S0-DJE-493

414-94-5538

call 974-5964

North Carrick Hall

Plasma Engineering

Plasma engineering

M. ALI KETTANI,
Associate Professor and Chairman,
Electrical Engineering Dept.,
College of Petroleum and Minerals,
Dhahran, Saudi Arabia

MAX F. HOYAUX,
Formerly Associate Professor,
Electrical Engineering Dept.,
University of Pittsburgh,
Pittsburgh, Pennsylvania

A HALSTED PRESS BOOK

JOHN WILEY & SONS
New York – Toronto

English edition first published in 1973
by Butterworth & Co (Publishers) Ltd
London; 88 Kingsway, WC2B 6AB

Published in the United States and Canada
by Halsted Press Division, John Wiley & Sons Inc.
New York

© Butterworth & Co (Publishers) Ltd, 1973

Library of Congress Cataloging in Publication Data
Kettani, M Ali.
Plasma engineering
Bibliography: p.
1. Plasma (Ionized gases). I. Hoyaux, Max
Florian, 1919- joint author. II. Title.
[DNLM: 1. Gases. QC 809.P5 K43p 1973]
QC718.K47 1973 530.4'4 72-7452
ISBN 0 470-47330-4

Printed in Hungary

Preface

Plasma physics is a rapidly expanding field of science. For a long time, it has coincided with the field of electrical discharges in gases, but recently, new fields of application of plasma physics have appeared. The conjunction of magnetohydrodynamics and plasma physics bears the promise of industrial applications, to convert directly thermal energy into electricity, without rotating machines. In the earth's atmosphere, the fast moving bodies produce plasmas which are likely to create new fields of interest in aeronautics and communications science. In astrophysics, the presence of a plasma is the general rule, not only in the interior of stars, but also in cosmic space itself, where important magnetohydrodynamic phenomena are occurring. Moreover, plasma guns seem to be a very promising way of propelling vehicles in space. In addition, plasma physics bears now the burden of the most formidable challenge ever faced by humanity: the controlled thermonuclear process. The importance of lasers in a very large variety of scientific and industrial applications cannot be overstated. More recently, the appearance of the field of solid-state plasmas opens completely new horizons.

The importance of applied plasma physics has been understood by many engineering schools, and plasma physics courses have been introduced in the curricula of those schools since the early 1960s, both at undergraduate and at graduate levels. This led to the appearance of many books in this field. Unfortunately, most of these books cannot be satisfactorily used as textbooks for a plasma engineering course. Some of them are too specialised or too theoretical and neglect the practical aspects and the possibility of application of primary interest to the engineer. We believe that there is a need for a textbook in plasma engineering and hence joined our efforts hoping to fulfil the demand.

This text results from the fusion of two courses offered at two different institutions: two-semester graduate course in Plasma Physics at the Carnegie Institute of Technology (presently Carnegie–Mellon University) and more recently at the University of Pittsburgh by Max Hoyaux; and a two-semester graduate course in Plasma Engineering formerly given by Ali Kettani at the School of Engineering of the University of Pittsburgh. The first five chapters are concerned with the physical principles of ionised gases, while each of the last seven chapters treats a plasma field of application separately. These last chapters are related to each other but are strongly dependent on the understanding of the first five chapters. Thus a plasma engineering course may be formed from Chapters 1–5 with a selection from the other chapters. The problems at the end of the book have been chosen to clarify and to extend the scope of the text itself with emphasis on triggering the curiosity of the student and encouraging his urge for originality. The material in this book has been carefully chosen from published and unpublished research work, to communicate the maximum information with a minimum of overspecialised details and to keep an acceptable evenness in the level and the unity of the text.

The authors wish to thank their students for contributing indirectly to this book, by their questions, their interest, and their encouraging hard work. They are grateful to the College of Petroleum and Minerals, Dhahran, Saudi Arabia, especially Ronald E. Scott, Dean of Engineering, and to the University of Pittsburgh, Pa., for their technical help in the preparation of this manuscript. Special thanks should be extended to Mr. Muhammad Aslam for the typing of the text.

M. Ali Kettani
Max F. Hoyaux

Contents

1 Introduction

1.1 DEFINITION OF A PLASMA

A plasma is a conglomeration of positively and negatively charged particles. It is on the average neutral, because the number density of the positive charges is equal to that of the negative ones. The plasma may also contain neutral particles or may be fully ionised, in which case all the particles are charged. The plasma will then contain two or more kinds of charge carriers: free electrons and positive ions. Sometimes it has also negative ions and more than one kind of positive ion. The ions may be singly charged or multiply charged; they may also be atomic or molecular.

An important parameter is the degree of ionisation, which is the percentage of those gaseous atoms or molecules initially present that has been decomposed into charge carriers. When the plasma contains neutral particles, the degree of ionisation is less than one. It is closer to unity when the plasma is strongly ionised and closer to zero when it is weakly ionised. Sometimes, multiply ionised atoms account conventionally for a degree of ionisation greater than one. The word plasma itself was introduced by Langmuir before any interest began to be devoted to fully ionised gases, and cannot be restricted to fully ionised gases.

The fully ionised plasma is sometimes considered as a particular state of matter. Indeed, matter can be found in four different states: solid, liquid, gas or plasma. It is well known that when enough energy is added to a solid, the solid will change its state and become a liquid which in turn with enough energy added to it will become a gas. The molecules of such a gas will assume a variety of degrees of freedom. If one keeps adding energy, the kinetic energy of the particles will increase and molecular impacts will become so intense that dissociation will result between some of the electrons and the rest of each molecule. The gas will become a plasma. The energy added is partly converted into random kinetic energy and partly into dissociation energy. The physical properties are different from one state of matter to the other.

The plasma state is actually by far the most common form of matter (up to 99 per cent in the universe). It is also the most energetic state; a body will require on average 10^{-2} eV*/particle to change its state from solid to liquid or

* 1 eV = $1{\cdot}602\times10^{-19}$ J.

from liquid to gas, whereas a change of state from gas to plasma will require from 1 to 30 eV/particle, depending on the material.

1.2 IMPORTANT CHARACTERISTICS

The two independent characteristics of a plasma are the charged particle *density n* and the *temperature T*. Since the plasma is usually neutral, there are as many positively charged particles as negatively charged ones and $n = n^+ = n^-$ where n^+ is the positive charge density and n^- is the negative charge density. Implicitly, we assume here the simplified case of a plasma made exclusively of electrons and of one species of singly charged—atomic or molecular—positive ions. Typically, for laboratory plasmas $n = 10^{19}$ particles/m^3 and $T = 10^5$K. It should also be noted that a practical plasma always has some charge separation, often exceedingly small, localised or temporary but not negligible.

When the plasma departs from neutrality, then $\Delta n = n^+ - n^-$. To determine the conditions of existence of a plasma, in spite of this departure from neutrality, two other characteristics are introduced. These are the Debye radius r_D (or Debye screening length) and the characteristic plasma frequency ω_p.

The Debye radius[18]

The Debye radius is defined as the radius of a sphere centred around any charged particle and as such, at its surface, the kinetic energy of the particles equals on the average the mutual electrostatic potential energy.

Consider a *systematic* departure from neutrality in an extended region of space. This departure will be governed by Poisson's law which can be written:

$$\nabla^2 V(r) = -\varrho_e/\varepsilon_0 \tag{1.1}$$

$V(r)$ being the space potential at a certain distance r, ε_0 the permittivity of the free space $\left(\varepsilon_0 = \frac{1}{36\pi} \times 10^{-9}\ \text{F/m}\right)^*$ and $\varrho_e = \Delta n . e$ is the space charge density, where e is the modulus of electronic charge ($e = 1{\cdot}602 \times 10^{-19}$ C). Considering only the effect of the electrons and assuming that the positive ions make it uniform, equation (1.1) becomes in spherical co-ordinates:

$$\nabla^2[V(r)] = \frac{1}{r^2}\frac{\mathrm{d}}{\mathrm{d}r}\left[r^2 \frac{\mathrm{d}V(r)}{\mathrm{d}r}\right] = \Delta n^- e/\varepsilon_0 \tag{1.2}$$

If for the sake of simplicity Δn^- is taken as n^- and is assumed to be constant in time and in space, integrating equation (1.2) gives:

$$V(r) = \frac{n^- e r^2}{6\varepsilon_0} \tag{1.3}$$

* or ε of the host-lattice for a solid-state plasma (Chapter 11).

Stating that the kinetic energy at radius r_D equals the potential energy (definition of the Debye radius) and expressing the potential V by its value in equation (1.3) gives:

$$e\frac{n^{-}er_{D}'^{2}}{6\varepsilon_0} = \frac{3}{2}kT$$

thus

$$r_D' = 3\left(\frac{\varepsilon_0 kT}{e^2 n^-}\right)^{1/2} \tag{1.4}$$

where k is Boltzmann's constant ($= 1{\cdot}38\times10^{-23}$ J/K). This is only a simplified derivation, the real value of the Debye radius is equal to the third of that given in equation (1.4), or $r_D = r_D'/3$.

The *first condition* for the existence of a plasma is that the Debye radius should have specified values, that is:

$$r_D > (n)^{-1/3} \tag{1.5}$$

Consequently, the Debye radius can also be defined as the minimum size of a system such that collective effects are dominant as compared to single particle effects.

The *second condition* of existence of a plasma is that the Debye radius be smaller than the overall length l of the plasma. This is also the condition for approximate overall charge neutrality. This condition can be illustrated in the following simple calculation. In a linear plasma, equation (1.1) becomes:

$$\Delta V = -l^2\, \Delta n\, e/\varepsilon_0$$

For $l = 0{\cdot}1$ m and $\Delta n = 10^{12}$ p./m^3, which is less than 1 ppm in most plasmas of interest, it can be seen that ΔV is already of the order of 200 V. Such a potential difference would be immediately compensated by suitable motions of the carriers. It is directly proportional to the carrier difference Δn. A dense plasma is generally a better conductor than a light one and consequently has a smaller Δn. However, for very small values of l, ΔV may be uncompensated, hence the concept of a minimum length given by the Debye radius.

The Debye radius is of the order of magnitude for the distance over which a significant departure from neutrality can be permanently maintained. This expression holds good for a number of phenomena in the steady state, in particular that described by Figure 1.1. Here we have assumed a stationary background of positive ions and we have also assumed that the plasma is bounded by an electrode biased with respect to the bulk of the plasma, say, negatively. This negative electrode rarefies the electrons in its vicinity thus creating a local positive space charge. The electron density is controlled by Boltzmann's special equation describing the rarefaction of a gas subjected to a retarding potential:

$$n_1/n_2 = \exp\left[e(V_1 - V_2)/kT\right] \tag{1.6}$$

This equation applies between the bulk of the plasma ($n = n_\infty$, $V = 0$ by definition) and a point close to the electrode (n, V); hence

$$n/n_\infty = \exp(eV/kT) \tag{1.7}$$

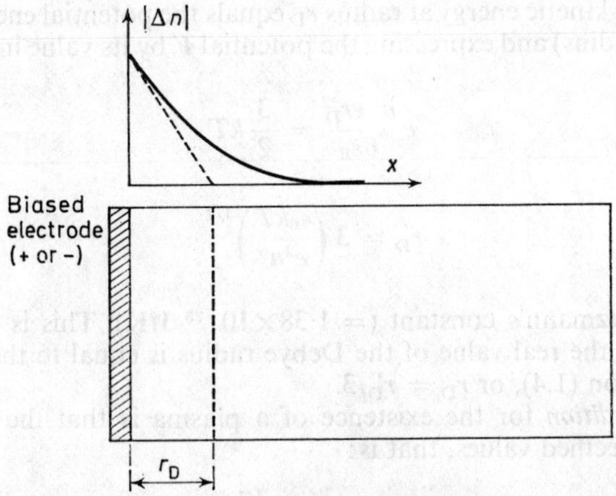

Fig. 1.1. Simplified model for the Debye (or Thomas–Fermi) screening length, also applicable to an n-type semiconductor. The influence of an electrode biased with respect to the bulk of the plasma is only 'felt' over an extremely small depth

In addition we have Poisson's equation:

$$\frac{d^2V}{dx^2} = -e(n_\infty - n)/\varepsilon_0 \tag{1.8}$$

Substituting n from equation (1.7) in equation (1.8), we obtain:

$$\frac{d^2V}{dx^2} = -en_\infty[1-\exp(eV/kT)]/\varepsilon_0 = e^2n_\infty V/\varepsilon_0 kT \tag{1.9}$$

the solutions of this equation are exponentials with a sub-tangent given by r_D as defined earlier.

In most cases of practical interest, r_D is small in comparison with the size of the plasma (for example 100 m in astrophysical applications, 10^{-6} m in low-pressure arcs, 10^{-8} m in high-pressure ones, and still less in solid-state plasmas). The charge separation imposed by biased electrodes is, in the steady state, confined to a thin layer. The bulk of the plasma stays equipotential and unaffected by the externally applied electric field. This does not apply, however, to fields already there when the plasma was created, as would be the case in an electrical discharge produced from an avalanche. This shielding effect will be taken into account in the Coulomb field of the individual charge carriers, notably in scattering phenomena.

Characteristic plasma frequency ω_p[42]

Consider a plasma region of thickness l as shown in Figure 1.2. In this region, the electrons are assumed to be displaced by a certain distance x in the direction of one of the sides of the plasma slab. At one end, the face has an excess of

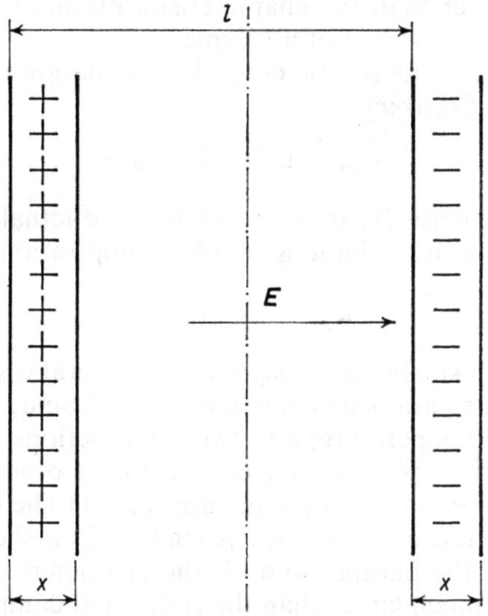

Fig. 1.2. A linear plasma departing from neutrality. The electrons are displaced by a distance x in the direction of one of the sides of the plasma slab of width l

electronic charge $-n^{-}ex$ per unit area and at the other one, there is left an excess of positive ion charge $+n^{-}ex$ per unit area.

Under these circumstances it would be extremely interesting to know how fast the plasma will react. To answer this question one must calculate the characteristic plasma frequency. Disregarding the ions (since the electrons will obviously react faster due to their smaller mass), the balance of forces becomes:

$$F = -eE = -M^{-}\,\mathrm{d}^2x/\mathrm{d}t^2 \tag{1.10}$$

where E is the electric field generated as a result of the displacement and M^{-} is the electronic mass ($M^{-} = 9{\cdot}108\times10^{31}$ kg), Poisson's law yields in one dimension:

$$E = -n^{-}ex/\varepsilon_0 \tag{1.11}$$

Replacing E by its value in equation (1.10) and rearranging:

$$x + \frac{M^{-}\varepsilon_0}{n^{-}e^2}\frac{\mathrm{d}^2x}{\mathrm{d}t^2} = 0 \tag{1.12}$$

This is the equation for an oscillating motion, the solution of which is $x = x_0 \exp(\mathrm{j}\omega_p t)$ where ω_p is defined as the *characteristic plasma frequency*; its value is:

$$\omega_p = (n^{-}e^2/M^{-}\varepsilon_0)^{1/2} \tag{1.13}$$

The plasma will oscillate with a frequency $f_p = \omega_p/2\pi$ around its equilibrium position. If collisions are neglected, this oscillation will continue indefinitely. In order that these oscillations may develop, the collisional frequency ν_c should be small, much smaller than the characteristic plasma frequency. This is the *third condition* for the existence of a plasma.

Actually, a second characteristic collective oscillation can also occur at the so-called *ion plasma frequency:*

$$\omega_{pi} = (n^+ e^2/M^+ \varepsilon_0)^{1/2}$$

where M^+ is the ionic mass. It can be proved that the actual plasma frequency is, taking into account both oscillations (in phase opposition):

$$\Omega_p = (\omega_p^2 + \omega_{pi}^2)^{1/2} \tag{1.14}$$

and since ω_{pi} is much smaller than ω_p, Ω_p is in general essentially equal to ω_p.

To summarise, four parameters characterise a plasma; two are independent (particle density and temperature) and two are functions of the others (Debye radius and characteristic plasma frequency). On the other hand, for a plasma to exist at all, three conditions should be satisfied: (1) The Debye radius should be larger than the distance between the particles; (2) it should be smaller than the overall length of the plasma; and (3) the characteristic plasma frequency should in general be much larger than the collisional damping frequency.

1.3 SHORT SURVEY OF THE APPLICATIONS OF PLASMA PHYSICS

The fields of science and engineering in which plasma physics can be applied are increasing more and more in number. Plasma physics applications have for long been identified only with those of electrical discharges in gases, since the electrical discharge has been used for its rectifying properties (mercury vapour rectifier, for instance) as a source of light as well as in control devices (thyratron, plasma magnetron, etc.). Special kinds of discharges were also used as particle detecting devices, for instance ionisation chambers, proportional counters and Geiger–Müller counters.

Electrical discharges have also been used extensively as sources of electrical carriers of both signs, principally for accelerating purposes or in mass spectrography. The possibility of creating very high temperatures in electrical discharges has led to the 'plasma torch', which is used for cutting and welding purposes. The plasma torch can in fact be considered as a first step towards the application of plasma devices to the propulsion of rockets in space.

Independently of this application, the plasma gun is becoming a field of interest in itself. One of the major applications of plasma guns is the injection of plasmas into 'magnetic bottles' in view of the feasibility studies of controlled thermonuclear processes. Thermonuclear devices in themselves have created a tremendous revival in the interest of scientists and engineers towards plasma physics. They offer unique possibilities of producing power in amounts never thought of before.

The behaviour of plasmas submitted to magnetic fields is of paramount importance. The notions of magnetic constriction of electrical discharges, of magnetic confinement in thermonuclear devices, and those applications to plasma physics belonging to the vast class of magnetohydrodynamic phenomena, are soon likely to be important fields of application in science and engineering. Indeed, a large amount of work is being done in the field of power generation and in that of propulsion in space and for submarines.

Plasma physics becomes applicable in aeronautics, in studies at high speeds and high altitudes. In astrophysics, much more matter in the universe, both in stars and in interstellar space, is in the plasma state than in any other state, either solid, liquid or non-ionised gas. The three traditional states of matter appear, on the cosmic scale, as rather exceptional.

More recently, plasma phenomena were used for building microwave amplifiers (masers). The study of the possibility of extending the research toward the optical region led to the construction of the first light amplifier (laser) in

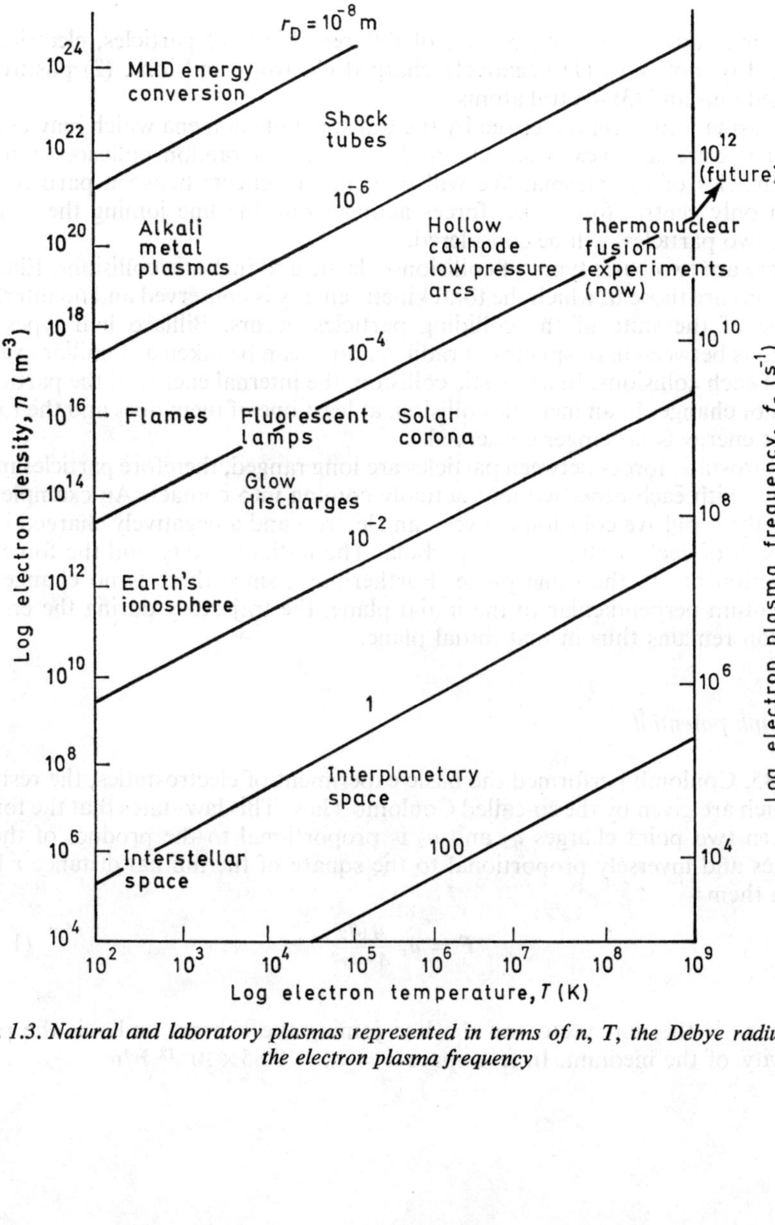

Fig. 1.3. Natural and laboratory plasmas represented in terms of n, T, the Debye radius and the electron plasma frequency

1960. Interest in solid-state plasmas is even more recent. Its theory is still in its infancy and its applications in science and industry cannot yet be predicted with accuracy.

The fields of application of plasma physics can be represented in a diagram in which the logarithm of plasma temperature is plotted along the abscissa, the ordinate being logarithm of the electron density (see Figure 1.3). Values of the characteristic plasma frequency and the Debye screening length are also represented on the same diagram.

1.4 INTERACTIONS BETWEEN PARTICLES

A plasma consists, as seen before, of different kinds of particles, electrically charged or not, viz., (1) negatively charged electrons and ions, (2) positively charged ions and (3) neutral atoms.

A plasma is also characterised by the transport phenomena which may occur in it. These are due to external fields and braked by the random collisions among the particles of the plasma. We will treat binary effects between particles in which only central forces, i.e. forces acting along the line joining the centre of the two particles, will be considered.

There are two basic types of collisions, elastic and inelastic collisions. Elastic collisions are those in which the total kinetic energy is conserved and no internal change of the state of the colliding particles occurs. Billiard ball types of collisions between hard spheres of radii r_1 and r_2 can be taken as familiar examples of such collisions. In an elastic collision, the internal energy of the particles does not change. In an inelastic collision, at least one of them does and the total kinetic energy is no longer conserved.

Electrostatic forces between particles are long ranged, therefore particles may interact with each other without actually coming into contact. An example of this is the repulsive collision between an electron and a negatively charged ion. The path of each particle is a hyperbola. The initial velocity and the force of interaction are in the same plane. Furthermore, since there is no change in momentum perpendicular to the initial plane, the trajectory during the entire collision remains thus in that initial plane.

Coulomb potential

In 1785, Coulomb performed the basic experiment of electrostatics, the results of which are given by the so-called Coulomb's law. This law states that the force between two point charges q_1 and q_2 is proportional to the product of these charges and inversely proportional to the square of the mutual distance r between them:

$$\boldsymbol{F} = \boldsymbol{a}_r \frac{q_1 q_2}{4\pi\varepsilon r^2} \tag{1.15}$$

where $\boldsymbol{a}_r$ is the unit vector on the line joining the charges and ε is the permittivity of the medium. In free space $\varepsilon = \varepsilon_0 = 8{\cdot}85 \times 10^{-12}$ F/m.

Contingent on whether the charges are of the same sign or of opposite signs, there will be a repulsion or an attraction. The electric field due to q_1 is

$$\boldsymbol{E} = \boldsymbol{F}/q_1 \tag{1.16}$$

and Poisson's law yields for the potential,

$$V(r) = -\int E(r)\,\mathrm{d}r \tag{1.17}$$

and with $\mathcal{C}_2 = q_2/4\pi\varepsilon$,

$$V(r) = \pm\mathcal{C}_2/r \tag{1.18}$$

The Coulomb potential given by equation (1.18) is represented in Figure 1.4.

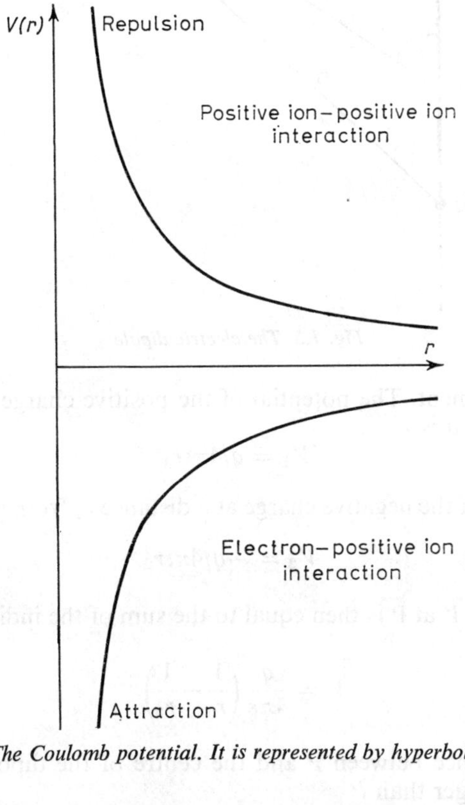

Fig. 1.4. The Coulomb potential. It is represented by hyperbolic functions

Polarisation

When an electric field is applied to a dielectric, the electrons are displaced with respect to their nuclei. The atoms will therefore, behave as dipoles. The dielectric is said to be polarised with a moment $\mathcal{D} = \varepsilon_0\psi\boldsymbol{E}$, $\varepsilon_0\psi$ being the polarisability of the molecule and ψ the dielectric susceptibility. Thus, $\boldsymbol{D} = \varepsilon_0(1+\psi)\,\boldsymbol{E}$ is the total electric flux density. Some molecules are already dipoles in their normal state, that is, the charge centroid of the electrons and that of the positive ions do not coincide.

Consider the electric dipole shown in Figure 1.5. It consists of two point charges $+q$ and $-q$ separated by a distance l. The product ql is called the

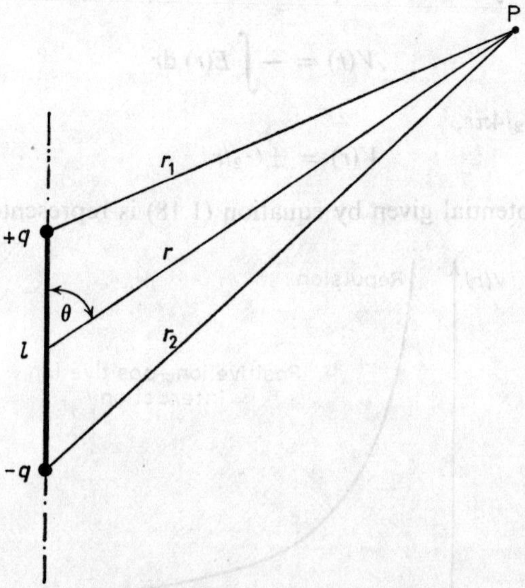

Fig. 1.5. The electric dipole

electric dipole moment. The potential of the positive charge at a point P at a distance r_1 from $+q$ is:

$$V_1 = q/4\pi\varepsilon r_1$$

and the potential of the negative charge at a distance r_2 from point P is:

$$V_2 = -q/4\pi\varepsilon r_2$$

The total potential V at P is then equal to the sum of the individual potentials; that is:

$$V = \frac{q}{4\pi\varepsilon}\left(\frac{1}{r_1} - \frac{1}{r_2}\right) \tag{1.19}$$

Calling r the distance between P and the centre of the dipole, one can write when r is much larger than l:

$$r_1 = r - \frac{l}{2}\cos\theta \tag{1.20}$$

and

$$r_2 = r + \frac{l}{2}\cos\theta \tag{1.21}$$

where θ is the angle between the axis of the dipole and the line of length r. Replacing in equation (1.19), r_1 and r_2 by their values in equations (1.20) and (1.21) it is found that:

$$V = \frac{ql \cos \theta}{4\pi\varepsilon_0 r^2} \tag{1.22}$$

As a vector the dipole moment is—by definition—equal to ql. Thus one can write:

$$V = \frac{p}{4\pi\varepsilon_0} \frac{\cos \theta}{r^2} \tag{1.23}$$

The applied field is given by equation (1.16), and the dipole moment becomes:

$$p = \frac{\psi q}{4\pi} \frac{1}{r^2} \tag{1.24}$$

Substituting in equation (1.23), one finds:

$$V(r) = \frac{\psi q}{16\pi^2\varepsilon_0} \cos \theta \frac{1}{r^4} \sim 1/r^4 \tag{1.25}$$

When a positive ion approaches an atom of high atomic number Z, the potential will be $V(r) = \mathscr{C}_1/r$ with $\mathscr{C}_1 = Z\mathscr{C}_2$. Since Z is large, $\mathscr{C}_1$ becomes large. The potential is then short-ranged and repulsive. Due to these short-range effects, the incoming ion encounters a rapidly increasing repulsion as it penetrates the atomic cloud. Since this overlap repulsion is steep, one can readily approximate it by a repulsive barrier at range r_0 determined by the particle energy $\mathscr{E}$ as shown in Figure 1.6.

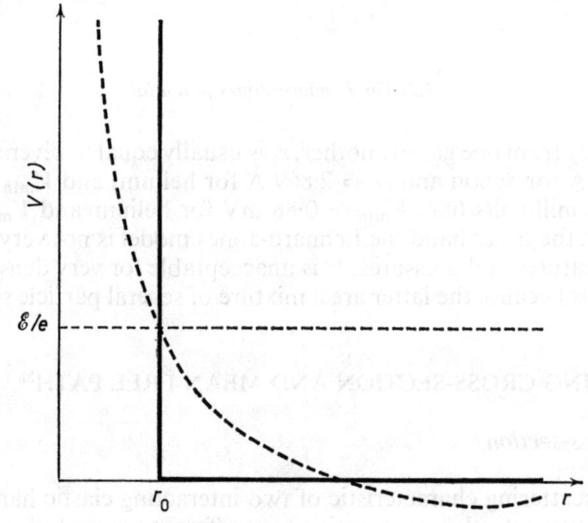

Fig. 1.6. Hard-sphere approximation. It is represented by a repulsive barrier at range r_0 determined by the energy of the particle

Overall effect

The overall effect can be represented by a more realistic model than the hard-sphere one. There are several proposals, among them the Lennard-Jones[32] potential is the most used in gas dynamics. This potential is defined as:

$$V(r) = 4V_{min}[(r_0/r)^{12} - (r_0/r)^6] \tag{1.26}$$

In this equation, $(r_0/r)^6$ represents the attractive term and $(r_0/r)^{12}$ represents the repulsive term, V_{min} corresponds to the maximum energy of attraction which happens at $r_{min} = 2^{1/6}r_0$ where r_0 is the radius of the molecule.

The Lennard-Jones potential is represented in Figure 1.7. It describes in a satisfactory manner most of the spherical non-polar molecules. The parameters

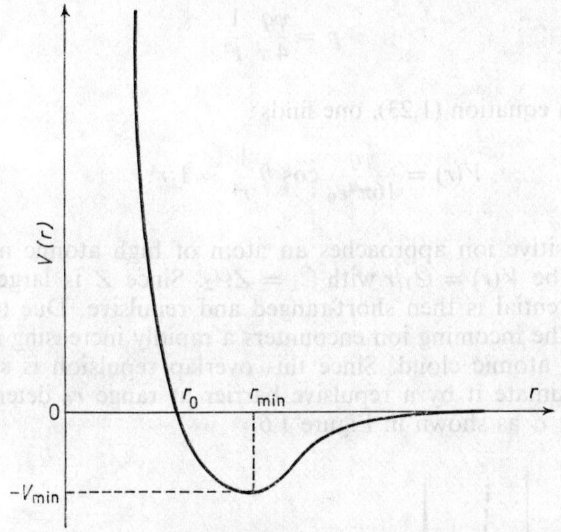

Fig. 1.7. The Lennard-Jones potential

V_{min} and r_0 vary from one gas to another. r_0 is usually equal to several angstroms (e.g. $r_0 = 4{\cdot}6$ Å for xenon and $r_0 = 2{\cdot}869$ Å for helium) and V_{min} varies from one to several millivolts (e.g. $V_{min} = 0{\cdot}88$ mV for helium and $V_{min} = 19$ mV for xenon). On the other hand the Lennard-Jones model is not very satisfactory at high temperatures and pressures. It is unacceptable for very dense and highly ionised plasmas because the latter are a mixture of several particle species.

1.5 SCATTERING CROSS-SECTION AND MEAN-FREE PATH[35]

Scattering cross-section

Consider the scattering characteristic of two interacting elastic hard spheres of radii r_1 and r_2 respectively, representing two different types of plasma particles. In order that these two spheres collide, their centres must be at a distance

(r_1+r_2) apart as shown in Figure 1.8. Assume now that the particles of type 2 are stationary and of uniform density and that a given particle of type 1 moves toward the others. The problem is usually simplified by representing the moving particle by a material point and the stationary ones by spheres of radius

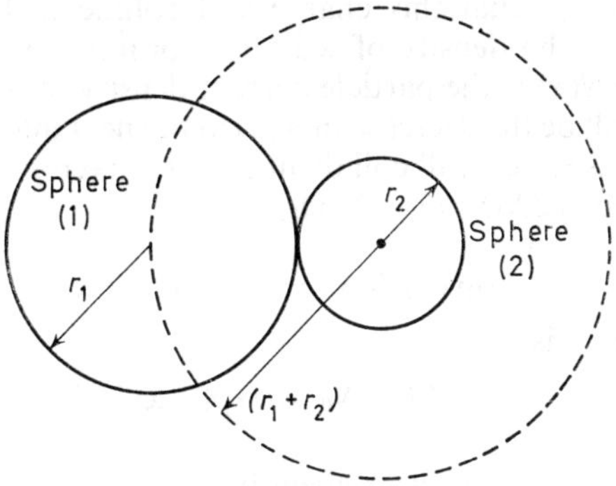

Fig. 1.8. Two colliding hard spheres

(r_1+r_2). The scattering cross-section Q_t is by definition the area presented by the scatterer to the incoming particles. In the relevant case, it is:

$$Q_t = \pi(r_1+r_2)^2$$

If the radius of the projectile particles is much smaller than the radius of the scatterer, the scattering cross-section becomes:

$$Q_{to} = \pi r_2^2 \tag{1.27}$$

This, of course, is under the assumption that the target molecules have a negligible motion. If, now, the incident particles are identical to the scatterers, the scattering cross-section becomes equal to four times the value given by equation (1.27). It is not possible, in an elementary calculation, to account correctly for the motion of the largest molecules. It is conceivable, as the molecules are three-dimensional bodies, that this motion somewhat increases the probability of interception, thus the scattering cross-section. A rigorous calculation[26] shows that, if the incident particles are identical to the target molecules and move at the same speed, the scattering cross-section becomes $Q_{tv} = 4\sqrt{2}\pi r^2$. This equation can be checked experimentally, for instance from the value of the viscosity of a gas.

The differential scattering cross-section $Q(\theta, \phi)$ is still defined as the area presented by a scatterer but this time exclusively scattering into a unit solid angle in the direction θ, ϕ. It is evidently related to the total cross-section by:

$$Q_t = 2\pi \int_0^{\pi} Q(\theta, \phi).\sin\theta.d\theta \tag{1.28}$$

Mean-free path and frequency of collision

Let us follow the motion of the point charge as it travels a distance dx in the gas. The probability p_r that this charge will collide with a type 2 target is $Q_t n_{sc}\, dx$, where n_{sc} is the density of scatterers or number of relevant particles per unit volume. If $N(x)$ is the particle current density of the point charges at a distance x, $dN(x)$ will be the decrease in $N(x)$ in the next interval dx. It represents the number of points which will collide in dx after having travelled a distance x without collision; thus $dN(x) = -N(x)p_r$ and

$$dN(x)/N(x) = -(n_{sc}Q_t).dx \tag{1.29}$$

The solution of which is:

$$N(x)/N_0 = \exp(-x/\lambda_t)$$

where λ_t is the *total mean free path* given by:

$$\lambda_t = 1/n_{sc}Q_t \tag{1.30}$$

Here again, it is assumed that the velocity of the target particles is negligible compared to that of the incident particles.

Nov, if v_i is the average speed of the incident particles, then, during a time dt, the particles travel the distance $dx = v_i\, dt$ and replacing dx by its value in equation (1.29), one obtains:

$$dN(t)/N(t) = -(n_{sc}Q_t v_i)\, dt$$

the solution of which is:

$$N(t)/N_0 = \exp(-\nu_c t) \tag{1.31}$$

where ν_c is the frequency of collision here given by:

$$\nu_c = n_{sc}Q_t v_i \tag{1.32}$$

A quantity called the *average time of flight* τ_c may also be introduced. Since ν_c is the average number of collisions per unit time, it can be seen without difficulty that $\tau_c = 1/\nu_c$.

1.6 MOMENTUM TRANSFER CROSS-SECTION

The momentum transfer cross-section is of the utmost importance in the study of transport phenomena. Consider an incident particle of velocity v_2, and mass M encountering a wall of potential of infinite mass as shown in Figure 1.9. The velocity of the incident particle can be divided into two components, namely v_r (radial) orthogonal to the wall of potential, and v_t (tangential) perpendicular to the former. After the collision, obviously v_t keeps its direction, whereas v_r is completely reversed (as may be seen in Figure 1.9). Therefore, the relevant transfer of momentum $\Delta(Mv)$ from the centre of potential toward the particle will be $-2Mv_r$. In the case of a 'grazing' collision (where $v_r = 0$), it will be zero, and in the case of a 'head-on' collision (where $v_t = 0$) one has $v_r = v_i =$ incident velocity, and therefore $\Delta(Mv) = -2Mv_i$. More generally, the change in momentum along the initial direction is $\Delta(Mv) = Mv_i(1-\cos\theta)$, where θ is

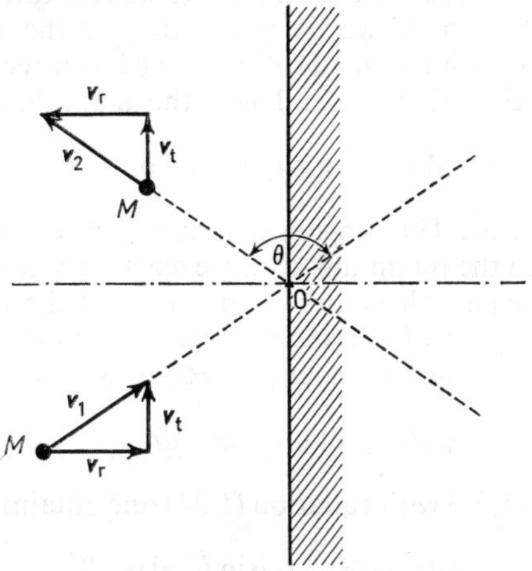

Fig. 1.9. Particle interacting with a potential of infinite mass

defined in Figure 1.9; accordingly, going back to the case of a scattering centre, we define the 'momentum transfer cross-section' as:

$$Q_m = \iint Q(\theta, \phi)(1-\cos\theta)\cdot\sin\theta \, d\theta \, d\phi \quad (1.33)$$

Where $Q(\theta, \phi)$ is the differential scattering cross-section defined above and θ is the angle of incidence. If $Q(\theta, \phi)$ is independent of both θ and ϕ

$$Q_m = Q\int_{\phi=0}^{2\pi}\int_{\theta=0}^{\pi}(1-\cos\theta)\sin\theta \, d\theta \, d\phi = 4\pi Q$$

The value of the total scattering cross-section Q_t is given by equation (1.28), thus for a constant differential scattering cross-section, the total scattering cross-section and the cross-section for momentum transfer are equal to $4\pi Q$.

1.7 ENERGY DEPENDENCES OF SCATTERING CROSS-SECTIONS

For the hard-sphere model, the total scattering cross-section given by equation (1.27) is independent of energy and velocity. The hard-sphere model being only a crude approximation, in general, this is not the case. Consider, for instance, a parallel beam of monoenergetic particles and let Φ_1 be its flux, i.e. the number of particles in the beam which in unit time cross a surface of unit area placed orthogonally. Suppose now, that this beam encounters a fixed scattering centre, e.g. heavy ion, due to the target the incident particles will be scattered in various directions.

If the incident flux is homogeneous, then the impact parameters of the incident particles will be distributed uniformly around the target particle. Therefore, the angular distribution of the scattered particles is characteristic of the law of force between the incident particles and the target. This can be specified by using a differential cross-section. If we designate dN_1/dt the number of particles scattered per unit time within a small solid angle $d\Omega$ centred on a given direction at θ and ϕ where θ is the colatitude and ϕ is the azimuth, its value will be:

$$dN_1/dt = \Phi_1 Q(\theta, \phi)\, d\Omega \tag{1.34}$$

where $d\Omega$ is $\sin\theta . d\theta . d\phi$. The 'impact parameter' x_b of an incident particle is equal, by definition, to the minimum distance existing between an axis along the incident velocity of the particle before interaction and the target particle. Consider now all the particles of the incident beam having an impact parameter between x_b and $x_b + dx_b$; their number per second will be:

$$\phi_1 2\pi x_b . dx_b = dN_1/dt \tag{1.35}$$

Comparing equation (1.35) with equation (1.34) one obtains:

$$Q(\theta, \phi) = [x_b/\sin\theta] . dx_b/d\theta \tag{1.36}$$

The total cross-section is 2π times the integral of $Q(\theta, \phi)$ extended over all the possible values of θ. The factor 2π implies the integration in the azimuth direction. For physical reasons, it is evident that such an integral can easily diverge. Indeed, it corresponds to the surface area of a circle of radius x_0 for which θ goes to zero. However, if $V(r)$ goes asymptotically to zero, as does the Coulomb potential and many others, such a circle simply does not exist; the hard-sphere is the only model satisfying such a condition. In practical problems, the divergence is eliminated by the use of factors related to the phenomenon of 'shielding' seen in Section (1.2).

In the case of a *Coulomb potential*, the interaction potential is given by equation (1.18), or

$$V(r) = e/4\pi\varepsilon r \tag{1.37}$$

This problem was solved by Rutherford apropos of alpha-particle deflection in connection with the discovery of the atomic nucleus. The trajectory of the particle is here a hyperbola, and the calculation of the angle of deflection can be easily performed from the geometric properties of hyperbolas (see Figure 1.10). Geometrically, it is known that the hyperbola is the locus of points such that the difference between their distances to the two loci F_1 and F_2 is a constant $2d$. Considering, first, the point of closest approach, its distance to F_1 is:

$$r_m = F_1O + d = [x_b/\cos(\theta/2)] + d \tag{1.38}$$

Considering, next, the point at infinite distance on the left it is found that:

$$d = x_b . \tan(\theta/2) \tag{1.39}$$

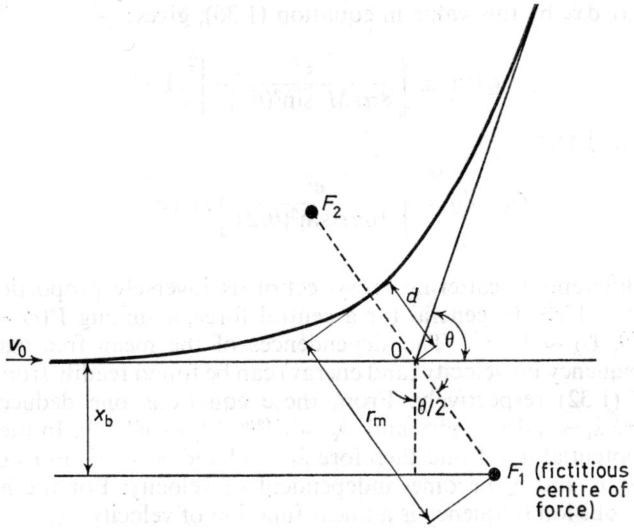

Fig. 1.10. The trajectory of an incident particle under the effect of a Coulomb potential. Points F_1 and F_2 are the foci of the hyperbolic locus

By substituting for d in equation (1.38) and expressing the cosine as a function of the tangent, one finds:

$$r_m = x_b[\tan(\theta/2)+(1+\tan^2(\theta/2))^{1/2}] \quad (1.40)$$

On the other hand, from the laws of conservation of energy and angular momentum, one finds:

$$v^2 = (dr/dt)^2+\frac{v^2x_b^2}{r^2}+\frac{2eV}{M}$$

and for $r = r_m$, $dr/dt = 0$, thus, introducing the value of V_m obtained from equation (1.37) in this relation, one finds:

$$r_m^2-\frac{e^2}{2\pi\varepsilon Mv^2}r_m-x_b^2 = 0 \quad (1.41)$$

Solving this second-order equation leads to the value of r_m. A comparison of this result and equation (1.40) shows that:

$$\tan(\theta/2) = e^2/4\pi\varepsilon Mv^2x_b \quad (1.42)$$

After differentiation and rearrangement one obtains:

$$x_b\,dx_b = \frac{2\pi\varepsilon Mv^2x_b^3\,d\theta}{e^2\cos^2(\theta/2)}$$

Replacing $x_b\, dx_b$ by this value in equation (1.36), gives:

$$Q(\theta) = \left[\frac{e^2}{8\pi\varepsilon M . \sin^2(\theta/2)}\right]^2 . 1/v^4$$

or since $\mathcal{E} = \frac{1}{2}Mv^2$,

$$Q(\theta, \mathcal{E}) = \left[\frac{e^2}{16\pi\varepsilon \sin^2(\theta/2)}\right] . 1/\mathcal{E}^2 \tag{1.43}$$

thus, the differential scattering cross-section is inversely proportional to $\mathcal{E}^2$, or $(Q(\theta, \mathcal{E}) \approx 1/\mathcal{E}^2$. In general for a central force, assuming $V(r) \sim 1/r^n$, one obtains $Q(\theta, \mathcal{E}) \sim 1/\mathcal{E}^{2/n}$. The dependences of the mean free path and the collision frequency on velocity (and energy) can be found readily from equations (1.30) and (1.32) respectively. From these equations one deduces that for $V(r) \sim 1/r^n$, $\lambda_t \sim \mathcal{E}^{2/n} \sim v^{4/n}$ and $\nu_c \sim \mathcal{E}^{(2/n-1/2)} \sim v^{(4/n-1)}$. In the case of a Coulomb potential $n = 1$ and therefore $\lambda_t \sim v^4$ and $\nu_c \sim v^3$. For polarisation, $n = 4$, $\lambda_t \sim v$, and ν_c becomes independent of velocity. For the hard-sphere model, the collision frequency is a linear function of velocity.

1.8 DISTRIBUTION OF PARTICLE VELOCITIES[36, 37]

The kinetic theory of gases aims at interpreting the macroscopic properties (such as temperature, pressure, viscosity and others) from the motions of individual molecules or atoms. In a simplified analysis the gas is first considered non-ionised. The smallest macroscopic sample of a gas is still composed of a tremendous number of molecules and the consideration of the individual motions based on dynamics would be an impossible task. Moreover, the initial conditions must generally be considered as unknown. In such conditions, the motion of the molecules can only be studied in a statistical way and average initial conditions are assumed. The fundamental hypothesis is that of the so-called 'molecular chaos'. It is a well known paradox in probabilities that from the most complete chaos, some order always appears. This is particularly true in kinetic theory.

When a gas is enclosed in a container at a constant temperature, its velocity distribution function tends rapidly towards a unique and perfectly determined function isotropic in the velocity, called the Maxwell–Boltzmann distribution function.

$$f(v) = n(M/2\pi kT)^{3/2} \exp(-Mv^2/2kT) \tag{1.44}$$

where M is the mass of the molecules concerned, n is the particle density and $k = 1{\cdot}380 \times 10^{-23}$ J/K is Boltzmann's constant. The derivations of equation (1.44) are numerous and most of them have been bitterly criticised. The best, however, seems to be the one given by Boltzmann himself.

In this book, a rather crude proof will be given on basis of a simplified version of the hypothesis of molecular chaos mentioned above. Motions of molecules in a gas will be considered as random with respect to an ideal state in which all the molecules are at complete rest. Those random motions (errors) are assumed to conform to the well-known *Gaussian* function which should be applied to the three dimensions.

Consider a molecule of velocity $\boldsymbol{v}$, such that v_x, v_y and v_z are the components of $\boldsymbol{v}$ respectively in the x-, y- and z-directions. For each of the directions, a Gaussian distribution function is assumed; thus:

$$\left.\begin{aligned} f(v_x)\,\mathrm{d}v_x &= \mathcal{A}\exp(-v_x^2/v_0^2)\,.\,\mathrm{d}v_x \\ f(v_y)\,\mathrm{d}v_y &= \mathcal{A}\exp(-v_y^2/v_0^2)\,.\,\mathrm{d}v_y \\ f(v_z)\,\mathrm{d}v_z &= \mathcal{A}\exp(-v_z^2/v_0^2)\,.\,\mathrm{d}v_z \end{aligned}\right\} \tag{1.45}$$

The probabilities $f(v_x)$, $f(v_y)$ and $f(v_z)$ are assumed to be completely independent of each other, allowing thus the use of the theorem of composed probabilities. In 'hodographic space', where v_x, v_y and v_z are plotted in cartesian co-ordinates (Figure **1.11**) the probability that the extremity of the vector-velocity

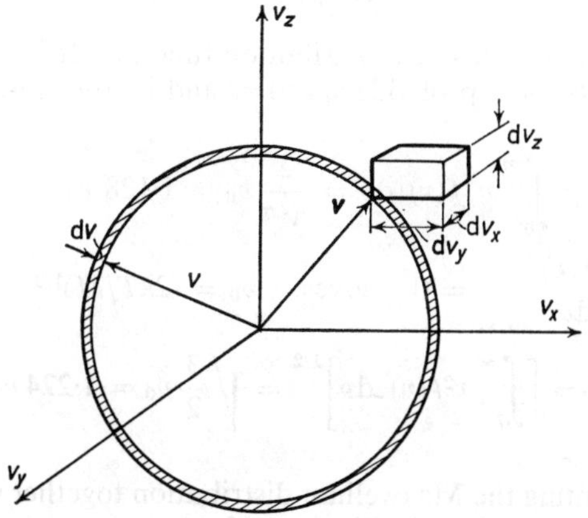

Fig. 1.11. The two elementary volumes considered $\mathrm{d}v_x\cdot\mathrm{d}v_y\cdot\mathrm{d}v_z$ *and* $4\pi v^2\,\mathrm{d}v$ *in the hodographic space*

$\boldsymbol{v}$ comes within a given volume $\mathrm{d}v_x.\mathrm{d}v_y.\mathrm{d}v_z$ in the neighbourhood of the point of co-ordinates v_x, v_y, v_z is equal to the product of all the individual probabilities given by the three equations (1.45). Thence:

$$f(\boldsymbol{v})\,\mathrm{d}\boldsymbol{v} = \mathcal{A}^3\exp(-v^2/v_0^2)\,.\,\mathrm{d}v \tag{1.46}$$

where $\mathrm{d}\boldsymbol{v} = \mathrm{d}v_x.\mathrm{d}v_y.\mathrm{d}v_z$ is the elementary volume, and $v^2 = v_x^2+v_y^2+v_z^2$.

However, the primary interest is often in the magnitude of the velocity or speed regardless of its direction. Therefore, a volume element between two spheres of radii v and $v+\mathrm{d}v$ should be considered in equation (1.46), rather than the parallelepiped volume element. Its magnitude is $4\pi v^2\,\mathrm{d}v$; thus (see Figure 1.11):

$$f(v) = \mathrm{const}\times v^2\exp(-v^2/v_0^2) \tag{1.47}$$

This probability must satisfy the two conditions, that $\int_{-\infty}^{+\infty} f(v)\,\mathrm{d}v = 1$ and that

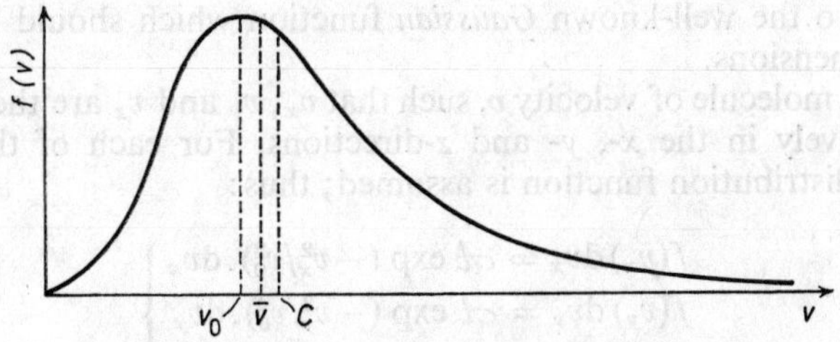

Figure 1.12. Maxwell's distribution function

v_0 must be a function of the temperature of the gas by the relation $\frac{1}{2}Mv_0^2 = kT$. Under these conditions, equation (1.47) becomes:

$$f(v) = \frac{4}{\sqrt{\pi}}\frac{v^2}{v_0^3}\exp\left(-Mv^2/2kT\right) \tag{1.48}$$

This is the well-known Maxwell distribution function. It is characterised by its average speed $\bar{v}$, its most probable speed v_0 and its root mean square speed C defined as[14]:

$$\left.\begin{aligned}\bar{v} &= \int_0^\infty v.f(v).\mathrm{d}v = \frac{2}{\sqrt{\pi}}v_0 = 1{\cdot}128\ v_0\\ \left.\frac{\mathrm{d}f(v)}{\mathrm{d}v}\right|_{v=v_0} &= 0 \quad \text{gives} \quad v_0 = (2kT/M)^{1/2}\\ C &= \left[\int_0^\infty v^2 f(v).\mathrm{d}v\right]^{1/2} = \sqrt{\frac{3}{2}}v_0 = 1{\cdot}224\, v_0\end{aligned}\right\} \tag{1.49}$$

The curve representing the Maxwellian distribution together with the particular speeds defined above is given in Figure 1.12.

1.9 TRANSPORT PARAMETERS[5]

Because of a combination of the Brownian motion and the effect of collision, particles are randomised in a plasma. In the absence of any external force, the random motion of the plasma particles will result in a flow of particles from one region to another only if there exists a gradient of particle density, of pressure or of temperature in the plasma. The net effect of an applied external force is to cause the particles to move in a direction determined by the force. The existence of such motions results in mass and charge transports which are elementary examples of transport phenomena. These phenomena are simply described by five transport coefficients: diffusion $\mathcal{D}$, mobility b, electrical conductivity σ, thermal conductivity $\varkappa$, and viscosity v.

Diffusion[17]

The process by which a net flow of particles results from the random motion of the individual particles is called *diffusion*. Consider a sphere of unit radius, with two meridians of longitude ϕ and $\phi+d\phi$ and two parallels of colatitude θ and $\theta+d\theta$. The trapezoidal surface dS in between is $\sin\theta.d\phi.d\theta$. Starting from the centre of the sphere, if a direction is selected at random, its probability $f(\theta, \phi)$ of crossing the sphere through a given portion is given by the surface area of the sphere of unit radius, thus:

$$f(\theta, \phi).d\theta.d\phi = \frac{1}{4\pi}\sin\theta.d\theta.d\phi$$

Consider, now, a plane dividing the space into two parts, and in this plane a unit surface area. In the half-space on the left of the assumed vertical plane, consider a class A medium of molecules whose velocity v aims at the unit surface. Those molecules crossing the surface between instants t and $t+dt$, are, at time t, inside a prism based on the unit surface and having four sides parallel to $\boldsymbol{v}$ and of length $v\,dt$. Thus the volume of the prism is $v.\cos\theta.dt$ where θ is the angle between the direction of $\boldsymbol{v}$ and the normal to the dividing plane. If n is the number density of molecules, then the total number of molecules falling in class A is

$$dN_A = v.\cos\theta.dt.\frac{1}{4\pi}.\sin\theta.d\theta.d\phi.n.f(v).dv \qquad (1.50)$$

where $n.f(v).dv$ is the density of molecules having a speed between v and $v+dv$ and $f(v)$ is the distribution function of these velocities.

The *random flux* of particles is, then, defined as the total number of particles crossing the unit surface from the left per unit time. It is equal to

$$\Gamma_r = \int_{\phi=0}^{2\pi}\int_{\theta=0}^{\pi/2}\int_{v=0}^{\infty} dN_A/dt$$

The result of these integrations is

$$\Gamma_r = n.\bar{v}/4 \qquad (1.51)$$

where $\bar{v}$ is given in equation (1.49). It can be shown that the random current from the right is equal to that from the left, when the plane separating the two regions of space is 'virtual' (i.e. not a solid surface).

In the case of a solid surface this is not true any longer. There will be a net flow by diffusion such as $\Gamma_{diff} = \Gamma_{right} = \Gamma_{left}$, and replacing the random currents by their values from equation (1.51), it is found:

$$\Gamma_{diff} = [n(+x)-n(-x)].\bar{v}/4 \qquad (1.52)$$

Expanding $n(+x)$ and $n(-x)$ in Taylor's series in the vicinity of $x = 0$, equation (1.52) becomes:

$$\Gamma_{diff} = 2x.(\partial n/\partial x).\bar{v}/4 \qquad (1.53)$$

Since the mean free path λ is equal to $2x$, equation (1.53) becomes (in three dimensions):

$$\Gamma_{\text{diff}} = \lambda.\bar{v}.\nabla n/4 \tag{1.54}$$

The diffusion coefficient $\mathcal{D}$ is defined as the coefficient of ∇n; thus

$$\Gamma_{\text{diff}} = -\mathcal{D}.\nabla n \tag{1.55}$$

and consequently:

$$\mathcal{D} = \lambda.\bar{v}/4 \tag{1.56}$$

On the other hand if diffusion is considered as the result of a force, itself resulting from a pressure gradient ∇p_r, one might write:

$$\nabla p_r = -\Gamma_{\text{diff}}\,\nu_c M \tag{1.57}$$

where ν_c is the collision frequency and M is the mass of the particles. For perfect gases the equation of state is:

$$p_r = nkT \tag{1.58}$$

Thus, combining equations (1.57) and (1.58):

$$\Gamma_{\text{diff}} = -\frac{kT}{M\nu_c}\nabla n \tag{1.59}$$

For a Maxwellian distribution function:

$$\tfrac{1}{2}Mv^2 = \tfrac{3}{2}kT \tag{1.60}$$

$$\bar{v} = \nu_c \lambda$$

Thus

$$\mathcal{D} = \lambda.\bar{v}/3 \tag{1.61}$$

Rigorous mathematical analysis of the molecular motions leads to the same result as was shown by Jeans[26], Kennard[28] and Loeb[34].

Mobility[16]

Under the effect of an externally applied force, the particles acquire a drift velocity v_d and are said to have a mobility $\mathscr{b}$. If the external force is due to an applied electric field $\boldsymbol{E}$, the resulting drift velocity of the particles can readily be obtained from the balance of forces; it is given by:

$$v_d = e\langle 1/\nu_c \rangle E/M \tag{1.62}$$

where $\langle 1/\nu_c \rangle$ is the average of the inverse of the collision frequency over the distribution function of the particle velocities. The mobility is defined as the ratio of the drift velocity to the electric field, thus

$$\mathscr{b} = e\langle 1/\nu_c \rangle/M = v_d/E \tag{1.63}$$

Electrical conductivity[15]

One of the most important transport coefficients is the electrical conductivity microscopically defined as the new flow of charges in response to an applied field. Macroscopically, it is found experimentally that the current density produced by an electric field is proportional to it, the constant of proportionality being by definition the electrical conductivity σ, or since $\boldsymbol{E} = -\nabla V$, then:

$$\boldsymbol{J} = -\sigma \nabla V = \sigma \boldsymbol{E}$$

where V is the potential. In a plasma where both positive and negative particles are present, the electrical conductivity is given by:

$$\sigma = n^+ q^+ v_d^+ + n^- q^- v_d^- \tag{1.64}$$

where q^+ and q^- are the electric charges, of the positive and the negative ions, respectively. Taking account of equation (1.63) to eliminate v_d^+ and v_d^- from equation (1.64), one obtains:

$$\sigma = \frac{n^+ q^{+2}}{M^+} \langle 1/v_c^+ \rangle + \frac{n^- q^{-2}}{M^-} \langle 1/v_c^- \rangle \tag{1.65}$$

One can write in general $\sigma = \sigma^+ + \sigma^-$ and the electronic contribution is:

$$\sigma^- = n^- e^2 \langle 1/v_c^- \rangle / M^-$$

Introducing equation (1.63)

$$\sigma^- = n^- e b \tag{1.66}$$

For a fully ionised gas in which the plasma consists only of simple ionised atoms and an equal number of electrons, v_c^- is given by equation (1.32) with $n_{sc} = n^+$ since the scatterers are the ions and the impinging beam is formed by the electrons in this case. For a fully ionised neutral plasma the contribution of the positive ions can be neglected, and equation (1.65) becomes:

$$\sigma = e^2 \langle 1/Q_t v^- \rangle / M^- \tag{1.67}$$

In this case, the electrical conductivity is not a function of the number density of carrier pairs. For a Coulomb interaction $Q_t \sim v^{-4}$, therefore $\sigma \sim v^3 \sim T^{3/2}$ or more exactly $\sigma = 3 \times 10^7 (kT_e)^{3/2}$. This neglects screening effects. When they are taken into account, they make the conductivity a function of the number density of carrier pairs as shown in Figure 1.13.

Thermal conductivity[12]

When there is a temperature gradient ∇T in a gas, a heat flux Q results which is directly proportional to the temperature gradient, the constant of proportionality is by definition the 'thermal conductivity' $\varkappa$, and

$$Q = -\varkappa . \nabla T \tag{1.68}$$

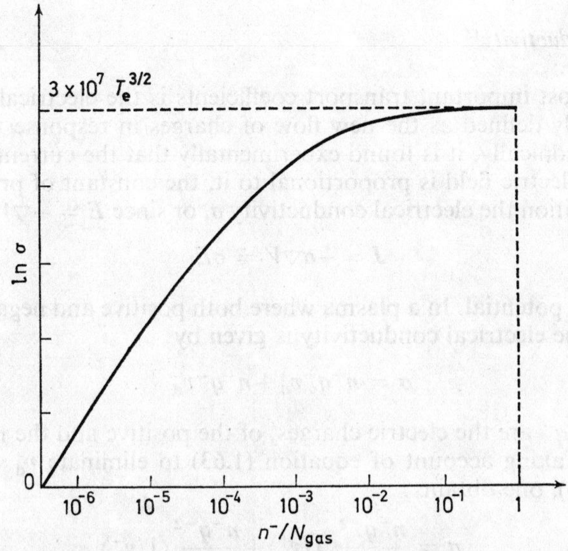

Fig. 1.13. The electrical conductivity as a function of the fractional ionisation. $(n^-/N_{gas})T^-$ *is the electron temperature in* keV

This thermal conductivity can be interpreted in terms of kinetic energy transport from one point to another. Where the temperature is greater, the average kinetic energy is also greater, so that a molecule passing from a region of higher temperature towards one of lower temperature contributes to 'heating' the latter. It can be shown that

$$\varkappa = k.n.\lambda.\bar{v}/2 \tag{1.69}$$

Viscosity[9, 12]

In some media, the component of the pressure tensor Ψ_{xy} in the plane xy is proportional to the gradient of the velocity v in the y-direction. The constant of proportionality is defined as the viscosity v; thus

$$\Psi_{xy} = -v\,\mathrm{d}v_y/\mathrm{d}x \tag{1.70}$$

This concept of viscosity can be interpreted in terms of momentum exchange among molecules pertaining to different layers. It can be shown here again that

$$v = n.M.\lambda.\bar{v}/3 \tag{1.71}$$

This viscosity is called 'static viscosity' in fluid mechanics.

Einstein's relation

The diffusion coefficient is given by equation (1.61), whereas the mobility is given by equation (1.63). Consequently, one can write:

$$\mathcal{D}/\mathscr{b} = \frac{M}{3e}.\lambda.\bar{v}/\langle 1/\nu_c \rangle \tag{1.72}$$

Replacing in this equation, $\bar{v}$ by its value as given by equation (1.60), one finds:

$$\mathcal{D}/\mathscr{b} = kT/e \tag{1.73}$$

which is called Einstein's relation. Thus, the ratio of the diffusion coefficient to the mobility is a linear function of the temperature.

1.10 THERMODYNAMICS OF GASES[4, 21]

In thermodynamics, a gas is said to be 'perfect' when it obeys the following equation of state $p_r\mathcal{V} = \mathcal{R}T$, in which $\mathcal{V}$ and $\mathcal{R}$ correspond to a mole. In this equation, $\mathcal{V}$ is the volume, $\mathcal{R}$ is the gas constant, p_r is the pressure and T is the temperature. According to Avogardo's hypothesis, the number of molecules per mole is given and is independent of the particular gas considered. This hypothesis has been checked experimentally by some fourteen different methods, which all agree within the experimental limits, and the experimental value of Avogadro's number $\mathcal{N}$ is approximately $60{\cdot}2\times10^{25}$. Boltzmann's constant is defined as $k = \mathcal{R}/\mathcal{N} = 1{\cdot}3\times10^{-23}$ J/K and if n is the number of molecules per unit volume, then $\mathcal{V} = \mathcal{N}/n$, leading to equation (1.58). If the gas is exposed to high energy radiations or is submitted to an extremely high temperature or to a high pressure, it will not any more obey the above equation of state satisfactorily. The equation of state of such a 'real' gas can be written $p_r = \alpha_s n.kT$, where α_s is called the departure coefficient. This coefficient can be smaller than, equal to or larger than unity, and it is generally a function of the temperature and of the pressure. At high pressures and low temperatures, the effect of the Van der Vaals intermolecular forces becomes important, thus affecting the value of α_s. At high temperatures and low pressures, some ionisation may take place, changing the number of particles in the gas and therefore the value of α_s.

A perfect gas can be interpreted as a collection of perfectly elastic spherical molecules of negligible size. The molecules as well as the containing walls being assumed perfectly smooth, the only interactions that can occur between them are elastic collisions. This is hardly the case in a plasma. However, under the assumption of thermal equilibrium, the plasma can be considered as a mixture of gases of different species, each of them obeying the equation of the perfect gas [equation (1.58)]. The total pressure of the plasma will then follow Dalton's law which states that the total pressure in a gas mixture is the sum of the partial pressures of its components.

Actually, the pressure is not only a function of the temperature and of the particle density, it is also a function of the degree of ionisation of the gas and of its chemistry. The degree of ionisation α is by definition the ratio of the number of ions existing in a given volume to the total number of ions and neutrals existing in the same volume. However, if an adequate relation for the equilibrium value of α is found, then the pressure can be expressed as a function of the particle density and the temperature only. This introduces the notion of the 'partition function' Z, a concept of statistical physics.

By assuming that the plasma is a neutral continuum, the Maxwell–Boltzmann statistics[31] can be applied. It states that the number of particles n_i in the ith cell of phase space is, in the state of maximum thermodynamic probability:

$$n_i = \frac{g_i}{\beta} \exp(-\mathcal{E}_i/kT) \tag{1.74}$$

where β is a temporarily undefined constant, g_i is the degeneracy (multiplicity) of the ith level and $\mathcal{E}_i$ is its energy. g_i is an integer equal to the number of states having exactly the same energy $\mathcal{E}_i$. The total number of particles is evidently $n = \sum_i n_i$, or

$$n = \frac{1}{\beta} \sum_i g_i \exp(-\mathcal{E}_i/kT)$$

thus

$$\beta = \frac{1}{n} \sum_i g_i \exp(-\mathcal{E}_i/kT) \tag{1.75}$$

The 'partition function' Z is then defined as

$$Z = \sum_i g_i \exp(-\mathcal{E}_i/kT) \tag{1.76}$$

The energy $\mathcal{E}$ of a non-ionised gas is equal to the sum of the energies of rotation, translation and vibration as well as of the electronic excitation energies of its atoms or molecules. Therefore, the total partition function of the gas will be equal to the product of the individual partition functions.

All the thermodynamic functions can, then, be obtained in terms of the partition function. For instance, the internal energy U of a system can be written: $U = \sum_i \mathcal{E}_i n_i$, and replacing n_i by its value from equation (1.74) and differentiating equation (1.76), one finds:

$$U = n.k.T^2.[\partial(\ln Z)/\partial T]_{\mathcal{V}} \tag{1.77}$$

and similarly for Helmholtz free energy:

$$\mathcal{F} = -n.k.T.[\ln(Z/n)+1] \tag{1.78}$$

the entropy:

$$\mathcal{S} = n.k.[1+\ln(Z/n)+T[\partial(\ln Z)/\partial T]_{\mathcal{V}}] \tag{1.79}$$

and the pressure:

$$p_r = n.k.T.[\partial(\ln Z)/\partial \mathcal{V}]_T \tag{1.80}$$

It is of interest to recall the relations between the internal energy U, the enthalpy $\mathcal{H}$ and the specific heats at constant pressure C_p and constant volume C_v. They are:

$$U-U_0 = \int_{T_0}^{T} C_v \, \mathrm{d}T \tag{1.81}$$

$$\mathcal{H}-\mathcal{H}_0 = \int_{T_0}^{T} C_p \, \mathrm{d}T \tag{1.82}$$

The specific heats are themselves related by $C_p - C_v = \mathcal{R}$, where $\mathcal{R}$ is the universal constant of gases ($\mathcal{R} = \mathcal{N}k = 8{\cdot}314$ J/mol K). The ratio of specific heats is:

$$\gamma = C_p/C_v = (n_f+2)/n_f \tag{1.83}$$

where n_f is the number of degrees of freedom in a gas. This number is involved in the principle of 'equipartition of energy'. The latter states that provided classical mechanics is applicable within a sufficient degree of aproximation each degree of freedom is in the average endowed with the same energy for all the particles involved. In other words, the average kinetic energies of the molecules in two gases at the same temperature and pressure are identical. To show this, consider a solid plane dividing a plasma into two regions, and consider the impulsion given by the particles to the wall per unit surface per unit time. This impulsion corresponds to a force per unit surface, i.e. to a pressure. Assuming that each particle rebounds elastically on the surface and that the latter is perfectly smooth, it can be seen that the impulsion thus given up is doubled. For a particle of speed v coming under an angle θ with respect to the normal to the plane, the impulsion is $2\,M.v.\cos\theta$. Following the same argument as given in Section 1.9, one finds that the number density of molecules bouncing on the wall per unit area per unit time as:

$$\mathrm{d}N/\mathrm{d}t = \frac{1}{4\pi}(v.\cos\theta)\sin\theta.\mathrm{d}\theta.\mathrm{d}\phi(2.n.M.v.\cos\theta)\,f(v).\mathrm{d}v \tag{1.84}$$

and the pressure will be:

$$p_r = \int_{\phi=\theta}^{\pi/2}\int_{\theta=0}^{\pi/2}\int_{v=0}^{\infty} \mathrm{d}N/\mathrm{d}t$$

and after integration:

$$p_r = nMC^2/3 \tag{1.85}$$

If there are two types of molecules, by Avogadro's hypothesis $n_1 = n_2$, and for $p_1 = p_2$, one will have then:

$$M_1C_1^2/2 = M_2C_2^2/2 \tag{1.86}$$

Note, finally, that the speed of sound in a continuum is given by:

$$v_s = (\partial p_r/\partial \varrho)^{1/2} \tag{1.87}$$

where ϱ is the mass density of the gas ($\varrho = nM$). For a perfect gas, equation (1.87) becomes:

$$v_s = (\gamma . kT/M)^{1/2} \tag{1.88}$$

Since γ is never very different from unity one sees that the speed of sound is of the same order of magnitude as the molecular speeds v_0, $\bar{v}$ and C.

REFERENCES AND BIBLIOGRAPHY

1. ARZIMOVICH, L. A., *Elementary Plasma Physics*, Blaisdell, New York (1965).
2. BACHYNSKI, M. P., 'Plasma Physics: an Elementary Review', *Proc. Inst. Radio Engrs*, **49**, 1751 (1961).
3. BATES, D. R., *Atomic and Molecular Processes*, Academic Press, New York (1962).
4. BENSON, S. W., BUSS, J. H. and MYERS, H., 'The Thermodynamic Properties of Ionised Gases', IAS paper 59–95, *IAS International Summer Meeting*, Los Angeles, Calif. June 16 (1959).
5. BIRD, R. B., STEWART, W. E. and LIGHTFOOT, E. N., *Transport Phenomena*, Wiley, New York (1960).
6. BROWN, S. C., *Basic Data of Plasma Physics*, Wiley, New York (1959).
7. BROWN, S. C. (Ed.), 'Outline of a Course in Plasma Physics', *Am. J. Phys*., **31**, No. 8, 637 (1963).
8. CAMBEL, A. B., *Plasma Physics and Magnetofluidmechanics*, McGraw-Hill, New York (1963).
9. CHANDRASEKHAR, S., 'Dynamical Friction', *Astrophys. J.*, **97**, 255, 263 (1943).
10. CHANDRASEKHAR, S., *Plasma Physics*, University of Chicago Press (1962).
11. CHAPMAN, S. and COWLING, T. G., *The Mathematical Theory of Nonuniform Gases*, Cambridge University Press (1953).
12. CHAPMAN, S., 'Viscosity and Thermal Conductivity of a Completely Ionized Gas', *Astrophys. J.*, **120**, No. 1, 151 (1954).
13. CLAUSER, F. N., *Plasma Dynamics*, Addison-Wesley, Reading, Mass. (1960).
14. COBINE, J. D., *Gaseous Conductors*, McGraw-Hill, New York (1941).
15. COHEN, R. S., SPITZER, L. Jr. and MCROUTLY, P., 'The Electrical Conductivity of an Ionized Gas', *Phys. Rev.*, **80**, 230 (1950).
16. COMPTON, K. T., 'Mobilities of Electrons in Gases', *Phys. Rev.*, **22**, Nov. (1923).
17. CRANK, J., *The Mathematics of Diffusion*, Oxford University Press (1956).
18. DEBYE, P. and HUCKEL, W., *Physikal Z.*, **24**, 183, 305 (1923).
19. DELCROIX, J. L., *Introduction to the Theory of Ionized Gases*, Interscience, New York (1960).
20. DRUMMOND, J. E. (Ed.), *Plasma Physics*, McGraw-Hill, New York (1961).
21. FLUGGE, S., 'Thermodynamik der Gase, *Handbuch der Physik*, Vol. 12, Springer-Verlag, Berlin (1958).
22. GARTENHAUS, S., *Elements of Plasma Physics*, Holt, Rinehart and Winston, New York (1964).
23. HELLUND, E. J., *The Plasma State*, Reinhold, New York (1961).
24. HOLT E. H. and HASKELL, R. E., *Plasma Dynamics*, MacMillan, New York (1965).
25. HOYAUX, M. F., *Notes on Plasma Physics*, Carnegie Institute of Technology, Pittsburgh, Pa. (1964).
26. JEANS, J., *An Introduction to the Kinetic Theory of Gases*, Cambridge University Press (1962).
27. JOST, W., *Diffusion in Solids, Liquids and Gases*, Academic Press, New York (1952).
28. KENNARD, E. H., *Kinetic Theory of Gases*, McGraw-Hill, New York (1938).
29. KETTANI, M. A., *Notes on Plasma Engineering*, University of Pittsburgh, Pittsburgh, Pa. (1967).

30. KUNKEL, W. B. (Ed.), *Plasma Physics in Theory and Application*, McGraw-Hill, New York (1966).
31. LANDAU, L. D. and LIFSHITZ, E. M., *Statistical Physics*, Addison-Wesley, Reading, Mass. (1958).
32. LENNARD-JONES, J. E., 'On the Determination of Molecular Fields from the Equations of State of a Gas', *Proc. R. Soc.*, **A 106,** 463 (1924).
33. LINHART, J. G., *Plasma Physics*, Interscience, New York (1960).
34. LOEB, L. B., *Basic Processes of Gaseous Electronics*, University of California Press, Berkeley, Calif. (1955).
35. MEIENS, C., 'Diffusion and Elastic Collision Losses of Fast Electrons in Plasmas', *J. appl. Phys.*, 29, 903 (1958).
36. MORSE, P. M., ALLIS, W. P. and LAMAR, E. S., 'Velocity Distribution for Elastically Colliding Electrons', *Phys. Rev.*, **48,** 412 (1935).
37. PRESENT, R. D., *Kinetic Theory of Gases*, McGraw-Hill, New York (1958).
38. SEKIGUCHI, T. and HERNDON, R. C., 'Thermal Conductivity of an Electron Gas in a Gaseous Plasma', *Phys. Rev.*, **112,** 1 (1958).
39. SENGUPTA, D. L., 'The Electrical Conductivity of a Partially Ionized Gas', *Proc. Inst. Radio Engrs*, **49,** 1872 (1961).
40. SPITZER, L., Jr., *Physics of Fully Ionized Gases*, Interscience, New York (1962).
41. TANENBAUM, B. S., *Plasma Physics*, McGraw-Hill, New York (1967).
42. TONKS, L. and LANGMUIR, I., 'Oscillations in Ionized Gases', *Phys. Rev.*, **33,** 195 (1929).
43. UMAN, M. A., *Introduction to Plasma Physics*, McGraw-Hill, New York (1964).
44. VON ENGEL, A., *Ionized Gases*, Oxford University Press (1955).

2 Fundamental plasma processes

2.1 PARTICLE ORBITS IN ELECTRIC AND MAGNETIC FIELDS[16]

One of the most important phenomena occurring in plasmas is the interaction between the fields and the moving charges. This phenomenon exists in many applications such as space charge control tubes, electron tube devices (cathode-ray tubes, electron lenses, etc.), and magnetic confinement in thermonuclear approaches.

According to the kinetic theory of gases, the motion of charge carriers can be considered as a succession of free paths between 'collisions' or 'deflections'. Along a free path, the charge carrier behaves as if it were in a vacuum. In the case of the hard-sphere model, the collisions are distinct events, well separated from each other. For the potential model, from a certain density upward, it can no longer be said that a given charge carrier is interfering with scattering centres one at a time.

To determine the trajectory of a charged particle moving in a static electromagnetic field, one needs to know the forces present, resulting from *electromagnetic* fields, and the laws of *mechanics*. The electromagnetic fields are of two different kinds: 'macrofields', which are generally created by external electrodes or currents, or by the space charge or the motion of large fragments of plasma, and 'microfields' which result from individual particles. In the hard-sphere model, the latter is neglected. In the potential model, particles can be treated separately only if the density of the scattering centres is not too high.

The force due to the 'macroscopic' electromagnetic field is the sum of two individual forces, an electric force $\boldsymbol{F}_E = q\boldsymbol{E}$, due to the electric field $\boldsymbol{E}$ on the charge q and a magnetic force $\boldsymbol{F}_B = q\boldsymbol{v}\times\boldsymbol{B}$, due to the magnetic field $\boldsymbol{B}$ on a charge moving with a velocity $\boldsymbol{v}$. Thus, neglecting the force of gravity,

$$\boldsymbol{F} = q(\boldsymbol{E}+\boldsymbol{v}\times\boldsymbol{B}) \tag{2.1}$$

The motion of charge carriers will also be governed by the laws of classical mechanics, with a constant mass and no associated wave except in two cases: when relativistic speeds are reached and when the plasma has such a density that the factor of degeneracy* is not negligible with respect to unity, e.g. in the

* When there is more than one state on the same energy level, these states are called degenerate.

white dwarf stars. In classical mechanics Newton's first law states:

$$F = M\,\mathrm{d}\boldsymbol{v}/\mathrm{d}t \tag{2.2}$$

where M is the mass of one particle.

Equations (2.1) and (2.2) are the two fundamental equations for studying the motion of charged particles. The field forces should equal Newton's force and if cartesian co-ordinates are chosen such that z is in the $\boldsymbol{B}$-direction, then:

$$M\,\mathrm{d}v_x/\mathrm{d}t = q(E_x+v_yB) \tag{2.3a}$$

$$M\,\mathrm{d}v_y/\mathrm{d}t = q(E_y-v_xB) \tag{2.3b}$$

$$M\,\mathrm{d}v_z/\mathrm{d}t = qE_z \tag{2.3c}$$

where subscripts x, y and z refer to the components in the x-, y-, and z-directions, respectively.

Motion along the magnetic field

Along the z-co-ordinate, equation (2.3c) yields after integration:

$$v_z = v_{z0}+\frac{q}{M}\int_0^t E_z\,\mathrm{d}t$$

For a constant E_z, this is an accelerated motion in the z-direction. One can also write that $v_z = \mathrm{d}z/\mathrm{d}t$ and that the energy in the $\boldsymbol{B}$-direction is $\mathcal{E}_{\parallel} = Mv_z^2/2$. Therefore after taking into consideration equation (2.3c), one finds after integrating the electric field,

$$\mathcal{E}_{\parallel}-\mathcal{E}_{\parallel 0} = q[V(z)-V(o)] \tag{2.4}$$

where $V(z)$ is the electric potential which depends on the co-ordinate z; from equation (2.4), $\mathcal{E}_{\parallel}(z) = qV(z)$. This equation states that the electric potential at a point is the amount of energy per unit charge spent by a moving charged particle at that point.

Motion perpendicular to the magnetic field

It is assumed that the magnetic and electric fields are homogeneous and constant in space and time. The motion perpendicular to the magnetic field is described by equations (2.3a) and (2.3b). If a new frame of reference is defined such as:

$$\begin{aligned} v'_x &= -\frac{E_y}{B}+v_x \\ v'_y &= \frac{E_x}{B}+v_y \end{aligned} \tag{2.5}$$

and since $v'_x = \mathrm{d}x'/\mathrm{d}t$ and $v'_y = \mathrm{d}y'/\mathrm{d}t$, then,

$$\mathrm{d}^2x'/\mathrm{d}t^2 = \frac{q}{M} B\, \mathrm{d}y'/\mathrm{d}t$$

$$\mathrm{d}^2y'/\mathrm{d}t^2 = -\frac{q}{M} B\, \mathrm{d}x'/\mathrm{d}t \tag{2.6}$$

or in vectorial notation,

$$\mathrm{d}\boldsymbol{v}'/\mathrm{d}t = \frac{q}{M}\boldsymbol{v}'\times\boldsymbol{B} \tag{2.7}$$

whereby equations (2.5) become:

$$\boldsymbol{v}-\boldsymbol{v}' = \boldsymbol{v}_0 = \boldsymbol{E}\times\boldsymbol{B}/B^2 \tag{2.8}$$

There is no electric field in the new frame of reference. The motion is due to the magnetic field only. The velocity of the frame itself can be identified with an electric drift of the particles. Consider now, a complex plane where y is the imaginary axis and x is the real axis (see Figure 2.1). In this frame, the complex co-ordinate is:

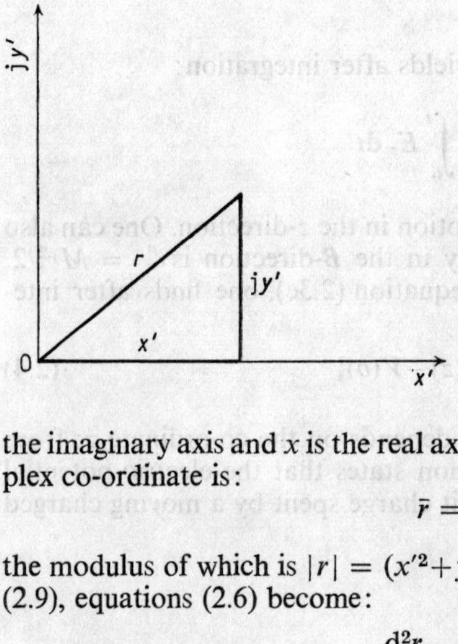

Fig. 2.1. The complex frame of reference describing the motion of an electron perpendicularly to a magnetic field. This is a cyclotron motion of gyroradius r_0 and gyrofrequency ω_c

$$\bar{r} = x'+\mathrm{j}y' \tag{2.9}$$

the modulus of which is $|r| = (x'^2+y'^2)^{1/2}$. Taking into consideration equation (2.9), equations (2.6) become:

$$\frac{\mathrm{d}^2r}{\mathrm{d}t^2}+\mathrm{j}\frac{qB}{M}\frac{\mathrm{d}r}{\mathrm{d}t} = 0$$

This is a linear differential equation, the solution of which is in the form:

$$r = r_0 \exp(-\mathrm{j}\omega_c t)+r'_0 \tag{2.10}$$

where $\omega_c = qB/M$ is the gyrofrequency of the particle, and $r_0 = Mv/qB$ is its gyroradius. Thus, in the frame where the electric field is equal to zero, the path of the particle is a circle of radius r_0 (gyroradius) on which the particle moves with a frequency ω_c (gyrofrequency). This is a cyclotron motion.

With the existence of both the electric and magnetic fields, the motion of the particle will be a drift of its guiding centre upon which a rotation is superimposed. This is, in general, a helicoidal motion as shown in Figure 2.2.

The general problem of the motion of charge carriers through a non-uniform magnetic field is complicated. It can be simplified under some assumptions. For instance, the space variation of the magnetic field can be assumed to be

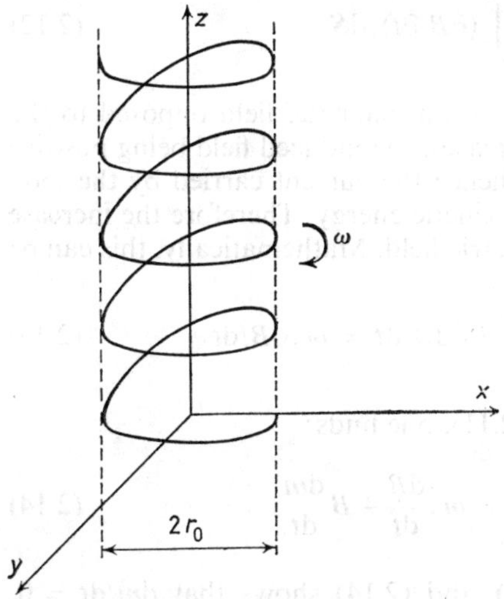

Fig. 2.2. The helicoidal orbit of an electron moving in the presence of both an electric and a magnetic field. The motion of the particle is a drift of its guiding centre upon which a rotation is superimposed

small if the distance involved is comparable to the gyroradius, or also the variation of the magnetic field can be assumed to be slow with respect to the duration of one revolution in the cyclotronic motion.

2.2 ADIABATIC INVARIANTS[34]

Orbits are described by three adiabatic invariants: the magnetic moment ***m***, the flux Φ within a guiding surface, and the action $\mathcal{J}$.

Magnetic moment

In the case of a homogeneous and constant magnetic field, the particle has a cyclotron motion in a plane perpendicular to the magnetic field. It should be noted that, both for positive and negative particles, the current carried by the particles creates a magnetic field which is opposed to the external magnetic field. It can thus be concluded that an ionised gas is diamagnetic.

In all the cases where the magnetic induction is non-uniform, the notion of magnetic moment becomes important. For a loop, the magnetic moment ***m*** is defined as the product of the current I times the loop area S. Since for a 'cyclotron orbit' S is πr_0^2 and I is $q\omega_c/2\pi$, one finds, after replacing ω_c and r_0 by their values:

$$m = \mathcal{E}_\perp / B \tag{2.11}$$

where $\mathscr{E}_\perp = Mv^2/2$ is the kinetic energy in the direction perpendicular to the magnetic field.

It can be shown[7] that the magnetic moment is constant for slowly time-varying fields. Assume that $\boldsymbol{B}$ is slowly varying with time, i.e. that the relative change of the magnetic field is small during a single gyration period. An e.m.f. will be induced such as:

$$\int \boldsymbol{E}.\mathrm{d}\boldsymbol{l} = -\iint (\partial \boldsymbol{B}/\partial t).\mathrm{d}\boldsymbol{S} \tag{2.12}$$

This e.m.f. carries a current which generates a magnetic field opposed to the applied one. When the magnetic field increases, the induced field being positive by definition, the speed of the particle, hence the current carried by the loop is increased and the particle increases its kinetic energy. Therefore the increase of kinetic energy is due to the induced electric field. Mathematically, this can be written:

$$\mathrm{d}(\mathscr{E}_\perp)/\mathrm{d}t = -I\int \boldsymbol{E}.\mathrm{d}\boldsymbol{l} = IS\,\mathrm{d}B/\mathrm{d}t = \boldsymbol{m}.\mathrm{d}\boldsymbol{B}/\mathrm{d}t \tag{2.13}$$

and taking into consideration equation (2.11), one finds:

$$\mathrm{d}(\mathscr{E}_\perp)/\mathrm{d}t = \frac{\mathrm{d}(mB)}{\mathrm{d}t} = \boldsymbol{m}.\frac{\mathrm{d}\boldsymbol{B}}{\mathrm{d}t} + B\frac{\mathrm{d}m}{\mathrm{d}t} \tag{2.14}$$

A comparison between equations (2.13) and (2.14) shows that $\mathrm{d}m/\mathrm{d}t = 0$, therefore m is a constant of the motion.

It can also be shown that the magnetic moment remains constant for a magnetic field varying with position. The theory to be outlined corresponds to a magnetic field pattern known as the 'magnetic mirror'. Consider the motion of a particle spinning in the xy-plane, the $\boldsymbol{B}$-field being function of the r- and z-co-ordinates. From Maxwell's equations:

$$\nabla.\boldsymbol{B} = 0 \tag{2.15}$$

Taking cylindrical co-ordinates, neglecting the variation of $\boldsymbol{B}$ in the azimuthal direction and assuming that the z-component of the magnetic field is much larger than the r-component, one obtains in the neighbourhood of the z-axis:

$$B_r = -\frac{r_0}{2}\frac{\partial B}{\partial t} \tag{2.16}$$

This means that an axial force is created, the value of which is $F_\parallel = qvB_r$, or after replacing B_r, v and r_0 by their values one obtains $F_\parallel = -m\,\partial B/\partial z$. Due to this force the particle orbit cannot stay in the same plane. It shifts in the axial direction. If the magnetic field is independent of time, then, after deriving the constant total kinetic energy,

$$\frac{\mathrm{d}\mathscr{E}_\perp}{\mathrm{d}t} + \frac{\mathrm{d}\mathscr{E}_\parallel}{\mathrm{d}t} = 0 \tag{2.17}$$

On the other hand, since $F_{\|} = M\,\mathrm{d}v_{\|}/\mathrm{d}t = -m\,\partial B/\partial z$, then:

$$Mv_{\|}\,\mathrm{d}v_{\|}/\mathrm{d}t = -mv_{\|}\,\partial B/\partial z = -m\,\mathrm{d}B/\mathrm{d}t \tag{2.18}$$

From equation (2.11), replacing $\mathscr{E}_{\perp}$ by its value in equation (2.17), one obtains:

$$\frac{\mathrm{d}\mathscr{E}_{\|}}{\mathrm{d}t} + \frac{\mathrm{d}}{\mathrm{d}t}(mB) = 0$$

and from equation (2.18):

$$\frac{\mathrm{d}\mathscr{E}_{\|}}{\mathrm{d}t} + m\frac{\partial B}{\partial t} = 0 \tag{2.19}$$

Therefore, one should necessarily have m constant.

Magnetic flux Φ

The magnetic flux is equal, by definition, to the product of the magnetic induction times the surface πr_0^2 of the loop. Replacing r_0 and B by their values as functions of m, one obtains:

$$\Phi = 2\pi Mm/q^2 \tag{2.20}$$

Therefore, the magnetic flux across the orbit is a constant of the motion. The actual orbit follows a flux tube as shown in Figure 2.3.

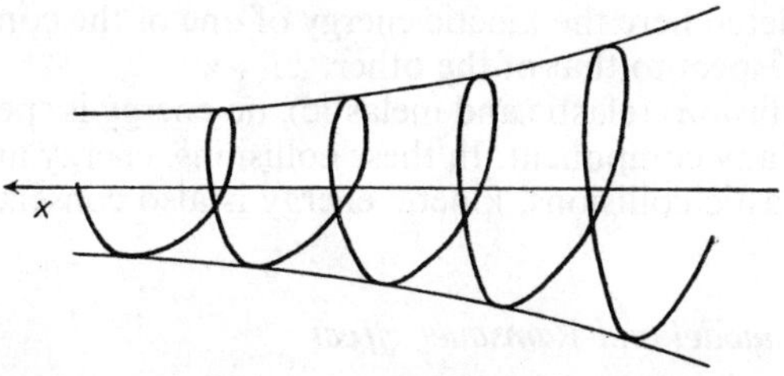

Fig. 2.3. Orbit in an increasing magnetic field. It is a helicoid of variable radius and pitch included in a flux tube

Action $\mathscr{J}$

The action is another adiabatic invariant; it is defined as $\mathscr{J} = \int_{z_1}^{z_2} v_{\|}\,\mathrm{d}z$, and it is constant when the magnetic fields vary slowly. If $\theta(z)$ is the angle of magnetic reflection, one can write $v_{\|} = v\cos[\theta(z)]$ and thus:

$$\mathscr{J} = v\int_{z_1}^{z_2} [1 - \sin^2\theta(z)]^{1/2}\,\mathrm{d}z \tag{2.21}$$

since v is constant. The condition of reflection for the particles states that:

$$\sin[\theta(z)] > [B(z)/B_{\max}]^{1/2}$$

and substituting $[B(z)/B_{\max}]^{1/2}$ in equation (2.21), one finds:

$$\mathcal{J} = \frac{v}{B_{\max}} \int_{z_1}^{z_2} [B_{\max} - B(z)]^{1/2}\, dz = m^{1/2} \times \text{constant} \tag{2.22}$$

Therefore the action is a constant of the motion.

The principle of the magnetic mirror is used in 'magnetic bottle' configurations to confine plasmas artificially. Natural magnetic mirrors exist also, and the Van Allen belts around the earth provide an example.

2.3 ELASTIC COLLISIONS[11]

Apart from emission of light by ordinary excited atoms and molecules, particles generally do not change state during their free paths but only at the moment of their collisions with other particles or under impact of light quanta. Thus, almost all elementary processes occurring in a plasma are collisions. If the density is not very high, which is generally the case in practical applications, only binary collisions need to be considered. Moreover, the temperature is not always the same for all the components of a plasma. In low-pressure electrical discharges in gases, the electron temperature may be several decimal orders of magnitude above the gas or the ionic temperature. In thermonuclear approaches, the ionic compound is generally the hottest. For this reason, in collisions considered here the kinetic energy of one of the components has been neglected with respect to that of the other.

In classical collisions (elastic and inelastic), no energy is spent in changing the internal state of any component. In these collisions, energy and momentum are conserved. In elastic collisions, kinetic energy is also conserved in the process.

The hard-sphere model and Ramsauer effect

In 1903, Lenard determined experimentally the attenuation of a beam of monoenergetic electrons by a gas. When the beam was passed into a field-free space, electrons were scattered out of the beam by collision with the gas atoms. The simple apparatus of Lenard ignored only the deflected electrons. Electrons which suffered energy changes without being deflected by the beam would still be detected as if they did not suffer any collision. The experiment of Lenard led to a constant cross-section Q_t of collision corresponding to the hard-sphere model and equal to πr_a^2 where r_a is the radius of the atom.

However, the atom is actually a system of charges with their resultant electric fields. It does not agree entirely with the hard-sphere mathematical model. An electron passing by the atom will be deflected by these fields. The difficulty of Lenard's apparatus was overcome by Ramsauer. Ramsauer's method can be described as follows. Electrons are emitted by a photoelectric source and they are bent in a uniform magnetic field so that a collimated monoenergetic beam is formed at a source slit. Residual electrons are measured in a collector and collisions in a field-free box. If an electron suffers any angular deflection greater than the angular aperture of the detector, it will be lost to the beam. Ramsauer

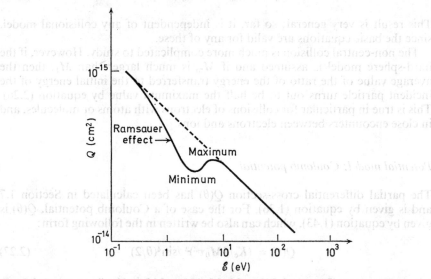

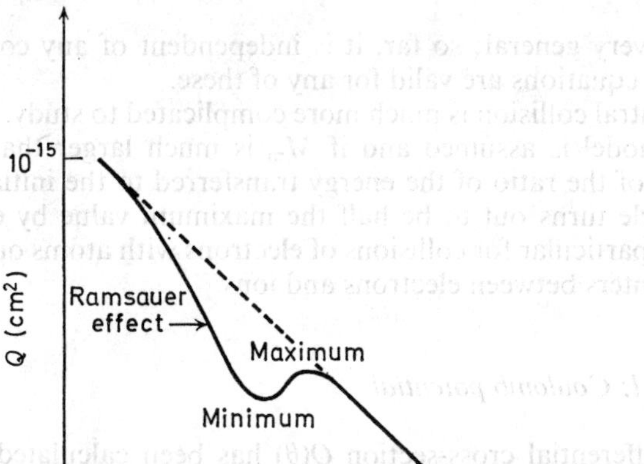

Fig. 2.4. Ramsauer effect. It is the extremely low value of the total cross-section at low electron energies

found that the scattering cross-section for elastic collisions varies with the energy of electrons. Some gases exhibit extremely low values of Q_t at low electron energies; this is known as the Ramsauer effect. It is shown in Figure 2.4.

Energy exchange in an elastic collision

The total kinetic energy is here an invariant and the only existing transfer of kinetic energy is between the two particles involved in an elastic collision. Consider the case of a *central collision* of a moving particle with a molecule at rest, that is, the case where the initial velocity of the particle is directed towards the centre of the molecule. The problem is one-dimensional. Let M_m be the mass of the molecule, M_i that of the incident particle, v the initial velocity of the latter, v_1' its final velocity after collision, and v_2' the recoil velocity of the molecule. From the equation of momentum conservation:

$$M_i v = M_m v_2' + M_i v_1' \tag{2.23}$$

The equation of kinetic energy conservation states:

$$\tfrac{1}{2} M_i v^2 = \tfrac{1}{2} M_m v_2'^2 + \tfrac{1}{2} M_i v_1'^2 \tag{2.24}$$

Once this system is solved for v_1' and v_2', it can be shown that:

$$M_m v_2'^2/2 = \chi_m M_i v^2/2 \tag{2.25}$$

with

$$\chi_m = 4M_m M_i/(M_m+M_i)^2 \tag{2.26}$$

This result is very general; so far, it is independent of any collisional model, since the basic equations are valid for any of these.

The non-central collision is much more complicated to study. However, if the hard-sphere model is assumed and if M_m is much larger than M_i, then the average value of the ratio of the energy transferred to the initial energy of the incident particle turns out to be half the maximum value by equation (2.26). This is true in particular for collisions of electrons with atoms or molecules, and in close encounters between electrons and ions.

Potential model: Coulomb potential

The partial differential cross-section $Q(\theta)$ has been calculated in Section 1.7 and is given by equation (1.36). For the case of a Coulomb potential, $Q(\theta)$ is given by equation (1.43), which can also be written in the following form:

$$Q(\theta) = (K_c/4M_0 v^2)^2 /\sin^4(\theta/2) \tag{2.27}$$

where $K_c = q_1 q_2/2\pi\varepsilon_0$ and $M_0 = M_1 M_2/(M_1+M_2)$ in the most general case; the subscripts 1 and 2 refer to the two colliding particles. From equation (2.27), it is apparent that the scattering is highly anisotropic and that the integral over all the values of θ diverges, a result easily understood because of the long range of Coulomb interaction. The farther away the particle passes, the less it is deflected, but the deflection angle never reaches zero. This divergence is a weakness in the model and several attempts to eliminate it have been made.

Potential model: large angle deflection

A 'large angle deflection' is by definition one corresponding to a value of θ at least equal to 90°. To calculate the 'deflection part' of the Coulomb cross-section, first set in equation (1.42) $\tan(\theta/2)$ equal to one. This gives, in the most general form mentioned above,

$$x_{bc} = K_c/M_0 v^2 = K_c/2\mathcal{E} \tag{2.28}$$

where $\mathcal{E}$ is the initial kinetic energy of the fictitious particle having a mass M_0. In such a case, a cross-section can be defined with the same accuracy as in the hard-sphere model, since any x_b smaller than that given by equation (2.28) gives a deflection at least equal to 90°. Hence, by definition, the collision part of the cross-section is:

$$Q_c = \pi x_{bc}^2 = \pi K_c^2/4\mathcal{E}^2 \tag{2.29}$$

Insertion of numerical values shows that, in most practical cases, this is already greater than any 'mechanical' collision cross-section, so that consideration of the Coulomb field alone is generally sufficient.

Potential model: small angle deflection

If θ is smaller than 90° only a small error is made when $\theta/2$ is substituted for $\tan(\theta/2)$. In the process of deflection, the particle acquires a component of velocity $v_\perp$ perpendicular to the initial velocity and such as $v_\perp = v\theta$. If, in a time interval τ, dN deflections occur, with an impact parameter in the range x_b, $x_b + \mathrm{d}x_b$, one can expect a variation in v_1^2 equal to dN times that resulting from v_1, thus:

$$\mathrm{d}v_\perp^2 = v^2\theta^2\,\mathrm{d}N \tag{2.30}$$

with

$$\mathrm{d}N = 2\pi n v \tau x_b\,\mathrm{d}x_b \tag{2.31}$$

where n is the number of scattering centres and τ is the time interval of collision. Replacing v by its value in equation (1.42), one finds:

$$v = (2K_c/M_0 x_b \theta)^{1/2} \tag{2.32}$$

Taking into consideration equations (2.30), (2.31) and (2.32), one obtains:

$$\mathrm{d}v^2 = \frac{8\pi K_c^2 n\lambda}{M_0^2 v^2}\frac{\mathrm{d}x_b}{x_b} \tag{2.33}$$

where $\lambda = \tau v$. To integrate this equation, it is usually assumed that the lower limit of x_b is x_{bc} and that the upper limit is the Debye screening length r_D. For a large number of small-angle scatterings $v_\perp$ is on the average equal to v; furthermore, λ can be considered as the mean free path such as $n\lambda = 1/Q_d$, where Q_d is the cross-section of the small angle-scattering; after taking these simplifying assumptions into consideration one finally finds:

$$Q_d = 2\pi K_c^2 \ln(r_D/x_{bc})/\mathscr{E}^2 \tag{2.34}$$

The weakness of this result is that the interaction potential is assumed to be a Coulomb potential up to the Debye screening length with the interaction abruptly going to zero at that distance. Actually the Coulomb potential energy K_c/r should be replaced by a Yukawa potential energy of the form:

$$\mathscr{E}(r) = K_c \exp(-r/r_D)/r$$

But this does not significantly change the final result.

Gvosdover effect[24]

The long-distance Coulomb interaction of the electrons in the microfield of the positive ions is known in the theory of high-current gas discharges as the Gvosdover effect. It conforms essentially to the same theory mentioned above, with the simplification that M_0 is the mass of the electron and that the relative kinetic energy $\mathcal{E}$ can be replaced by $3kT/2$, T being the electron temperature. The coefficients thus obtained are slightly different from those obtained in Gvosdover's expression:

$$Q_d = \frac{\pi}{2}(e^2/4\pi\varepsilon_0 kT)^2 \ln(6\pi\varepsilon_0 kT/e^2N^{1/3}) \tag{2.35}$$

It can be concluded that the scattering cross-section of one particle depends upon the state of the entire plasma in the neighbourhood. It should also be noticed that as soon as the relative amount of atoms or molecules ionised is above a few per cent, the electron mean free path is more and more governed by the ionic density, rather than the neutral gas density.

Induced dipole potential

The interaction between a charge carrier and a neutral molecule is sometimes studied by an 'induced dipole' model. If the molecule is polarisable, the presence in its vicinity of a charge carrier induces a dipole, which in turn attracts the charge carrier. As discussed in Section 1.4 the law of attraction is $1/r^4$.

2.4 INELASTIC COLLISIONS[2]

If the collisions are inelastic, equation (2.24) is no longer valid. When both particles have a certain velocity v before interaction and v' after interaction, the balance of kinetic energy yields for an inelastic collision:

$$M_1v_1^2 + M_2v_2^2 = M_1v_1'^2 + M_2v_2'^2 + 2\Delta\mathcal{E}$$

where $\Delta\mathcal{E}$ is the 'heat of reaction' per interaction. It can always be written that $\Delta\mathcal{E} = eV_r$ where V_r is a 'potential of reaction'. The 'heat of reaction' can be negative as well as positive. A collision in which the total kinetic energy decreases ($\Delta\mathcal{E} > 0$) is designated as a 'collision of the first kind'; a collision in which it increases is a 'collision of the second kind'.

An atom consists of a central positive charge, called a nucleus, which is surrounded by a cloud of revolving electrons. The whole structure is electrically neutral and almost the entire mass of the atom is concentrated in the nucleus. The chemical properties of an atom depend on the number Z of the electrons and on their arrangement. An electron has a negative charge ($-e$), therefore the atom will have a total negative charge ($-Ze$) and consequently, because of the overall neutrality, the charge of the nucleus will be ($+Ze$). Nuclei are formed of two fundamental particles: protons of one atomic unit mass and of charge $+e$ and neutrons of one atomic unit mass and of charge zero. If A is the mass number of an atom, the number of neutrons in the nucleus will be ($A-Z$). These particles are bound together in a very small space by short range attractive forces that are much stronger than the electrostatic repulsion between the protons. This picture of the atom was first proposed by the Danish scientist

Bohr, the proton–neutron structure of the nucleus having been added later by Heisenberg.

The two most important types of inelastic collisions are the excitation and the ionising collisions. They are possible only when the speed of the colliding particle is at least equal to the orbital speed of an atomic electron. Therefore heavy particles will not cause much ionisation by impact unless their kinetic energy is very high. Also, collisions of this type follow the Heisenberg Uncertainty Principle which states that $\Delta\mathcal{E}\,\Delta t > h/4$, where $h = 6{\cdot}624 \times 10^{-34}$ J s is Planck's constant. This means that experiment cannot fix both $\mathcal{E}$ and t to a limited precision but energy is determined only within a range $\Delta\mathcal{E}$ and time within a range Δt, the product of the two ranges being equal to a constant.

Excitation collisions[23]

An atom A is said to be excited when an electron is lifted from a lower to a higher energy level. This may be due to a mechanical collision with an electron, an ion, or a fast neutral particle. Figure 2.5(a) shows the case of an excitation by an electron in which:

$$\mathrm{A+e \rightarrow A^{*}+e}$$

where A* is the excited atom. Figure 2.5(b) shows the case of excitation by absorption of a photon. A photon is considered as a particle of zero mass and

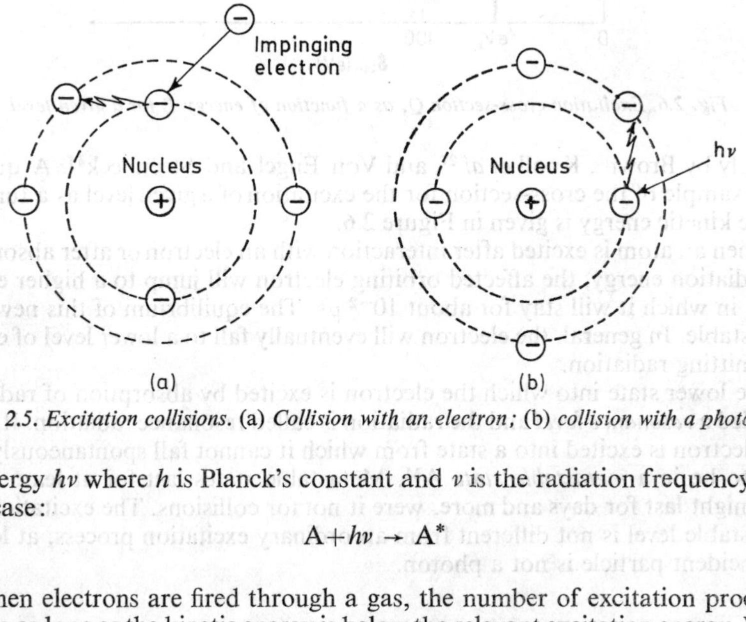

Fig. 2.5. Excitation collisions. (a) *Collision with an electron;* (b) *collision with a photon*

of energy $h\nu$ where h is Planck's constant and ν is the radiation frequency. For this case:

$$\mathrm{A}+h\nu \rightarrow \mathrm{A}^{*}$$

When electrons are fired through a gas, the number of excitation processes is zero as long as the kinetic energy is below the relevant excitation energy. When the excitation energy is exceeded, the excitation process suddenly commences; as the energy increases, the process rises to a maximum and then declines slowly.

A cross-section can thus be defined for the excitation of this particular level. If several excitation levels are involved, the different excitation cross-sections simply add up to give a total excitation cross-section.

It should be noted that excitation by electron collision is very different from photoexcitation. If one were able to throw photons of increasing energy through a gas, the curve of excitation cross-section would consist of a series of well defined, widely separated and roughly Gaussian peaks, each corresponding to one of the excitation levels. Examples of cross-sectional characteristics are given

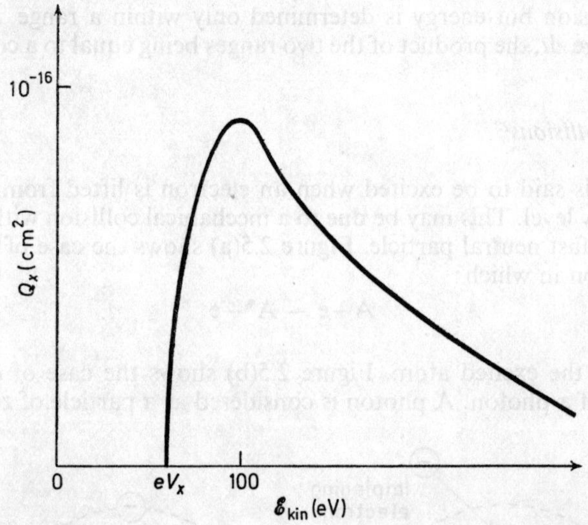

Fig. 2.6. Excitation cross-section Q_e as a function of energy $\mathcal{E}$ for a given level

notably by Brown[5], Knoll *et al.*[27], and Von Engel and Steenbeck[44]. A qualitative example of the cross-section for the excitation of a given level as a function of the kinetic energy is given in Figure 2.6.

When an atom is excited after interaction with an electron or after absorption of radiation energy, the affected orbiting electron will jump to a higher energy level, in which it will stay for about 10^{-2} μs. The equilibrium of this new state is unstable. In general, the electron will eventually fall to a lower level of energy by emitting radiation.

The lower state into which the electron is excited by absorption of radiation is called a *resonance level* and the radiation is called resonance radiation. If, now, the electron is excited into a state from which it cannot fall spontaneously, it is said to be in a *metastable state* A^M. Metastable states last for at least 10 ms and might last for days and more, were it not for collisions. The excitation of a metastable level is not different from an ordinary excitation process, at least if the incident particle is not a photon.

Ionising collisions[1]

Ionisation can be regarded as an extreme case of excitation where the electron is given an energy greater than the highest excited levels of the atom. The electron escapes from the attraction of its parent nucleus with some kinetic energy, and the atom becomes an *ion*, having one positive charge. This is shown in Figure 2.7. In this case the remaining kinetic energy after collision is split into three parts,

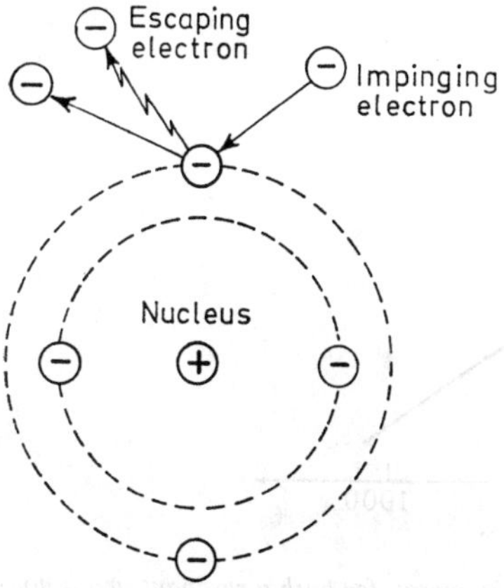

Fig. 2.7. Ionisation by collision of an atom with an electron. An electron of the outer energy level escapes from the influence of its parent nucleus

the recoil energy of the atom or molecule, the remaining energy of the ionising agent, and the kinetic energy of the electron detached. The reaction is:

$$A+e \rightarrow A^{+}+e+e$$

where A^{+} is the positively charged ion.

Conspicuous data on the curve giving the ionisation cross-section as a function of energy or speed exist when the incident particles are electrons. The shape of the curves is slightly different from those for excitation. The maximum is flatter, it occurs for higher energies; and the descent is slower. Moreover, from data related to cosmic rays, it is known that, at higher energies, the curve rises again. The experimental curve is frequently approximated analytically, either by a rising straight line or a step function.

Ionisation may be obtained by a variety of processes: (1) by collision with an electron, a positive ion or a fast neutral atom; (2) by 'collision' with a photon; (3) by electron attachment; (4) by collisions of the second kind; (5) step ionisation; (6) thermal ionisation; (7) by high-frequency electric fields; (8) surface ionisation.

When an ion collides with a neutral atom, the charge may be transferred from the former to the latter, provided the energy balance of the process can be satisfied. This phenomenon is important in gas mixtures where the two colliding particles may have different chemical natures. The different reactions leading to charge transfer are, the resonant reaction:

$$A^{+}+A \rightarrow A+A^{+}$$

and the two non-resonant reactions:

$$A^+ + B \rightarrow A + B^+$$
$$A^{++} + A \rightarrow A^+ + A^+$$

The charge transfer cross-sections are shown in Figure 2.8 for both resonant and non-resonant reactions. It should be noted that charge transfer can cause

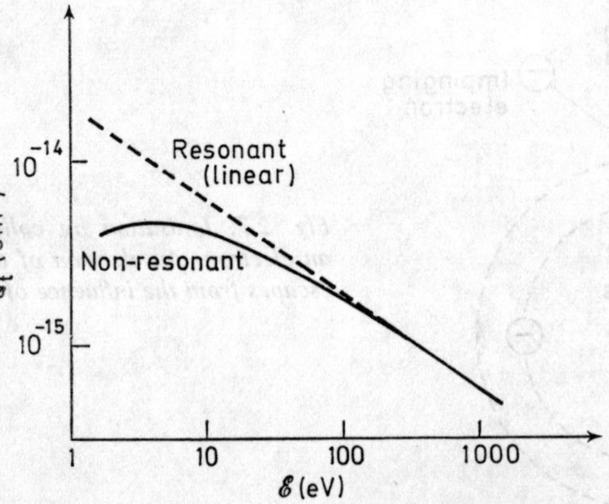

Fig. 2.8. Charge transfer cross-section Q_t as a function of energy for both a resonant and a non-resonant reaction. For a resonant reaction the characteristic is linear

serious losses of hot ions in contained plasmas if impurity atoms stream in at low energies.

When a neutral atom collides with a photon of high energy, the valence electron* will eventually escape from the atom, following the reaction:

$$A + h\nu \rightarrow A^+ + e$$

where A^+ is the positively charged ion. Figure 2.9 shows the variation of the excitation and ionisation cross-sections as functions of the energy of the photon. An excitation level corresponds to each energy of the orbiting electron. The ionisation level can be considered as the excitation level of the highest energy level of the atom (valence electron). The excitation cross-section functions of the radiation energy are Gaussian.

When electrons travel at low speed through a gas, they exhibit a high probability of being trapped by neutral atoms. There are two important ways of forming a negative ion by electron attachment: radiation attachment and dissociative attachment. In the first case, an electron will recombine with a neutral atom at low temperature to form a negative ion following the reaction:

$$A + e \rightarrow A^- + h\nu$$

* Electron situated in the outermost energy shell of the atom.

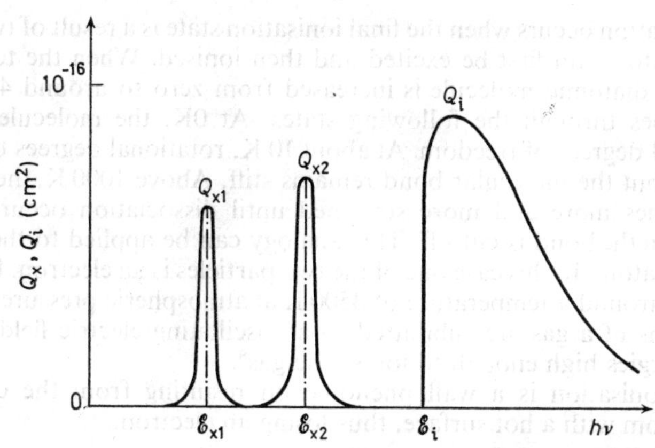

Fig. 2.9. Cross-sections for atom–photon interactions. Q_{e1} and Q_{e2} are the excitation cross-sections corresponding to two energy levels. Q_i is the ionisation cross-section

The released energy goes into emitted photons. The radiative cross-section at 1 eV energy is equal to about 10^{-23} cm^2. In the second case, an electron will recombine with a neutral molecule to form an unstable negatively charged molecular ion. This will in turn be dissociated into a negative ion and a neutral atom. The energy goes into kinetic energy of the dissociating fragments of the unstable molecular ion. The dissociation cross-section at 1 eV energy is equal to about 10^{-17} cm^2 in this case. The reaction is:

$$AB+e \rightarrow (AB)_u^- \rightarrow A^- + B$$

Curves giving the variation of the cross-section as a function of speed are given notably by Brown[5].

A collision of the second kind resulting in ionisation is that of two metastable atoms or molecules. In the collision process part or all of the excitation energy is transferred from one to the other, and if the total energy available is sufficient, ionisation may result. The reaction is:

$$A^M + A^M \rightarrow A^+ + A + e$$

For instance, mercury has a metastable level at 5·4 eV, whereas the ionisation level is at 10·4 eV. In a collision between two such metastable atoms, enough energy is then available to ionise one of them, provided the other one goes to normal. It is a collision of the second kind since 0·4 eV becomes available in kinetic energy. In gas mixtures metastable atoms or molecules of one gas may collide with a normal atom of the other gas, when the energy of the metastable state in the former exceeds the ionisation energy of the latter. This can happen in noble gas mixtures or mercury vapour–noble gas components. A third possibility is the interaction between a metastable atom and an electron:

$$A^M + e \rightarrow A^+ + e + e$$

This reaction, as well as the former one, has large cross-sections.

Step ionisation occurs when the final ionisation state is a result of two or more stages. An atom can first be excited and then ionised. When the temperature of a typical diatomic molecule is increased from zero to around 4000 K, the molecule goes through the following states. At 0K, the molecule has three translational degrees of freedom. At about 10 K, rotational degrees of freedom are excited but the molecular bond remains stiff. Above 1000 K, the molecular bond becomes more and more stretched until dissociation occurs at about 3000 K when the bond is cut off. This analogy can be applied to thermal ionisation of an atom. In this case one of the two particles is an electron. For oxygen this occurs around a temperature of 4500 K at atmospheric pressure. When the free electrons of a gas are submitted to an oscillating electric field, they may acquire energies high enough to ionise the gas[8].

Surface ionisation is a wall phenomenon resulting from the collision of a neutral atom with a hot surface, thus losing an electron.

Degree of ionisation and Saha's equation[12, 17]

The degree of ionisation α is the ratio of the number of ions existing in a volume $\mathcal{V}$ over the total number of ions and neutrals existing in that volume; thus:

$$\alpha = n_{x+}/(n_x + p_{x+}) \tag{2.36}$$

where n_{x+} is the ion density, and n_x is the neutral density. The total pressure of a plasma is equal to the sum of the pressure of its components. Thus, if thermal equilibrium is assumed:

$$p_r = \frac{kT}{\mathcal{V}}(n_x + n_{x+} + n_e)$$

where n_e is the electron density. For a neutral plasma, $n_e = n_{x+}$, and consequently the pressure becomes:

$$p_r = \frac{kT}{\mathcal{V}} n_{x+}(1+\alpha)/\alpha \tag{2.37}$$

The Indian physicist Saha derived an equation relating the degree of ionisation to temperature, pressure and the ionisation potential V_i, by using classical thermodynamics concepts:

$$\alpha^2/(1-\alpha^2) = \mathcal{A}\frac{T^{5/2}}{p_r}\exp(-eV_i/kT) \tag{2.38}$$

where $\mathcal{A}$ is a constant. Replacing p_r in equation (2.37) by its value, one obtains:

$$(n_{x+})^2/n_x = \mathcal{A}' T^{3/2}\exp(-eV_i/kT)$$

for $n_{x+} \ll n_x$.

Plasma deionisation processes[1]

It is conceivable that charge carriers of opposite sign must recombine in one way or another, since in laboratory experiments ionised gases lose their conducting properties almost as soon as the cause of ionisation is removed. However, such an event is rather rare. There are two important types of deionisation processes: electron–ion recombination and electron attachment.

There are three types of electron–ion recombinations, depending on the product of the reaction. These are radiative recombination, three-body recombination, and dissociative recombination. In the first case, a positively charged ion becomes an excited atom by absorbing an electron. Radiative energy is emitted:

$$A^+ + e \rightarrow A^* + h\nu$$

where A^* is the excited atom. This reaction is found notably in astrophysical plasmas. The radiation cross-section is about 10^{-21} cm^2 at 1 eV. In the second case, a positively charged ion becomes an excited atom by absorbing an electron. The energy goes to another colliding electron:

$$A^+ + e + e \rightarrow A^* + e$$

This kind of reaction occurs notably in stellarators† and can be considered as the inverse of ionisation of an excited atom by electron impact. In the third case a positively charged molecular ion absorbs an electron and becomes an excited molecule. This molecule is very unstable; it dissociates almost instantaneously in an excited atom and a neutral atom:

$$(AB)^+ + e \rightarrow (AB)_u^* \rightarrow A^* + B$$

The energy goes into kinetic energy of the dissociating fragments of the unstable excited molecule. This kind of interaction occurs in the ionosphere. As for the dissociative cross-section, its value at 1 eV is about 10^{-16} cm^2.

A negative ion can be formed either by radiation attachment or by dissociative attachment. In the first case, a slow electron may become associated with a neutral atom to form a negative ion:

$$A + e \rightarrow A^- + h\nu$$

The released energy goes into emitted photons. The radiative cross-section at 1 eV energy is equal to about 10^{-23} cm^2. In the second case, an electron will become associated with a neutral molecule to form an unstable negative molecular ion. This will in turn be dissociated into a negative ion and a neutral atom. The energy goes into kinetic energy of the dissociating fragments of the unstable molecular ion. The dissociation cross-section at 1 eV energy is equal to approximately 10^{-17} cm^2. The reaction is:

$$AB + e \rightarrow (AB)_u^- \rightarrow A^- + B$$

† Configurations for the confinement of plasma in thermonuclear approaches (Chapter 7).

2.5 MOBILITY OF CHARGE CARRIERS MOVING IN A GAS

Mobility was defined with the other transport parameters in Section 1.9. Equation (1.63) was the result of an elementary calculation. In the most general case this equation can be written:

$$\mathscr{b} = K_1 q \langle 1/\nu_c \rangle M$$

where K_1 is a factor which may be a function of the temperature. In the case of electrons, this variation changes from one gas to another. In the case of ions $\mathscr{b}$ varies approximately as T^{-2}. Values of $\mathscr{b}$ for different cases are given by Brown[5]. It can also be shown that, if p_0 is the reduced pressure, the product $\mathscr{b}p_0$ is a function of temperature.

Mobility in a temperature gradient[25]

In the presence of an electric field, particles drift with a velocity v_{d1} equal to $\mathscr{b}E$. In the presence of a concentration gradient, the velocity of the particles is given by equation (1.55), which yields:

$$v_{d2} = -\mathscr{D}\nabla n/n$$

By using equation (1.73), this equation becomes:

$$v_{d2} = -\frac{\mathscr{b}kT}{nq}\nabla n \tag{2.39}$$

If there is furthermore a temperature gradient, it will create a drift velocity equal to:

$$v_{d3} = -\frac{\mathscr{b}k}{q}\nabla T \tag{2.40}$$

In the presence of both a temperature gradient and a concentration gradient as well as an electric field, the particles will have a velocity equal to the sum of the velocities v_{d1}, v_{d2}, and v_{d3}. This will give after some rearrangements:

$$v_d = -\mathscr{b}\left[\nabla V + \frac{k}{q}\frac{1}{n}\nabla(nT)\right]$$

where V is the electrical potential such that $\boldsymbol{E} = -\nabla V$. The drift current density for the relevant carrier is evidently $\boldsymbol{J} = nqv_d$; then:

$$J = -\mathscr{b}[nq\nabla V + k\nabla(nT)] \tag{2.41}$$

In order to obtain the total current density, one has to take the sum of the contributions arising from all the different kinds of charge carriers.

Mobility in a magnetic field[18, 40]

If the temperature gradient is neglected and only the electric field, magnetic field, and concentration gradient are considered, the equation of momentum conservation of the partially ionised gas becomes:

$$\boldsymbol{v}_{\mathrm{d}} = -\mathcal{D}\frac{\nabla n}{n} + \mathscr{b}(\boldsymbol{E} + \boldsymbol{v}_{\mathrm{d}} \times \boldsymbol{B}) \tag{2.42}$$

with a uniform magnetic field in the z-direction such that $\boldsymbol{B} = \boldsymbol{k}B$, $\boldsymbol{E}$ in the x-direction such that $\boldsymbol{E} = \boldsymbol{j}E_x$, and n function of x only, one finds from equation (2.42):

$$\begin{aligned} v_x &= -\frac{\mathcal{D}}{1+(\mathscr{b}B)^2}\,\frac{1}{n}\,\frac{\mathrm{d}n}{\mathrm{d}x} + \frac{\mathscr{b}B}{1+(\mathscr{b}B)^2}\,\frac{E_x}{B} \\ v_y &= \frac{\mathcal{D}\mathscr{b}B}{1+(\mathscr{b}B)^2}\,\frac{1}{n}\,\frac{\mathrm{d}n}{\mathrm{d}x} - \frac{(\mathscr{b}B)^2}{1+(\mathscr{b}B)^2}\,\frac{E_x}{B} \end{aligned} \tag{2.43}$$

This velocity v_y gives rise to the Hall current in the y-direction.

Mobility in time dependent fields

If collisions are considered and if the only field present is an a.c. electric field, the balance of forces on the particle yields:

$$M\frac{\mathrm{d}v_{\mathrm{d}}}{\mathrm{d}t} - qE + Mv_{\mathrm{d}}\nu_{\mathrm{c}} = 0 \tag{2.44}$$

where ν_{c} is the frequency of collision. For an a.c. field, equation (2.44) yields for the mobility $\mathscr{b} = v_{\mathrm{d}}/E$:

$$\mathscr{b} = \frac{q}{M}\langle 1/\nu_{\mathrm{c}} + \mathrm{j}\omega \rangle \tag{2.45}$$

If, further, a field slowly variable with time is applied, the balance of force equation becomes:

$$M\frac{\mathrm{d}\boldsymbol{v}_{\mathrm{d}}}{\mathrm{d}t} - q(\boldsymbol{E} + \boldsymbol{v}_{\mathrm{d}} \times \boldsymbol{B}) + M\boldsymbol{v}_{\mathrm{d}}\nu_{\mathrm{c}} = 0$$

In this case $\mathscr{b}$ becomes a tensor such that $\boldsymbol{v}_{\mathrm{d}} = \mathbf{b}\boldsymbol{E}$. In the particular case where $\boldsymbol{B}$ is in the z-direction, this tensor becomes:

$$\mathbf{b} = \frac{q}{M}\begin{vmatrix} \dfrac{\nu_{\mathrm{c}}+\mathrm{j}\omega}{(\nu_{\mathrm{c}}+\mathrm{j}\omega)^2+\omega_{\mathrm{c}}^2} & \dfrac{-\omega_{\mathrm{c}}}{(\nu_{\mathrm{c}}+\mathrm{j}\omega)^2+\omega_{\mathrm{c}}^2} & 0 \\ \dfrac{\omega_{\mathrm{c}}}{(\nu_{\mathrm{c}}+\mathrm{j}\omega)^2+\omega_{\mathrm{c}}^2} & \dfrac{\nu_{\mathrm{c}}+\mathrm{j}\omega}{(\nu_{\mathrm{c}}+\mathrm{j}\omega)^2+\omega_{\mathrm{c}}^2} & 0 \\ 0 & 0 & \dfrac{1}{\nu_{\mathrm{c}}+\mathrm{j}\omega} \end{vmatrix}$$

where ω_{c} is the cyclotron frequency defined in Section 2.1.

2.6 DIFFUSION AND AMBIPOLAR DIFFUSION

The diffusion coefficient was defined in Section 1.9, and equation (1.55) was deduced. Here again, it can be shown that the product $\mathcal{D}p_0$ is a function of temperature. Diffusion can be considered as the result of a force, itself the result of a pressure gradient arising from the density or temperature gradient or both. Indeed, in such a case, the force $\boldsymbol{F}$ per particle in a volume element is $-\nabla p_r/n$ where p_r is the pressure of the relevant kind of carriers. By taking equation (1.58) into account, this force becomes:

$$\boldsymbol{F} = -\frac{k}{n}\nabla(nT) \tag{2.46}$$

and the drift velocity is then:

$$\boldsymbol{v}_d = \mathscr{b}\boldsymbol{F}/q = -\frac{\mathscr{b}k}{qn}\nabla(nT)$$

Now using equation (1.73), it is found:

$$\boldsymbol{v}_d = \frac{\mathcal{D}}{Tn}\nabla(nT) = \frac{\mathcal{D}}{p_r}\nabla p_r \tag{2.47}$$

Ambipolar diffusion in a field-free space

A plasma does not tolerate the presence of large-scale or long-lived uncompensated charges. When this occurs, the electrons tend to diffuse much more rapidly than the positive ions in the same gradient of concentration. This charge separation tends to give rise to very high electric fields, which accelerate the ions and decelerate the electrons in order to compensate for the unbalance. Thus, in the steady state, positive ions and electrons will move together with exactly the same drift speed. This is the definition of an ambipolar flow.

From the equation of momentum conservation, the particle currents Γ^+ and Γ^- become after inclusion of the space charge:

$$\begin{aligned} \Gamma^+ &= -\mathcal{D}^+\nabla n^+ + n^+\mathscr{b}^+E \\ \Gamma^- &= -\mathcal{D}^-\nabla n^- - n^-\mathscr{b}^-E \end{aligned} \tag{2.48}$$

For the case of an ambipolar flow $\Gamma^- = \Gamma^+ = \Gamma$. By taking into account the plasma neutrality, equations (2.48) yield:

$$\begin{aligned} E &= \frac{\mathcal{D}^+ - \mathcal{D}^-}{\mathscr{b}^+ + \mathscr{b}^-}\frac{\nabla n}{n} \\ \Gamma &= -\frac{\mathscr{b}^+\mathcal{D}^- + \mathscr{b}^-\mathcal{D}^+}{\mathscr{b}^+ + \mathscr{b}^-}\nabla n \end{aligned} \tag{2.49}$$

The ambipolar diffusion coefficient $\mathcal{D}_a$ is by definition equal to the ratio of the current density over the gradient of the number density. Thus:

$$\mathcal{D}_a = \frac{b^+\mathcal{D}^- + b^-\mathcal{D}^+}{b^+ + b^-} \tag{2.50}$$

Taking into account equation (1.73) and the fact that the electronic mobility is much larger than the ionic mobility, one finds:

$$\mathcal{D}_a/\mathcal{D}^+ = 1 + \frac{T^-}{T^+}$$

where T^- and T^+ are respectively the electronic and the ionic temperatures.

Ambipolar diffusion in an electric field[21]

In general, the electric field arising from the charge separation described above superposes itself upon any external electric field. In this analysis one should take account of equation (2.46), and the equation of momentum conservation becomes:

$$\begin{aligned} \Gamma^+ &= -b^+\left[\nabla V + \frac{k}{q^+ n^+}\nabla(n^+T^+)\right] \\ \Gamma^- &= -b^-\left[\nabla V + \frac{k}{q^- n^-}\nabla(n^-T^-)\right] \end{aligned} \tag{2.51}$$

For ambipolar diffusion in a neutral plasma $\Gamma^+ = \Gamma^- = \Gamma$ and $n^+ = n^- = n$. The total electric field $-\nabla V$ can be considered as consisting of two parts, an external electric field $E_p = -\nabla V'$, and an electric field $E_a = \nabla(V' - V)$, specifically created by the mechanism of ambipolar diffusion explained above. It is unclear whether such an assumption can always be made, but it is certain that the formalism described helps to solve practical problems. For an ambipolar neutral flow one obtains from equations (2.51):

$$\nabla V' = \nabla V + \frac{k}{en}\,\frac{b^-\nabla(nT^-) + b^+\nabla(nT^+)}{b^+ - b^-} \tag{2.52}$$

with $e = q^+ = -q^-$. Since T^- is much larger than T^+ and b^- is much larger than b^+, equation (2.52) can be simplified and yields after integration:

$$V' \approx V - \frac{kT^-}{e}\ln\left(nT^-/n_0T_0^-\right) \tag{2.53}$$

(where $n_0T_0^-$ is a constant of integration), provided that the gradient of T^- is much smaller, in relative values, than that of n, which is generally the case in practice. After a calculation similar to the preceding, one finds an equation identical to equation (2.50) which becomes for $T^- \gg T^+$ and $b^- \gg b^+$:

$$\mathcal{D}_a = b^+kT^-/e \tag{2.54}$$

Ambipolar diffusion in a magnetic field[3]

In this case, the effect of the magnetic field should be included in the equations, thus:

$$\begin{aligned} \Gamma^{+} &= -\mathcal{D}^{+}\nabla n^{+}+n^{+}b^{+}(\boldsymbol{E}+\boldsymbol{v}^{+}\times\boldsymbol{B}) \\ \Gamma^{-} &= -\mathcal{D}^{-}\nabla n^{-}-n^{-}b^{-}(\boldsymbol{E}+\boldsymbol{v}^{-}\times\boldsymbol{B}) \end{aligned} \tag{2.55}$$

In a plasma column of finite length in a magnetic field, equations (2.55) become, when only the density gradient perpendicular to the magnetic field is considered:

$$\begin{aligned} \Gamma^{+} &= -\mathcal{D}^{+}\nabla_{\perp}n^{+}+n^{+}b^{+}E_{\perp}+n^{+}b^{+}v_{\perp}^{+}B \\ \Gamma^{-} &= -\mathcal{D}^{-}\nabla_{\perp}n^{-}-n^{-}b^{-}E_{\perp}-n^{-}b^{-}v_{\perp}^{-}B \end{aligned} \tag{2.56}$$

These equations can be written in the form:

$$\begin{aligned} \Gamma^{+} &= -\chi^{+}\mathcal{D}^{+}\nabla_{\perp}n^{+}+n^{+}\chi^{+}b^{+}E_{\perp} \quad \text{(a)} \\ \Gamma^{-} &= -\chi^{-}\mathcal{D}^{-}\nabla_{\perp}n^{-}-n^{-}\chi^{-}b^{-}E_{\perp} \quad \text{(b)} \end{aligned} \tag{2.57}$$

where

$$\chi^{\pm} = 1/(1+b^{\pm 2}B^{2})$$

Assuming plasma neutrality and equating equation (2.57a) to equation (2.57b) for an ambipolar flow, one obtains:

$$\Gamma = -\chi_{a}\mathcal{D}_{a}\nabla_{\perp}n \tag{2.58}$$

where

$$\chi_{a} = 1/(1+b^{+}b^{-}B^{2})$$

and the ambipolar diffusion coefficient $\mathcal{D}_a$ is identical to that given by equation (2.50). On the other hand, if ν_i is the frequency of ionisation then $\nabla.\Gamma = -n^{-}\nu_{i}$, and taking equation (2.58) into account, it is found that:

$$\nabla_{\perp}^{2}n+\frac{\nu_{i}}{\chi_{a}\mathcal{D}_{a}}n = 0 \tag{2.59}$$

This differential equation can be written in cylindrical co-ordinates, by assuming cylindrical symmetry and after considering the boundary conditions $n = n_0$ for $r = 0$ and $n = 0$ for $r = r_0 =$ radius of the column, the solution of equation (2.59) becomes:

$$\nu_{i}/\chi_{a}\mathcal{D}_{a} = (2{\cdot}405)^{2}/r_{0}^{2}$$

As the magnetic field increases, χ_a decreases and therefore the frequency of ionisation decreases. However, at a critical value B_c of the magnetic field the plasma becomes unstable as shown in Figure 2.10.

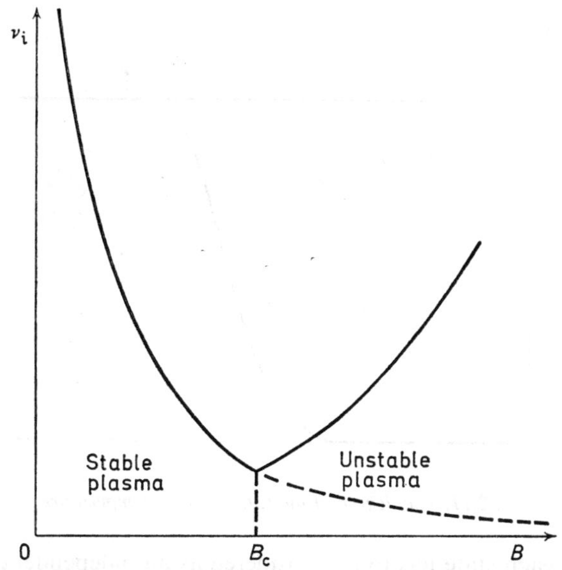

Fig. 2.10. Ionisation frequency versus magnetic field. Above a critical value of the magnetic field the plasma becomes unstable

2.7 GENERATION AND EXCITATION OF CHARGE CARRIERS IN A GAS[13, 32]

Thermal ionisation can be described by Saha's equation[37] as given by equation (2.38). If the degree of ionisation is sufficiently small, this equation can be written:

$$\alpha = \mathcal{A}'' \frac{T^{5/4}}{p_r^{1/2}} \exp(-eV_i/2kT)$$

The general behaviour of α as a function of T is shown in Figure 2.11 which represents equation (2.38). The degree of ionisation remains negligible until a certain temperature is attained, rises steeply afterwards, and becomes practically equal to unity over a range of less than a factor of two in temperature. An ionisation degree of 50 per cent is attained for $T = 9500$ K if $V_i = 7{\cdot}5$ V and for $T = 16\,100$ K if $V_i = 15$ V. The variation is little less than proportional.

The thermal excitation process can be studied as a reversible chemical equation. Thus, if α_e is the relative proportion of excited molecules, one has, within a good degree of approximation,

$$\alpha_e/(1-\alpha_e) = \exp(-eV_e/kT)$$

V_e being the excitation potential. However, in some cases called *multiplets*, the same value of V_e corresponds to several excited configurations, indistinguishable from the energetic point of view of, but different from the standpoint of the Schrödinger ψ function, hence different by the shape of the electron cloud.

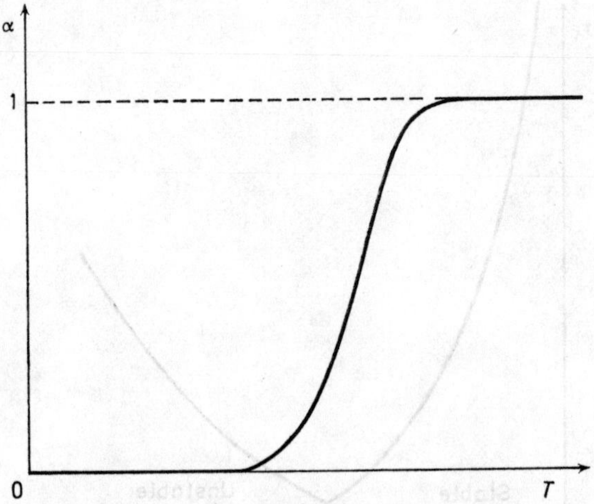

Fig. 2.11. The degree of ionisation versus temperature

In these cases, each state has to be considered as an independent excited level, disregarding the identity of the excitation potentials. This is best taken into account by modifying each value of α_e by a statistical weight g which is equal to the number of configurations which correspond to the same energy. Hence, the final expression is:

$$\alpha_e/(1-\alpha_e) = g \exp(-eV_e/kT) \tag{2.60}$$

Equations (2.38) and (2.60) are valid only at thermal equilibrium.

When light of appropriate spectral composition falls upon a gas and the pressure of the latter is not very small, considerable ionisation can occur in the gas. An approximate treatment can be given if the spectral composition is known. Von Engel and Steenbeck[44] considered a Planck spectrum; in this case, the number of photons of frequency close to ν is:

$$\mathrm{d}n = \frac{2\pi}{c^2}\nu^2 \exp(-h\nu/kT)\,\mathrm{d}\nu \tag{2.61}$$

c being the velocity of light in a vacuum and h Planck's constant. The number of photons having an energy greater than eV_i is obtained by a simple integration from $\nu = eV_i/h$ to infinity and emerges as:

$$n = (2\pi/c^2)(kT/h)^3[(eV_i/kT)^2+2(eV_i/kT)+2]\exp(-eV_i/kT)$$

In photoexcitation, the energy brought by the incident photon has to be equal to that required for excitation. The major causes of spectral line broadening are here Heisenberg's uncertainty principle and the Doppler effect associated with thermal motion.

In a quiescent plasma, where the carriers display a given velocity distribution, ionisation can occur by the electronic component. The probability of ionisation by a carrier is a function of its velocity. It frequently happens that only the very rare carriers at the far end of the high velocity tail of a distribution like the Maxwellian one, have enough energy to produce ionisation. For instance, in a mercury-vapour arc, the average energy of the electrons is about a tenth of that required for ionisation. If ν_v is the number of collisions per unit time in the gas and $\bar{\tau}_v$ is the mean free flight time at speed v then:

$$f_e(v/v_0)\, dv/v_0 = \nu_v \bar{\tau}_v \tag{2.62}$$

where f_e is the electronic distribution function of velocities. On the other hand,

$$f_c(v/v_0)\, dv/v_0 = \nu_c/\nu \tag{2.63}$$

where $f_c(v/v_0)$ is the collision probability and ν is the inverse of the general average of free flight times. Eliminating ν_v from these two equations, one obtains:

$$f_c(v/v_0) = f_e(v/v_0)/\nu \bar{\tau}_v$$

Let $f_i(v/v_0)$ be the probability for a collision at the velocity v to be ionising. $f_i(v/v_0)$ is zero as long as $Mv^2/2 < eV_i$ and positive for higher values. The total frequency of ionising collisions is given by:

$$\nu_i = \int_0^\infty v f_c(v/v_0) f_i(v/v_0)\, d(v/v_0) \tag{2.64}$$

By taking into account equation (2.63),

$$\nu_i/p_r = \int_0^\infty \frac{v f_i(v)}{f_1(v)} f_c(v/v_0)\, d(v/v_0)$$

where f_1 is a function of v depending only upon the nature of the neutral gas. Thus, it can be deduced that ν_i/p_r is a function of temperature. The best known treatment of this equation is that of Killian, who assumes a Maxwellian distribution and the statistical independency of the mean free path from the velocity (f_1 = constant). He obtains[26]:

$$\nu_i/p_r = \frac{2\mathcal{A}M}{e\sqrt{\pi}} \bar{v}^3 \exp(-eV_i/kT)\left(1 + \frac{eV_i}{2kT}\right) \tag{2.65}$$

in which $\mathcal{A}$ is the slope of f_i as a function of kinetic energy.

Excitation by the electronic component of a quiescent plasma is similar to the phenomenon of ionisation since the impinging electron is always there to carry the excess energy.

2.8 ANNIHILATION OF CHARGE CARRIERS IN A GAS[31]

The presence of negative ions in a discharge is generally synonymous with high pressure; in such a case, mobility and diffusion coefficients are small, and the motion of the ions is necessarily slow. A positive and a negative ion, either arising from the same atom such as Cl^+ or Cl^- or originating from the division of an unsymmetrical molecule like H_2O into H^+ and OH^-, can come into such proximity that they are influenced by the microfield of each other and begin to move according to the laws of mobility. If the circumstances are favourable, they can collide. Since the initial kinetic energy was essentially zero, the total momentum is zero and the centre of gravity of the system is practically at rest.

If the collision leads to recombination of the electric charges, several circumstances can arise.

(1) The extra electron of the negative ion detaches itself and becomes fixed on the positive ion; thus, two neutral atoms appear. The recombination energy appears mainly as heat.

(2) The kinetic energy developed in the reaction above is not necessarily equal to the recombination energy eV_i since part of the latter may be emitted as a light quantum or as several light quanta. In the limiting case, all the recombination energy will be emitted as light and the two neutral products of the reaction come to rest.

(3) In the case of the ion pair arising from one molecule (e.g. H_2O, Cl_2), the final product of the reaction is one neutral molecule, whether or not in an excited state. All the energy liberated will have been developed in the form of photons.

If the positive and negative ions are equal in number, the recombination process is quadratic; that is:

$$dn/dt = -\varrho_c n^2$$

hence:

$$1/n = (1/n_0) + \varrho_c t$$

where ϱ_c is the recombination coefficient. The average lifetime of a pair is obtained by:

$$\tau_r = 1/\varrho_c n \tag{2.66}$$

The recombination coefficient can be found as a function of the mobility coefficient by using Poisson's law. Consider that the positive and the negative ion both move towards each other in their mutual Coulomb microfield; if r is the distance between the two and assuming singly charged ions:

$$E = e/4\pi\varepsilon_0 r^2 \tag{2.67}$$

The ions travel toward each other with a drift velocity:

$$v_d = (b^+ + |b^-|)E \tag{2.68}$$

There are n ion pairs per unit volume, hence $2n$ particles distributed at random in space. The average distance between two particles is then $(2n)^{-1/3}$. The average lifetime can be defined as:

$$\tau_r = \int_0^{(2n)^{-1/3}} dr/v_d$$

By using equations (2.67) and (2.68), this equation can be solved, and equating it to equation (2.66), one finds:

$$\varrho_c = \frac{3e}{2\pi\varepsilon_0}(b^+ + |b^-|) \tag{2.69}$$

Equation (2.69) is relatively well satisfied experimentally at high pressur

In the case of the recombination of positive ions and electrons, the positive ions are practically at rest, while the electrons travel at high speeds; the ratio of speeds is then well below the ratio of masses, so that the total momentum is practically zero. Therefore, the *singly* neutral particle, product of the reaction can keep no energy in kinetic form; everything must be emitted as light quanta; the total energy is then equal to the sum of the ionisation energy eV_i plus the kinetic energy of the electron. If only one photon is emitted per recombination, the spectrum of the recombining plasma appears as a continuum at the short wavelength side of the characteristic spectral line corresponding to $h\nu = eV_i$. If the descent involves an intermediate stopping on an excited level V_e, the spectrum comprises a sharp line at $h\nu = eV_e$ plus a continuum on the short wavelength side of $h\nu = e(V_i - V_e)$. Direct recombination of positive ions and electrons is a rare event. However, this recombination is more common as a three-body reaction with the intervention of a neutral molecule. The first step is the attachment of an electron to a neutral molecule, thus resulting in a negative ion; next, the positive and the negative ions recombine as above.

Since metastable atoms are unable to liberate their potential energy spontaneously by photon emission, their annihilation in the gas occurs only by collision among themselves, with electrons and with neutral molecules, the latter case being the major source of metastable annihilation. Von Engel and Steenbeck[44] give for the rate of annihilation of metastables in the gas:

$$\frac{1}{n^*}\frac{\mathrm{d}n^*}{\mathrm{d}t} = 2\mathcal{C}_\mu(r+r^*)^2 n\left[2\pi kT_g\left(\frac{1}{M}+\frac{1}{M^*}\right)\right]$$

in which r and r^* are, respectively, the radius of the normal and the metastable molecule, M and M^* are their masses, n and n^* are their concentrations, T_g is the temperature of the neutrals, and $\mathcal{C}_\mu$ is the probability that a single collision is an annihilation process; it is small with respect to unity in many cases.

2.9 WALL AND ELECTRODE EFFECTS

Surface emission is a complex phenomenon, involving more than one process. Scientific opinion has considerably changed in time concerning the kind of phenomenon prevailing in various cases, as well as about the numerical values of the coefficients. The most conspicuous source of carrier emission is generally the cathode, although the anode or sometimes the inert walls can also play an appreciable role in some instances.

Under the effect of temperature, incandescent metals and oxides emit electrons freely. This phenomenon, known as *thermionic emission*[14], is studied either from the point of view of velocity statistics of the semi-free electrons inside the metal or on a thermodynamic basis. From the second method the following expression for the produced current density J is deduced:

$$J = \mathcal{A}T^2 \exp(-eV_i/kT) \qquad (2.70)$$

in which $\mathcal{A} = 0{\cdot}006\,02$ A/m^2 if T is in Kelvin, and V_i is the potential required to transform a semi-free electron inside the metal into a free electron in space. The expression seems to be satisfied in metallic emitters. The value of V_i changes from one metal to another and numerical values are given by Brown[5] and others. If the field configuration above the emitting surface is such that the electrons stay in its neighbourhood for some time, a negative space charge accumulates and prevents the emission of further electrons. The electron density in front of the electrode surface must be kept below a limiting value given by:

$$n_{\text{lim}} = (J/e)(2\pi M/kT)^{1/2}$$

which acts in the same sense as the saturation pressure of a vapour.

Hertz showed that visible ultra-violet light of a sufficiently short wavelength can cause emission of electrons from metals. The energy per photon has to be larger than a certain energy eV_i. Taking into account the relations between energy, frequency and wavelength in angstroms, one finds $\lambda_c \leq 12\,340/V_i$. The emitted electron has an initial kinetic energy equal to the difference between the photon energy and the extraction work, so that:

$$Mv^2/2 = h\nu - eV_i$$

The ratio of the number of electrons emitted to the number of incident photons is designated as m_ϕ. In the neighbourhood of λ_c, it is given approximately by:

$$m_\phi = \mathscr{C}_\phi \frac{\lambda_c}{\lambda}\left(1 - \frac{\lambda}{\lambda_c}\right)$$

where $\mathscr{C}_\phi$ is a constant. Figure 2.12 shows m_ϕ as a function of λ for several pure metals[6].

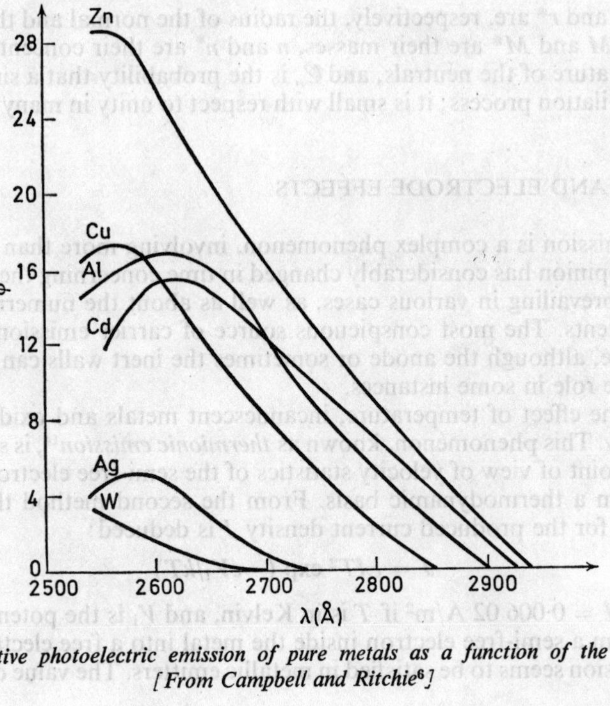

Fig. 2.12. Relative photoelectric emission of pure metals as a function of the wavelength. [From Campbell and Ritchie[6]]

When an electron impinges on a solid surface with a kinetic energy larger than eV_i, it extracts one or more other electrons. The number of secondary electrons including the rebounding one is generally designated by δ which may be either smaller or larger than unity. In the first case, it means that a high percentage of the primaries are absorbed. δ is a function of V_i and there are one or two critical values of V_i for which $\delta = 1$. For instance, for aluminium, the maximum value of δ is 1·90 at 220 V and $\delta = 1$ for $V_i = 35$ V and $V_i = 45$ V; for tungsten, $\delta = 1{\cdot}45$ for $V_i = 700$ V and $\delta = 1$ for $V_i = 200$ V and $V_i = 240$ V. Numerical tables are given by Engel and Steenbeck[44] as well as Thomson and Thomson[42].

The impact of positive ions on solid surfaces can also result in electron emission. The positive ion comes to within a few angstroms from the metal surface, one of the semi-free electrons crosses the potential barrier between the bulk of the material and the positive ion by the 'tunnel effect'*; the positive ion is neutralised and rebounds as a neutral. The energy thus liberated is communicated to the lattice and may serve to extract one or several electrons.

When an electric field of convenient sign and strength is applied normally to a solid surface, electrons can be emitted (field emission). This field emission is generally considered as a correction to thermionic emission. For low fields equation (2.70) becomes:

$$J = \mathcal{A}T^2 \exp\left(-eV_i/kT\right) \exp\left[(e/k)\sqrt{(eE/4\pi\varepsilon_0)}\right]$$

and for high fields and/or low temperatures:

$$J = \mathcal{A}(T+\mathcal{C}E)^2 \exp\left[-\frac{eV_i}{k(T+\mathcal{C}E)}\right]$$

where $\mathcal{C}$ is a constant of the order of 10^{-4} K m/V.

If a metastable atom collides with a solid surface, electron emission can occur if $V_e > V_i$. The phenomenon does not differ from that occurring with positive ions.

Positive ions can be emitted by solid surfaces under three different mechanisms: (1) by thermionic emission in the presence of an electric field, (2) when the work function of the metal is larger than the ionisation potential of the gas (e.g. Cs), and (3) in the Kunsman anode which is a hollow palladium tube with a high pressure inside and an electric field outside.

At low pressures, the recombination of carriers occurs almost exclusively at the walls. In general, a negative ion striking a wall has a high probability of being decomposed into a neutral molecule and an electron. Also, any positive ion striking the wall has a good probability of meeting an electron. In metals, this probability seems indistinguishable from unity, whereas in insulators it is significantly smaller but different from zero because of the existence of an electron layer on the insulating wall.

The probability of de-excitation of metastable atoms by collision with the wall seems to be practically unity in all cases. The rate of de-excitation is controlled by the rate of diffusion.

* When the electron penetrates through the barrier of potential instead of moving over it.

2.10 CREATION OF A PLASMA[4, 30]

Consider n_0 electrons emitted by the cathode. These electrons move towards the anode under the action of an external electric field and ionise the gas. Each electron will create λ_t new electrons per unit distance in the field direction. Thus, the increase of electrons dn produced by n electrons in a distance dx will be $\lambda_t n\,dx$, the solution of which is:

$$n = n_0 \exp(\lambda_t x) \tag{2.71}$$

λ_t is known as the *first Townsend coefficient*. To calculate this coefficient note that $\Gamma = \mathit{b}En$ and that $\nabla.\Gamma = n\nu_i$, thus:

$$\nabla.\Gamma = \frac{\nu_i}{\mathit{b}E}\Gamma$$

and

$$\Gamma = \Gamma_0 \exp\left(\frac{\nu_i}{\mathit{b}E}x\right) \tag{2.72}$$

Comparing equation (2.72) with equation (2.71) and taking into account the relation between Γ and n, one finds for the first Townsend coefficient:

$$\lambda_t = \nu_i/\mathit{b}E \tag{2.73}$$

This coefficient is given per unit distance. An electron falling through a potential difference of one volt will create η_E new electrons; thus:

$$dn = \eta_E n\,dV$$

the solution of which is:

$$n = n_0 \exp(\eta_E V) \tag{2.74}$$

Since $Ex = V$, one finds after comparison with equations (2.71) and (2.73):

$$\eta_E = \nu_i/\mathit{b}E^2 = \lambda_t/E$$

where η_E is the first Townsend coefficient per volt. Figure 2.13 illustrates the variation of the ionisation coefficient η_E as a function of E/p_r for several gases, p_r being the pressure.

In argon and neon mixtures, large variations of the first Townsend coefficient per volt are observed. This is a result of the interaction between the metastable neon atoms and the neutral argon atoms. Since the neon metastable level lies above the argon ionisation potential, collisions between the two result in ionisation of the argon and return of the neon to its ground level. This phenomenon has large variations of η_E and is known as the *Penning effect*. It is illustrated in Figure 2.14 for several percentages of argon–neon.

A *second Townsend coefficient* γ_t is defined as the number of electrons ejected for each incident positive ion hitting the wall or an electrode. This coefficient is

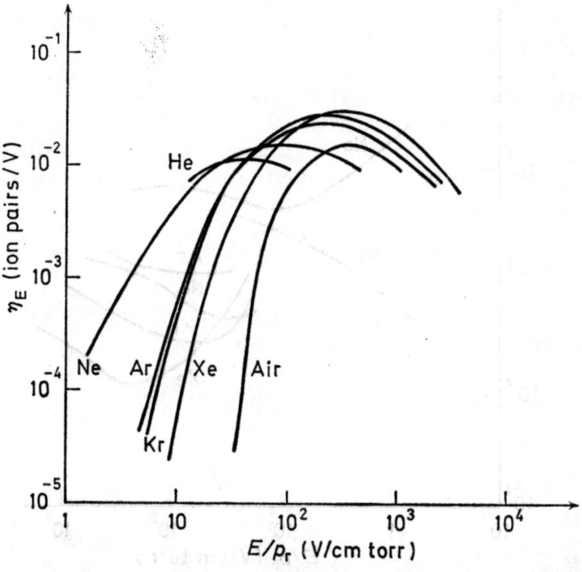

Fig. 2.13. Ionisation coefficient η_E as a function of E/p_r for air and noble gases

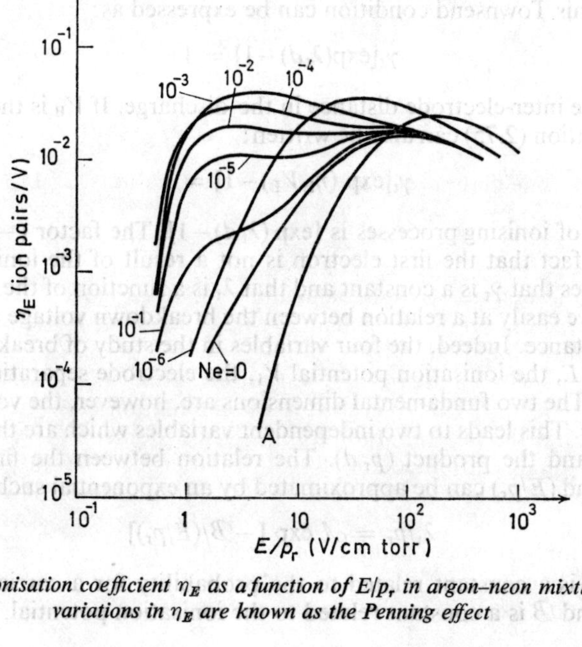

Fig. 2.14. The ionisation coefficient η_E as a function of E/p_r in argon–neon mixtures. The large variations in η_E are known as the Penning effect

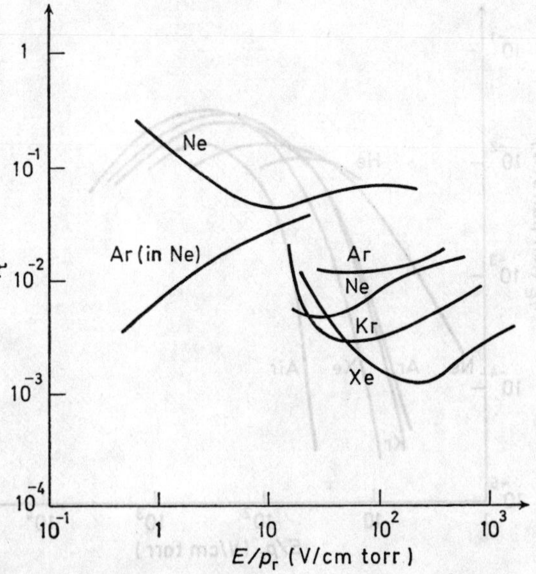

Fig. 2.15. The second Townsend coefficient as a function of E/p_r for noble gases

also a function of the ratio of the electric field over pressure in the gas. It is shown in Figure 2.15 for several noble gases.

Breakdown occurs when a sufficient number of ion pairs are formed in a discharge avalanche so that at least one new electron is produced to maintain the discharge. This Townsend condition can be expressed as:

$$\gamma_t[\exp(\lambda_t d)-1] = 1 \tag{2.75}$$

where d is the inter-electrode distance in the discharge. If V_B is the breakdown voltage, equation (2.75) can also be written:

$$\gamma_t[\exp(\eta_E V_B)-1] = 1 \tag{2.76}$$

The number of ionising processes is $[\exp(\lambda_t d)-1]$. The factor (-1) takes into account the fact that the first electron is not a result of the ionising process. If one assumes that γ_t is a constant and that λ_t is a function of the electric field, one can arrive easily at a relation between the breakdown voltage and the inter-electrode distance. Indeed, the four variables in the study of breakdown are the electric field E, the ionisation potential V_i, the electrode separation d, and the pressure p_r. The two fundamental dimensions are, however, the voltage and the unit distance. This leads to two independent variables which are the breakdown voltage V_B and the product $(p_r d)$. The relation between the first Townsend coefficient and (E/p_r) can be approximated by an exponential such as:

$$\lambda_t/p_r = \mathcal{A} \exp[-\mathcal{B}/(E/p_r)]$$

in which $\mathcal{A}$ is a constant related to the probability for a carrier to produce ionisation and $\mathcal{B}$ is a constant related to the ionisation potential. Replacing λ_t

by this value in equation (2.76), assuming that γ_t is constant and noticing that $V_B = Ed$, one arrives after some rearrangement to:

$$V_B/V_{Bm} = \frac{(p_r d)/(p_r d)_m}{1+\ln[(p_r d)/(p_r d)_m]} \tag{2.77}$$

which is known as *Paschen's law*. This law is illustrated in Figure 2.16 for several gases. The curve rises to the right of the minimum, since for an increasing pressure E/p_r will decrease, and therefore E will decrease for each mean free

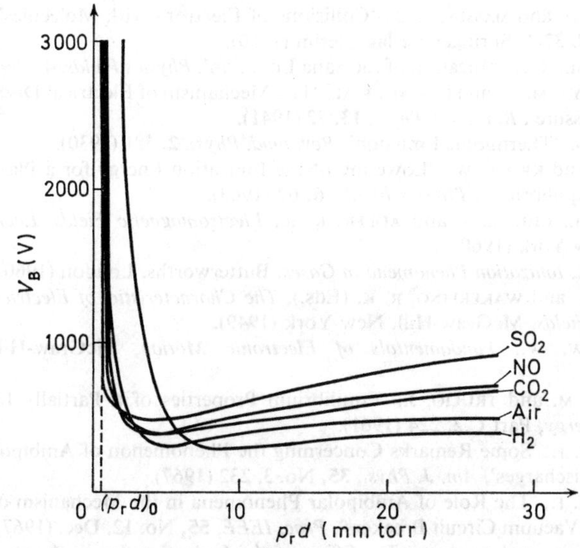

Fig. 2.16. Paschen's curves for several gases. They represent the breakdown voltage as a function of $p_r d$. For very small values of $(p_r d)$ no discharge is possible. This is the Hittorf effect

path λ. This requires that the voltage is increased to produce breakdown. At the minimum, the electrode separation is approximately equal to the mean free path, so that below this point, the decreasing number of collisions of the electrons with the gas atoms makes the ionisation probability fall drastically, requiring a rapidly increasing breakdown voltage. For $(p_r d)$ below a certain $(p_r d)_0$, however, no discharge is possible. This phenomenon is known as the *Hittorf effect*.

REFERENCES AND BIBLIOGRAPHY

1. BARGER, R. L., *Ionization and Deionization Processes in Low Density Plasma Flows*, NASA Tech. Note D-740 (1961).
2. BIONDI, M. A., 'Atomic Collisions Involving Low Energy Electrons and Ions', *Adv. Electronics Electron Phys.*, **18**, (1963).
3. BLACKMAN, M., 'Self-Magnetic Field in High Current Discharges', *Proc. phys. Soc.*, **B64**, 1039 (1951).

4. BROWN, S. C., 'Breakdown in Gases, Alternating and High Frequency Fields', *Encyclopaedia of Physics* (ed. by S. FLUGGE), Vol. 22, Springer-Verlag, Berlin, 531 (1956).
5. BROWN, S. C., *Basic Data in Plasma Physics*, MIT Press (1959).
6. CAMPBELL, N. R. and RITCHIE, D., *Photoelectric Cells*, Pitman, New York (1934).
7. CHANDRASEKHAR, S., *Plasma Physics*, University of Chicago Press (1960).
8. CHUAN, R. L., *A Note on the Addition of Heat to a Gas Through Electrical Discharge*, USCEC Rept. 68–202, Jan. (1957).
9. COMPTON, K. T. and LANGMUIR, I., 'Electrical Discharges in Gases, Part I, Survey of Fundamental Processes', *Rev. Mod. Phys.*, **2,** 123 (1930).
10. COMPTON, K. T. and LANGMUIR, I., 'Electrical Discharges in Gases, Part II, Fundamental Phenomena in Electrical Discharges', *Rev. Mod. Phys.*, **3,** 191 (1931).
11. CRAGGS, J. D. and MASSEY, H. S., 'Collisions of Electrons with Molecules', *Handbuch der Physik*, Vol. 37/1, Springer-Verlag, Berlin (1950).
12. DEWAN, E. M., 'Generalization of the Saha Equation', *Physics Fluids*, **4,** 759 (1961).
13. DRUYVESTEYN, M. J. and PENNING, F. M., 'The Mechanism of Electrical Discharges in Gases at Low Pressure', *Rev. mod. Phys.*, **13,** 72 (1941).
14. DUSHMAN, S., 'Thermionic Emission', *Rev. mod. Phys.*, **2,** 381 (1930).
15. ECKER, G. and KROLL, W., 'Lowering of the Ionization Energy for a Plasma in Thermodynamic Equilibrium', *Physics Fluids*, **6,** 62 (1963).
16. FANO, R. M., CHU, L. J. and ADLER, R. B., *Electromagnetic Fields, Energy and Forces*, Wiley, New York (1960).
17. FRANCIS, G., *Ionization Phenomena in Gases*, Butterworths, London (1960).
18. GUTHRIE, A. and WAKERLING, R. K. (Eds.), *The Characteristics of Electrical Discharges in Magnetic Fields*, McGraw-Hill, New York (1949).
19. HARMON, W. W., *Fundamentals of Electronic Motion*, McGraw-Hill, New York (1953).
20. HARRIS, G. M. and TRULIO, J., 'Equilibrium Properties of a Partially Ionized Plasma', *J. nucl. Energy*, Part C.2, 224 (1961).
21. HOYAUX, M. F., 'Some Remarks Concerning the Phenomenon of Ambipolar Diffusion in Gaseous Discharges', *Am. J. Phys.*, **35,** No. 3, 232 (1967).
22. HOYAUX, M. F., 'The Role of Ambipolar Phenomena in the Mechanism of the Post-Zero Current in Vacuum Circuit Breakers', *Proc. IEEE*, **55,** No. 12, Dec. (1967).
23. HOYAUX, M. F. and GANS, P., *The Effect of the Ionization due to Atoms in a Metastable State upon the Characteristics of a Positive Column with Axial Symmetry*, Rpt. EOARDC-TN-53-3, Contract AF61(514)630-C.
24. HOYAUX, M. F. and GANS, P., *The Influence of the Gvosdover Effect upon the Characteristics of a Positive Column with Axial Symmetry in the Domain in which the Self-Magnetic Field is Important*, Rept. EOARDC-TN-55-4, Contract AF61(514)-630-C.
25. KERREBROCK, J. L., 'Conduction in Gases with Elevated Electron Temperature', in MANNAL, C. and MATHER, N. W. (Eds.), *Proc. 2nd Symp. Eng. Aspects of MHD*, Columbia University Press, New York, 327 (1962).
26. KILLIAN, T. J., 'Thermionic Phenomena Caused by Vapors of Rubidium and Potassium', *Phys. Rev.*, **27,** 578 (1926).
27. KNOLL, M., OLLENDORF, F. and ROMPE, R., *Gasentladungstabellen*, Edwards Brothers, Ann Arbor, Mich. (1944).
28. LANDAU, L. D. and LIFSHITZ, E. M., *Electrodynamics of Continuous Media*, Pergamon Press, Oxford (1960).
29. LANDSHOFF, R., 'Transport Phenomena in a Completely Ionized Gas in Presence of a Magnetic Field', *Phys. Rev.*, **76,** 904 (1949).
30. LLEWELLYN-JONES, F., *Ionization and Breakdown in Gases*, Methuen, London (1957).
31. LOEB, L. B., The Recombination of Ions', *Handbuch der Physik*, Vol. 21, Springer-Verlag, Berlin (1956).
32. MEEK, J. M. and CRAGGS, J. D., *Electrical Breakdown of Gases*, Clarendon Press, Oxford (1953).

33. NORTHROP, T. G., *The Adiabatic Motion of Charged Particles*, Interscience, New York (1963).
34. PANOFSKY, W. and PHILLIPS, M., *Classical Electricity and Magnetism*, Addison-Wesley, Reading, Mass. (1962).
35. PAPOULAR, R., *Electrical Phenomena in Gases*, Iliffe, London (1965).
36. PENNING, F. M., *Electrical Discharges in Gases*, MacMillan, New York (1957).
37. SAHA, M. N., 'Ionization in the Solar Chromosphere', *Phil. Mag.*, **40,** 472, Oct. (1920).
38. SAHA, M. N. and SRIVASTAVA, B. N., *Treatise on Heat*, The Indian Press, Allahabad (1958).
39. SODHA, M. S. and VARSHNI, Y. P., 'Transport Phenomena in Completely Ionized Gas Considering Electron-Electron Scattering', *Phys. Rev.*, **111,** 1203 (1958).
40. SODHA, M. S. and VARSHNI, Y. P., 'Dependence of Electron Mobility on Magnetic Field in a Fully Ionized Gas', *Phys. Rev.*, **114,** 946 (1953).
41. STRATTON, J. A., *Electromagnetic Theory*, McGraw-Hill, New York (1941).
42. THOMSON, J. J. and THOMSON, G. P., *Conduction of Electricity Through Gases*, Cambridge University Press, London (1928).
43. TONKS, L., 'Theory of the Magnetic Effects in the Plasma of an Arc', *Phys. Rev.*, **56,** 360 (1939).
44. VON ENGEL, A. and STEENBECK, M., *Elektrische Gasentladungen*, Edwards Brothers, Ann Arbor, Mich. (1944).

3 Plasma equations

3.1 MAXWELL'S EQUATIONS[32]

Maxwell's equations express the relations between electric and magnetic fields in a medium. Since these fields are functions of space and time, there will be a need for four relations.

Electric currents create magnetic fields. Consider, for instance, a current I flowing in an infinitesimal element of length $\mathrm{d}l$. The magnetic flux density $\mathrm{d}B$ produced by this current at a point P, a distance r from the element, and making an angle θ with its axis is:

$$\mathrm{d}B = KI\,\mathrm{d}l \sin\theta/r^2$$

or in vectorial form

$$\mathrm{d}\boldsymbol{B} = KI\frac{\mathrm{d}\boldsymbol{l}\times\boldsymbol{r}}{r^3}$$

known as Ampère's law, where K is a constant of proportionality. This constant is defined as $\mu/4\pi$, where μ is the permeability of the medium. For vacuum $\mu = \mu_0 = 4\pi\times10^{-7}$ H/m. The total flux density $\boldsymbol{B}$ produced at point P by the current flowing in a long conductor will then be:

$$\boldsymbol{B} = \frac{\mu I}{4\pi}\int\frac{\sin\theta}{r^2}\,\mathrm{d}l$$

If the conductor is infinite and linear, $\boldsymbol{B} = \mu I/2\pi r$ where r is the radial distance from P to the linear conductor (see Figure 3.1). If $\boldsymbol{B}$ is now integrated around a path that encloses the wire once, then:

$$\int \boldsymbol{B}.\mathrm{d}\boldsymbol{l} = \mu I \tag{3.1}$$

This result is valid in all cases where the integration is around a closed path. To make this equation independent of the medium, a magnetic field vector $\boldsymbol{H}$ is introduced and defined such that:

$$\boldsymbol{B} = \mu\boldsymbol{H} \tag{3.2}$$

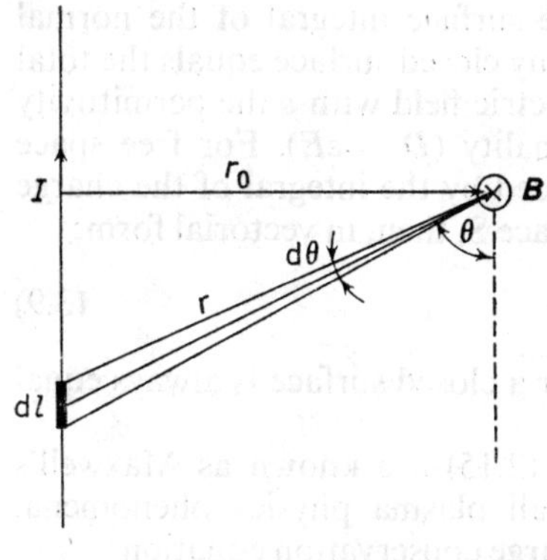

Fig. 3.1. Flux density near a long straight wire in which a current I is flowing. $\sin\theta\, dl = r\, d\theta$ *and* $r_0 = r \sin\theta$

Introducing this relation in equation (3.1) and replacing the current I by the surface integral of the conduction current density $\boldsymbol{J}$ over the area described by the path of integration of $\boldsymbol{H}$, one finds:

$$\int \boldsymbol{H}.\mathrm{d}\boldsymbol{l} = \iint_S \boldsymbol{J}.\mathrm{d}\boldsymbol{S} \tag{3.3}$$

Since the total current density is equal to the conduction current density $\sigma\boldsymbol{E}$ where σ is the conductivity of the wire and $\boldsymbol{E}$ is the electric field, plus the displacement current density $\partial\boldsymbol{D}/\partial t$, where $\boldsymbol{D}$ is the electric flux density, equation (3.3) can be written in a more general form, after applying Stokes' theorem,

$$\nabla\times\boldsymbol{H} = \boldsymbol{J}+\frac{\partial\boldsymbol{D}}{\partial t} \tag{3.4}$$

Faraday's law gives the total e.m.f. (V) induced in a closed loop as a function of the total flux Φ_m producing this e.m.f. following the relation:

$$V = -\mathrm{d}\Phi_m/\mathrm{d}t \tag{3.5}$$

For a loop of n turns enclosing the same flux, the total e.m.f. will be equal to n, the value given by equation (3.5), or if the total flux linkage Φ_t is $n\Phi_m$, then $V = -\mathrm{d}\Phi_t/\mathrm{d}t$. The total flux through a circuit is equal to the integral of B over the area bounded by the circuit; thus:

$$V = -\frac{\mathrm{d}}{\mathrm{d}t}\iint_S \boldsymbol{B}.\mathrm{d}\boldsymbol{S} \tag{3.6}$$

This changing magnetic field produces an electric field $\boldsymbol{E}$ such that

$$V = \int \boldsymbol{E}.\mathrm{d}\boldsymbol{l} \tag{3.7}$$

or $\boldsymbol{E} = -\nabla V$. Combining equation (3.7) with equation (3.6), one finds equation (2.12), and after use of Stokes' theorem:

$$\nabla\times\boldsymbol{E} = -\partial\boldsymbol{B}/\partial t \tag{3.8}$$

In electrostatics, Gauss' law states that the surface integral of the normal component of the electric flux density $\boldsymbol{D}$ over any closed surface equals the total enclosed charge q. $\boldsymbol{D}$ is proportional to the electric field with ε the permittivity of the medium as the constant of proportionality ($\boldsymbol{D} = \varepsilon\boldsymbol{E}$). For free space $\varepsilon = \varepsilon_0 = 8{\cdot}855\times10^{-12}$ F/m. If now q is replaced by the integral of the charge density ϱ_e over the volume enclosed by the surface S, then, in vectorial form:

$$\nabla.\boldsymbol{D} = \varrho_e \tag{3.9}$$

For the magnetic field, the integral of $\boldsymbol{B}$ over a closed surface is always equal to zero, thus equation (2.15).

The four equations (3.4), (3.8), (3.9), and (2.15) are known as Maxwell's equations. These equations are satisfied in all plasma physics phenomena. To these equations, one should also add the charge conservation equation:

$$\nabla.\boldsymbol{J}+\frac{\partial\varrho_e}{\partial t} = 0 \tag{3.10}$$

3.2 LIOUVILLE'S THEOREM

When the statistical properties of the gas become complex, the orbital approach either becomes too intricate or approximate. The Boltzmann approach is more general and more abstract; it yields results that are in agreement with the facts to within the statistical errors.

In connection with the Boltzmann approach, most of the problems of statistical mechanics are best studied in abstract, multidimensional spaces called

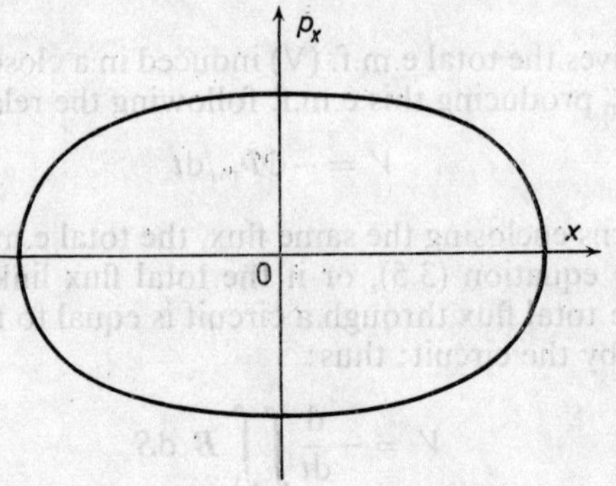

Fig. 3.2. The trajectory of an oscillating point in the μ-space. It is an ellipse

'phase spaces'. Two of them are important. The first one, the μ-space, is six-dimensional, and makes use, as co-ordinates, of the three cartesian co-ordinates plus the three components of momentum x, y, z, p_x, p_y, p_z or any set of Lagrangian co-ordinates for a point plus the associated generalised momenta. In this space each particle of the plasma is represented by a point. If there were only one degree of freedom, this μ-space could be physically represented on a plane (Figure 3.2). For instance, the trajectory in this phase space (dynamic trajectory) of an oscillating point is an ellipse. If several points oscillate along the same line, they will be represented by an equal number of ellipses.

The second important phase space is the Γ-space. If there were only one particle under consideration, both the first and the second phase spaces would be identical, but if there are several particles involved, instead of each of them being represented by a different point in a six-dimensional space, all of them are represented by *one point* in a $6N$-dimensional space, N being the relevant number of particles. In this space, the dynamic trajectory of N particles oscillating linearly is one complex $2N$-dimensional characteristic, instead of N two-dimensional separate characteristics.

In cartesian co-ordinates, it is more practical in non-relativistic problems to use the components of velocity instead of the components of momenta $\boldsymbol{p}$. If no magnetic force is involved, it suffices to divide the relevant momenta by M. Whenever magnetic forces are involved, $\boldsymbol{p}$ is no longer $M\boldsymbol{v}$ but $M\boldsymbol{v}+qA$ where A is the magnetic vector potential.

In Chapter 1, use was made of a velocity-distribution function, but only in a simplified manner; it was considered only as a function of speed. In some problems, the velocity-distribution function is also dependent upon the space co-ordinates or upon time. It is now appropriate to define a distribution valid in all cases. This distribution is a function of *seven* variables x, y, z, p_x, p_y, p_z and t, or x, y, z, v_x, v_y, v_z and t such that, in a six-dimensional volume element $dx\,dy\,dz\,dp_x\,dp_y\,dp_z$, the number of relevant particles at the instant t is $f(x, y, z, p_x, p_y, p_z, t)\,dx\,dy\,dz\,dp_x\,dp_y\,dp_z$ or a similar expression with the vs.

In the μ-space, the distribution function f is a function of seven variables. Thus, the probability of finding a particle in a given volume element depends only upon the co-ordinates and momenta of *this* particle, not upon those of the other particles. This theorem is a simplified version of Liouville's theorem for a large number of *non-interacting* particles.

In the six-dimensional μ-space, the particles are conservative just as they are in the ordinary space. Hence, the conservation theorem may be applied in the phase space; it yields:

$$\frac{\partial f}{\partial t}+\nabla_6.(f\boldsymbol{v}_6) = 0 \tag{3.11}$$

where ∇_6 is the six-dimensional divergence and $\boldsymbol{v}_6$ is a six-dimensional velocity vector, the components of which are the exact time derivatives of the six co-ordinates of the μ-space. By writing temporarily x_1, x_2, x_3 instead of x, y, z and using a similar convention for the ps, equation (3.11) becomes:

$$\frac{\partial f}{\partial t}+\sum_1^3\left[\frac{\partial(f\dot{x}_j)}{\partial x_j}+\frac{\partial(f\dot{p}_j)}{\partial p_j}\right] = 0 \tag{3.12}$$

where the dot represents the operator $\partial/\partial t$, or:

$$\frac{\partial f}{\partial t}+\sum_1^3\left[\frac{\partial f}{\partial x_j}\dot{x}_j+\frac{\partial f}{\partial p_j}\dot{p}_j+f\left(\frac{\partial \dot{x}_j}{\partial x_j}+\frac{\partial \dot{p}_j}{\partial p_j}\right)\right] = 0 \tag{3.13}$$

Hamilton's canonical equations yield:

$$\frac{\partial q_j}{\partial q_j}+\frac{\partial \dot{p}_j}{\partial p_j}=\frac{\partial}{\partial q_j}(\partial \mathcal{q}/\partial p_j)-\frac{\partial}{\partial p_j}(\partial \mathcal{q}/\partial q_j)=\frac{\partial^2 \mathcal{q}}{\partial p_j \partial q_j}-\frac{\partial^2 \mathcal{q}}{\partial q_j \partial p_j}=0$$

where $\mathcal{q}$ is the Hamiltonian. Thus, equation (3.13) becomes:

$$\frac{\partial f}{\partial t}+\sum_{1}^{3}[(\partial f/\partial x_j)(\mathrm{d}x_j/\mathrm{d}t)+(\partial f/\partial p_j)(\mathrm{d}p_j/\mathrm{d}t)]=\mathrm{d}f/\mathrm{d}t=0 \qquad (3.14)$$

known as Liouville's equation. Hence, the total six-dimensional 'hydrodynamic' time derivative of f is zero. This means that, if one travels with the particles in the phase space, one will find that their density in the neighbourhood of a given mobile point does not vary with time.

Instead of considering the μ-space representation of a collection of N identical non-interacting particles orbiting in the same portion of the physical space, one could as well consider one particle at N different, discretely separated, instants of its life. Since the Hamiltonian function is the same for any of the N non-interacting particles, or for this single particle, nothing is changed in the preceding argument. If the particles are interacting, neither f nor $\mathcal{q}$ can be expressed as functions of seven variables only. The probability of finding a particle somewhere in the μ-space is influenced by the location of all the other particles. On the other hand, the Hamiltonian function comprises terms corresponding to external potentials and to potentials resulting from the interactions between particles. For instance, in the case of a Coulomb interparticle interaction, the Hamiltonian comprises terms such as $q_i q_j/2\pi\varepsilon r_{ij}$ with $r_{ij}=[(x_i-x_j)^2+(y_i-y_j)^2+(z_i-z_j)^2]^{1/2}$ and the total number of such terms is equal to the total number of interacting pairs of particles. This number is $N(N-1)/2$, i.e. virtually $N^2/2$ if all particles are electrically charged. Similarly, magnetic interactions include additional terms in which both the space co-ordinates and the momenta are involved[16].

Therefore, in the case of interacting particles, both f and $\mathcal{q}$ are functions of $(6N+1)$ variables, one Lagrangian co-ordinate and one associated momentum per degree of freedom, plus time. In the Γ-space the canonical equations have the same form as in the μ-space, it being understood that the sums of 3 terms must be replaced by sums of $3N$ terms in equations (3.12), (3.13) and (3.14). With such formulation, the particles no longer need to be identical to each other and the entire plasma can be represented by a single point in the Γ-space. Liouville's theorem is then valid for interacting particles, provided the Γ-space is used instead of the μ-space.

3.3 BOLTZMANN'S EQUATION[4, 6]

In the case of identical non-interacting particles, the Liouville theorem can also be written in the Γ-space. The problem can be simplified by introducing two properties of identical non-interacting particles; the distribution function f and the Hamiltonian function $\mathcal{q}$.

The distribution function, written as a function of $6N$ variables plus time, factorises to a product of N functions, each involving only the co-ordinates and momenta of one particle, plus time. This is because the probability of a compound event made of independent parts is the product of the probabilities of the separate parts. If the particles are identical, the factors are also identical.

The Hamiltonian function of N non-interacting particles is the sum of N terms, each of them involving only the co-ordinates and momenta of one particle. If the particles are identical, the terms of the Hamiltonian are identical as well.

In the case of weakly interacting particles, or when the particles interact only occasionally, these properties are 'approximately true'. In other words, such a decomposition of f into a product of factors and of $\mathcal{H}$ into a sum of terms, all identical if the particles are identical, will be 'approximately possible'. As a result an expression of Liouville's theorem in the μ-space will be 'approximately correct'.

Several methods can be used to improve the accuracy of this simplified version of the theorem. First, an important part of the interaction between the particles would be taken into account if, instead of considering the scalar, vector, and gravitational potential of the external sources only, one introduces the influence of the plasma itself, but considering only global densities of electricity, electric current, and matter. For instance, instead of considering the true discontinuous density of electricity $\varrho_e(r, t) = \sum q_j \delta(r-r_j)$ (where δ is the Dirac function and q the charge), and similar expressions for the current and mass densities J and ϱ, one considers an average density $\varrho_e(r, t) = qn(r, t)$, where n is the number of particles per unit volume; the latter is calculated by integration of f over the three p co-ordinates. If more than one group of identical particles is involved, a summation should be made. Such an approximation, which disregards the microfield of the particles but takes into account the macrofield of the plasma as a whole, has been in fact widely used throughout the orbital approach. Second, the hitherto neglected part of the interactions could be taken into account by writing in the right-hand side of equation (3.14) an hitherto unspecified $(\partial f/\partial t)_{\text{int}}$ instead of zero. Of course, the only part of the interactions to be taken into account in such an expression is that of the microfield only, which has not been taken into consideration in the new expression of the Hamiltonian. This solves, in principle, the problem, for the new Hamiltonian can be expressed as a function of the three co-ordinates of the physical space, their associated momenta and time only, and, in the demonstration of Liouville's theorem, the exact expression of the Hamiltonian is irrelevant; it suffices that the laws of analytical dynamics are satisfied. Therefore, for a collection of N identical interacting particles, Liouville's theorem can be written in the μ-space in the form:

$$\frac{\partial f}{\partial t} + \sum_1^3 [(\partial f/\partial x_j)(\mathrm{d}x_j/\mathrm{d}t) + (\partial f/\partial p_j)(\mathrm{d}p_j/\mathrm{d}t)] = (\partial f/\partial t)_{\text{int}} \tag{3.15}$$

It is possible to attempt an evaluation of the right-hand side of equation (3.15) using the exact form of the interaction terms in the $6N+1$ variable Hamiltonian function described above. Assume, for a while, that the microfield part of the interaction is completely negligible. Assume also that the expression of the

Hamiltonian involves no magnetic terms. Under such conditions $p = Mv$. If the system of co-ordinates in the μ-space is changed from x, y, z, p_x, p_y, p_z to x, y, z, v_x, v_y, v_z then the corresponding volume elements will be in the ratio M^3. Hence, if the density of figurative points in the (r, p) space is constant according to equation (3.14), it will also be constant in the (r, v) space. Thus:

$$\frac{\partial f}{\partial t}+\sum_1^3 [(\partial f/\partial x)(dx_j/dt)+(\partial f/\partial v_j)(dv_j/dt)] = df/dt = 0 \tag{3.16}$$

If magnetic terms are involved in the Hamiltonian function, then $p = Mv + qA$. In any change of co-ordinates, the ratio of the corresponding volume elements can be calculated by using the Jacobian. The Jacobian is a determinant calculated by taking the partial derivative of any co-ordinate of one system with respect to all the co-ordinates of the second system. For instance, in the physical space, the Jacobian is:

$$\text{Jacobian} = \begin{vmatrix} \dfrac{\partial x_1}{\partial x_2} & \dfrac{\partial x_1}{\partial y_2} & \dfrac{\partial x_1}{\partial z_2} \\ \dfrac{\partial y_1}{\partial x_2} & \dfrac{\partial y_1}{\partial y_2} & \dfrac{\partial y_1}{\partial z_2} \\ \dfrac{\partial z_1}{\partial x_2} & \dfrac{\partial z_1}{\partial y_2} & \dfrac{\partial z_1}{\partial z_2} \end{vmatrix}$$

and the value of the determinant is equal to the ratio of the corresponding volume elements. This method is easily applied to the change of co-ordinates under consideration, and it shows that the ratio of the volume elements is constant and equal to M^3 even when magnetic forces are present.

Note that $dv_j/dt = a_j = F_j/M$ where a is the acceleration and F the external force. By taking this into account, and introducing a hodographic operator ∇_v such as:

$$\nabla_v = i\frac{\partial}{\partial v_x}+j\frac{\partial}{\partial v_y}+k\frac{\partial}{\partial v_z} \tag{3.17}$$

where i, j and k are the unit vectors in the v_x, v_z and v_y directions respectively. Equation (3.16) becomes, in vectorial form:

$$\frac{\partial f}{\partial t}+v.\nabla f+\frac{F}{M}.\nabla_v f = 0 \tag{3.18}$$

known as Vlasov's equation or the collisionless Boltzmann equation. F is the sum of the electric, magnetic and gravitational forces resulting from external fields as well as the macroscopic forces resulting from the plasma itself. If viscous-like forces were to be taken into consideration, then Vlasov's equation would be:

$$\frac{\partial f}{\partial t}+v.\nabla f+\nabla_v.\left[\frac{F}{M}f\right] = 0$$

This is rarely the case in plasma physics except when the influence of the neutral component is approximated by a viscous force on the charge carriers.

If, now, the microscopic interaction term is taken into consideration, then equation (3.18) becomes:

$$\frac{\partial f}{\partial t}+\boldsymbol{v}.\nabla f+\frac{\boldsymbol{F}}{M}.\nabla_{v} f=(\partial f/\partial t)_{\text{inter}} \tag{3.19}$$

Boltzmann was concerned with *collisions* in the sense of the kinetic theory of gases, and equation (3.19) is generally quoted in the more restrictive form:

$$\frac{\partial f}{\partial t}+\boldsymbol{v}.\nabla f+\frac{\boldsymbol{F}}{M}.\nabla_{v} f=(\partial f/\partial t)_{\text{coll}} \tag{3.20}$$

known as Boltzmann's equation.

If collisions in the sense of the kinetic theory of gases are involved, the phase trajectory of a particle ceases to be continuous. Indeed, a hard-sphere collision introduces no important modification in the space location of the particle, but can introduce serious variations in the velocity. In the μ-space, the particle seems suddenly to leave the volume element in which it was located and to reappear instantly in another volume element. Still more specifically, Boltzmann had in mind the *constant radius* hard-sphere model, which is so successful in the kinetic theory of gases. In that particular case, $(\partial f/\partial t)_{\text{coll}}$ can be calculated by statistical considerations of the collisions. Boltzmann's equation is almost ideally accurate as far as the necessities of the kinetic theory of gases are concerned. But it is integro-differential in the seven relevant variables, and extremely difficult to solve except in very simple cases. Its adequacy in plasma physics is far from being as good.

In plasma physics, it is uncertain that a Boltzmann-like collisional term is the correct answer to most of the problems because of the 'long-range' character of the Coulomb interaction. A deflection of, say, an electron in the Coulomb microfield of a positive ion is often too different from a hard-sphere collision for $(\partial f/\partial t)_{\text{coll}}$ to be of any utility. This is because the Coulomb microfield has a range much greater than the average distance between close particles. Therefore, the concept of a particle describing free paths separated by discrete collisions is inadequate; the picture of the particle interacting all along its path is much more realistic in many cases. Even if the deflections can be considered as discrete events, the cumulative effect of numerous small-angle deflections can be much greater than the effect of the less numerous large-angle deflections, more nearly resembling collisions (Gvosdover effect).

In the hodographic space of co-ordinates v_x, v_y and v_z, the motion of a particle undergoing such numerous small angle deflections resembles the motion in the physical space of a particle in the Brownian motion, and transpositions of the methods used in the latter case have been successfully effected. The best known is the Fokker–Planck method[43], leading to a particular form of equation (3.19) known as the Fokker–Planck equation. A rigorous demonstration of this equation has been given by Chandrasekhar[7]. An entirely different account is given by Kittel[25]. Fokker–Planck's equation can be written for a plasma as:

$$\frac{\partial f}{\partial t}+\boldsymbol{v}.\nabla f+\frac{\boldsymbol{F}}{M}.\nabla_v f = \nabla_v.(\mathcal{A}f)+\nabla_v.\nabla_v:(\mathcal{B}f) \tag{3.21}$$

with $\mathcal{A} = \langle \Delta \boldsymbol{v}/\tau \rangle$ and $\mathcal{B} = \langle \Delta \boldsymbol{v}.\Delta \boldsymbol{v}/\tau \rangle$.

The motion, in the hodographic space, of the particle is considered here for a lapse of time τ long enough to comprise a large number of deflections, and displacements are averaged in this space per unit time. The term $\Delta_v.(\mathcal{A}f)$ is often smaller than the other terms of equation (3.21) and it is equal to zero in symmetrical configurations. The term $\nabla_v.\nabla_v:(\mathcal{B}f)$ is never zero because it involves quadratic displacements combined in a dyadic. The sum of these two terms involves only operations in the hodographic part of the μ-space since the deflections influence the velocity of the particle considerably more than its position in space.

Equation (3.19) is valid disregarding the exact nature of the interaction between the particles, and only the magnitude of the coefficients involved in vector $\mathcal{A}$ and tensor $\mathcal{B}$ depend upon this nature. If such coefficients could be determined experimentally, the problem would be solved. Some theoreticians calculated those coefficients in particular cases, for instance in the case of the inverse-square law force of the Coulomb interaction.

Calculations based on the Fokker–Planck equation are often more accurate than those based on the original Boltzmann equation; however, the best results are obtained by combining both types of interaction terms, since in many cases large angle deflections play a non-negligible role and are best treated as collisions in the Boltzmann sense of the word.

Since most plasmas comprise three or more kinds of particles, an equation such as equation (3.19) must be written for each species. If neutral particles are involved, the electromagnetic part of the force is zero, and in general the other forces are negligible.

On the right-hand sides of equations (3.19), (3.20) and (3.21) the interaction terms should be decomposed into as many factors as there are species of particles. For instance, in a plasma comprising electrons, one species of positive ions and one species of neutral particles, the interaction term on the right-hand side of the equation relevant to the electrons should comprise an (e–e) factor (electrons–electrons), an (e–i) factor (electrons–ions), and an (e–n) factor (electrons–neutrals). Quite often, however, only one, and more exceptionally two factors are predominant. In most electrical discharges in gases, the (e–n) factor is largely predominant; at very high current intensities, the (e–i) factor (Gvosdover effect) is taken into account; and in strongly and fully ionised plasmas, the (e–n) factor is zero and the (e–i) interaction is predominant.

3.4 MACROSCOPIC RELATIONS

Consider a distribution function $f(x, y, z, v_x, v_y, v_z, t) = f(\boldsymbol{r}, \boldsymbol{v}, t)$ in which the second expression is shorthand for the first expression. Similarly the hypervolume element is $\mathrm{d}x\ \mathrm{d}y\ \mathrm{d}z\ \mathrm{d}v_x\ \mathrm{d}v_y\ \mathrm{d}v_z = \mathrm{d}\boldsymbol{r}.\mathrm{d}\boldsymbol{v}$, where $\mathrm{d}\boldsymbol{r}$ and $\mathrm{d}\boldsymbol{v}$ are volume elements. The total number of particles in a volume element $\mathrm{d}\boldsymbol{r}$ close to the end of the vector $\boldsymbol{r}$, disregarding the velocities, is obtained by integration of the three velocity components; hence:

$$n(r, t) = \iiint_{-\infty}^{+\infty} f(r, v, t)\,\mathrm{d}v \tag{3.22}$$

The drift velocity is the average value of the velocity $\boldsymbol{v}$ considered as a vector. Hence, it is the sum of all the velocity vectors, divided by the number of particles:

$$\boldsymbol{v}_\mathrm{d}(\boldsymbol{r}, t) = \frac{1}{n(\boldsymbol{r}, t)} \iiint_{-\infty}^{+\infty} \boldsymbol{v} f(r, v, t)\,\mathrm{d}v \tag{3.23}$$

More generally, consider an unspecified quantity $\mathcal{Y}(\boldsymbol{r}, \boldsymbol{v}, t)$. Its average value near the point $\boldsymbol{r}$ at time t is:

$$\overline{\mathcal{Y}}(r, t) = \frac{1}{n(\boldsymbol{r}, t)} \iiint_{-\infty}^{+\infty} \mathcal{Y}(\boldsymbol{r}, \boldsymbol{v}, t)\, f(\boldsymbol{r}, \boldsymbol{v}, t)\,\mathrm{d}v \tag{3.24}$$

Equation (3.24) has been written as if the quantity $\mathcal{Y}$ were a scalar, but in fact, it is valid also for vector quantities or tensors of any order. $\mathcal{Y}$ does not have to be a function of all the variables $\boldsymbol{r}$, $\boldsymbol{v}$ and t. For instance, equations (3.22) and (3.23) are special cases of equation (3.24) respectively for $\mathcal{Y} = 1$ and $\mathcal{Y} = \boldsymbol{v}$.

The macroscopic quantities in a plasma are always expressed by using average values of the microscopic quantities, so that equation (3.20) must be multiplied by $\mathcal{Y}$ and integrated over all the values of $\boldsymbol{v}$. For simplicity, only those quantities which are functions of $\boldsymbol{v}$ alone will be dealt with; the general case is not much different in principle[39].

First, integrate equation (3.20):

$$\iiint_{-\infty}^{+\infty} \mathcal{Y}(\boldsymbol{v})\frac{\partial f}{\partial t}\,\mathrm{d}v + \sum_1^3 \iiint_{-\infty}^{+\infty} \mathcal{Y}(\boldsymbol{v}) v_i \frac{\partial f}{\partial x_i}\,\mathrm{d}v + \sum_1^3 \iiint_{-\infty}^{+\infty} \mathcal{Y}(\boldsymbol{v}) \frac{F_i \partial f}{M \partial v_i}\,\mathrm{d}v$$
$$= \iiint_{-\infty}^{+\infty} \mathcal{Y}(\boldsymbol{v})(\partial f/\partial t)_\mathrm{coll}\,\mathrm{d}v$$

The left-hand side of this equation comprises seven terms which are of three kinds:

$$\iiint_{-\infty}^{+\infty} \mathcal{Y}(\boldsymbol{v})\frac{\partial f}{\partial t}\,\mathrm{d}v = \frac{\partial}{\partial t}\iiint_{-\infty}^{+\infty} \mathcal{Y}(\boldsymbol{v})\, f(\boldsymbol{r}, \boldsymbol{v}, t)\,\mathrm{d}v = \partial(n\overline{\mathcal{Y}})/\partial t \tag{3.25}$$

$$\iiint_{-\infty}^{+\infty} \mathcal{Y}(\boldsymbol{v})\frac{\partial f}{\partial x_i}\,\mathrm{d}v = \frac{\partial}{\partial x_i}\iiint_{-\infty}^{+\infty} \mathcal{Y}(\boldsymbol{v}) v_i f(\boldsymbol{v}, \boldsymbol{r}, t)\,\mathrm{d}v = \partial(n\overline{\mathcal{Y}v_i})/\partial x_i \tag{3.26}$$

$$\iiint_{-\infty}^{+\infty} \mathcal{Y}(\boldsymbol{v})\frac{F_i(\boldsymbol{r}, \boldsymbol{v})}{M}\frac{\partial f}{\partial v_i}\,\mathrm{d}v = \left[\mathcal{Y}(\boldsymbol{v})\frac{F_i(\boldsymbol{r}, \boldsymbol{v})}{M} f(\boldsymbol{r}, \boldsymbol{v}, t)\,\mathrm{d}v\right]_{-\infty}^{+\infty}$$
$$- \iiint_{-\infty}^{+\infty} \frac{f(\boldsymbol{r}, \boldsymbol{v}, t)}{M}\frac{\partial}{\partial v_i}[\mathcal{Y}(\boldsymbol{v})F_i(\boldsymbol{r}, \boldsymbol{v})]\,\mathrm{d}v = -\frac{n}{M}\partial(\overline{\mathcal{Y}F_i})/\partial x_i \tag{3.27}$$

The integrated term in equation (3.27) is zero since logically $f(-\infty) = f(+\infty) = 0$.

It is of interest to group terms like equation (3.26) by using the operator ∇. The latter will describe the divergence of a vector if $\mathscr{Y}$ is a scalar quantity; if $\mathscr{Y}$ is a vector, ∇ will still correspond to a divergence, but this time, a dyadic. The case of tensors of higher degree leads to similar conclusions. Sums of terms like equation (3.27) can also be grouped by introducing the 'hodographic divergence' defined by equation (3.17). In using this operator, one should recall that all the forces in plasma physics either do not depend upon velocity (e.g. electric and gravitational forces) or depend upon it, but in such a manner that $\partial F_i/\partial v_i = 0$ (e.g. magnetic forces). In both cases, the hodographic divergence of $\boldsymbol{F}$ is always zero.

By taking all the partial results into account, equation (3.24) becomes:

$$\frac{\partial}{\partial t}[n(\boldsymbol{r}, t)\overline{\mathscr{Y}(\boldsymbol{r}, t)}] + \nabla.[n(\boldsymbol{r}, t)\overline{\boldsymbol{v}\mathscr{Y}(\boldsymbol{r}, t)}] - \frac{n(\boldsymbol{r}, t)}{M}\boldsymbol{F}(\boldsymbol{r}, t).\nabla_{\boldsymbol{v}}\mathscr{Y}(\boldsymbol{v})$$
$$= \iiint_{-\infty}^{+\infty} \mathscr{Y}(\boldsymbol{v})(\partial f/\partial t)_{\text{coll}}\, \mathrm{d}\boldsymbol{v} \qquad (3.28)$$

In this equation if $\mathscr{Y}$ is a scalar, the 'hodographic divergence' development leads to a 'hodographic gradient'.

Three practical cases are of special importance. If, in equation (3.28), $\mathscr{Y}$ is set equal to unity, a particle balance equation is formed; if it is set equal to $M\boldsymbol{v}$, a momentum transport equation is obtained, and if $\mathscr{Y}$ is set equal to $Mv^2/2$ an energy balance equation is obtained.

3.5 CONTINUITY EQUATION

For $\mathscr{Y} = 1$, equation (3.28) becomes:

$$\frac{\partial n}{\partial t} + \nabla.(n\boldsymbol{v}_\text{d}) = (\partial n/\partial t)_{\text{coll}} \qquad (3.29)$$

since $\boldsymbol{v}_\text{d} = \langle \boldsymbol{v} \rangle$ and

$$\iiint_{-\infty}^{+\infty} (\partial f/\partial t)_{\text{coll}}\, \mathrm{d}\boldsymbol{v} = \left.\frac{\partial}{\partial t}\right|_{\text{coll}} \iiint_{-\infty}^{+\infty} f(\boldsymbol{r}, \boldsymbol{v}, t)\, \mathrm{d}\boldsymbol{v} = (\partial n/\partial t)_{\text{coll}}$$

The exact expression for $(\partial n/\partial t)_{\text{coll}}$ depends upon the given case. For instance, if the expression concerns the molecules of a gas in which no chemical effects occur, the particles are conservative and the right-hand side of equation (3.29) has to be set equal to zero; this is the case of a fully ionised plasma. On the other hand, in a partly ionised plasma as well as in a gas in which chemical transformations occur, or in any similar case, the right-hand side of equation (3.29) is not zero and has to be computed by a special statistical approach using all the velocities; thus implying a partial return to the orbital theory. Here the results of Sections 2.4 and 2.7 should be used by writing $(\partial n/\partial t)_{\text{coll}}$ as the difference between the contribution of all the ionising and recombining processes in the plasma. For instance, for a quiescent plasma one obtains $(\partial n/\partial t)_{\text{coll}} = \nu_i n$, where ν_i is the frequency of ionisation processes. Equation (3.29) becomes then:

$$\frac{\partial n}{\partial t} + \nabla.(n\boldsymbol{v}_d) = \nu_i n$$

This is a microscopic relation describing a given species of plasma particles. In a plasma containing several species, each has its own microscopic continuity equation. Consider now a two-component plasma where the particles present are electrons and ions; the equations of continuity for ions and electrons, respectively, will be:

$$\frac{\partial n^+}{\partial t} + \nabla.(n^+\boldsymbol{v}_d^+) = 0 \tag{3.30a}$$

$$\frac{\partial n^-}{\partial t} + \nabla.(n^-\boldsymbol{v}_d^-) = 0 \tag{3.30b}$$

Multiplying these two equations by the electric charge e and subtracting equation (3.30b) from equation (3.30a), one obtains:

$$\frac{\partial}{\partial t}[e(n^+ - n^-)] + \nabla.e(n^+\boldsymbol{v}_d^+ - n^-\boldsymbol{v}_d^-) = 0 \tag{3.31}$$

Noting that the current density $\boldsymbol{J}$ is $e(n^+\boldsymbol{v}_d^+ - n^-\boldsymbol{v}_d^-)$ and the charge density ϱ_e is $e(n^+ - n^-)$, equation (3.31) leads to equation (3.10), the charge conservation equation describing macroscopically the entire plasma.

If now equation (3.30a) is multiplied by M^+, the ionic mass, and equation (3.30b) by M^-, the electronic mass, and equation (3 30a) is added to equation (3.30b), one obtains:

$$\frac{\partial}{\partial t}(n^+M^+ + n^-M^-) + \nabla.(M^+n^+\boldsymbol{v}_d^+ + M^-n^-\boldsymbol{v}_d^-) = 0 \tag{3.32}$$

Note that the mass density is:

$$\varrho = n^+M^+ + n^-M^- \tag{3.33}$$

and that the total macroscopic drift velocity $\boldsymbol{v}_d$ of the plasma is such that:

$$\varrho\boldsymbol{v}_d = n^+M^+\boldsymbol{v}_d^+ + n^-M^-\boldsymbol{v}_d^- \tag{3.34}$$

Equation (3.32) then becomes:

$$\frac{\partial \varrho}{\partial t} + \triangle.(\varrho\boldsymbol{v}_d) = 0 \tag{3.35}$$

This is the macroscopic fluid continuity equation of the plasma.

For plasmas with more than two components, similar reasoning leads to the same continuity equations [i.e. equations (3.30) and (3.35)]. These equations are valid even if the continuity equations for the various species contain sinks and sources representing ionisation, recombination, or chemical reactions. Indeed, in the process of subtraction and summation all the collisional terms must cancel each other. However, equation (3.10) cannot now be considered as related to the motion of the plasma in its entirety, since there will be diffusion of different species with respect to each other without the necessity of a net charge transport. The quasi-neutrality of the plasma is therefore not sufficient in itself to simplify the continuity equation. For instance, in a three-component plasma containing neutral atoms, electrons, and singly charged positive ions, the charge carriers (electrons and ions) will transfer momentum to each other and to the neutral particles. The velocity $\boldsymbol{v}_d$ in equation (3.35) will be the mean velocity of the ionised fraction of the plasma, and it is essential to evaluate the relative motion between ions and neutrals. This motion is known as *ion slip*. For a weakly ionised gas the frame in which the neutral component of the plasma is at rest can be taken as the frame of reference. In this case, the concepts of mobility and ambipolar diffusion seen above should be taken into account by assuming a momentum transfer to fixed scattering centres.

3.6 MOMENTUM TRANSPORT EQUATION

If, in equation (3.28), $\mathscr{U}$ is set equal to the vector quantity $M\boldsymbol{v}$, the second term in the equation gives rise to the divergence of a dyadic:

$$\frac{\partial(nM\boldsymbol{v}_d)}{\partial t}+\nabla.(nM\langle\boldsymbol{v}.\boldsymbol{v}\rangle)-n\langle\boldsymbol{F}\rangle = \iiint_{-\infty}^{+\infty} M\boldsymbol{v}(\partial f/\partial t)_{coll}\,\mathrm{d}v \qquad (3.36)$$

This equation is not equivalent to any of the equations developed by the orbital approach. However, a number of equations obtained by that approach correspond to particular, simplified cases of equation (3.36).

The orbital 'random' velocity $\boldsymbol{v}$ generally consists of two terms: the drift velocity $\boldsymbol{v}_d$ which is the average value of $\boldsymbol{v}$ and a 'pure random' velocity $\boldsymbol{v}-\boldsymbol{v}_d$, say $\boldsymbol{u}$. In the orbital approach, it is assumed that $\boldsymbol{v}_d$ is so small with respect to $\boldsymbol{v}$ that there is no need to make a distinction between $\boldsymbol{v}$ and $\boldsymbol{u}$. This will not be the case here; thus:

$$\langle\boldsymbol{v}.\boldsymbol{v}\rangle = \langle(\boldsymbol{v}_d+\boldsymbol{u}).(\boldsymbol{v}_d+\boldsymbol{u})\rangle = \langle\boldsymbol{v}_d.\boldsymbol{v}_d\rangle+\langle\boldsymbol{v}_d.\boldsymbol{u}\rangle+\langle\boldsymbol{u}.\boldsymbol{v}_d\rangle+\langle\boldsymbol{u}.\boldsymbol{u}\rangle \qquad (3.37)$$

But $\boldsymbol{v}$ is not a statistical variable; it is itself an average value, so that equation (3.37) is equivalent to:

$$\boldsymbol{v}_d.\boldsymbol{v}_d+\boldsymbol{v}_d.\langle\boldsymbol{u}\rangle+\langle\boldsymbol{u}\rangle.\boldsymbol{v}_d+\langle\boldsymbol{u}.\boldsymbol{u}\rangle = \boldsymbol{v}_d.\boldsymbol{v}_d+\langle\boldsymbol{u}.\boldsymbol{u}\rangle$$

$\langle\boldsymbol{u}\rangle$ is zero, since $\boldsymbol{u}$ is a purely random component. By developing the divergence of the symmetric dyadic, one finds:

$$\nabla:(nM\boldsymbol{v}_d.\boldsymbol{v}_d) = (nM\boldsymbol{v}_d.\nabla)\boldsymbol{v}_d+(\boldsymbol{v}_d.\nabla)(nM\boldsymbol{v}_d)$$

In addition, let $\boldsymbol{\Psi} = nM\langle\boldsymbol{u}.\boldsymbol{u}\rangle$. After expanding the time derivative of equation (3.36) and making use of equation (3.29), one obtains:

$$nM\left(\frac{\partial}{\partial t}+\boldsymbol{v}_d.\nabla\right)\boldsymbol{v}_d = n\langle\boldsymbol{F}\rangle-\nabla:\boldsymbol{\Psi}+\iiint_{-\infty}^{+\infty} M\boldsymbol{v}(\partial f/\partial t)_{coll}\,\mathrm{d}v+M\boldsymbol{v}_d(\partial n/\partial t)_{coll} \qquad (3.38)$$

The left-hand side of this equation involves the product of the species density $\varrho = nM$. The first term on the right-hand side is the density of external forces; the force $\langle F \rangle$ being the sum of all the forces due to the electric, magnetic, and gravitational fields present. The second term is the divergence of the dyadic $\boldsymbol{\Psi}$. The third term can be interpreted as a kind of friction hampering the motion of the relevant particles and the last term represents a component attributable to the momentum introduced by the particles when they were created, disappearing with them when they are annihilated; it is seldom important in practice.

The second term $\nabla : \boldsymbol{\Psi}$ deserves some explanation. First, note that the only missing term in equation (3.38) is a pressure gradient, and that the nine components of $\boldsymbol{\Psi}$ have the dimensions of pressure. If $\boldsymbol{u}$ is isotropic, then the tensor degenerates into a scalar times a unit dyadic since $\langle u_x^2 \rangle = \langle u_y^2 \rangle = \langle u_z^2 \rangle = C^2/3$, while $\langle u_x u_y \rangle = \langle u_y u_z \rangle = \langle u_z u_x \rangle = 0$. The divergence of the tensor is replaced by the gradient of the scalar, and as $nMC^2/3$ is the pressure p_r, the divergence of $\boldsymbol{\Psi}$ is then replaced by the gradient of p_r, since each of the average squares is one-third of the mean square velocity. In general, this degeneracy does not occur, since, when the magnetic field is significant, the plasma is known to become anisotropic; hence the pressure ceases to be a scalar. Only for zero and weak magnetic fields can the simplification be introduced in equation (3.38).

The overall momentum equation of the plasma is obtained by summation of all the species present. The sum of all the terms $\int\int\int_{-\infty}^{+\infty} M\boldsymbol{v}(\partial f/\partial t)_{\text{coll}}\, \mathrm{d}v$ is equal to zero since the total momentum is conserved during a collision. By taking account of this, the macroscopic equation [equation (3.38)] of conservation of momentum for the entire plasma becomes:

$$\varrho\left(\frac{\partial}{\partial t}+\boldsymbol{v}_{\mathrm{d}}.\nabla\right)\boldsymbol{v}_{\mathrm{d}} = -\nabla:\boldsymbol{\Psi}+\boldsymbol{F}\sum_{s} n_{s} \tag{3.39}$$

where $\boldsymbol{v}_{\mathrm{d}}$ is the drift velocity of the plasma, $\boldsymbol{\Psi}$ is the pressure tensor of the entire plasma, $\boldsymbol{F}$ is the sum of external forces, and n_s is the number density of a particle species. If the electromagnetic field only is considered, the force will be given by equation (2.1). When $e\sum_s n_s = \varrho_e$ the charge density, and $\boldsymbol{J} = e\boldsymbol{v}_{\mathrm{d}}$ the current density, equation (3.39) becomes:

$$\varrho\left(\frac{\partial}{\partial t}+\boldsymbol{v}_{\mathrm{d}}.\nabla\right)\boldsymbol{v}_{\mathrm{d}}+\nabla:\boldsymbol{\Psi} = \varrho_{e}\boldsymbol{E}+\boldsymbol{J}\times\boldsymbol{B} \tag{3.40}$$

The pressure tensor can be written as the sum of a stress tensor $\boldsymbol{X}$ and the hydrostatic pressure p_r times a unit tensor; equation (3.40) becomes then:

$$\varrho\frac{\mathrm{D}\boldsymbol{v}_{\mathrm{d}}}{\mathrm{D}t}+\nabla:\boldsymbol{X}+\nabla p_{r} = \varrho_{e}\boldsymbol{E}+\boldsymbol{J}\times\boldsymbol{B} \tag{3.41}$$

where $\mathrm{D}/\mathrm{D}t = \partial/\partial t+\boldsymbol{v}_{\mathrm{d}}.\nabla$, the hydrodynamic derivative of the drift velocity. In this equation the forces resulting from polarisation and particle magnetic moments have been neglected and it is assumed that there are no sinks or sources in the plasma. To find an expression for $\boldsymbol{X}$, the collision term $(\partial f/\partial t)_{\text{coll}}$

in Boltzmann's equation is usually replaced approximately by an expression involving only the distribution functions. This yields s Boltzmann equations with s unknown distribution functions, where s is the number of species in the plasma. To solve this system it is further assumed for instance that all the distribution functions are quasi-Maxwellian about a single temperature T, such as:

$$f_s = f_{0s} + \beta f_{1s} + \beta^2 f_{2s} + \ldots$$

where f_{0s} is the Maxwellian distribution function and β is a small parameter which is linear in pressure gradients, concentration gradients and forces. By this method one obtains expressions for f_{1s}, f_{2s}, etc., from which all the transport coefficients can be deduced. This yields for the shear stress tensor X_{ij}[20]:

$$X_{ij} = -\upsilon\left(\frac{\partial v_i}{\partial x_j} + \frac{\partial v_j}{\partial x_i}\right) + \frac{2}{3}\upsilon\delta_{ij}\nabla . \boldsymbol{v}_{\mathrm{d}}$$

where υ is the viscosity coefficient of the plasma.

3.7 ENERGY CONSERVATION EQUATION[35]

If the plasma is isotropic and Maxwellian, the conservation of energy equation takes a simple form. Five algebraic quantities are sufficient to define the state of one component of the plasma; for instance, the density n of the carriers, the three components of the drift velocity $\boldsymbol{v}_{\mathrm{d}}$ and the carrier temperature T. Since four equations are already known—the carrier balance and the three components of momentum transfer—only one extra scalar equation is needed. Such an equation can be obtained if, in equation (3.28), $\mathscr{U}$ is set equal to the scalar quantity $Mv^2/2$. If $\boldsymbol{F}$ is a purely electromagnetic force, taking into account that the magnetic induction does no work, the result can be stated as:

$$\frac{\partial}{u)}\partial t M\langle v^2\rangle/2) + \nabla . (nM\langle v^2 \boldsymbol{v}\rangle) = \boldsymbol{J}.\boldsymbol{E} + \iiint_{-\infty}^{+\infty} (Mv^2/2)\,(\partial f/\partial t)_{\mathrm{coll}}\,\mathrm{d}v \qquad (3.42)$$

The first term of this equation is the local increase in time of the particle kinetic-energy; the second is a convective term; the third is the power density of the electromagnetic force; the fourth represents the extent to which the balance of the first three is altered by collisions.

Whenever the velocity distribution results from the superposition of an isotropic part, not necessarily Maxwellian, and a drift velocity, we have:

$$nM\langle v^2\rangle/2 = nM\frac{v_{\mathrm{d}}^2}{2} + nM\frac{C^2}{2} = \frac{1}{2}nMv_{\mathrm{d}}^2 + \frac{3}{2}p_{\mathrm{r}}$$

where p_{r} is the pressure, also:

$$\langle v^2 \boldsymbol{v}\rangle = \langle(\boldsymbol{v}_{\mathrm{d}}+\boldsymbol{u})^2 . (\boldsymbol{v}_{\mathrm{d}}+\boldsymbol{u})\rangle = \langle v_{\mathrm{d}}^2 \boldsymbol{v}_{\mathrm{d}}\rangle + 2\langle(\boldsymbol{v}_{\mathrm{d}} . \boldsymbol{u})\,\boldsymbol{v}_{\mathrm{d}}\rangle + \langle u^2 \boldsymbol{v}_{\mathrm{d}}\rangle + \langle v_{\mathrm{d}}^2 \boldsymbol{u}\rangle$$
$$+ 2\langle(\boldsymbol{v}_{\mathrm{d}} . \boldsymbol{u})\boldsymbol{u}\rangle + \langle u^2 \boldsymbol{u}\rangle$$

Since the distribution of $\boldsymbol{u}$ is purely isotropic, all the odd functions of $\boldsymbol{u}$ vanish on the average. Furthermore, the third and fifth terms of the expansion can be grouped together; indeed:

$$(\boldsymbol{v}_d.\boldsymbol{u})\boldsymbol{u} = (v_{dx}u_x+v_{dy}u_y+v_{dz}u_z)(\boldsymbol{i}u_x+\boldsymbol{j}u_y+\boldsymbol{k}u_z)$$

Consider only the x-component; in the averaging process, the factors in u_xu_y and u_xu_z vanish, hence:

$$\langle(\boldsymbol{v}_d.\boldsymbol{u})\boldsymbol{u}\rangle_x = \langle v_{dx}u_x^2\rangle = v_{dx}u_x^2 = v_{dx}C^2/2 = (C^2\bar{\boldsymbol{v}}_d)_x/2$$

whereas:

$$\langle u^2\boldsymbol{v}_d\rangle = \langle u^2\rangle\langle\boldsymbol{v}_d\rangle = C^2\bar{\boldsymbol{v}}_d$$

hence:

$$\langle v^2\bar{\boldsymbol{v}}\rangle = v_d^2\bar{\boldsymbol{v}}_d+\frac{5}{3}C^2\bar{\boldsymbol{v}}_d = v_d^2\bar{\boldsymbol{v}}_d+5\frac{p_r\boldsymbol{v}_d}{nM}$$

By substituting these terms in equation (3.42), one obtains:

$$nM\boldsymbol{v}\frac{\partial\boldsymbol{v}}{\partial t}+\frac{1}{2}Mv_d^2\frac{\partial n}{\partial t}+\frac{3\partial p_r}{2\partial t}+\frac{1}{2}Mv_d^2.(n\boldsymbol{v}_d)+\frac{1}{2}nM\boldsymbol{v}_d.\nabla v_d^2+\frac{5}{2}\boldsymbol{v}_d\nabla p_r$$

$$+\frac{5}{2}p_r\nabla.\boldsymbol{v}_d = \boldsymbol{J}.\boldsymbol{E}+\iiint_{-\infty}^{+\infty}(Mv^2/2)(\partial f/\partial t)_{\text{coll}}\,\mathrm{d}\boldsymbol{v}$$

From this equation, subtract the particle-balance equation multiplied by $Mv_d^2/2$ and the momentum-transfer equation written for an isotropic plasma (i.e. with $\triangle p_r$ instead of $\nabla{:}\boldsymbol{\Psi}$) dot-multiplied by $\boldsymbol{v}_d$. Using the operator D/Dt the final result can be rearranged as:

$$3\frac{\mathrm{D}p_r}{\mathrm{D}t}-\frac{5p_r}{n}\frac{\mathrm{D}n}{\mathrm{D}t} = M\iiint_{-\infty}^{+\infty}\left(u^2-\frac{5}{2}C^2\right)(\partial f/\partial t)_{\text{coll}}\,\mathrm{d}\boldsymbol{v}$$

where all the collisional terms are grouped together.

There are many cases in which the right-hand side is identically zero or can be neglected for the relevant purpose; in such cases:

$$\mathrm{d}p_r/\mathrm{d}n = \tfrac{5}{3}p_r/n \qquad (3.43)$$

or $p_rn^{-5/3}$ is constant in a frame of reference moving with the bulk of the relevant kind of particles. This is seen to be identical with the law of the ordinary adiabatic compression or expansion, with γ being given the value corresponding to a monatomic gas.

When a plasma is immersed in a very strong magnetic field, the theory above is inadequate since it does not take into account the anisotropy of the ionised gas. More specifically, the concepts of temperature, thermal conductivity, and pressure become tensor concepts. For instance, the energy transfer is described by a third-order tensor, and the electric conductivity acquires the properties of a second-order tensor; hence $\boldsymbol{J}$ has to be calculated as a function of $\boldsymbol{E}$.

The proper way of obtaining the relevant relation consists in setting in equation (3.28) the condition that $\mathscr{U}$ is equal to the second-order tensor $\boldsymbol{\Psi}/n$. The mathematical treatment is the same as for $\mathcal{Y} = Mv$ and for $\mathcal{Y} = Mv^2/2$, with Ψ having at least nine components. A particular treatment has been given by Delcroix[14]. In addition to the second-order Q tensor Ψ, Delcroix's fundamental equation involves a third-order tensor which is the heat transfer. The latter intervenes only through its divergence.

If the presence of a strong magnetic field motivates the anisotropy, there is still an equivalence between the two directions perpendicular to it. If the frame of reference is chosen such that the z-axis is parallel to the magnetic field, the tensor Ψ becomes diagonal. Furthermore, $\Psi_{xx} = \Psi_{yy} = p_{r\perp}$ and $\Psi_{zz} = p_{r_{||}}$. Since the temperature is proportional to the pressure, there is still equipartition among the two directions perpendicular to the magnetic field. In some problems, it can be interesting to relate the presence of the magnetic field to the effect of collisions and the gyration radius to a mean free path. In such conditions, it is possible to generalise equation (3.43); it suffices to recall that γ, the coefficient in the second term adiabatic equation, is related to the number of degrees of freedom n_f by equation (1.83), and in the ordinary space, for non-rotating molecules, $n_f = 3$. When a strong magnetic field is applied and the collisions are scarce, $p_{r\perp}$ must obey a similar law with two degrees of freedom only, hence $\gamma = 2$; as for $p_{r_{||}}$ under the same conditions, there is only *one* degree of freedom and γ takes the value 3.

However, this is not the only way of writing the equations of adiabatic motion in an anisotropic plasma; the derivation of the above equations neglects the condition that the magnetic field and the plasma are in general very strongly coupled; hence any variation of the plasma density modifies also the value of the magnetic field. The relevant equations are designated as the Chew, Goldberger and Low[9, 12] (CGL) approximation.

In the anisotropic plasma, the random motions can be different, especially during fast transients, in the direction of the magnetic field and in the direction parallel to it. Energy exchanges exist between the two, but they are not instantaneous. Consider, for instance, an adiabatic compression. If it takes place in a given direction, it may increase only the relevant component of the tensor temperature, in relation to the relevant component of the tensor pressure. Later, heat can be transmitted anisotropically to the surroundings and temperature equalisation in one or two directions can proceed much faster than component equalisation on the spot.

Next, consider an alternate one-dimensional, nearly adiabatic compression. Only one component of temperature is affected as a first approximation. However, energy can flow with a certain delay between this component and the other ones on the spot. Thus, by an increase of pressure, the instantaneous temperature in the relevant component may be slightly decreased, whereas a decrease in pressure increases it slightly with respect to the steady-state value. Hence, there is a hysteresis and, therefore, energy can be communicated to the plasma in that way. The mathematical description of all these phenomena requires that the energy transfer equation be treated in the tensor form.

If the rigorous treatment is performed, the consequences of Boltzmann's equation form an infinite sequence which cannot be solved absolutely. The momentum of a particle is introduced as a quantity $\mathcal{Y}$, and the equation involves the divergence of a second-order tensor, which is the pressure tensor. The treatment of energy transfer involves the introduction of a quantity $\mathcal{Y}$ which has the same tensor nature as the pressure tensor. The final equation involves a heat-transfer tensor which is of the third order. The process can be repeated several times and each equation will introduce a new unknown quantity, which is a tensor of one unit greater than the order of the tensor introduced as quantity $\mathcal{Y}$. In each state, the system of equations involves one more unknown tensor than the number of equations already written. The only way to escape from this situation is to assume that, in the highest order equation, the highest order tensor can be neglected. For instance, in the adiabatic approximation the energy transfer equation is written by neglecting the third-order tensor describing the heat transfer from point to point. Unfortunately, this neglects at the same time the heat transfer between different components of the tensor temperature at a given point.

A more general energy equation is obtained by considering both the kinetic and internal energy U of the particle. For this case, in equation (3.28) $\mathcal{Y}$ is set equal to $\frac{1}{2}Mv^2+U$. Using the same averaging method of Section 3.6 and noting that in a direction i, one has:

$$Q_i/n = \langle u_i(\tfrac{1}{2}Mu^2+U)\rangle$$

where Q_i is the relevant component of the heat flux vector, the pressure is $p_{rik} = nM\langle u_i u_k\rangle$, the mass density is $\varrho = nM$, the average energy per unit mass of the particle is $e_s = \langle U\rangle/M$, and the electromagnetic force is given by equation (2.1). The conservation of energy equation now becomes:

$$\frac{\partial}{\partial t}\varrho\left(\frac{1}{2}v_d^2+e_s\right)+\nabla.\varrho\left(\frac{1}{2}v_d^2+e_s\right)\boldsymbol{v}_d+\nabla.\boldsymbol{Q}+\nabla.(\boldsymbol{v}_d p_{rik}) = \boldsymbol{E}.\boldsymbol{J} + \int U(\partial f/\partial t)_{\text{coll}}\,\mathrm{d}\boldsymbol{v} \quad (3.44)$$

By taking into consideration equation (3.35), equation (3.44) becomes:

$$\varrho \boldsymbol{v}_d \frac{\mathrm{D}\boldsymbol{v}_d}{\mathrm{D}t}+\varrho\frac{\mathrm{D}e_s}{\mathrm{D}t}+\nabla.\boldsymbol{Q}+\nabla.(\boldsymbol{v}_d p_{rik}) = \boldsymbol{E}.\boldsymbol{J}+\int U(\partial f/\partial t)_{\text{coll}}\,\mathrm{d}\boldsymbol{v} \quad (3.45)$$

The equation of overall conservation of energy in the plasma is obtained by summing up all equations relevant to the different species present; one obtains:

$$\varrho \boldsymbol{v}_d \frac{\mathrm{D}\boldsymbol{v}_d}{\mathrm{D}t}+\varrho\frac{\mathrm{D}e_s}{\mathrm{D}t}+\nabla.\boldsymbol{Q}+\nabla.(\boldsymbol{v}_d p_{rik}) = \boldsymbol{E}.\boldsymbol{J} \quad (3.46)$$

where the quantities are here macroscopic and related to the plasma as a whole. The internal energy is given by equation (1.81). The heat flux vector is:

$$\boldsymbol{Q} = -\varkappa\nabla T+\sum_s \boldsymbol{v}_s\varrho_s\mathcal{H}_s$$

where $\varkappa$ is the thermal conductivity coefficient, $\boldsymbol{v}_s$ is the diffusion velocity of a species, and $\mathcal{H}_s$ is the enthalpy of the species.

Equation (3.46) can be written in a more convenient form by noting that:

$$\mathscr{H} = \frac{1}{2}v_d^2 + e_s + \frac{p_r}{\varrho}$$

and

$$\nabla.(\boldsymbol{v}_d p_{rik}) = \nabla.(p_r \boldsymbol{v}_d) + \Pi$$

where

$$\Pi = \nabla(\boldsymbol{v}_d : \boldsymbol{X})$$

After using equation (3.41), one obtains:

$$\varrho\frac{De_s}{Dt} + \varrho p_r \frac{D(\varrho^{-1})}{Dt} = -\Pi - \nabla.\boldsymbol{Q} + \boldsymbol{J}.(\boldsymbol{E} + \boldsymbol{v}_d \times \boldsymbol{B}) \tag{3.47}$$

the term $\boldsymbol{J}.(\boldsymbol{E}+\boldsymbol{v}_d\times\boldsymbol{B}) = \boldsymbol{J}.\boldsymbol{E}^*$ is often called the Joule heating. In a plasma, the conduction current density $\boldsymbol{J}$ is usually the dot product of a tensor with the electric field in the moving frame of reference such as $\boldsymbol{J} = \boldsymbol{\sigma} : \boldsymbol{E}^*$, where $\boldsymbol{\sigma}$ is the electric conductivity tensor.

Finally, it can be shown that the equation of energy becomes, when expressed as a function of the entropy $\mathscr{S}$,

$$\varrho T\frac{D\mathscr{S}}{Dt} + \Pi + \nabla.\boldsymbol{Q} - \boldsymbol{E}^*.\boldsymbol{J} = 0 \tag{3.48}$$

3.8 BOLTZMANN'S EQUATION TREATMENT OF DIFFUSION AND MOBILITY

Mobility and diffusion are anisotropic phenomena. The external force in the first case and the concentration gradient in the second one impose anisotropies in the velocity distribution, so that the latter may be expanded as a sum of two terms:

$$f(\boldsymbol{r}, \boldsymbol{v}) = f_0(\boldsymbol{r}, \boldsymbol{v}) + f_1(\boldsymbol{r}, \boldsymbol{v}) \tag{3.49}$$

A steady state is assumed; hence, the disappearance of t from the brackets and of the time derivative from the left-hand side of Boltzmann's equation. In equation (3.49), f_0 represents a purely isotropic distribution, such that:

$$n = \iiint_{-\infty}^{+\infty} f_0(\boldsymbol{r}, \boldsymbol{v})\, d\boldsymbol{v} \tag{3.50}$$

whereas the same integral for f_1 becomes zero. Equation (3.20) can be written:

$$\boldsymbol{v}.\nabla(f_0+f_1) + \frac{\boldsymbol{F}}{M}\nabla_v.(f_0+f_1) = (\partial f_0/\partial t)_{coll} + (\partial f_1/\partial t)_{coll}. \tag{3.51}$$

The term $(\partial f_0/\partial t)_{coll}$ can be set equal to zero, since f_0 is an equilibrium distribution that the collisions tend to establish, but do not modify any longer when it is established. On the other hand, in spite of the assumed steady state, $(\partial f_1/\partial t)_{coll}$ is not to be set equal to zero, since the role of collisions is essential in diffusion and mobility phenomena. For instance, in the case of mobility, if there were no collisions, the carriers would be uniformly accelerated, hence the perturbation would be an ever-increasing function of time. Also, f_1 is assumed to be a small perturbation, which can be neglected with respect to f_0.

The concentration gradient, or the external force, tends to perturb without any limit, the distribution function. In equation (3.51), $(\partial f_1/\partial t)_{\text{coll}}$ is a restoration term which would tend, if the concentration gradient or the external force were suppressed, to re-establish the unperturbed distribution function more or less rapidly; thus an acceptable approximation is $(\partial f_1/\partial t)_{\text{coll}} = -f_1/\tau_m$ where τ_m is a relaxation time, the physical meaning of which depends upon the relevant problem. For instance, if an electron gas drifts among heavy molecules, the motion is completely randomised after *one* collision, hence τ_m is the average mean-free-flight time $\bar{\tau}$. If now positive ions drift among molecules, a single collision is not enough to randomise the motion since the scattering is only isotropic in the centre of a mass system and the positive ion retains an important fraction of its drift motion. Hence, τ_m is equal to several $\bar{\tau}$, e.g. three times.

When these considerations are introduced, equation (3.51) becomes:

$$\boldsymbol{v}.\nabla f_0 + \frac{\boldsymbol{F}}{\boldsymbol{M}}.\nabla_v f_0 = -f_1/\tau_m \tag{3.52}$$

The drift velocity is given by:

$$v_d = \frac{1}{n}\iiint_{-\infty}^{+\infty} f\boldsymbol{v}\, \mathrm{d}v = \frac{1}{n}\iiint_{-\infty}^{+\infty} f_1\boldsymbol{v}\, \mathrm{d}v \tag{3.53}$$

f is replaced by f_1 since f_1 is the only anisotropic component, and is alone involved in a drift phenomenon. In equation (3.53), n is given by equation (3.50). By replacing f_1 by its value from equation (3.52) in equation (3.53):

$$v_d = -\frac{1}{n}\iiint_{-\infty}^{+\infty} \boldsymbol{v}[\tau_m\boldsymbol{v}.\nabla f_0 + \tau_m(\boldsymbol{F}/\boldsymbol{M}).\nabla_v f_0]\, \mathrm{d}v$$

The first term in the bracket represents the diffusion; the second corresponds to the mobility. Since the integration can be performed simply only if $\boldsymbol{F}$ does not depend on $\boldsymbol{v}$, $\boldsymbol{F}$ will be assumed for clarity to be purely electric.

In order to estimate the diffusion coefficient $\nabla n = \iiint_{-\infty}^{+\infty} \nabla f_0\, \mathrm{d}v$; the coefficient of ∇n may be written:

$$\mathcal{D} = C^2\tau_m/3 = \langle(\boldsymbol{v}.\boldsymbol{i})^2\tau_m\rangle$$

where $\boldsymbol{i}$ is a unit vector in the direction of ∇n. Since the square of the projection of $\boldsymbol{v}$ in any direction is one-third of the mean-square velocity, the coefficient of ∇n is by definition the diffusion coefficient. For electrons among heavy molecules, one obtains:

$$\tau_m = \lambda/\bar{v}$$

In order to estimate the mobility coefficient, let $\boldsymbol{i}$ be, this time, parallel to the force $\boldsymbol{F}$. Noting that:

$$(\partial f_0/\partial v_n) = \frac{v_n}{v}(\partial f_0/\partial v)$$

and setting $\boldsymbol{F} = q\boldsymbol{E}$, the coefficient of $\boldsymbol{E}$ becomes:

$$b = -\frac{1}{n}\iint\int_{-\infty}^{+\infty} (q\tau_{\mathrm{m}}/Mv)\,(\boldsymbol{v}.\boldsymbol{i})^2\,(\partial f_0/\partial v)\,\mathrm{d}\boldsymbol{v}$$

Recalling the statement above concerning the square of the scalar product $(\boldsymbol{v}.\boldsymbol{i})$, and replacing the volume integral of $\mathrm{d}v = \mathrm{d}v_x\,\mathrm{d}v_y\,\mathrm{d}v_z$ by a linear integral of $\mathrm{d}\boldsymbol{v}$ through a passage to spherical co-ordinates v, ϕ, θ, one obtains:

$$b = -\frac{4\pi q\tau_{\mathrm{m}}}{3nM}\int_0^{\infty} v^2(\partial f_0/\partial v)\,\mathrm{d}v$$

This result takes into account the fact that f_0 is isotropic in order to perform the integrations on the angles. This integration is calculated by parts. One has to assume, for f_0, a form such that the integrated term is zero. It is possible to verify that this is true in particular for a Maxwellian distribution. By extracting the value of n from the integral and substituting average brackets for the remaining part, the calculation leads to the classical value of mobility. If a Maxwellian distribution and a mean-free-path independent of the velocity are assumed, one obtains for one particle:

$$b = K_1 q\tau_{\mathrm{m}}/M$$

where K_1 is a constant coefficient. In any case, equation (1.73) is satisfied when the velocity distribution is Maxwellian. This is a consequence of the implicit assumption that τ_{m} is the same for a perturbation f_1 induced by external forces or by a concentration gradient. If the temperature is assumed to be space dependent, one finds following the same reasoning:

$$\boldsymbol{F}_{\mathrm{c}} = -\nabla p_{\mathrm{rc}}/n$$

where $\boldsymbol{F}_{\mathrm{c}}$ is the force per particle in a unit volume and p_{rc} is the pressure of the relevant kind of carrier. This result is obtained without assuming as we did in the orbital theory that the pressure gradient divided by the concentration acts as a field force. Finally, the problem of the mobility in a magnetic field can be treated in the same way and the result that the transverse mobility is equal to the longitudinal mobility divided by $[1+(\omega_{\mathrm{c}}/\nu)^2]$ also emerges directly.

3.9 FULLY IONISED PLASMAS[44]

The fully ionised plasma differs from the weakly ionised plasmas in the following respects.

(1) There are only two components: electrons and ions. In the weakly ionised plasma, interactions between the carriers and the neutral particles are predominant and only exceptionally interactions of the carriers among themselves have been considered. Here, this kind of interaction becomes the rule.

(2) In the fully ionised plasma, the generation and annihilation processes are entirely disregarded, the conditions of temperature and confinement being assumed to be such that recombination occurs neither in the volume nor at the walls so that there is no opportunity for ionising processes.

(3) Similarly no excitation process is generally considered in the fully ionised plasma, so that energy-exchange phenomena are governed by elastic collisions. Thus, a fully ionised plasma emits almost no radiation.

(4) Since the mobility phenomena are governed by collisions between carriers, the number of scattering centres is proportional to the number of carriers as in a metal. Hence, the concept of resistivity, which is seldom in use in low pressure weakly ionised plasmas, is a useful concept for fully ionised plasmas.

The theory of fully ionised plasmas developed by Spitzer rests on a system of two equations similar to equation (3.38), one for the electrons of charge $-e$ and the other for the ions of charge $+Ze$. The force includes not only the electromagnetic terms, but also a gravitational term $-\varrho\nabla G$ added in the case of astrophysical applications. The following assumptions are also introduced.

(1) The plasma is assumed to be quasi-isotropic. Not only is the divergence of the tensor $\boldsymbol{\Psi}$ replaced by the gradient of a scalar p_r, but also all drift velocities, current densities, etc., are assumed to be small perturbations with respect to this isotropic state, so that expressions involving squares or products of such quantities can be neglected.

(2) The plasma is neutral and the electrons are assumed to be of negligible mass with respect to the ions.

(3) Each species of particles is conservative and the only effect of collisions is an exchange of momentum between the two components.

The current density is given by:

$$\boldsymbol{J} = e(n^+ Z\boldsymbol{v}_d^+ - n^-\boldsymbol{v}_d^-)$$

whereas the mass density and the plasma drift velocity are given by equations (3.33) and (3.34). Because of the neutrality one has $n^- = Zn^+$. The two equations of the type of equation (3.38) are, taking into account such simplifications and definitions, added after multiplication respectively by Ze/M^+ and $-e/M^-$. This gives the two Spitzer equations:

$$\varrho\, \partial\boldsymbol{v}_d/\partial t = \boldsymbol{J}\times\boldsymbol{B} - \nabla p_r - \varrho\nabla G \tag{3.54}$$

and

$$\frac{M^-\,\partial\boldsymbol{J}}{n^- e^2 \partial t} = \boldsymbol{E} + \boldsymbol{v}_d\times\boldsymbol{B} + \frac{1}{en^-}\nabla p_{re} - \frac{1}{en^-}\boldsymbol{J}\times\boldsymbol{B} - \frac{\boldsymbol{J}}{\sigma} \tag{3.55}$$

p_{re} being the electron pressure and p_r the total pressure while σ is the conductivity of the plasma.

In equation (3.54), which is an overall equation of motion, the collision term disappears since there is only exchange of momentum between the two components; the electric force disappears too, since matter as a whole is assumed to be electrically neutral. The acceleration of matter results from the magnetic force, the pressure gradient, and the gravitational force.

The left-hand side of equation (3.55) represents inductive effects which, owing to the high conductivity of the plasma, are generally very important. $\boldsymbol{E}^* = (\boldsymbol{E} + \boldsymbol{J}\times\boldsymbol{B})$ is equivalent to an electromotive force in the frame of reference moving with the plasma. The gradient of the electron pressure tends to result in an increase of current since the current arises almost exclusively from the motion of electrons. Eventually, the gradient of electron pressure combines with the electromotive force in a kind of ambipolar diffusion. The next term is the 'dynamo' effect for conductors in motion, and the last one is the resistive effect. Since the gravitational field accelerates both components in the same way it induces no electric current.

To solve a practical problem, these equations must be combined with conservation equations and a simplified energy balance identifying the Joule effect J^2/σ with the heating; we need eventually a complete description of the electromagnetic field.

In the steady state, it is usual, in addition to setting the time derivatives equal to zero, to eliminate the 'dynamo' term between equations (3.54) and (3.55) so that they become respectively:

$$\nabla p_r = \boldsymbol{J}\times\boldsymbol{B} - \varrho\nabla G \tag{3.56}$$

and

$$\boldsymbol{E}^* = \frac{\boldsymbol{J}}{\sigma} + \frac{1}{\overline{n}\,\overline{e}}(\sigma\nabla G + \nabla p_{ri}) \tag{3.57}$$

with $p_{ri} = p_r - p_{re}$.

The magnetic confinement in its non-tensor form can be studied by setting G equal to a constant. Then, taking into account equation (2.15), multiplying equation (3.56) vectorially by $\boldsymbol{B}$, expanding the double cross-product and setting $\boldsymbol{J} = \boldsymbol{J}_{\parallel} + \boldsymbol{J}_{\perp}$, where $\boldsymbol{B}$ is the reference, one obtains:

$$\boldsymbol{J}_{\perp} = \boldsymbol{B}\times\nabla p_r/B^2$$

A similar treatment given to equation (3.57) yields:

$$\boldsymbol{v}_{\perp} = \left[\boldsymbol{B}\times\left(\frac{1}{e\overline{n}}\nabla p_{ri} - \boldsymbol{E}\right)\right]\Big/ B^2$$

In addition, from equation (3.56), ∇p_r is perpendicular to both $\boldsymbol{B}$ and $\boldsymbol{J}$. The conservation of electricity gives $\nabla.\boldsymbol{J} = 0$ and the conservation of matter $\nabla.(\varrho \boldsymbol{v}_d) = 0$. It can be logically assumed that all the pressure gradients have the same direction at a given point; hence in equation (3.57) ∇p_{ri} is also perpendicular to both $\boldsymbol{B}$ and $\boldsymbol{J}$. But in the general case, $\boldsymbol{B}$ and $\boldsymbol{J}$ are not perpendicular to each other. The case when $\boldsymbol{B}$ and $\boldsymbol{J}$ are mutually parallel is excluded, since ∇p_r would then be zero, which does not conform to the condition of confinement.

The condition $\nabla.\boldsymbol{B} = 0$ requires that the magnetic lines of force are continuous. However, these lines do not generally close on themselves; a given line circles around and around and at the limit tends to occupy all the surface of a figure topologically equivalent to a torus (Figure 3.3). From equation (3.56),

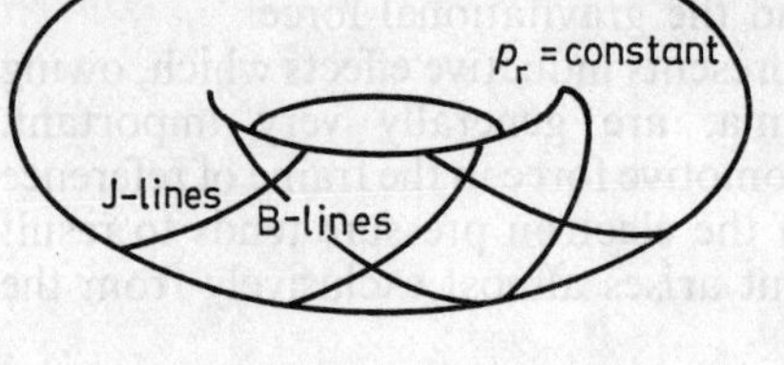

Fig. 3.3. The isobaric surface (p_r = constant) on which the magnetic and current density lines of force are situated. It has the shape of a torus

it is seen that such a surface coincides with an isobaric surface (p_r = constant). Since $\boldsymbol{J}$ obeys the same divergence condition, the lines of current are also on the surface of constant p_r, but do not close on themselves. The conservation of matter necessitates that the lines of hydrodynamic flow constitute a third pattern on the surface of constant p_r; in general, this surface does not coincide with that which is constituted on the lines of current since the transport of matter is mainly due to ions, whereas the transport of electricity is mainly due to electrons.

Finally, if in equation (3.56) $\boldsymbol{J}$ is replaced by $\nabla\times\boldsymbol{B}/\mu$ [equation (3.4)], a magnetic tensor concept can be introduced. If an intrinsic system of co-ordinates is used locally to describe this tensor, the non-diagonal components vanish, whereas the diagonal components take approximately the values $-B^2/2\mu$, $-B^2/2\mu$, and $+B^2/2\mu$. This would be rigorous if the magnetic lines of force were straight and parallel. Thus, the pressure exerted by the magnetic field orthogonally to its lines of force becomes a constant equal to $(p_r+B^2/2\mu)$ when one travels along a line everywhere parallel to ∇p_r. The value of the constant is approximately given by the value of the magnetic energy where the plasma density has become negligible.

The concept of *frozen magnetic fields*[1, 11] is of paramount importance for fully ionised plasmas, and more generally for all the highly conducting mobile media. In an infinitely conducting medium, $\partial\boldsymbol{B}/\partial t$ is necessarily zero since any variation of induction induces currents which tend to oppose exactly any variation of $\boldsymbol{B}$. For cases where the medium is mobile, it has been stated that 'magnetic lines of force travel with it' and thus are 'frozen' into it. In equation (3.57), assume the gravitational, pressure, and displacement current terms to have negligible effects. One obtains from equations (2.15), (3.4), and (3.8):

$$\partial\boldsymbol{B}/\partial t = \nabla\times(\boldsymbol{v}_d\times\boldsymbol{B})+\frac{1}{\sigma\mu}\nabla^2\boldsymbol{B} \tag{3.58}$$

If $\boldsymbol{v}_d$ is zero, this is a diffusion equation. The diffusion coefficient is inversely proportional to the conductivity σ, hence the smaller σ, the faster a local magnetic field tends to diffuse into the surrounding medium. As $1/\sigma\mu$ is dimensionally a surface per unit time, relaxation time can be calculated by dividing characteristic surfaces involved in the problem by the diffusion coefficient. The relaxation times turn out to be of the order of seconds for metals and weakly ionised plasmas, of hours or days for fully ionised plasmas and superconducting materials, of years in stellar interiors and of billions of years in interstellar space. If σ is infinity, the equation takes the same form as a hydrodynamic equation with $\boldsymbol{B}$ replacing the so-called vortex vector; it is known that, in non-viscous fluids, vortices are neither created nor destroyed and travel with the fluid.

3.10 THE MHD APPROXIMATION

Magnetohydrodynamics (MHD) is the science that brings together the concepts of electromagnetism and those of hydrodynamics. Therefore, both sets of equations are used: electromagnetic field theory equations given by Maxwell's relations of Section 3.1 and the flow equations given by the conservation of mass, momentum and energy. This complete system of equations is usually referred to as the MHD equations. The flow equations obtained from Boltzmann's equation can also be derived by postulating a continuous plasma with appropriate characteristics.

The plasma is assumed to be a continuous medium present in a magnetic field. This medium may be composed of several distinct species, some of which may be electrically charged so that currents can flow within it. As a result of this electric conductivity of the medium, the electromagnetic field will give rise to two important effects, creation of body forces acting on the field and exchange of energy between the field and the fluid.

Since the interest is rather in engineering MHD than in questions relating to astrophysics, some simplifications can be made. This set of approximations is generally referred to as the MHD approximation. In such an approximation the following assumptions are made:

(1) In Maxwell's equations the displacement current $\partial \boldsymbol{D}/\partial t$ is neglected compared to the conduction current,

(2) the current flow $\varrho_e \boldsymbol{v}_d$ due to the transport of excess charge is neglected compared to the conduction current, and

(3) the electrostatic body force $\varrho_e \boldsymbol{E}$ is neglected in the equation of motion.

If $\varrho = \varrho(\boldsymbol{r}, t)$ is the mass density of the plasma fluid at a point $\boldsymbol{r}$ at a time t and $\boldsymbol{v}_d = \boldsymbol{v}_d(\boldsymbol{r}, t)$ is its drift velocity at that same point and in the same time, the rate of increase of mass within an arbitrary fixed volume $\mathcal{V}$ is:

$$\frac{\mathrm{d}}{\mathrm{d}t}\int_{\mathcal{V}} \varrho \,\mathrm{d}\mathcal{V} = \int_{\mathcal{V}} \frac{\partial \varrho}{\partial t}\,\mathrm{d}\mathcal{V}$$

Mass will be lost by passing through the boundary surface at the rate $\varrho \boldsymbol{v}_d . \mathrm{d}\boldsymbol{S}$. For mass conservation, the rate of mass increase should be equal to the rate of decrease; thus:

$$\int_{\mathcal{V}} \frac{\partial \varrho}{\partial t}\,\mathrm{d}\mathcal{V} = -\int_S \varrho \boldsymbol{v}_d . \mathrm{d}\boldsymbol{S}$$

Applying Gauss' theorem to this equation, one obtains, after an obvious simplification, equation (3.35).

Consider now a moving volume whose surface always encloses the same amount of fluid. Newton's second law states that mass times acceleration is equal to the sum of forces; thus:

$$\int_{\mathcal{V}} \varrho \frac{\mathrm{D}\boldsymbol{v}_d}{\mathrm{D}t}\,\mathrm{d}\mathcal{V} = \int_S \dot{\boldsymbol{F}}\,\mathrm{d}\boldsymbol{S} + \int_{\mathcal{V}} \boldsymbol{F}\,\mathrm{d}\mathcal{V} \tag{3.59}$$

where the first term on the right-hand side represents the surface force and the second term the volume force. The surface force $\dot{\boldsymbol{F}}$ can be considered in terms of a stress tensor with a scalar viscosity. The volume force $\boldsymbol{F}$ is mainly due to the motion of free charges. Taking account of this and the MHD approximation, equation (3.59) leads to equation (3.41) without the electric term.

The rate of increase of energy in the fluid is equal to the rate of increase of kinetic energy and of internal and thermal energy; that is $\frac{1}{2}\int_{\mathcal{V}} \varrho(\mathrm{D}v_d^2/\mathrm{D}t)\,\mathrm{d}\mathcal{V}$ $+\int_{\mathcal{V}} \varrho(\mathrm{D}e_s/\mathrm{D}t)\,\mathrm{d}\mathcal{V}$. This increase must certainly be equal to all the energy inputs per unit time from all the possible sources. These are electromagnetic energy, energy brought by heat conduction and by diffusion of particles, and the rate at which surface forces are doing work. Doing this, an equation identical to equation (3.45) is obtained after some obvious rearrangements.

It would be of interest to look in more detail into the magnetic body force $\boldsymbol{J}\times\boldsymbol{B}$ appearing in the equation of motion. If the MHD approximation is taken into account, one obtains from equations (3.2) and (3.4):

$$\boldsymbol{J}\times\boldsymbol{B} = \frac{1}{\mu}(\nabla\times\boldsymbol{B})\times\boldsymbol{B} = \frac{1}{\mu}\left[(\boldsymbol{B}.\nabla)\boldsymbol{B}-\frac{1}{2}\nabla B^2\right]$$

Seen in its equation of motion context, the quantity $B^2/2\mu$ appears to have the same effect as a pressure, hence the concept of 'magnetic pressure'. Because of equation (2.15), one also has:

$$(\boldsymbol{B}.\nabla)\boldsymbol{B}/\mu = \nabla(\boldsymbol{B}.\boldsymbol{B})/\mu$$

and the dyadic $\boldsymbol{B}.\boldsymbol{B}/\mu$ becomes analogous to a mechanical tensile stress.

The frozen magnetic field concept can also be shown through a hydrodynamic demonstration. Consider a closed contour enclosing the plasma fluid. The magnetic flux through this contour would be:

$$\Phi = \iint \boldsymbol{B}.\mathrm{d}\boldsymbol{S}$$

The contour is assumed to travel with the fluid and to enclose always the same amount of fluid. The total time derivative of Φ comprises two parts; the first one exists for a fixed contour in a variable induction and the second one results from the motion of the contour and would exist in the case of a time independent induction; thus:

$$\mathrm{d}\Phi/\mathrm{d}t = \iint (\partial\boldsymbol{B}/\partial t).\mathrm{d}\boldsymbol{S} - \int \boldsymbol{v}_d\times\boldsymbol{B}.\mathrm{d}\boldsymbol{l}$$

For an infinite conductivity $\mathrm{d}\Phi/\mathrm{d}t$ equals zero for any contour travelling with the fluid. Since one can imagine a situation with the magnetic induction different from zero only in the neighbourhood of a single magnetic line of force, the conclusion is that this line of force moves with the fluid. Since the equations are linear, any complicated distribution of magnetic field can always be visualised as a superposition of similar configurations, i.e. everything happens as if the magnetic lines of force travelled with the fluid.

REFERENCES AND BIBLIOGRAPHY

1. ALFVÉN, H., *Cosmical Electrodynamics*, Clarendon, Oxford (1950).
2. ARIS, R., *Vectors, Tensors and the Basic Equations of Fluid Mechanics*, Prentice-Hall, Englewood Cliffs, N.J. (1962).
3. BINDER, R. C., *Fluid Mechanics*, Prentice-Hall, Englewood Cliffs, N.J. (1962).
4. BOLTZMANN, L., *Lectures on Gas Theory*, University of California Press (1964).
5. BOYD, R. L. and TWIDDY, N. D., 'Electron Energy Distribution in Plasmas', *Proc. R. Soc.*, **250A**, 53 (1959).
6. BRUEKNER, K. A. and WATSON, K. M., 'Use of the Boltzmann Equation for the Study of Ionized Gases at Low Density', *Phys. Rev.*, **102,** 19 (1956).
7. CHANDRASEKHAR, S., 'Stochastic Problems in Physics and Astronomy', *Rev. mod. Phys.*, **15,** No. 1, 1 (1943).
8. CHAPMAN, S. and COWLING, T. C., *The Mathematical Theory of Non-Uniform Gases*, Cambridge University Press (1953).
9. CHEW, C. F., GOLDBERGER, M. L. and LOW, F. E., 'The Boltzmann Equation and the One-Fluid Hydromagnetic Equations in the Absence of Particle Collisions', *Proc. R. Soc.*, **236,** 112 (1956).
10. CLAUSER, F. H. (Ed.), *Plasma Dynamics*, Addison-Wesley, Reading, Mass. (1960).
11. COWLING, T. G., *Magnetohydrodynamics*, Interscience, New York (1957).
12. DAVIDSON, R. C., *An Invariant of the Higher Order CGL Theory*, Princeton University Rept., MATT. 480, October (1966).
13. DAVYDOV, B., 'On the Theory of the Motion of Electrons in Gases and Semiconductors', *Sov. Phys.*, **12,** 269 (1937).
14. DELCROIX, D. L., *Introduction to the Theory of Ionized Gases*, Interscience, New York (1960).
15. FERRARO, V. C. and PLUMPTON, C., *An Introduction to Magnetofluid Mechanics*, Oxford University Press, Fair Lawn, N.J. (1961).
16. GARTENHAUS, S., *Elements of Plasma Physics*, Holt, Rinehart and Winston, New York (1964).
17. GASIOROWICZ, S., NEUMANN, M. and RIDELL, R. J., 'Dynamics of Ionized Media', *Phys. Rev.*, **101,** 922 (1956).
18. GRAD, H., 'Principles of the Kinetic Theory of Gases', *Handbuch der Physik*, Vol. 12, Springer-Verlag, Berlin (1958).
19. GREEN, H. S., 'Ionic Theory of Plasmas and Magnetohydrodynamics', *Physics Fluids*, **2,** 341 (1959).
20. HIRSHFELDER, J. O., CURTISS, L. F. and BIRD, R. B., *Molecular Theory of Gases and Liquids*, Wiley, New York (1954).
21. HUGHES, W. F., 'Relativistic MHD and Irreversible Thermodynamics', *Proc. Camb. phil. Soc.*, **57,** 878 (1961).
22. HUGHES, W. F. and YOUNG, F. J., *Electromagnetodynamics of Fluids*, Wiley, New York (1966).
23. JANCEL, M. and KAHAN, T., *Electrodynamique des Plasmas*, Dunod, Paris (1960).
24. JEANS, J., *Kinetic Theory of Gases*, Cambridge University Press, Greenwich, Con. (1946).
25. KITTEL, C., *Elementary Statistical Physics*, Wiley, New York (1958).
26. KORNOWSKI, E., 'On some Unsteady Free Molecular Solutions to the Boltzmann Equation', *Gen. Elec. Rept.*, R 59 SD 463, Nov. (1959).
27. LANDAU, F. M. and LIFSHITZ, E. M., *Fluid Mechanics*, Addison-Wesley, Reading (1959).
28. LANDAU, F. M. and LIFSHITZ, E. M., *Electrodynamics of Continuous Media*, Addison-Wesley, Reading, Mass. (1960).
29. LANDSHOFF, R., 'Convergence of the Chapman-Enskog Method for a Completely Ionized Gas', *Phys. Rev.*, **82,** 442 (1951).
30. LIEPMANN, H. W. and ROSHKO, A., *Elements of Gas Dynamics*, Wiley, New York (1957).

31. LÜST, R. and SCHLÜTER, A., 'Angular Momentum Transport by Magnetic Fields', *Z. Astrophys.*, **38,** 190 (1955).
32. MAXWELL, J. C., *A Treatise on Electricity and Magnetism*, (1891) Dover, New York (1954).
33. MILNE-THOMSON, L. M., *Theoretical Hydrodynamics*, MacMillan, New York (1960).
34. MORSE, P. M., ALLIS, W. P. and LAMAR, E. S., 'Velocity Distribution for Elastically Colliding Electrons', *Phys. Rev.*, **48,** 412 (1935).
35. PAI S.-I., 'Energy Equation of Magnetogasdynamics', *Phys. Rev.*, **105,** No. 5, 1424 (1957).
36. PAI S.-I., *Magnetogasdynamics and Plasma Dynamics*, Prentice-Hall, Englewood Cliffs, N.J. (1962).
37. POST, E. J., *Formal Structure of Electromagnetics*, North-Holland, Amsterdam (1962).
38. ROCARD, Y., *Thermodynamique*, Masson, Paris (1952).
39. ROSE, D. J. and CLARK, M., *Plasmas and Controlled Fusion*, MIT Press, Cambridge, Mass. (1961).
40. ROSENBLUTH, M. N. and ROSTOKER, N., 'Theoretical Structure of Plasma Equations', *Physics Fluids*, **2,** 23 (1953).
41. SAMARAS, D. G., *Theory of Ion Flow Dynamics*, Prentice-Hall, Englewood Cliffs, N.J. (1962).
42. SLEPIAN, J., 'Hydromagnetic Equations for Two Isotopes in a Completely Ionized Gas', *Phys. Rev.*, **112,** No. 5, 1441 (1958).
43. SPITZER, L., Jr. and HARM, R., 'Transport Phenomena in a Completely Ionized Gas', *Phys. Rev.*, **89,** 977 (1953).
44. SPITZER, L., Jr., *Physics of Fully Ionized Gases*, Interscience, New York (1956).
45. TWIDDY, N. D., 'Electron Energy Distribution in Plasmas', *Proc. R. Soc.*, London, **275A,** 338 (1963).
46. WATSON, K. M., 'Use of the Boltzmann Equation for the Study of Ionized Gases of Low Density', *Phys. Rev.*, **102,** 12 (1956).

4 Radiations and waves in plasmas

4.1 RADIATIONS IN A PLASMA[27, 37]

Radiation is emitted after excitation of electrons in the bound energy states of ions, atoms and molecules. The excitation is a result of the particle interactions in the plasma as discussed in Chapter 2. The radiations may also be a result of the free charge motions in their respective Coulomb fields, in the fields of atoms and molecules, and in externally applied fields. Two types of radiations are of the outmost importance: the *brake radiation*, often designated by the German name bremsstrahlung, and *cyclotron radiation*[44]. Bremsstrahlung is a result of electron–ion collision and cyclotron radiation is emitted by electrons accelerated in an applied magnetic field.

Transport of radiation (Figure 4.1)

Let $\boldsymbol{r}$ be the direction of the radiation energy flow. If dP_r is the power flowing through an area dS perpendicular to $\boldsymbol{r}$ within the cone of solid angle $d\Omega$, then the radiation intensity at a point O within dS, in the radian frequency interval between ω and $\omega + d\omega$, is defined as:

$$dI_r = dP_r/dS\, d\omega\, d\Omega \tag{4.1}$$

The emission coefficient is defined as:

$$J_r = dP_r/dS\, dr\, d\omega\, d\Omega \tag{4.2}$$

and the absorption coefficient is defined by:

$$\alpha_r = -dI_r/I_r\, dr \tag{4.3}$$

Comparing equations (4.1) and (4.2) one notices that the radiation intensity J'_r per unit path is dI_r/dr and taking account of the loss of energy expressed by equation (4.3) one finds:

$$dI_r/dr = J_r - \alpha_r I_r \tag{4.4}$$

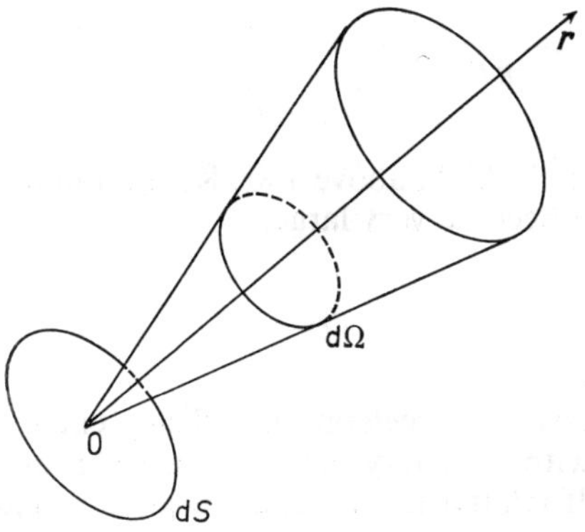

Fig. 4.1. A beam radiating in the r-direction from a point O on the surface dS *and within the solid angle* dΩ

For a uniform plasma J_r and α_r are constants and equation (4.4) can be integrated readily, leading to:

$$I_r = (J_r/\alpha_r)\,[1 - \exp(-\alpha_r l)] \tag{4.5}$$

where l is the length of the ray within the plasma. This equation is represented in Figure 4.2. If the radiating particles have a Maxwellian distribution of velocities, the theory of thermal radiation from hot bodies can be used, which leads to*:

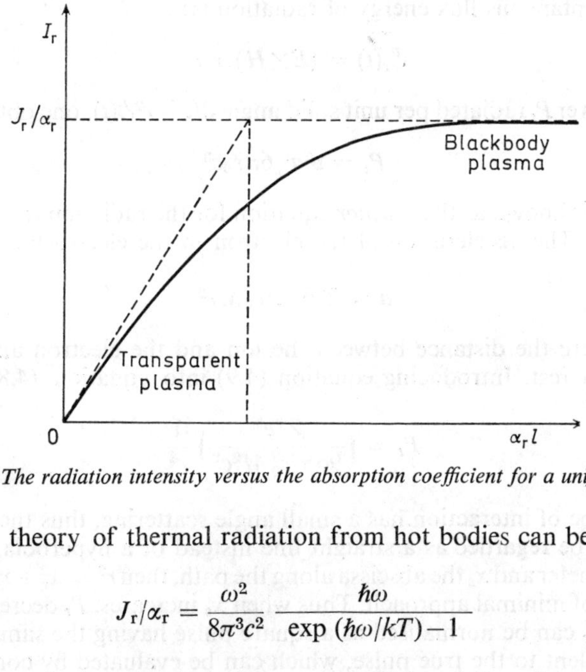

Figure 4.2. The radiation intensity versus the absorption coefficient for a uniform plasma

$$J_r/\alpha_r = \frac{\omega^2}{8\pi^3 c^2}\,\frac{\hbar\omega}{\exp(\hbar\omega/kT) - 1} \tag{4.6}$$

where c is the velocity of light in free spacc and $\hbar = h/2\pi$, where h is Planck's constant. For low frequencies, when $\hbar\omega \ll kT$, equation (4.6) becomes after

* This equation is often called the Kirchhoff–Planck law.

simplification:

$$J_r/\alpha_r = \frac{\omega^2}{8\pi^3 c^2} kT$$

Since the electronic temperature T is usually high, above 1000 K, the radiant energy loss from a blackbody plasma can become very large.

Bremsstrahlung

An electric charge undergoing an electrostatic acceleration radiates energy. Consider an electron of charge e moving with a velocity $\boldsymbol{v}(t)$ and an acceleration $\boldsymbol{a}(t)$ at a distance $\boldsymbol{r}(t)$ from the observer. If relativistic effects are negligible and if $\boldsymbol{r}(t)$ is very large; one can write:

$$\boldsymbol{E} = \frac{e}{4\pi\varepsilon_0 c^2 r^3} \boldsymbol{r}\times(\boldsymbol{r}\times\boldsymbol{a})$$

$$\boldsymbol{B} = \frac{e}{4\pi\varepsilon_0 c^3 r^2} (\boldsymbol{a}\times\boldsymbol{r}) \tag{4.7}$$

and the instantaneous flux energy of radiation is:

$$\mathscr{E}_r(t) = (\boldsymbol{E}\times\boldsymbol{H}).\boldsymbol{r}/r$$

Since the power P_r radiated per unit solid angle $d\Omega$ is $r^2\mathscr{E}(t)$, one obtains readily:

$$P_r = e^2a^2/6\pi\varepsilon_0 c^3 \tag{4.8}$$

This relation is known as the *Larmor* equation for the radiation from an accelerated charge. The acceleration of the electron in the electrostatic field of the ion is:

$$a = Ze^2/4\pi\varepsilon_0 M r^2 \tag{4.9}$$

where r is here the distance between the ion and the electron and the ion is considered at rest. Introducing equation (4.9) into equation (4.8) one finds:

$$P_r = \left(\frac{Z^2e^6}{96\pi^3\varepsilon_0^3 M^2c^2}\right)\frac{1}{r^4} \tag{4.10}$$

The main type of interaction has a small angle scattering, thus the orbit of the electron can be regarded as a straight line instead of a hyperbola. If x_b is the impact parameter and x_s the abscissa along the path, then $r^2 = x_b^2 + x_s^2$ and $x_s^2 = 0$ at the point of minimal approach. Thus when x_s increases, P_r decreases rapidly. Therefore, P_r can be normalised as a square pulse having the same height and width equivalent to the true pulse, which can be evaluated by considering the points at which P_r is equal to half the maximum value. The pulse width is then:

$$l_p = \sqrt{\frac{3}{2}} x_b$$

and

$$\tau = l_p/v = \sqrt{\frac{3}{2}}\frac{x_b}{v}$$

Therefore, the energy radiated by one electron in one collision is:

$$\mathcal{E}_1 = P_r/\tau = (3/2)^{1/2}\frac{Z^2e^6}{96\pi^3\varepsilon_0^3c^3x_b^3M^2v} \tag{4.11}$$

The number of interactions per unit volume and per unit time between n^- missile electrons of speed v with n^+ target ions of cross-section $2\pi x_b$ is:

$$d\nu = n^-n^+v2\pi x_b\,dx_b = n^{+2}Zv2\pi x_b\,dx_b$$

The power density of the interactions corresponding to all the possible values of x_b is $dP_b = \mathcal{E}_b\,d\nu$, thus:

$$P_b = \mathcal{C}\int_{x_{b\,min}}^{\infty} dx_b/x_b^2 \tag{4.12}$$

where $\mathcal{C}$ is a constant value. The upper limit is the Debye radius, but since the integral converges rapidly, infinity has been taken. The lower limit, $x_{b\,min}$, is the electron Compton radius $\hbar/Mv$. If a Maxwellian distribution function of the velocities is assumed one obtains readily after integrating equation (4.12):

$$P_b = \left(\frac{e^6(3k)^{1/2}}{12\pi\varepsilon_0^3c^3M^{3/2}h\pi^{1/2}}\right)Z^3n^{+2}T^{1/2}$$

$$\cong 10^{-38}Z^3n^{+2}(kT)^{1/2} \qquad \text{(MKSC units)} \tag{4.13}$$

Because of the straight-line approximation of the hyperbola, the result is the same whether the accelerating field is attractive or repulsive. This may be true at low frequencies only, but at higher frequencies the bremsstrahlung is different for the two cases. Note also that the bremsstrahlung is isotropic, i.e. it is independent of direction.

Cyclotron radiation

Cyclotron radiation occurs in the interval between collisions under the effect of a magnetic field accelerating an electron. The power radiated by a gyrating particle is found[42] to be:

$$P_{c1} = \frac{2}{3}\frac{e^4Z^4B^2v_\perp^2}{4\pi\varepsilon_0c^3M^2}$$

or since $Mv_\perp^2/2 = kT$, one obtains for n particles in a unit plasma volume:

$$P_c = \frac{e^4}{3\pi\varepsilon_0c^3M^3}Z^4B^2(kT)n$$

$$= 10^{-19}Z^4B^2(kT)n \qquad \text{(in MKSC units)} \tag{4.14}$$

In the case of an ideal confinement, with no magnetic field inside the confined plasma, the centre fringe of the plasma would be in a region of strongly heterogeneous magnetic field; it is not yet clear whether or not this would give less cyclotron radiation. Equation (4.14) is valid only for non-relativistic and reasonably homogeneous plasmas. Note also that cyclotron radiation is anisotropic, and it is elliptically polarised.

Other radiations

Consider a system of two colliding particles of charges q_1 and q_2 and of masses M_1 and M_2 respectively. The instantaneous dipole moment in the centre of mass of the system can be calculated easily and is found to be, when the particles are not moving with a relativistic speed,

$$\mathbf{p} = \frac{M_1 M_2}{M_1+M_2}\left(\frac{q_1}{M_1}-\frac{q_2}{M_2}\right)\mathbf{r}$$

where $\mathbf{r}$ is the separation between the two interacting particles. This equation shows that when the two particles are identical, $\mathbf{p}$ is equal to zero. However, for relativistic speeds of the particle, electron–electron radiation can be of the order of bremsstrahlung, at about 10^9K, that is for $kT \cong Mc^2$.

Bremsstrahlung is a result of collisions between electrons and ions, atoms, or molecules. Cyclotron radiation is emitted by cold electrons, warm electrons, and highly relativistic electrons. Co-operative motions in plasmas lead also to the emission of radiation. An example is plasma oscillations that cause the emission of electromagnetic waves in inhomogeneous plasmas.

4.2 PLASMA SHEATHS[17, 40]

Sheaths are dark regions in otherwise bright plasmas, with a significant departure from neutrality. Most often, the departure always has the same sign. In the case of narrow constrictions, however, experiment, as well as theory, shows that the inlet of the constriction on the cathodic side of the discharge is occupied by a double sheath, a positive one on the cathodic side and a negative one on the anodic side. Those regions with significant departures from space charge neutrality have thicknesses of the same order of magnitude as the Debye screening length. The conditions in which sheaths are formed in plasmas have for long remained obscure, until the 1950s when they received some attention.

The Debye screening length is given by equation (1.4). Now, consider a plasma in which the electronic temperature equals the ionic temperature. Such a condition can be considered as satisfied in most of the high pressure plasmas, but significant departures are encountered as soon as the overall pressure (neutral pressure and electronic pressure plus ionic pressure) falls below, say, one tenth of an atmosphere.

In the relevant circumstances, i.e. a punctual 'test' charge in an otherwise homogeneous and equipotential plasma, equation (1.1) should be considered in spherical co-ordinates, assuming spherical symmetry, hence:

$$\frac{1}{r^2}\frac{\mathrm{d}}{\mathrm{d}r}\left(r^2\frac{\mathrm{d}V}{\mathrm{d}r}\right) = -\frac{\varrho_e}{\varepsilon_0} \tag{4.15}$$

ϱ_e can be estimated by remarking that in the neighbourhood of a positive charge, negative charges tend to accumulate according to equation (1.7), in which n_∞ is the plasma density at large distances from the positive charge and V is the voltage drop between the equipotential plasma and the relevant point. If eV/kT is not too large, then equation (1.7) becomes:

$$n^- = n_\infty\left(1+\frac{eV}{kT}\right) \tag{4.16}$$

If both electronic and ionic temperatures are equal, the accumulation of positive ions is given by:

$$n^+ = n_\infty\left(1-\frac{eV}{kT}\right) \tag{4.17}$$

Hence, ϱ_e is given as a function of V by subtracting equation (4.16) from equation (4.17) and multiplying by e:

$$\varrho_e = -2e^2 n_\infty V/kT$$

If this value is inserted into equation (4.15), an equation in V only is obtained, the solution of which is known as the Yukawa potential; it is:

$$V = \frac{e}{4\pi\varepsilon_0 r}\exp\left(-r/r_D\right) \tag{4.18}$$

where r_D has the value given by equation (1.4). If the central charge is negative, the same equation is obtained, except that the sign of the voltage is reversed.

Bohm[13] has developed a theory which is more realistic, in the case of low pressure plasmas, than the one stated above. He considers a one-dimensional case in which the cause of the sheath formation is a plane wall. Experimental measurements suggest that, in most of the cases of sheath formation, either along a wall or near an electrode or at the mouth of a constriction, the neighbouring plasma is in such a condition that the electron random velocity is much larger than its drift velocity whereas for the ions the reverse is true. Such circumstances occur very widely as the result of such phenomena as ambipolar diffusion in a strong concentration gradient. If such is the case, both the ions and the electrons are subjected to two drift velocities, one resulting from the concentration gradient, and the other from the electric field of ambipolar diffusion. In the case of electrons, these drift velocities almost cancel each other, whereas in the case of ions they add up. Therefore, Bohm neglects the drift velocity of the electrons and the random velocity of the ions, thus obtaining a simple model. The electron density still obeys equation (1.7), but the ions are accelerated as a beam; if their drift velocity in the plasma is already v_0^+, its value v^+ at a point where the voltage with respect to the equipotential plasma is V, is given by:

$$M^+v^{+2} = 2eV + M^+v_0^{+2} \tag{4.19}$$

Bohm considers implicitly potentials accelerating the ions and creating a rarefaction of the electrons. Since the flow of positive ions must remain constant, $n_\infty v_0^+ = n^+v^+$. In this equation v^+ is substituted from equation (4.19), thus obtaining:

$$n^+ = n_\infty\left(1 - \frac{eV}{M^+v_0^{+2}}\right) \tag{4.20}$$

The electron density is obtained from equation (4.16). If both equation (4.16) and equation (4.20) are introduced into equation (1.1), one obtains in a linear configuration:

$$\mathrm{d}^2V/\mathrm{d}x^2 = \frac{en_\infty}{\varepsilon_0 kT^-}\left(1 - \frac{kT^-}{M^+v_0^{+2}}\right)V \tag{4.21}$$

This equation has the interesting property that only departures from neutrality having the shape of waves are permitted, if the drift kinetic energy of the positive ions is less than $kT/2$; thus, no real sheath could be formed in such circumstances. On the other hand, if the kinetic energy of the positive ions is larger than this value, equation (4.21) has only exponential solutions, corresponding to the apparition of sheaths. If the characteristic rising length of those exponentials is deduced from equation (4.21), values of the same order as the Debye screening length are obtained.

4.3 PLASMA OSCILLATIONS[4, 9]

The analysis of oscillations and waves in plasmas is exceptionally intricate and a considerable amount of literature has been devoted to it during recent years[2, 6, 41]. The number of possible wave-like phenomena in a plasma is exceptionally large, and when the simplifying assumptions are eliminated, new kinds of oscillations and propagations appear. When drastic simplifications are introduced in the models, three major classes of waves appear: *electrostatic*, *magnetohydrodynamic* and *electromagnetic*, in addition to the ordinary sound waves in the neutral component of the plasma. If broader hypotheses are introduced, the clear-cut distinction between these three classes of phenomena tends to disappear, and continuous transitions from one kind to the other are observed when changes are made in the two key variables, the frequency and the angle with respect to the privileged direction (e.g. $\boldsymbol{B}$). At this degree of approximation, a much more complete description is needed, comprising equations describing accurately the motion of the charge carriers on one side and Maxwell's equation on the other side. Electrostatic oscillations are the simplest example.

Electron *electrostatic oscillations* were detected experimentally by Penning[39] in 1926 and demonstrated theoretically by Tonks and Langmuir[43] three years later. In this theory small amplitude disturbances are assumed for the electron density and the electronic velocity. The plasma is, further, assumed spatially uniform with no steady fields and no steady velocity; the ion motions are neglected. One can thus write:

$$n^- = n_0^- + \tilde{n}$$
$$\boldsymbol{v}_\mathrm{d} = \tilde{\boldsymbol{v}}$$
$$\boldsymbol{E} = \tilde{\boldsymbol{E}}$$

where n_0^- is the steady state value and $\tilde{n}$, $\tilde{\boldsymbol{v}}$, $\tilde{\boldsymbol{E}}$ are the disturbances, $\tilde{n}$ is assumed much smaller than n_0^-. By taking account of this, equation (3.41) of motion becomes after neglecting collisions:

$$\frac{\partial}{\partial t}(n^- M^- \boldsymbol{v}_\mathrm{d}^-) = -\nabla p_\mathrm{r}^- - n^- e\boldsymbol{E} \tag{4.22}$$

After developing and neglecting the second order quantities this equation becomes:

$$M^- n_0^- \partial\tilde{\boldsymbol{v}}/\partial t = -\nabla\tilde{p} - n_0^- e\tilde{\boldsymbol{E}} \tag{4.23}$$

The equation of continuity for the electrons is equation (3.30a). Its right-hand side is set equal to zero because the oscillation is quite fast and the probability of having a generation or annihilation phenomenon during one oscillation can be completely disregarded. This equation becomes after using assumptions in equation (4.22):

$$\frac{\partial\tilde{n}}{\partial t} + n_0\nabla.\tilde{\boldsymbol{v}} = 0 \tag{4.24}$$

On the other hand,

$$\nabla\tilde{p} = \frac{\partial\tilde{p}}{\partial\tilde{\varrho}}\nabla\tilde{\varrho} \tag{4.25}$$

with $\tilde{\varrho} = \tilde{n}M^-$. Taking into consideration equations (4.23), (4.24) and (4.25), one obtains easily:

$$\left(\frac{\partial^2}{\partial t^2} + \omega_\mathrm{p}^2 - v_\mathrm{s}^2\nabla^2\right)\tilde{n} = 0 \tag{4.26}$$

where v_s is the velocity of sound, in the medium, given by equation (1.87), and ω_p is the electron plasma frequency defined by equation (1.13). Equation (4.26) is the equation for the electron density oscillations.

For a cold plasma or long wavelength disturbances this equation is reduced to

$$\frac{\partial^2\tilde{n}}{\partial t^2} + \omega_\mathrm{p}^2\tilde{n} = 0$$

In this case there will be no wave propagation and there are only oscillations of $\tilde{n}$ with a unique angular frequency ω_p.

More refined calculations show that the positive ions are not immobile, but oscillate so that on the average the centre of mass of a carrier pair is at rest. This modifies slightly the characteristic frequency which becomes Ω_p as given by equation (1.14).

4.4 ELECTROACOUSTIC WAVES

If the plasma is not cold and if the disturbances are not of the long wavelength type the term $v_s^2\nabla^2 n$ cannot be disregarded in equation (4.26). In this case, the solution of the equation becomes:

$$n^- = n_0^- \exp(\mathrm{j}\omega t) \exp(\pm \mathrm{j}kx) \tag{4.27}$$

This is the equation of two propagating waves, in which ω is the frequency of the wave and $\pm k$ are constants of propagation of the two waves. The dispersion relation of these waves is obtained by replacing $\tilde{n}$ in equation (4.26) by its value in equation (4.27); thus:

$$k^2 = (\omega^2 - \omega_p^2)/v_s^2$$

The group and the phase velocities can be obtained from their definitions, since $v_p = \omega/k$ and $v_g = \partial\omega/\partial k$; thus,

$$v_p/v_s = \left(1 - \frac{\omega_p^2}{\omega^2}\right)^{-1/2} \tag{4.28}$$

$$v_g/v_s = \left(1 - \frac{\omega_p^2}{\omega^2}\right)^{1/2} \tag{4.29}$$

A comparison between equations (4.28) and (4.29) shows that:

$$v_g v_p = v_s^2 \tag{4.30}$$

Equations (4.28), (4.29) and (4.30) are represented in Figure 4.3. This figure shows that there is a low frequency cut-off at the plasma frequency. Also, for frequencies large with respect to the cut-off frequency, the term $\omega_p^2\tilde{n}$ in equation (4.26) approaches zero, and the equation becomes identical to that of sound waves in the electron gas, the phase and group velocities tending toward the common value v_s.

In a perfect gas equation (1.87) becomes:

$$v_s = (\gamma\tilde{p}/\tilde{\varrho})^{1/2} \tag{4.31}$$

where γ is the ratio of the specific heats at constant pressure and constant volume. Since $\tilde{p}^- = \tilde{n}^- kT^-$ and $\tilde{\varrho}^- = \tilde{n}^- M^-$, equation (4.31) becomes identical to equation (1.88). For a collisionless plasma, the adiabatic compression is unidimensional and $\gamma = 3$. It can happen, however, that the resistivity of the plasma, resulting from collisions, gives rise to a negligible damping and the compression becomes isotropic. In such a case, γ has the usual value for a monatomic gas, i.e. 5/3.

To consider the effect of oscillating ions, both the equations of continuity for electrons and for ions should be written as well as the two equations of motion for the two species of particles. After some obvious rearrangements, one obtains

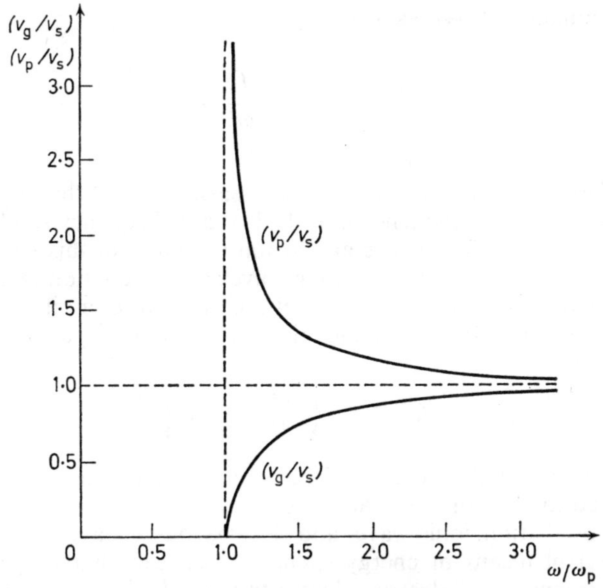

Figure 4.3. Phase and group velocities versus frequency for electroacoustic waves

two coupled equations for the perturbations $\tilde{n}^-$ and $\tilde{n}^+$ in electron and ion densities respectively; these are:

$$\left(-\frac{\partial^2}{\partial t^2}+v_{s-}^2\nabla^2\right)\tilde{n}^- = -\frac{n_0^- e}{M^-}\nabla.\tilde{\boldsymbol{E}} = \omega_p^2(\tilde{n}^- - \tilde{n}^+)$$

$$\left(-\frac{\partial^2}{\partial t^2}+v_{s+}^2\nabla^2\right)\tilde{n}^+ = +\frac{n_0^+ e}{M^+}\nabla.\tilde{\boldsymbol{E}} = -\omega_{pi}^2(\tilde{n}^- - \tilde{n}^+) \qquad (4.32)$$

The solutions of these equations are of the form given by equation (4.28); thus one obtains easily:

$$(\omega^2-\omega_p^2-k^2v_{s-}^2)\tilde{n}^- = -\omega_p^2\tilde{n}^+ \qquad (4.33a)$$

$$(\omega^2-\omega_{pi}^2-k^2v_{s+}^2)\tilde{n}^+ = -\omega_{pi}^2\tilde{n}^- \qquad (4.33b)$$

where v_{s-} is the speed of sound in the electronic gas and v_{s+} is the speed of sound in the ionic gas. Solving this system by eliminating $\tilde{n}^-$ from equations (4.33) leads to the dispersion equation:

$$\left(1-\frac{\omega^2}{\omega_{pi}^2}+\frac{k^2v_{s+}^2}{\omega_{pi}^2}\right)\left(1-\frac{\omega^2}{\omega_p^2}+\frac{k^2v_{s-}^2}{\omega_p^2}\right) = 1 \qquad (4.34)$$

For ion oscillations to occur, ω should be much smaller than ω_p; taking account of the definitions of the thermal velocities v_{s-} and v_{s+} and of the frequencies

ω_p and ω_{pi}, equation (4.34) becomes:

$$\frac{\omega^2}{\omega_{pi}^2} = \frac{k^2\left(1+\dfrac{T^+}{T^-}\right)}{k^2+\dfrac{\omega_p^2}{v_{s-}^2}} \tag{4.35}$$

Under this form, it is evident that ω is necessarily smaller than ω_{pi}, hence the latter constitutes a high frequency cut-off. For any frequency smaller than ω_{pi} there is a propagation. The phase and group velocities of this wave are easily obtained from equation (4.35). For long wavelengths equation (4.35) takes the form of a relation of proportionality between ω and k as it is the case for ordinary sound waves or electromagnetic waves in non-dispersive media; thus $\omega = kv_{s\text{-}}\omega_{pi}/\omega_p$ and:

$$v^- = v_g^- = v_{s-}\sqrt{\frac{M^-}{M^+}} = v_{s+}\sqrt{\frac{T^-}{T^+}}$$

The ion waves travel with velocities of the order of the ion thermal speed and are slow compared to electron thermal speed.

Electronic oscillation, ionic waves and electroacoustic waves have been found to be a powerful means of energy exchange between charged particles in a plasma. The frequency of electron electrostatic oscillations is several orders of magnitude above any kind of frequency collision, and, as such, could involve much shorter relaxation times in the randomisation of the plasma. Besides, the neutral component is entirely omitted and energy is exchanged between the charge carriers directly. This energy exchange is one of the possible reasons

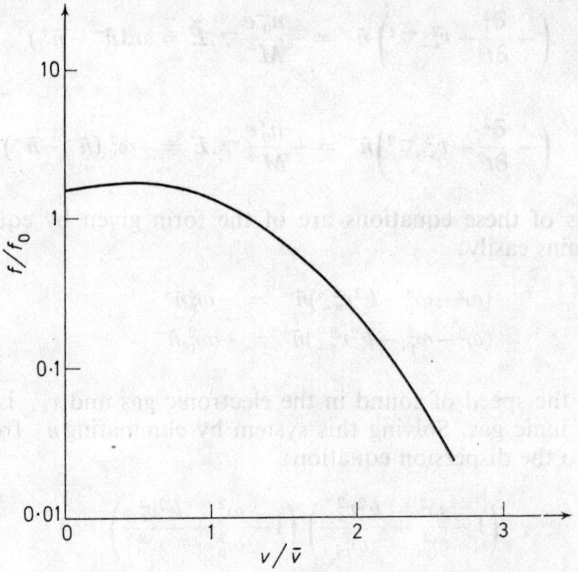

Fig. 4.4. Fast randomisation of a plasma by trapped electroacoustic waves, f being the distribution function of the electronic velocities and f_0 the Maxwellian distribution function. $\bar{v}$ is the average velocity of the electrons

why so many plasmas are Maxwellian (randomised) and for which departures from the Maxwellian state are wiped out frequently in a much shorter time than that predicted by a mere consideration of collisions. Hoyaux and Williams[28] reported experimental observation of the randomisation of the plasma by trapped electroacoustic waves. Partial randomisation of electron velocities in an electron beam crossing an oscillating sheath has been observed by Looney and Brown[35]. Kettani and Hoyaux[31, 32] developed a simplified theoretical model for the phenomenon of fast randomisation by electroacoustic waves. They assumed that, in certain plasma regions, electroacoustic waves can be trapped in resonant-like 'cavities'. They treated the interaction between those waves and the electron gas by a pseudo-collisional method leading to a generalisation of Boltzmann's H-theorem[30]. The velocity distribution which is essentially Maxwellian with a significant deficiency in the fast electron tail was deduced as shown in Figure 4.4.

4.5 ELECTROMAGNETIC WAVES[1, 5]

In the presence of magnetic fields, electromagnetic waves can be generated as a result of fluctuations in the magnetic pressure and tensile stresses. When the magnetic flux lines are frozen, the plasma acquires elastic properties, i.e. it resists bending and stretching along the magnetic field.

Shear waves appear in the presence of small transverse magnetic disturbances such as:

$$\boldsymbol{B} = \boldsymbol{B}_0 + \tilde{\boldsymbol{B}} \tag{4.36}$$

where $\boldsymbol{B}_0$ is the mean value of the magnetic field which is independent of time and $\tilde{\boldsymbol{B}}$ is the variation imposed by the wave itself. The influence of such waves, also called *Alfvén waves*, on the magnetic lines of forces is represented in Figure 4.5. To find the characteristics of the Alfvén waves the following assumptions are usually made: (1) the medium is conducting, and to obtain non-damped

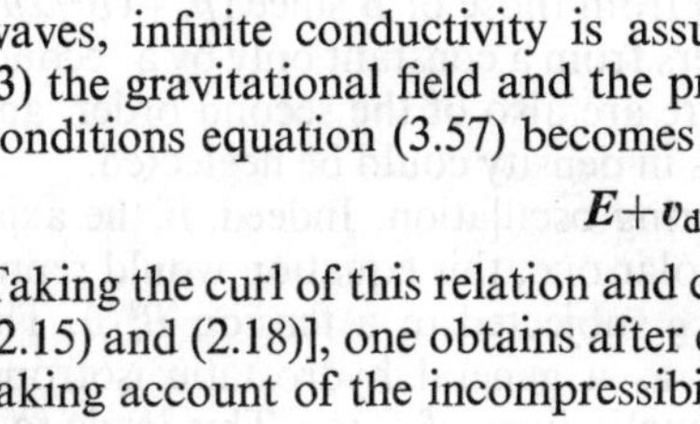

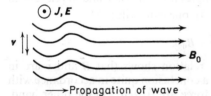

Fig. 4.5. Magnetic lines of force in the presence of Alfvén waves. The lines resist bending and stretching

waves, infinite conductivity is assumed; (2) the medium is incompressible; (3) the gravitational field and the pressure gradient are negligible. Under such conditions equation (3.57) becomes:

$$\boldsymbol{E} + \boldsymbol{v}_d \times \boldsymbol{B} = 0 \tag{4.37}$$

Taking the curl of this relation and considering Maxwell's equations [equations (2.15) and (2.18)], one obtains after expanding the curl of the cross product and taking account of the incompressibility of the medium:

$$-\frac{\partial \boldsymbol{B}}{\partial t} - (\boldsymbol{v}_d . \nabla)\, \boldsymbol{B} + (\boldsymbol{B} . \nabla)\, \boldsymbol{v}_d = 0 \tag{4.38}$$

The velocity $\boldsymbol{v}_d$ is itself a perturbation $\tilde{\boldsymbol{v}}$; thus introducing equation (4.36) into equation (4.38) and neglecting the second order perturbations, one obtains:

$$-\frac{\partial \tilde{\boldsymbol{B}}}{\partial t}+(\boldsymbol{B}_0.\nabla)\tilde{\boldsymbol{v}} = 0 \tag{4.39}$$

After differentiating equation (4.39) with respect to time and introducing equation (3.54) with the assumption that the mass density ϱ is constant, equation (4.39) becomes:

$$-\frac{\partial^2 \tilde{\boldsymbol{B}}}{\partial t^2}+\frac{1}{\varrho}(\boldsymbol{B}_0.\nabla)(\boldsymbol{J}\times\boldsymbol{B}) = 0 \tag{4.40}$$

Replacing $\boldsymbol{J}$ by its value in equation (3.4) where the displacement current has been neglected, equation (4.40) yields:

$$-\frac{\partial^2 \tilde{\boldsymbol{B}}}{\partial t^2}+\frac{B_0^2}{\mu\varrho}\nabla^2\tilde{\boldsymbol{B}} = 0 \tag{4.41}$$

This is the equation of a propagating wave, the phase and group velocities of which are equal to each other and to:

$$v_A = B_0/\sqrt{(\mu\varrho)} \tag{4.42}$$

known as the Alfvén speed. In most practical cases in plasma physics, this turns out to be at least three orders of magnitude below the velocity of light in vacuum. The general structure of the simple Alfvén wave can be deduced by going back to the basic equations. For instance, equation (3.4) yields for a plane wave having $\boldsymbol{k}$ as constant of propagation:

$$\mu\tilde{\boldsymbol{J}} = -\mathrm{j}.\boldsymbol{k}\times\tilde{\boldsymbol{B}} \tag{4.43}$$

showing that $\tilde{\boldsymbol{J}}$ is a perturbation at right angles both in space and in phase with respect to $\tilde{\boldsymbol{B}}$. Similarly, from equation (3.54), one can write:

$$\mathrm{j}\omega\varrho\tilde{\boldsymbol{v}} = \tilde{\boldsymbol{J}}\times\boldsymbol{B}_0$$

When compared to equation (4.43), this relation shows that $\tilde{\boldsymbol{B}}$ and $\tilde{\boldsymbol{v}}$ are in phase and have the same spatial direction, a conclusion quite in accordance with the hypothesis that the magnetic lines of force are frozen. On the other hand, the variations of pressure can be calculated from those of $\boldsymbol{B}$ since $[p_r+(B^2/2\mu)]$ is a constant; but $B^2 = B_0^2+\tilde{B}^2$, and B^2 differs from a constant only by a second-order term. Hence the variations in pressure are also of the second order, and this justifies, *a posteriori*, that the variations in density could be neglected.

Alfvén's waves can be compared to a string oscillation. Indeed, if the axial vector in equation (4.41) is replaced by a polar one, this equation would represent the oscillation of a string of density ϱ subjected to a tension B_0^2/μ. The magnetic stress tensor can be interpreted as a general hydrostatic isotropic pressure plus a tension parallel to the magnetic lines of force. This leads to a vivid interpretation of the Alfvén wave as an oscillation of a bundle of magnetic lines of force considered as strings. The magnetic lines of force have no mass density of their own, but since they are frozen they carry with them the mass density of the plasma (Figure 4.5).

If in equation (4.40), the displacement current is not neglected, the wave equation becomes:

$$\nabla^2 \tilde{\boldsymbol{B}} - \left(\frac{\mu \varrho}{B_0^2} + \varepsilon \mu \right) \frac{\partial^2 \tilde{\boldsymbol{B}}}{\partial t^2} = 0 \tag{4.44}$$

Under usual conditions, the first term in the bracket is several orders of magnitude larger than the second one as stated above ($v_p \ll c$). Equation (4.44) gives $\triangle^2 \tilde{\boldsymbol{B}}$ as a sum of two terms; the first term specific to Alfvén's wave and the second term corresponding to an electromagnetic wave in vacuum (for $\varepsilon = \varepsilon_0$ and $\mu = \mu_0$). Assume, now, that the plasma becomes more and more tenuous all other conditions being the same. The second term will become more and more significant, until eventually it becomes predominant. Thus, the Alfvén wave will have acquired the properties of an electromagnetic wave. Conversely, an electromagnetic wave as such cannot propagate in a medium in which the conductivity is too large. It actually becomes replaced by an Alfvén wave. This shows that the Alfvén wave is the form taken by an electromagnetic wave at low frequency in a highly conducting medium, if there is a magnetic field parallel to the propagation vector $\boldsymbol{k}$.

Another type of wave, closely related to the Alfvén wave, can propagate across the magnetic lines of force (Figure 4.6). The principle of propagation of this wave, designated as a magnetosonic wave, is simple. Consider a highly

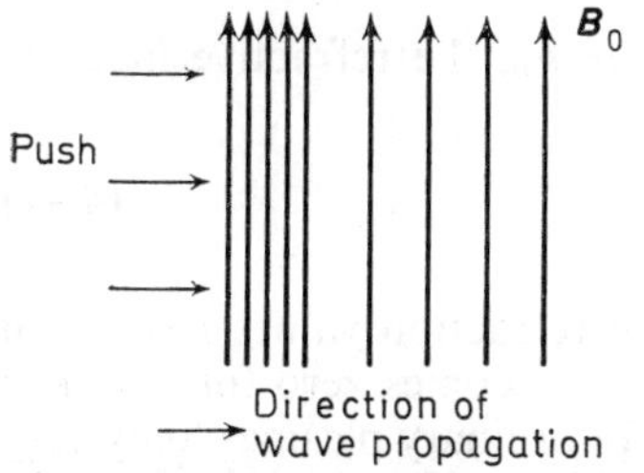

Fig. 4.6. Magnetic lines of force in the presence of magnetosonic waves. The medium is alternately compressed and stretched

conductive medium with magnetic lines of force frozen-in, and a sound wave propagating in the direction perpendicular to the magnetic lines of force. In the sound wave, the medium is alternately compressed and expanded, and if the magnetic lines of force are frozen-in, the value of the magnetic induction will vary in proportion with the field density. Since in a problem like this, the total pressure is equal to the sum of the kinetic and the magnetic pressures, it is normal to consider that the velocity of propagation of the wave will be:

$$v_p = v_A \left(\frac{\gamma}{2} + \frac{\gamma p_r}{\varrho v_A^2} \right)^{1/2} \tag{4.45}$$

where v_A is the Alfvén speed given by equation (4.42). With a magnetic field perpendicular to the direction of compression, provoking a gyration of the particles in a plane perpendicular to the plane of the wave, it should be considered that there is equipartition of energy in two of the dimensions of space, hence $\gamma = 2$. By taking into account equation (4.31), equation (4.45) becomes $v_p^2 = v_A^2 + v_s^2$.

If an electromagnetic wave falls on an ionised medium in the absence of a magnetic field, the electrons are set in oscillation by the electric field of the wave. This motion gives rise to the emission of secondary wavelets, which combine with the incident wave in a fashion equivalent to a modification of the refractive index ν_R. Indeed, in the case of electromagnetic waves, it can be assumed, *a priori*, that the electric charge density is zero. The difference between a plasma and a vacuum is the existence of a conduction current. In a collisionless plasma with no magnetic field, it is given by a simplified form of equation (3.55):

$$\partial \boldsymbol{J}/\partial t = \frac{n^- e^2}{M^-} \boldsymbol{E} \tag{4.46}$$

This equation is introduced into equation (3.4) and the elimination of $\boldsymbol{J}$ is conducted in the same manner used to demonstrate the existence of electromagnetic waves in a vacuum, giving:

$$\nabla^2 \boldsymbol{E} - \frac{1}{c^2}\left(\omega_p^2 + \frac{\partial^2}{\partial t^2}\right) \boldsymbol{E} = 0$$

The relevant dispersion relation is:

$$k = \frac{\omega}{c}\left(1 - \frac{\omega_p^2}{\omega^2}\right)^{1/2}$$

Since the dispersion relation in vacuum is $\omega/c = k_0$, the refractive index is $\nu_R = k/k_0$ or:

$$\nu_R = \left(1 - \frac{\omega_p^2}{\omega^2}\right)^{1/2} \tag{4.47}$$

as it is known in ionospheric applications where refraction values have been found to be in agreement with this expression. ν_R becomes zero for $\omega = \omega_p$; ω_p is then interpreted as a low frequency cut-off. In ordinary plasmas, only very high frequency waves are above the cut-off, and this justifies the existence of a displacement current. On this degree of approximation, a wave having a frequency below the cut-off frequency cannot be propagated, and since a collisionless plasma cannot absorb energy, the only solution is total reflection. Such a phenomenon is well known to occur in the ionosphere (Figure 4.7).

Consider now the case of an electromagnetic wave in a collisionless plasma, with a permanent magnetic field of induction $\boldsymbol{B}_0$ parallel to the propagation vector $\boldsymbol{k}$. Under ordinary conditions, the magnetic force related to the wave itself is negligible with respect to the magnetic force relevant to the permanent field $\boldsymbol{B}_0$ so that equation (3.55) becomes:

$$\frac{\partial \boldsymbol{J}}{\partial t} = \frac{n^- e^2}{M^-} \boldsymbol{E} - \frac{e}{M^-} \boldsymbol{J} \times \boldsymbol{B} \tag{4.48}$$

The motion of the positive ions has been neglected. This is not always justified, because the presence of a longitudinal magnetic field of sufficient intensity

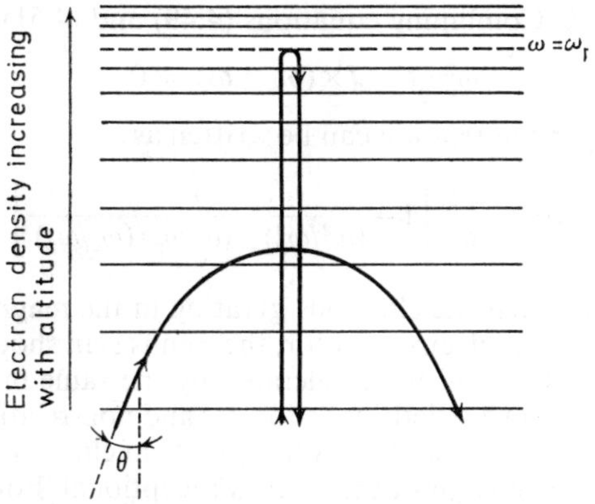

Fig. 4.7. Reflection of electromagnetic waves from the ionosphere

modifies the cut-off conditions and allows the propagation of waves of much lower frequencies. Besides, the positive ions can gyrate, resonating in the magnetic field and having very large motions. In order to render the elimination of $\boldsymbol{J}$ possible by using equation (3.4), the simple case of a circularly polarised wave

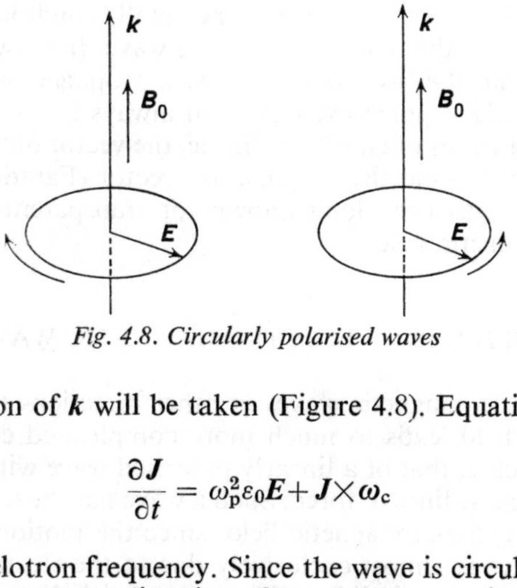

Fig. 4.8. Circularly polarised waves

around the direction of $\boldsymbol{k}$ will be taken (Figure 4.8). Equation (4.48) can now be written as:

$$\frac{\partial \boldsymbol{J}}{\partial t} = \omega_p^2 \varepsilon_0 \boldsymbol{E} + \boldsymbol{J} \times \boldsymbol{\omega}_c \tag{4.49}$$

where ω_c is the cyclotron frequency. Since the wave is circularly polarised and if Oz is taken as the common direction of $\boldsymbol{k}$ and the magnetic field, then:

$$\begin{aligned} J_x &= J \sin \omega t \\ J_y &= \pm J \cos \omega t \end{aligned} \tag{4.50}$$

in which one sign corresponds to the right polarisation and the other to the left polarisation. After differentiating equations (4.50) with respect to time, one obtains:

$$\frac{\partial \boldsymbol{J}}{\partial t} = \pm \boldsymbol{J} \times \boldsymbol{\omega} \tag{4.51}$$

with $\boldsymbol{\omega}$ parallel to $\boldsymbol{k}$. Combining equations (4.49) and (4.51), we find:

$$\omega_p^2 \varepsilon_0 \boldsymbol{E} + \boldsymbol{J} \times (\boldsymbol{\omega}_c \pm \boldsymbol{\omega}) = 0$$

From this, the dispersion relation can be written as:

$$k^2 = \frac{\omega^2}{c^2}\left[1 - \frac{1}{(\omega^2/\omega_p^2) \pm (\omega/\omega_p)(\omega_c/\omega_p)}\right] \tag{4.52}$$

If the wave is such that the electrons gyrating in the magnetic field rotate in the same direction as the electron vector, the minus sign should apply and there is a resonance; the electrons are accelerated by the same mechanism as in the cyclotron motion and, in a strictly collisionless and non-relativistic plasma, their energy would increase indefinitely. Several practical effects limit the energy to a finite value, but the behaviour of the wave is exceptional. For this reason, it has been termed the '*extraordinary*' *wave*. Nothing like this happens if the electric vector rotates in the opposite direction. There will be no resonance, only a modification of the numerical value of the refractive index [sign + in equation (4.52)]. This latter wave is termed the '*ordinary*' *wave*. Note that a wave which is ordinary for electrons is extraordinary for ions and vice versa. Also, when the frequency is very low with respect to both cyclotron resonances of the two plasma species, the conditions of propagation for both waves become those of an Alfvén wave. For ω very large or very small, conditions of propagation become independent of the polarisation of the wave. In between, the dispersion relation indicates that the two polarised waves propagate with different phase velocities. Since a plane polarised wave can always be decomposed into two circularly polarised waves of equal amplitude, the vector obtained by recombining the two, rotates around the propagation vector (Faraday rotation, Figure 4.8). This property has been long known for transparent anisotropic media immersed in a magnetic field.

4.6 GENERAL THEORIES OF ELECTROMAGNETIC WAVES[10, 23]

Propagation of electromagnetic waves in other directions than parallel to the external magnetic field leads to much more complicated equations. There is, however, a simple case, that of a linearly polarised wave with its electric vector parallel to the magnetic lines of force. Such a wave has the same properties as in the absence of an external magnetic field, since the motion of charge carriers induced by the wave is almost exclusively due to the electric field, hence it is polarised along the magnetic lines of force, and uninfluenced by the external magnetic field. This wave has a low frequency cut-off equal to the plasma frequency. For the other plane polarised wave, the situation is more complicated, since the charge carriers tend to rotate in a plane perpendicular to the direction of propagation. In this case, the wave loses its transverse character and can no longer be termed purely electromagnetic. However, at very high frequencies, well above the electron cyclotron and electrostatic resonances, the motion of the charge carriers is negligible. At low frequencies, the problem is simplified because the magnetic lines of force are trapped in the plasma, hence this wave tends toward the state described above as magnetosonic.

The preceding theories are interesting if the aim is to give a good physical insight into the different types of plasma waves. However, they are not fully satisfactory in that each type of wave comes out of a different model, created only for the requirements of the subject under consideration. It is sometimes difficult to state, and even more so to demonstrate, under which conditions each model is valid. In order to establish a general and rigorous theory of wave propagation in plasmas, it is fruitless to make, at the beginning, the distinction between electrostatic, Alfvén, magnetosonic and electromagnetic waves. In fact, all the waves are of the same nature; plasma waves, and the 'pure' cases already studied by means of oversimplified models, appear in general as limiting cases occurring only when very special conditions are met.

For instance, consider the propagation parallel to an external magnetic field. Electromagnetic waves, as such, occur only at very high frequencies, whereas in the low-frequency limit, they degenerate into Alfvén waves; in between, there exist the ion cyclotron wave, the electron cyclotron wave, and wide bands in which only evanescent waves can exist, assuming a collisionless plasma. Electro-acoustic waves have been described, assuming implicitly the absence of an external magnetic field. It is evident that they are essentially unaltered if there is a magnetic field parallel to $\boldsymbol{k}$, since the carrier motions would then be along the magnetic lines of force. However, if $\boldsymbol{k}$ and $\boldsymbol{B}_0$ are rotated with respect to each other, 'pure' electrostatic waves can no longer exist; the particles will start gyrating in the external magnetic field, and the wave ceases to be purely longitudinal. If $\boldsymbol{k}$ and $\boldsymbol{B}_0$ are perpendicular to each other, one obtains an unperturbed, plane-polarised electromagnetic wave; but in the other direction of polarisation, there is a continuous transition between a 'pure' electromagnetic wave at the high frequency extreme, and a 'pure' magnetosonic wave at the low frequency one. All the phenomena intervening in the intermediate range of frequencies, i.e. with the exception of frequencies well above and well below the four characteristic frequencies, and most of the phenomena relevant to propagation in other directions than that of the external magnetic field, can only be accurately known if a general model is introduced. There are two types of such general models: (1) models based entirely on the orbital approach[34, 35], and (2) models based on the Boltzmann approach[6, 22, 41].

There is a large number of general theories, since widely different assumptions can still be introduced. For instance, good results can be obtained if a scalar pressure is assumed, but in strong external magnetic fields, the concept of a tensor pressure is more valid. A usual approximation consists in assuming that the phenomena are adiabatic, but if this assumption is eliminated, other types of waves appear that are unrelated to any wave described above. The assumption of a collisionless plasma is often made, but in most cases, it omits phenomena of damping which, in practice, are not negligible.

4.7 WAVE DAMPING

At first sight, the only possible cause of damping in an electroacoustic wave should be the non-zero *resistivity* of the plasma. One should rewrite equation (4.22) with a finite conductivity term $-n^{-2}e^2v_d/\sigma$, resulting in the addition of a term $(\varepsilon_0/\sigma)\omega_p^2\,\partial\tilde{n}/\partial t$ to equation (4.26). The result of the addition of such a term would be the introduction of the imaginary symbol j in Equation (4.34), which is characteristic of damping. However, most of the time this is not the main cause of damping, the latter being completely ignored by the simplified model.

Landau damping[19]

The principal type of damping is designated as Landau damping[illegible]. The explanation as to why the above model excluded Landau damping is that it considers implicitly the electronic component of the plasma as a fluid. Somewhere in the argument, the physical contact with the particles has been lost. When a sine wave travels in a plasma, some particles, having velocities sufficiently close to the phase velocity of the wave, can be trapped and start travelling with it. A frame of reference for travelling with the phase velocity of the wave is a track with ups and downs, and balls thrown along it with various speeds. The balls can be divided into two classes: (1) in the first class, the speed is large enough to enable the ball to travel alternately uphill and downhill; this corresponds to the non-trapped part of the plasma; (2) in the second class, the balls have insufficient speeds to travel uphill; they cluster in the troughs and oscillate back and forth; these correspond to the trapped part of the plasma. Consider now the statistical distribution of the first class along the track: since the balls are accelerated in the troughs and slowed down at the crests, their statistical distribution is almost opposite to that of the second class. They tend to cluster around the crests, where their velocity is temporarily at a minimum. For the analogy to be complete with electrical particles, the curvature of the track is assumed, in some way, to be related to the density of the balls (Poisson's law). This feature is likely to modify the distribution of the balls between the first and second classes, because it modifies the height of the crests. Such a modification involves an energy exchange between the wave and the particles, which is the physical reason for the damping.

Damping of an Alfvén wave

Experiment shows that, in ordinary media Alfvén waves are strongly damped. There are two main causes of such damping: (1) finite conductivity, leading to the replacement in equation (4.37) of the zero right-hand side by $\boldsymbol{J}/\sigma$, and (2) viscosity, leading to the addition of complementary terms to equation (3.54); for example, a term $\nu_k \nabla^2 \boldsymbol{v}_d$ where ν_k is the kinematic viscosity.

For Alfvén waves in plasmas, the resistivity damping is largely predominant. After introducing the resistivity term, equation (4.41) becomes:

$$-\frac{\partial^2 \tilde{\boldsymbol{B}}}{\partial t^2} + \frac{B_0^2}{\mu \varrho} \nabla^2 \tilde{\boldsymbol{B}} = -\frac{1}{\sigma \mu} \frac{\partial}{\partial t} \nabla^2 \tilde{\boldsymbol{B}}$$

The dispersion relation for plane waves becomes, thus:

$$k = \omega \frac{\sqrt{(\varrho \mu)}}{B_0} \left(1 - \mathrm{j}\omega \frac{\varrho}{2\sigma B_0^2}\right)$$

an expression from which the damping attenuation length is deduced; it is approximately:

$$l = \frac{2B_0^3\sigma}{\omega^2\varrho^{3/2}\mu^{1/2}} \tag{4.53}$$

Thus, strong magnetic fields are required in order to obtain reasonably weak attenuations.

Attenuation of the electromagnetic wave

If the plasma is not collisionless, some energy is dissipated as heat, and the wave is attenuated. To calculate the amount of attenuation, one can either complete equation (4.46) by adding the resistivity term or replace it by a relation between the current density and the electric field, making use of the notion of complex mobility; in general,

$$\boldsymbol{J} = ne(b^+ - b^-)\boldsymbol{E}$$

In the absence of an external magnetic field, the mobilities are complex but scalar. Moreover, b^+ is negligible with respect to b^-. Introducing this in equation (3.4), one deduces:

$$\frac{1}{\mu_0}\nabla\times\boldsymbol{B} = \left(\varepsilon_0 - \frac{neb^-}{j\omega}\right)\frac{\partial\boldsymbol{E}}{\partial t} \tag{4.54}$$

This equation can also be written as:

$$\nabla\times\boldsymbol{B} = \mu_0\bar{\varepsilon}\frac{\partial\boldsymbol{E}}{\partial t}$$

with $\bar{\varepsilon}$ as complex permittivity and given by:

$$\bar{\varepsilon} = \varepsilon_0 - \frac{neb^-}{j\omega} \tag{4.55}$$

In the case of a collisionless plasma, the concept of mobility is physically meaningless. However, a purely imaginary mobility parameter can be formally defined; in such a case, the permittivity is real. A collisionless plasma can thus be considered as a non-absorbing medium with a refraction index of its own. When collisions are introduced, the mobility ceases to be a purely imaginary quantity, hence the permittivity is no longer a real quantity and the wave amplitude is necessarily variable. Physically, there is an attenuation. Thus, one can write formally as in a transparent medium:

$$\nabla^2\boldsymbol{E} = \mu_0\bar{\varepsilon}\frac{\partial^2\boldsymbol{E}}{\partial t^2} \tag{4.56}$$

The dispersion relation for equation (4.56) is then:

$$k^2 = \mu_0\left(\varepsilon_0 - \frac{neb^-}{j\omega}\right)\omega^2 \tag{4.57}$$

in which b^- is given by equation (2.45); thus, equation (4.57) becomes:

$$k^2 = \mu_0 \left[\varepsilon_0 + \frac{ne^2}{j\omega M^-(\nu_c + j\omega)} \right] \omega^2$$

an expression which after some elementary algebraic transformations, becomes for values of ν_c that are not excessively large compared to ω:

$$k^2 = \mu_0\varepsilon_0[1-(\omega_p^2/\omega^2)-j\,(\omega_p^2/\omega^2)\,(\nu_c/\omega)]\omega^2 \tag{4.58}$$

thus showing that the attenuation length is given by:

$$l = 2\frac{\omega^2}{\omega_p^2\nu_c}\left\{\frac{1-(\omega_p/\omega)^2}{\mu_0\varepsilon_0}\right\}^{1/2} \tag{4.59}$$

a result particularly important for plasma diagnostics (Figure 4.9).

If a longitudinal magnetic field is present in the plasma, a plane polarised wave can always be decomposed into two circularly polarised waves of equal

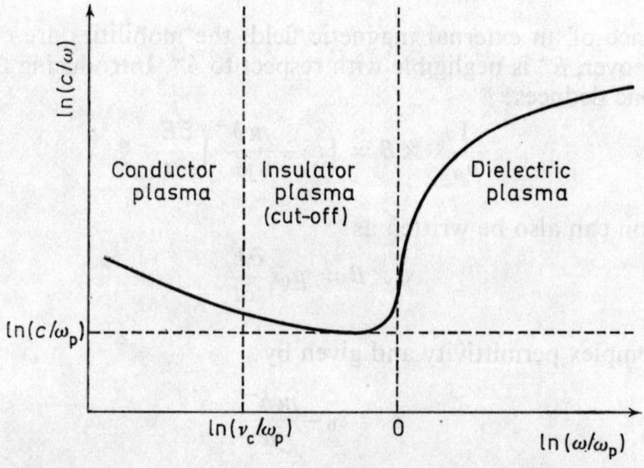

Fig. 4.9. Attenuation in a plasma for different frequencies of the electromagnetic wave. Depending on the frequency the plasma acts as a conductor, an insulator or a dielectric

amplitude. If damping is introduced, the two circularly polarised waves may display different attenuation lengths, especially near the cut-offs and the resonances. Such a situation is known to change the linear polarisation of the wave into an elliptic one; in extreme cases, only one of the circularly polarised waves may persist.

These results can be generalised by replacing the scalar mobility b by a tensor pseudo-mobility b' and by introducing the contribution of the positive ions. The latter, in fact, is always negligible except for the case of the ordinary wave at low frequencies. This generalisation leads to the replacement of equation (4.55) by:

$$\boldsymbol{\varepsilon} = \varepsilon_0\mathbf{I} + \frac{ne}{j\omega}(b^+ - b^-)$$

In the case of the collisionless plasma, this leads to the consideration of a real refractive index tensor, a situation similar to that encountered in crystal optics. Hence, the wave will exhibit properties of birefringence. If the external field is parallel to the z-axis, the tensor corresponding to a collisionless plasma has the following components:

$$\frac{\varepsilon_{xx}}{\varepsilon_0} = \frac{\varepsilon_{yy}}{\varepsilon_0} = 1 - \frac{\Omega_p^2(\omega^2 - \omega_c\omega_i)}{(\omega^2 - \omega_c^2)(\omega^2 - \omega_{ci}^2)}$$

$$\frac{\varepsilon_{xy}}{\varepsilon_0} = -\frac{\varepsilon_{yx}}{\varepsilon_0} = j\omega \frac{\Omega_p^2(\omega_i - \omega_c)}{(\omega^2 - \omega_c^2)(\omega^2 - \omega_{ci}^2)} \tag{4.60}$$

$$\varepsilon_{zz}/\varepsilon_0 = 1 - (\Omega_p^2/\omega^2)$$

all the other components being equal to zero. Ω_p is given by equation (1.14) and ω_{ci} is the ionic cyclotron frequency equal to eB/M^+. In the absence of a magnetic field, the tensor becomes diagonal with all the components equal to the z-zone. If, now, a circularly polarised wave propagating along the magnetic field is considered, then:

$$\varepsilon_{\pm}/\varepsilon_0 = 1 - \frac{\Omega_p^2}{(\omega \mp \omega_{ci})(\omega \pm \omega_c)} \tag{4.61}$$

This equation is similar to equation (4.52), but it takes account of the influence of the positive ions.

There are three major differences between the birefringence in crystalline optics and that in plasma physics.

(1) The components of the refractive index in a plasma are strongly frequency-dependent, with the relevant phenomena of cut-offs and resonances similar to those discussed in Section 4.6.
(2) In crystalline optics, the eigenvalues of the refractive index always correspond to plane polarised waves, whereas in plasma physics, they correspond to circularly polarised waves, or even more complicated situations.
(3) Damping is, in general, more conspicuous in plasmas, especially in the neighbourhood of resonance.

4.8 BEAM–PLASMA INTERACTION[15, 16]

Beam–plasma interaction phenomena were discovered experimentally by Langmuir more than thirty years ago and remained a laboratory curiosity until the late 1940s when they found interesting applications in the field of millimetre wave amplification. With the advent of thermonuclear projects, it became evident that manifestations of beam–plasma interactions can be present even if the beam is not introduced intentionally in the plasma. The experimental discovery of Langmuir is related closely to the more general aspect of the so-called 'Langmuir paradox'. In a plasma, when the velocity distribution of the electrons is artificially rendered non-Maxwellian, it tends, in general, towards a new Maxwellian state much more rapidly than a theory based exclusively upon the collisions would suggest. In this respect, the beam-plasma interaction is just a special aspect thereof, since the injection of a beam can be considered as a particular procedure for rendering the plasma non-Maxwellian.

Consider the case of a beam of particles injected into a gas. The beam is attenuated and the rate of attenuation can be used to determine the relevant cross-section resulting from the interaction of the individual particles with the individual atoms. However, once the gas is ionised, even weakly, it can no longer be said that the rate of attenuation of the beam is governed by the interactions of the particles constituting the beam with the particles of the plasma. In many cases, this cause of attenuation falls short by two to six orders of magnitude. This discrepancy is an example of a general plasma phenomenon, the complete understanding of which is the basis of many plasma applications: *collective phenomena.* In a collective phenomenon, an individual particle interacts with many particles of the plasma at a time in an orderly fashion. This is rendered possible by local fluctuations of the space charge. When an electron beam is injected into the plasma, it is not scattered only by the individual plasma particles but also by the local fields associated with such fluctuations. Furthermore, even in cases when the beam remains relatively ordered, its kinetic energy can be decreased and partially converted into electrostatic energy.

The elementary theory of the beam–beam interaction was developed almost simultaneously around 1950 by a number of different authors. The simplest way to introduce it is as a special case of the beam–plasma interaction. Thus, consider a one-dimensional configuration with both beams having an infinite extension perpendicular to their axis. Further, assume that all the space charges are compensated on average, with their density independent of space and time. Let n_1 and n_2 be the carrier densities in the two cold beams. As before, the relevant parameters are decomposed into two parts, one, corresponding to the average state, and the other, pertinent to the fluctuation. At this moment, no assumption is made about the origin of the fluctuation; thus:

$$n_{1,2} = n_{01,2} + \tilde{n}_{1,2}$$

$$v_{1,2} = v_{01,2} + \tilde{v}_{1,2}$$

where subscripts 1 and 2 correspond to each one of the two cold beams. The current densities can be divided similarly; hence for the first beam:

$$J_1 = J_{01} + \tilde{J}_1 = J_{01} - \tilde{n}_1 e v_{01} - n_{01} e \tilde{v}_1 \tag{4.62}$$

with a similar relation for the second beam.

Assume that the fluctuations are made of pure electrostatic waves, the field of which derives from an electrostatic potential such as:

$$\tilde{V} = V_0 \exp [\mathrm{j}(\omega t - kz)] \tag{4.63}$$

in which k is shorthand for k_z, the only component of the propagation vector assumed different from zero. In principle, either ω or k can be regarded as a complex quantity. For instance, if a fluctuation of definite frequency is imprinted on one of the beams, ω is real; the imaginary part of k is then relevant to the build-up or the attenuation of the fluctuation along the z-axis. Such a fluctuation can be considered as carried away by the particles and is termed 'convective'. On the other hand, one can think of some spatial periodicity imposed by the geometry, in which case k is real. Quite likely, in such a case, the true pattern will be some kind of standing wave rather than a travelling one. The imaginary part of ω is related to the local build-up or decay of such a wave, and the fluctuation is termed 'absolute' or 'non-convective'.

Using Poisson's equation and taking account of equation (4.63), one obtains:

$$k^2\tilde{V} = -e(\tilde{n}_1+\tilde{n}_2) \tag{4.64}$$

Assume also that the interaction between the two beams is collisionless; hence, a given particle always belongs to one or the other beam. Consequently, there is for each beam, an individual equation for charge conservation that yields, after using equation (4.63),

$$e\tilde{n}_{1,\,2} = -k\tilde{J}_{1,\,2} \tag{4.65}$$

The particles move under the influence of the wave electric field; thus:

$$-\omega\tilde{v}_{1,\,2}+v_{01,\,2}k\tilde{v}_{1,\,2} = ek\tilde{V}/M \tag{4.66}$$

Substituting in equation (4.65) the alternating part of the current density as given by equations (4.62), one obtains:

$$\tilde{v}_{1,\,2} = \frac{\tilde{n}_{1,\,2}}{kn_{01,\,2}}(\omega-kv_{01,\,2})$$

and after substitution into equation (4.66), a system of two equations in $\tilde{n}_1$ and $\tilde{n}_2$ is found. This system is solved for $\tilde{n}_1$ and $\tilde{n}_2$ and substituted into equation (4.64). Excluding the trivial solution $k^2\tilde{V} = 0$, the following relation is obtained:

$$1 = \frac{\omega_1^2}{(\omega-kv_{01})^2}+\frac{\omega_2^2}{(\omega-kv_{02})^2} \tag{4.67}$$

where ω_1 and ω_2 are the characteristic plasma frequencies of the two beams. This equation is the dispersion relation of the two interacting beams.

The preceding argument clearly shows that equation (4.67) can be generalised for more than two beams; it is thus obtained:

$$1 = \sum_{i=1}^{n}\frac{\omega_i^2}{(\omega-kv_{0i})^2} \tag{4.68}$$

A logical extension in the case of a continuous velocity distribution consists in considering it as made up of an infinite number of beams, thus replacing the sum in equation (4.68) by an integral. In such an integral, a differential $d\omega_i^2$ appears and must be interpreted as a combination of universal constants times dn_i, which can in turn be readily estimated from the velocity distribution. Similar considerations apply to the superposition of a random velocity to the drift velocity of the particles. Indeed, the interaction of a cold beam with a hot plasma is obtained by replacing in equation (4.67) the second factor in the right-hand side by a suitable integral. For a Maxwellian distribution without drift, this integral becomes $\omega_p^2/(\omega^2 - k^2C^2)$ where ω_p is the plasma frequency and C is the root mean square velocity in the plasma.

Extensive numerical calculations can be found in the literature for the mutual dependence of ω and k from equation (4.67)[16, 36]. Such calculations show that convective amplifying waves are most readily obtained in the neighbourhood of the plasma electrostatic frequency. Those calculations assume that the beam density is only a small fraction of the plasma density, the latter imposing its own characteristic frequency. One can deduce from equation (4.47) that just below the plasma frequency the dielectric constant is negative and extremely small since $v_R^2 = \varepsilon/\varepsilon_0$. When the dielectric constant is negative, the electric field changes to the opposite direction of the electric displacement. Thus, if such a medium is homogeneous, charges of the same sign would attract each other and charges of opposite sign would repel each other. Furthermore, if ε is very small, electric displacements of normal amplitude could correspond to very large fields.

Finally, if collisions are taken into account by modifying equation (4.66) and using the concept of complex mobility, they would limit the gain of the wave resulting from the interacting beams.

4.9 PLASMA–PLASMA INTERACTION

In some instances two plasmas will interact with each other without necessarily being combined in a single plasma. An obvious example is the interaction between two galaxies. Laboratory examples have been observed by many authors[3, 8, 26]. One example of non-Maxwellian probe characteristics below the space potential is several straight lines with decreasing slopes when the potential decreases. This can be interpreted as the result of the superposition of two or more independent plasmas, each having its own density, temperature and drift velocity. All the relevant parameters can easily be determined by extrapolation from the probe characteristics. However, the situation becomes more complicated when there is at least one large drift velocity involved.

A study of the plasma–plasma interaction has been made by Kettani[29]. He considered a fast plasma interacting with a slower one in an infinite medium. If one considers only the electronic components and neglects all the collisions, the three fundamental equations of motion, momentum transport, and state can be written for the two plasmas. Further, assuming that the electronic components are adiabatic, the theory of perturbation can be used by assuming that all the quantities are equal to the sum of a d.c. constant and a much smaller perturbation as seen in Section 4.8.

If it is further assumed that the drift velocity of the slow plasma is equal to zero and that perturbations follow relations similar to equation (4.63), after solving for the electric field, one finds:

$$\tilde{E} = \mathrm{j}\frac{\tilde{n}_s}{n_{0s}}\left(\frac{k\gamma}{e}\frac{p_{0s}}{n_{0s}} - \frac{\omega^2 M}{ke}\right) \tag{4.69}$$

and

$$\tilde{E} = \mathrm{j}\frac{\tilde{n}_f}{n_{0f}}\left(\frac{k\gamma}{e}\frac{p_{0f}}{n_{0f}} + \frac{2\omega M v_{0f}}{e} - \frac{kMv_{0f}^2}{e} + \frac{\omega^2 M}{ke}\right) \tag{4.70}$$

where subscript (s) stands for slow plasma and subscript (f) for fast plasma, p_0 is the average pressure, ω is the frequency of the perturbation, and k is its constant of propagation.

By using Poisson's law, assuming that the plasma as a whole is neutral, in the steady state, and neglecting the ionic perturbation, one finds for the electric field:

$$\tilde{E} = -\mathrm{j}\frac{e}{\varepsilon_0 k}(\tilde{n}_f + \tilde{n}_s) \tag{4.71}$$

By eliminating $\tilde{n}_f$ and $\tilde{n}_s$ from equations (4.69), (4.70) and (4.71), one obtains the following dispersion relation:

$$1 = \frac{1}{\left(\frac{\omega}{\omega_{ps}}\right)^2 - \frac{\eta^{*2}}{\mathcal{M}_f^2}\left(\frac{v_{ss}}{v_{sf}}\right)^2} + \frac{\left(\frac{\omega_{pf}}{\omega_{ps}}\right)^2}{\left(\frac{\omega}{\omega_{ps}} - \eta^*\right)^2 - \frac{\eta^{*2}}{\mathcal{M}_f^2}} \tag{4.72}$$

where ω_{ps} and ω_{pf} are the characteristic plasma frequencies of the slow and fast plasma respectively, $\mathcal{M}_f$ is the Mach number of the fast plasma, v_{ss} and v_{sf} are the speeds of sound in the slow and fast electronic plasmas respectively, and η^* is the gain of the slow plasma from the fast plasma defined as v_{0f}/ω_{ps}. The dispersion relation (4.72) describes the coupling between the space charge wave of the slow plasma and the space charge wave of the fast one. The waves of the slow (here still) plasma are given by:

$$1 - \frac{1}{(\omega/\omega_{ps})^2 - (\eta^*/\mathcal{M}_f)^2(v_{ss}/v_{sf})^2} = 0$$

which becomes after some mathematical rearrangements:

$$\omega^2 = \omega_p^2 + k^2 v_{ss}^2 \tag{4.73}$$

This relation was found by Bohm and Gross[14] and is the familiar form for electrostatic propagation as seen above. The waves of the fast plasma are given by:

$$1 - \frac{1}{[(\omega/\omega_{ps}) - \eta^*]^2 - (\eta^*/\mathcal{M}_f)^2} = 0$$

This relation leads to a second-degree algebraic equation in η^*, the solution of which becomes after some rearrangements:

$$k = \frac{-(\omega/\omega_{pf})\mathcal{M}_f \pm \sqrt{[(\omega/\omega_{pf})^2 - (1 - \mathcal{M}_f^2)]}}{r_{Df}\sqrt{\gamma(1 - \mathcal{M}_f^2)}}$$

where r_{Df} is the Debye radius of the electronic fast plasma and γ is the ratio of the specific heats. If the speed of sound in the fast plasma is equal to zero, the

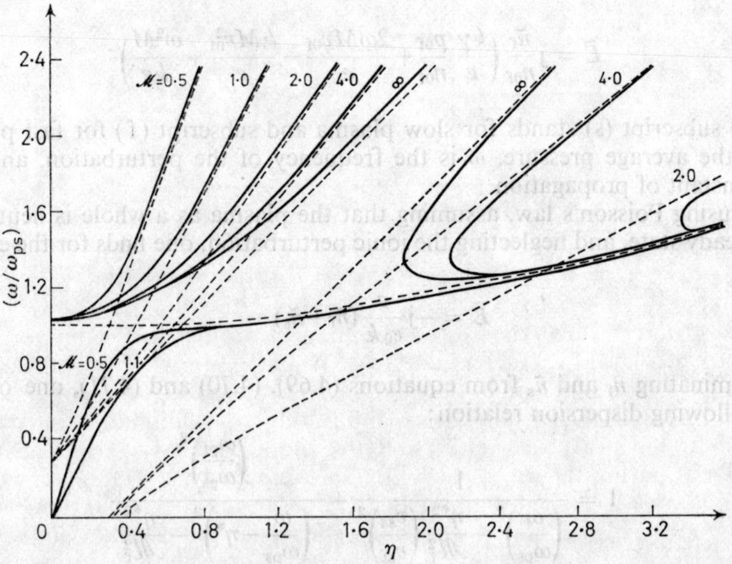

Fig. 4.10. Plasma–plasma interaction $\omega/\omega_{ps} = f(\eta^*)$ with the Mach Number as parameter. $(\omega_{pf}/\omega_{ps})^2 = 0{\cdot}1$, $(v_{ss}/v_{of})^2 = 0{\cdot}1$

Mach number will become infinite. This case corresponds to a cold fast plasma which was studied by Crawford[18].

If now the coupled equation (4.72) is considered, one can set $\omega/\omega_{ps} = f(\eta^*)$ with $(v_{ss}/v_{of}) = (v_{ss}/v_{sf})\mathscr{M}_f$, $(\omega_{pf}/\omega_{ps})$ and $\mathscr{M}_f$ as parameters. One can thus note that this equation is of the fourth degree in both η^* and ω. This equation is represented in Figure 4.10 from which four different regions can be recognised.

4.10 SHOCK WAVES[7, 11, 21]

A shock wave is a wave in which there is a discontinuity in the pressure, density and velocity. The occurrence of such phenomena can be interpreted as a possible result of the non-linearity of the magnetohydrodynamic equations. Indeed, shocks are often due to the non-linear character of the forces in the fluid causing a perturbation to travel with a velocity equal to the sum of the local maximum velocity and the velocity of sound in the medium. Thus, some regions of the perturbation will catch up with slower moving regions in such a way that the front of the travelling perturbation will become steeper and steeper until the shock takes place.

Consider, for instance, an accelerated piston in a confined fluid as shown in Figure 4.11. The region of the fluid near the piston must move with the velocity of the piston, whereas the part of the fluid far enough from the piston remains still. Therefore, the fluid velocity must vary with position from the velocity of the piston at the piston surface to zero far enough from that surface. Since the overall volume has decreased as a result of the motion of the piston, the fluid

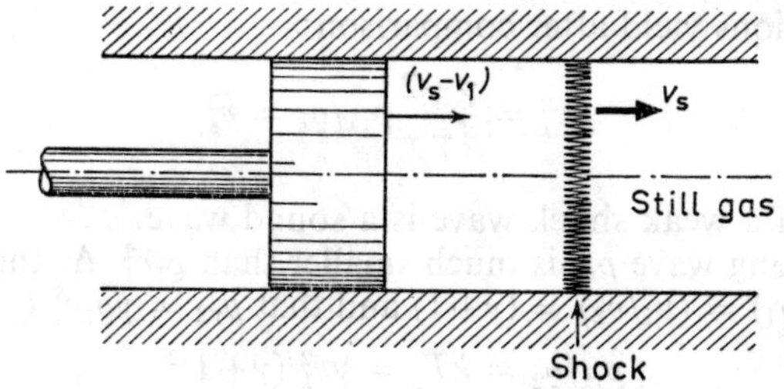

Fig. 4.11. Shock resulting from the motion of a piston in a confined gas

near the piston is thus compressed resulting in an increase of temperature. Therefore, the speed of sound will decrease from the piston surface to its minimum value at the still region of the fluid. If, now, the velocity profile is resolved into small steps, each step will correspond to the propagation of a sound wave moving with the local speed velocity relative to the fluid. At a certain point the high velocity part of the wave reaches the low velocity part resulting in the formation of a very steep front. At this front, viscous forces and thermal conduction become important and cause a steepening effect such that a steady state shock front is formed.

Consider now a stationary shock front moving in a channel of uniform cross-section. One can always find two planes parallel to the shock and sufficiently far from it so that the dissipative effects, viscous stresses, and flow gradients become negligible. Since there will be no accumulation of energy, mass, or momentum because of the steady state in the shock front, the flux of these quantities must remain the same across these two planes. Thus, from continuity equation (3.35) one obtains:

$$\varrho_1 v_1 = \varrho_2 v_2 \tag{4.74}$$

From equation (3.41) of conservation of momentum, if magnetic field and collisions are neglected, one obtains:

$$p_{r1} + \tfrac{1}{2}\varrho_1 v_1^2 = p_{r2} + \tfrac{1}{2}\varrho_2 v_2^2 \tag{4.75}$$

and from equation (3.46) of conservation of energy:

$$(\gamma+2)p_{r1} v_1 + \varrho_1 v_1^3 = (\gamma+2)p_{r2} v_2 + \varrho_2 v_2^3 \tag{4.76}$$

In these equations subscript (1) corresponds to conditions before the shock, whereas subscript (2) corresponds to conditions after the shock. Therefore, if the conditions before the shock are known, conditions after the shock can be deduced, and vice versa.

For a very weak wave the velocity of the fluid before the shock is almost equal to the velocity of the fluid after the shock; thus $(v_2/v_1) = 1-\delta_v$, and $(p_{r2}/p_{r1}) = 1+\delta_p$, where δ_v and δ_p are very small quantities compared to unity; thus equations (4.75) and (4.76) become respectively:

$$p_{r1}\delta_p - \tfrac{1}{2}\varrho_1 v_1^2 \delta_v = 0$$

$$\left(\frac{\gamma+2}{2}\right) p_{r1}(\delta_p - \delta_v) = \varrho_1 v_1^2 \delta_v$$

These two equations yield after comparison:

$$v_1^2 = \frac{\gamma+2}{\gamma} p_{r1}/\varrho_1 = v_s^2 \tag{4.77}$$

Thus the limit of a weak shock wave is a sound wave.

For a very strong wave p_{r1} is much smaller than $\varrho_1 v_1^2$. At the limit, it can be shown that $(\varrho_2/\varrho_1) = (v_1/v_2) = (\gamma+1)$ and that $p_{r2} = \gamma\varrho_1 v_1^2/(\gamma+1)$. Thus:

$$p_{r2}/\varrho_2 = kT_2 = \gamma v_1^2/(\gamma+1)^2 \tag{4.78}$$

After the shock the pressure as well as the temperature become functions of v_1^2, and the ratio of the densities before and after the shock becomes a function of γ. The thickness of a strong shock wave can be found qualitatively since the steepening effect on the shock front is limited by the viscous stress, thus:

$$p_{r2} = v\, dv/dx \tag{4.79}$$

where v is the viscosity coefficient. From equation (4.79) one can write for the shock thickness x_s approximately:

$$x_s = vv_1/p_{r2} = \lambda$$

where λ is the mean free path. For gases at low temperatures and densities, x_s is very small and the shock can be considered as a discontinuity. The mean free path becomes, however, long for high temperature plasmas and the discontinuity assumption is no longer valid. One should also note that, in a plasma, the viscosity stress is not always the determining factor for estimating the shock thickness.

If a magnetic field is applied, equations (4.75) and (4.76) should be modified to include the effect of the magnetic field. Considering the simple case where the shock wave plane is parallel to the applied magnetic field, equations (4.75) and (4.76) become respectively:

$$p_{r1} + \frac{1}{2}\varrho_1 v_1^2 + \frac{B_1^2}{2\mu} = p_{r2} + \frac{1}{2}\varrho_2 v_2^2 + \frac{B_2^2}{2\mu} \tag{4.80}$$

$$\frac{p_{r1}}{\varrho_1}\left(\frac{\gamma}{\gamma-1}\right) + \frac{v_1^2}{2} + \frac{B_1^2}{\mu\varrho_1} = \frac{p_{r2}}{\varrho_2}\left(\frac{\gamma}{\gamma-1}\right) + \frac{v_2^2}{2} + \frac{B_2^2}{\mu\varrho_2} \tag{4.81}$$

To equations (4.74), (4.80) and (4.81), Maxwell's equations should be added. In the steady state $\boldsymbol{E}+\boldsymbol{v}\times\boldsymbol{B} = 0$, and $\nabla\times\boldsymbol{E} = 0$. Thus $E_1 = E_2$, and:

$$v_1 B_1 = v_2 B_2 \tag{4.82}$$

Equations (4.74), (4.80), (4.81) and (4.82) form a system of four equations with four unknowns. By assuming that the conditions before the shock are known, v_2, ϱ_2, p_{r2}, and B_2 can be found in terms of v_1, ϱ_1, p_{r1}, and B_1. There is only one solution to this system that has only one shock wave. For instance, the shock velocity is found to be:

$$v_\phi = \left\{\frac{2}{\gamma+1}\,\frac{v_{s1}^2 + v_{A1}^2[1-(\gamma/2)]\,[(\varrho_2/\varrho_1)-\gamma/(\gamma-2)]}{(\varrho_1/\varrho_2)-[(\gamma-1)/(\gamma+1)]}\right\}^{1/2} \tag{4.83}$$

where v_{s1} is the speed of sound before the shock wave and v_{A1} is the Alfvén velocity before the shock wave such that $v_{A1} = (B_1^2/\mu \varrho_i)^{1/2}$. For weak shocks $\varrho_1/\varrho_2 = (1-\delta_v)$; thus equation (4.83) becomes at the limit $v_{\phi\ \mathrm{lim}} = (v_{s1}^2+v_{A1}^2)^{1/2}$. That is the superposition of a sonic wave and an Alfvén wave.

REFERENCES AND BIBLIOGRAPHY

1. ALFVÉN, H., 'Existence of Electromagnetic-Hydrodynamic Waves', *Nature, Lond.*, No. 3805, 405 (1942).
2. ALFVÉN, H., *Cosmical Electrodynamics*, Clarendon, Oxford, England (1950).
3. ALLEN, T. K., 'A Spectroscopic Study of Plasma-Electron Oscillations', *Proc. phys. Soc.*, **68A**, 676 (1955).
4. ALLIS, W. P., 'Electron-Plasma Oscillations', *Symp. on Electronic Waves*, Brooklyn Polytech. (1958).
5. ALLIS, W. P., 'Propagation of Waves in a Plasma in a Magnetic Field', *IRE Trans. on Microwave Theory and Techniques*, Vol. MTT-9, No. 1, 79 (1961).
6. ALLIS, W. P., BUSHBAUM, S. J. and BERS, A., *Waves in Anisotropic Plasmas*, MIT Press, Cambridge, Mass. (1963).
7. ANDERSON, J. E., *MHD Shock Waves*, MIT Press, Cambridge, Mass. (1963).
8. ARMSTRONG, E. B., 'Plasma-Electron Oscillations', *Nature, Lond.*, **160,** 713 (1947).
9. BAILEY, R. A. and EMELEUS, K. G., 'Plasma-Electron Oscillations', *Proc. R. Ir. Acad.*, **57A**, 53 (1955).
10. BANOS, A., 'MHD Waves', *Modern Physics for the Engineer* (edited by L. N. RIDENOUR and W. A. NIERENBERG), McGraw-Hill, New York (1961).
11. BAZAR, J. and FLEISCHMAN, O., 'Propagation of Weak Hydromagnetic Discontinuities', *Physics Fluids*, **2,** 366 (1959).
12. BEKEFI, G. and BROWN, S. C., 'Emission of Radio-Frequency Waves from Plasmas', *Am. J. Phys.*, **29,** 404 (1961).
13. BOHM, D., *Characteristics of Electrical Discharges in Magnetic Fields*, Symp. (edited by GUTHRIE and WAKERLING), McGraw-Hill, New York (1949).
14. BOHM, D. and GROSS, E. P., 'Theory of Plasma Oscillations', *Phys. Rev.*, **75,** 1851 (1949).
15. BOYD, G. D., FIELD, L. M. and GOULD, R. W., 'Interaction Between an Electron Beam and an Arc Discharge Plasma', *Symp. on Electronic Waveguides*, Brooklyn Polytech., 367, April (1958).
16. BOYD, G. D., GOULD, R. W. and FIELD, L. M., 'Interaction of Modulated Electron Beam with a Plasma', *Proc. IEEE*, **49,** 1906 (1961).
17. CRAWFORD, F. W. and CAMARA, A. B., 'Structure of the Double-Sheath in a Hot-Cathode Plasma', *J. appl. Phys.*, **36,** 3135 (1965).
18. CRAWFORD, F. W., *Beam-Plasma Interaction in a Warm Plasma*, Technical Report AEC Contract (04-3)326, May (1965).
19. DAWSON, J., 'On Landau Damping', *Physics Fluids*, **4,** 869 (1961).
20. DAWSON, J., 'Radiation From Plasmas', *Advances in Plasma Physics* (edited by A. SIMON), Interscience, New York (1968).
21. DEHOFFMAN, F. and TELLER, E., 'MHD Shocks', *Phys. Rev.*, **80,** 692 (1951).
22. DENISSE, J. and DELCROIX, J. L., *Plasma Waves*, Interscience, New York (1963).
23. DESIRANT, M. and MICHELS, J. L. (Eds.), *Electromagnetic Wave Propagation*, Academic Press, New York (1960).
24. ERICSON, W. B. and BAZAR, J., 'On Certain Properties of Hydromagnetic Shocks', *Physics Fluids*, **13,** 631 (1960).
25. GABOR, D., ASH, E. and DRACOTT, D., 'Langmuir's Paradox', *Nature, Lond.*, **176,** 916 (1955).
26. GARSCADDEN, A., 'Experiments on Plasma Oscillations', *J. Electron. Control*, **14,** 303 (1963).

27. HEITLER, W., *The Quantum Theory of Radiation*, Clarendon, Oxford (1954).
28. HOYAUX, M. F. and WILLIAMS, E. M., 'Fast Randomization Phenomena in an Unstable Plasma-Sac Configuration', *J. appl. Phys.*, **38**, No. 9, 3630 (1967).
29. KETTANI, M. A., 'Plasma-Plasma Interaction', *Instabilities in Highly Ionized Low Pressure Metal-Vapor Plasmas*, M. F. HOYAUX *et al.*, AEC Res. Dept. Carnegie Institute of Technology, Contract AT(301)3100, June (1965).
30. KETTANI, M. A., *Fast Randomization of an Electron Gas by Trapped Electroacoustic Waves*, Carnegie Institute of Technology, Ph.D. Thesis, May (1966).
31. KETTANI, M. A. and HOYAUX, M. F., 'Fast Randomization of an Electron Gas by Trapped Electroacoustic Waves', *Physics Fluids*, **11**, No. 1, 143 (1968).
32. KETTANI, M. A. and HOYAUX, M. F., 'Maxwellianisation Rapide d'un Gaz Electronique par des Ondes Electroacoustiques Confinées', *Onde elect.*, No. 492, 1 (1968).
33. LANDAU, L., 'On the Vibrations of the Electronic Plasma', *J. Phys. U.S.S.R.*, **10**, 25 (1946).
34. LONGMIRE, C. L., *Elementary Plasma Physics*, Interscience, New York (1963).
35. LOONEY, D. H. and BROWN, S. C., 'The Excitation of Plasma Oscillation', *Phys. Rev.*, **95**, 965 (1954).
36. MAHAFFEY, D. N., GULLOGH, G. M. and EMELEUS, K. G., 'Beam-Plasma Interaction', *Phys. Rev.*, **112**, 1052 (1958).
37. MITCHNER, M. (Ed.) *Radiation and Waves in Plasmas*, Stanford University Press (1961).
38. OSTER, L., 'Linearized Theory of Plasma Oscillations', *Rev. mod. Phys.*, **32**, 141 (1960).
39. PENNING, F. M., 'Scattering of Electrons in Ionized Gases', *Nature, Lond.*, **118**, 301 (1926).
40. SHERMAN, A., 'Viscous MHD Boundary Layer', *Physics Fluids*, **4**, 522 (1961).
41. STIX, T. H., *The Theory of Plasma Waves*, McGraw-Hill, New York (1961).
42. SUTTON, G. W. and SHERMAN, A., *Engineering MHD*, McGraw-Hill, New York (1965).
43. TONKS, L. and LANGMUIR, I., 'Oscillations in Ionized Gases', *Phys. Rev.*, **33**, 195 (1929).
44. TRUBNIKOV, B. A. and KUDRYATSOV, V. S., 'Plasma Radiation in a Magnetic Field', *Proc. 2nd U.N. Conf. on the Peaceful Uses of Atomic Energy*, Geneva, Switzerland, 93 (1958).
45. UMAN, M. A., *Introduction to Plasma Physics*, McGraw-Hill, New York (1964).

5 Plasma instabilities and turbulence

5.1 PLASMA EQUILIBRIUM[10, 11]

Instabilities in a plasma can be divided into two types : (1) instabilities due to the *microscopic* departure from equilibrium of the plasma characteristics, and (2) instabilities involving the plasma as a whole and consequently of a *macroscopic* nature. Consider a system described by a number of Lagrangian co-ordinates $q_1, q_2, \ldots, q_k$, in which k is the number of degrees of freedom n_f. The total energy $\mathscr{E}_t$ of the system will be equal to the sum of the potential enery $\mathscr{E}_{pot}$ and the kinetic energy $\mathscr{E}_{kin}$. At equilibrium the forces on the system should be equal to zero; thus, in terms of the Lagrangian co-ordinates, the condition of equilibrium becomes:

$$\partial \mathscr{E}_{pot}/\partial q_j = 0 \tag{5.1}$$

where $j = 1, 2, \ldots, k$.

For small deviations between point r and point x, the difference of potential energy $\Delta\mathscr{E}_{pot} = \mathscr{E}_{pot}(r) - \mathscr{E}_{pot}(x)$ would be given by a Taylor series, the first term of which will go to zero after applying equation (5.1), and when there is an equilibrium at point $r = x$ one obtains:

$$\Delta\mathscr{E}_{pot} = \frac{1}{2}(r-x)^2 \frac{\partial^2 \mathscr{E}_{pot}}{\partial q_j^2} + \ldots$$

If $\Delta\mathscr{E}_{pot} > 0$, then the system is stable, and if $\Delta\mathscr{E}_{pot} < 0$ the system is unstable. Thus, the condition of stability becomes:

$$\partial^2 \mathscr{E}_{pot}/\partial q_j^2 > 0 \tag{5.2}$$

This is explained as follows: since the sum of the potential and the kinetic energy is constant, any motion corresponding to an increase in potential energy corresponds to a decrease in kinetic energy. If the initial kinetic energy does not exceed a certain value (energy barrier), the system will eventually return towards the equilibrium position. In dynamics, it is shown that, in the linear approximation and for non-dissipative motions, sinusoidal oscillations occur. If equation (5.2) is not satisfied for any degree of freedom, the system departs from the equilibrium position with constant or increasing kinetic energy, and unless non-linear terms correct the situation the system never returns toward its equilibrium position.

As an example, consider a particle moving around a central field of force in a circular orbit. Assuming that the mass of the particle is unity, the two constants of the motion, i.e. the angular momentum and the total energy, can be written:

$$\dot{S} = r^2 \frac{d\theta}{dt} \tag{5.3}$$

and

$$\mathscr{E}_t = \frac{1}{2}\left(r\frac{d\theta}{dt}\right)^2 + \frac{1}{2}\left(\frac{dr}{dt}\right)^2 + \mathscr{E}_{pot}(r) \tag{5.4}$$

where r and θ are the two Lagrangian co-ordinates of the particle. After deriving equations (5.3) and (5.4) with respect to time, one obtains after some re-arrangements:

$$\frac{\partial \mathscr{E}_{pot}}{\partial r} = \frac{\dot{S}^2}{r^3} - \frac{d^2 r}{dt^2} \tag{5.5}$$

and then $(\partial\mathscr{E}_{pot}/\partial r)_{r_0} = (\dot{S}^2/r_0^3)$. Assuming that the orbit of the particle is the equilibrium orbit, one can study the stability of small motions about this orbit by letting $r = r_0(1+\delta)$ where $\delta \ll 1$. After expanding $\mathscr{E}_{pot}$ in powers of δr_0, one obtains:

$$(\partial\mathscr{E}_{pot}/\partial r) = (\partial\mathscr{E}_{pot}/\partial r)_{r_0} + \delta r_0(\partial^2\mathscr{E}_{pot}/\partial r^2)_{r_0} + \dots \tag{5.6}$$

Introducing equation (5.6) into equation (5.5), one obtains:

$$\frac{d^2\delta}{dt^2} = -\left[\frac{3}{r_0}\frac{\partial\mathscr{E}_{pot}}{\partial r} + \frac{\partial^2\mathscr{E}_{pot}}{\partial r^2}\right]_{r_0} \delta \tag{5.7}$$

or:

$$\frac{d^2\delta}{dt^2} + \omega^2\delta = 0 \tag{5.8}$$

The solution of this equation is:

$$\delta = \delta_0 \exp(j\omega t) \tag{5.9}$$

The orbit is stable if $\omega^2 > 0$, leading to an oscillatory orbit as shown by equation (5.9). It is unstable when $\omega^2 < 0$; in this case the distance δ grows exponentially.

Consider now the case of an attractive force varying as $\mathscr{C}/r^n$ near r_0, where $\mathscr{C}$ is the constant of proportionality; from equations (5.8) and (5.9), is obtained:

$$\omega^2 = \mathscr{C}(3-n)/r^{n+1} \tag{5.10}$$

It is thus seen that the motion is stable if the force decreases slower than $1/r^3$. Examples of stable and unstable equilibria are given in Figures 5.1 (a, b, c) showing a ball on a bump or in a trough. Figure 5.1 (c) shows that a particle may be stable for small displacements but becomes unstable when these displacements become larger.

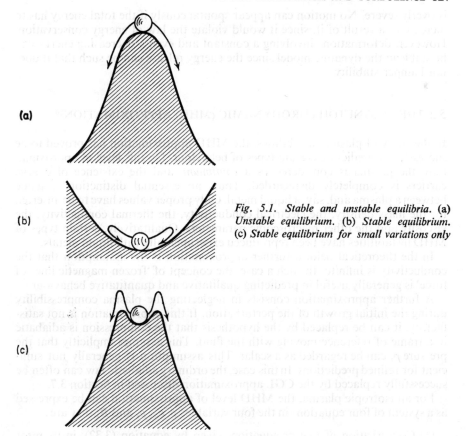

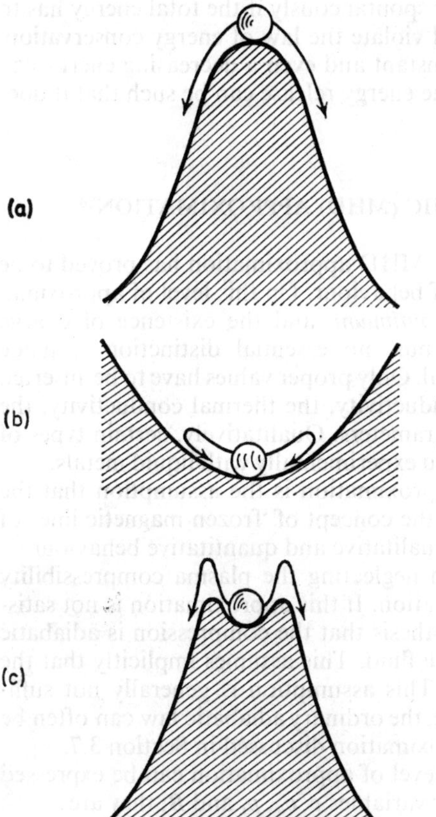

Fig. 5.1. Stable and unstable equilibria. (a) *Unstable equilibrium.* (b) *Stable equilibrium.* (c) *Stable equilibrium for small variations only*

Similar considerations apply to plasma equilibrium. In this case, the potential energy has to be replaced by the total energy, the latter including the kinetic energy of the particles (drift and random), the electromagnetic energy of the fields, taking into account modifications of the latter created by the plasma itself, and eventually, terms relevant to the electromagnetic radiation trapped within the plasma. Already on this degree of approximation, the situation is complicated since the number of degrees of freedom is extremely high. Simplifying assumptions must be introduced; collective instabilities involving deformations of the plasma as a whole, or motions of large groups of particles inside the plasma, are taken into account, but without experiments one can never be sure that all the possibilities have been explored. Furthermore, treating the plasma as a static system is not rigorous in many cases. The plasma is a dynamic system, and, as such, should be treated by the Lagrangian or Hamiltonian methods. The difference is especially sensible when the same particles travel back and forth along magnetic lines of force from a region where some kind of stability criterion is satisfied to a region where it is not. A purely static model would lead to the conclusion that the plasma is unstable, whereas in some cases the dynamic model proves the existence of an overall stability. The static model is overly severe. No motion can appear spontaneously if the total energy has to increase as a result of it, since it would violate the law of energy conservation. However, deformations involving a constant and even a decreasing energy can be stable in the dynamic model since the energy release can be such that it does not hamper stability.

5.2 THE MAGNETOHYDRODYNAMIC (MHD) APPROXIMATION[26]

In the study of plasma instabilities, the MHD approximation has proved to be successful in predicting certain types of behaviour. On this level of approximation, the plasma is considered as a *continuum*, and the existence of charge carriers is completely disregarded. Thus, no essential distinction is made between a plasma and, say, a liquid metal. Only proper values have to be inserted for the mass density, the electrical conductivity, the thermal conductivity, the viscosity and the other macroscopic parameters. Qualitatively, certain types of MHD instabilities have been reproduced experimentally with liquid metals.

In the theoretical field, a further approximation is the assumption that the conductivity is infinite. In such a case, the concept of 'frozen magnetic lines of force' is generally useful in predicting qualitative and quantitative behaviour.

A further approximation consists in neglecting the plasma compressibility during the initial growth of the perturbation. If this approximation is not satisfactory, it can be replaced by the hypothesis that the compression is adiabatic in a frame of reference moving with the fluid. This assumes implicitly that the pressure p_r can be regarded as a scalar. This assumption is generally not sufficient for refined predictions. In this case, the ordinary adiabatic law can often be successfully replaced by the CGL approximation discussed in Section 3.7.

For an isotropic plasma, the MHD level of approximation can be expressed as a system of four equations in the four variables ϱ, $\boldsymbol{v}_d$, p_r and $\boldsymbol{B}$; they are:

(1) Conservation of matter equation, given by equation (3.32), in its most general form.

(2) Equation of magnetic confinement, which is in fact an equation of motion in the perturbation form only. It is deduced from equation (3.54) by neglecting the gravitational term and by replacing $\boldsymbol{J}$ as a function of $\boldsymbol{B}$, the displacement current being neglected:

$$\varrho \frac{\partial \boldsymbol{v}_d}{\partial t} = -\nabla p_r - \frac{1}{2\mu}[\nabla B^2 - 2(\boldsymbol{B}.\nabla)\boldsymbol{B}] \tag{5.11}$$

Reverting to the method of derivation of equation (3.54), it is seen that, at the relevant level of approximation, $\partial \boldsymbol{v}_d/\partial t$ and $d\boldsymbol{v}/dt$ can be used indifferently. Equation (5.11) can be generalised significantly by introducing the viscosity of the plasma without modifying the isotropic character of p_r. A rigorous treatment would involve the equation with $\boldsymbol{\Psi}$ in which the viscosity terms appear as non-diagonal terms of the tensor. However, a simpler method consists of making use of results already obtained in fluid mechanics, and writing the viscosity terms in the same form as for a perfect gas, or even, in a cruder approximation, a liquid.

(3) Equation of magnetic field diffusion which is equation (3.58). In this equation, the last term is frequently neglected, and in this case the equation specifies that the magnetic field does not diffuse with respect to the plasma: frozen magnetic lines of force.

(4) Equation of state of the plasma, sometimes simplified as $\varrho = $ constant, and more generally as $p_r \varrho^{-\gamma} = $ constant in a frame of reference moving with the fluid. A major significant improvement is the CGL approximation as seen before.

In many cases, instead of a constant conductivity, a conductivity function of the state of the plasma, and namely of its temperature, is considered in equation (3.58). This introduces a coupling between this equation and the equation of state. The latter can also be modified to introduce thermal conductivity or convection; the simplest way to do this is to introduce gas dynamics relations, assuming that the proper coefficients are known or can be estimated by some independent method.

5.3 ENERGY PRINCIPLE

Bernstein *et al.*[6] have shown that, at the simplest level of approximation, a simple criterion of stability could be deduced, as if the plasma were subjected to the conditions of stability in statics (Section 5.1). In the above equations, the pressure is considered as a scalar, the resistivity and the viscosity are neglected, and the law of adiabatic compression is assumed in a frame of reference moving with the fluid; hence:

$$\frac{\partial \varrho}{\partial t} + \nabla . (\varrho \boldsymbol{v}_d) = 0 \qquad (5.12a)$$

$$\varrho \mathrm{d}\boldsymbol{v}_d/\mathrm{d}t = -\nabla p_r - \frac{1}{2\mu}[\nabla B^2 - 2(\boldsymbol{B}.\nabla)\boldsymbol{B}] \qquad (5.12b)$$

$$\partial \boldsymbol{B}/\partial t = \nabla \times (\boldsymbol{v}_d \times \boldsymbol{B}) \qquad (5.12c)$$

$$\mathrm{d}(p_r \varrho^{-\gamma})/\mathrm{d}t = 0 \qquad (5.12d)$$

Since the plasma is assumed not to dissipate any electrical or mechanical energy, there will be no energy coupling with the outside world. If a perfect equilibrium did exist, it could be maintained forever without any connection with the main energy supply. In such conditions, the existence of an energy principle can be regarded as evident, and a direct consequence of the most fundamental laws of physics. Furthermore, the conservation of the total energy—kinetic, elastic, and magnetic—can be deduced from equations (5.12). In equilibrium, the kinetic energy is assumed to be zero and if, for instance, the plasma is in the shape of a torus, this torus is assumed not to rotate around its axis. Further, a steady state is assumed, the equilibrium properties of which must be tested. It is characterised by the variables ϱ_0, p_0 and $\boldsymbol{B}_0$. In the light of the above assumptions v_0 is zero. The only meaningful equation remaining in the steady state is equation (5.11). This equation gives the condition of static magnetic confinement:

$$\nabla p_0 = -\frac{1}{2\mu}[\nabla B_0^2 - 2(\boldsymbol{B}_0 . \nabla)\boldsymbol{B}_0] \qquad (5.13)$$

A perturbation is defined by some displacement vector $\tilde{\boldsymbol{r}}(x, y, z, t)$ such that a point in the plasma which, in the steady state would be at the location $\boldsymbol{r}_0(x, y, z)$, is at the instant specified at the location $\boldsymbol{r}(x, y, z, t)$ with $\boldsymbol{r} = \boldsymbol{r}_0+\tilde{\boldsymbol{r}}$. The method is analogous to one of classical use in fluid mechanics. From the absence of mass motion in the steady state, it is deduced immediately that $\boldsymbol{v} = \mathrm{d}\tilde{\boldsymbol{r}}/\mathrm{d}t$. The other perturbed variables can be written as:

$$\varrho = \varrho_0+\tilde{\varrho}$$

$$p_r = p_0+\tilde{p} \tag{5.14}$$

$$\boldsymbol{B} = \boldsymbol{B}_0+\tilde{\boldsymbol{B}}$$

A linear approximation is considered; hence, squares and products of perturbations are neglected. By substituting the perturbed quantities in equations (5.12), one obtains for equation (3.32):

$$\tilde{\varrho} = -\nabla.\left(p_0\frac{\mathrm{d}\tilde{\boldsymbol{r}}}{\mathrm{d}t}\right) \tag{5.15}$$

Equation (5.11) yields, after taking into account equation (5.13) and after some elementary algebra:

$$\varrho_0\frac{\mathrm{d}^2\tilde{\boldsymbol{r}}}{\mathrm{d}t^2} = -\nabla\left(\tilde{p}+\frac{\boldsymbol{B}_0.\tilde{\boldsymbol{B}}}{\mu}\right)+\frac{1}{\mu}[\boldsymbol{B}\nabla.\tilde{\boldsymbol{B}}+(\tilde{\boldsymbol{B}}.\nabla)\boldsymbol{B}_0] \tag{5.16}$$

Equation (5.12c) becomes $\boldsymbol{B} = \nabla\times(\tilde{\boldsymbol{r}}\times\boldsymbol{B}_0)$, and equation (5.12d) can be written as $\tilde{\varrho} = (\tilde{p}/\gamma)(\varrho_0/p_0)$. If $\tilde{\varrho}$ is substituted in equation (5.15), one finds for $\tilde{p}$:

$$\tilde{p} = -\gamma p_0\nabla.\tilde{\boldsymbol{r}}-\tilde{\boldsymbol{r}}.\nabla p_0$$

By substituting in equation (5.16), one obtains after expanding the gradient of the scalar product:

$$\varrho_0\frac{\mathrm{d}^2\tilde{\boldsymbol{r}}}{\mathrm{d}t^2} = \nabla(\gamma p_0\nabla.\tilde{\boldsymbol{r}}+\tilde{\boldsymbol{r}}.\nabla p_0)-\frac{1}{\mu}[\boldsymbol{B}_0\times(\nabla\times\tilde{\boldsymbol{B}})+\tilde{\boldsymbol{B}}\times(\nabla\times\boldsymbol{B}_0)] \tag{5.17}$$

Equation (5.17) is an expression for a force per unit volume; if the configuration described by the mean variables corresponds to a stable equilibrium, $\varrho_0\mathrm{d}^2\tilde{\boldsymbol{r}}/\mathrm{d}t^2$ is then a restoring force tending to push back the plasma into its unperturbed state. The force per unit volume appears as a linear function of the displacement $\tilde{\boldsymbol{r}}$ in the perturbation, for each unperturbed configuration. It is not evident *a priori* that such a force density derives from a potential energy; however, in this particular case, this derivation appears as a direct consequence of the laws of physics. A more rigorous treatment consists in showing that the operator of $\tilde{\boldsymbol{r}}$ represented by the right-hand side of equation (5.17) is self-adjoint.

The quantity $-\frac{1}{2}\tilde{\boldsymbol{r}}.\mathrm{d}^2\tilde{\boldsymbol{r}}/\mathrm{d}t^2$ represents an energy density, and in order to calculate the total change in energy involved by the perturbation $\tilde{\boldsymbol{r}}(x, y, z, t)$, integration should be made over the entire volume; thus:

$$\tilde{\mathscr{E}} = -\frac{1}{2}\iiint_{\mathscr{V}}\boldsymbol{r}.\left\{\nabla(\gamma p_0\nabla.\tilde{\boldsymbol{r}}+\tilde{\boldsymbol{r}}\nabla p_0)-\frac{1}{\mu}[\boldsymbol{B}_0\times(\nabla\times\tilde{\boldsymbol{B}})+\tilde{\boldsymbol{B}}\times(\nabla\times\boldsymbol{B}_0)]\right\}\mathrm{d}\mathscr{V} \tag{5.18}$$

One can write:

$$\tilde{\boldsymbol{r}}.\nabla(\gamma p_0\nabla.\tilde{\boldsymbol{r}}+\tilde{\boldsymbol{r}}.\nabla p_0) = \nabla.[\tilde{\boldsymbol{r}}(\gamma p_0\nabla.\tilde{\boldsymbol{r}}+\tilde{\boldsymbol{r}}.\nabla p_0)]-(\gamma p_0\nabla.\tilde{\boldsymbol{r}}+\tilde{\boldsymbol{r}}.\nabla p_0)\nabla.\tilde{\boldsymbol{r}}$$

The integral of the first factor on the right-hand side can be transformed into a surface integral, which vanishes in many particular cases. For instance, p_0 goes asymptotically to zero at infinite distance, or $\tilde{\boldsymbol{r}}.\mathrm{d}\boldsymbol{S}$ is zero along a confining wall. It needs only be retained if the plasma is assumed to have a sharp boundary not corresponding to a wall: this is ideal magnetic confinement. The remaining part of equation (5.18) can be treated similarly:

$$\tilde{\boldsymbol{r}}.[\boldsymbol{B}_0\times(\nabla\times\tilde{\boldsymbol{B}})] = -\nabla\{[(\boldsymbol{r}\times\boldsymbol{B}_0).\tilde{\boldsymbol{B}}]-\tilde{\boldsymbol{B}}.[\nabla\times(\boldsymbol{r}\times\boldsymbol{B}_0)]\}$$

Again, the integral of the first factor on the right-hand side can be transformed for physical reasons into a vanishing surface integral; the second factor is $-\tilde{B}^2$, from the very definition of $\tilde{\boldsymbol{B}}$. In conditions that such both surface integrals vanish, the energy involved is:

$$\tilde{\mathscr{E}} = \frac{1}{2}\iiint_{(\mathcal{V})}\left\{\frac{1}{\mu}\{\tilde{B}^2-\tilde{\boldsymbol{r}}.[(\nabla\times\boldsymbol{B}_0)\times\tilde{\boldsymbol{B}}]\}+[\gamma p_0\nabla.\tilde{\boldsymbol{r}}+\tilde{\boldsymbol{r}}.\nabla p_0]\nabla.\tilde{\boldsymbol{r}}\right\}\mathrm{d}\mathcal{V} \qquad (5.19)$$

According to the principles given in Section 5.1, any departure from the ideal position in a stable equilibrium will involve an increment in potential energy. Hence a sufficient criterion of instability is the discovery of *one* function, $\tilde{\boldsymbol{r}}(x, y, z, t)$, for which the above expression is not positive. Strictly speaking, the existence of stability could never be proved, because absolutely all the possible configurations of perturbations would have to be tested. In practice, however, the situation is not that rigorous because of symmetry conditions. In a cylindrical configuration, for instance, any perturbation can be expressed as a sum or a series of eigenfunctions such as:

$$\tilde{\boldsymbol{r}} = \boldsymbol{r}_0(r)\cos k_z z\cos n_\theta\,\theta \qquad (5.20)$$

where k_z and n_θ are integral values. In a torus, an equivalent expression for the eigenfunctions is easily written. The situation can be simplified in any case where, for instance, the stability increases together with the indices k_z and n_θ so that only one or a few integral values have actually to be tested.

A method of interest in practice for the elimination of grossly unstable systems consists of imposing an arbitrary extra condition to $\tilde{\boldsymbol{r}}$. Since the existence of *one* unstable perturbation is sufficient for the system to be declared unstable, the existence of such a condition does not matter in theory, but can simplify the investigation in practice. Generally, the limiting condition used is $\nabla.\tilde{\boldsymbol{r}} = 0$. By deriving with respect to time, taking into account $\boldsymbol{v} = \mathrm{d}\tilde{\boldsymbol{r}}/\mathrm{d}t$, it can easily be interpreted physically. When this is satisfied, all the 'elastic' part of equation (5.19) vanishes; the 'magnetic' part is also simplified; further, one can obtain:

$$\tilde{\boldsymbol{B}} = (\boldsymbol{B}_0.\nabla)\tilde{\boldsymbol{r}}-(\tilde{\boldsymbol{r}}.\nabla)\boldsymbol{B}_0$$

Of course, if no perturbation satisfying this limiting condition leads to instability, the test is inconclusive.

Extensive application of this energy principle has been made to test certain configurations for gross hydromagnetic instabilities. In particular, the cylindrical configuration has been tested in many conditions to demonstrate the possibility of completely stabilising the pinch. This has led to the Suydam[42] criterion of stability, which can be written, with evident notations:

$$\frac{2\mu}{B_z^2}\frac{\mathrm{d}p_r}{\mathrm{d}r}+\frac{r}{4}\frac{\mathrm{d}}{\mathrm{d}r}\left[\ln\left(\frac{B_\theta}{rB_z}\right)^2\right]\geqslant 0 \qquad (5.21)$$

This is a *necessary* criterion for the perturbations of all orders described by equation (5.20) to be stable; from a strictly MHD point of view this criterion is insufficient, to say nothing about other types of instabilities.

The more frequent reasons for which the MHD model may not hold are all related to the corpuscular structure of the plasma, and fall into two main categories.

(1) Phenomena related to charge separation, as in the electrostatic plasma oscillations; owing to this restriction, the MHD model fails for time scales of the order of the electrostatic oscillation period and for wavelengths of the order of and below the Debye screening length.
(2) Phenomena related to the gyration of the particles in the magnetic field.

5.4 THE WAVE APPROACH[5, 7, 9,]

It is possible to overcome the deficiencies of category (1) mentioned above by regarding the instability as a self-growing wave. However, complications arise because the problem becomes related to that of wave propagation in guides, i.e. the consideration of conditions at the limits is necessary. The true conditions at a wall are often replaced by fictitious conditions at the 'plasma–vacuum' interface, i.e. conditions of ideal magnetic confinement are again assumed. In both cases, the major difference with respect to free propagation is that a convenient eigenfunction should replace the plane wave having a constant amplitude all over the space or having an amplitude modified only by the self-attenuation of the wave. Apart from this restriction, the method is identical to that outlined in Chapter 4, i.e. the introduction of Fourier components enables one to replace, after the usual substitution of $j\omega$ for $\partial/\partial t$ and $-j\boldsymbol{k}$ for ∇, the differential equation by corresponding algebraic equations. Again, the condition of non-trivial solution is that a certain determinant made with the coefficients of the known variables should vanish, thus yielding the dispersion relation. Instabilities are detected as spontaneously growing waves. For instance, some real values of the components of $\boldsymbol{k}$ are imposed, taking into account the configuration; the dispersion relation is solved for ω, which is generally a complex quantity. Positive imaginary parts correspond to growing waves. The solution of a given problem can take advantage of the properties of the direct and inverse Laplace transforms.

When there is plasma–vacuum interface in the model, it might seem that the perturbation results in a deformation of this boundary. Although this is often true physically, such deformation is not usually considered in the theoretical model, in which it might result in insurmountable mathematical difficulties. Instead, a change of the relevant parameters, similar to that expressed by equations (5.14), is assumed within the plasma, which is further assumed to keep its shape during the initial phase of the perturbation. Physically, a deformation of the fictitious boundary would eventually result, given a sufficient time, and in a manner similar to that of the simplified models.

The wave approach can be used together with equations (5.12). This enables the prediction of instabilities related to Alfvén waves and magnetosonic waves and the modifications thereof when the magnetic field is oblique with respect to the propagation vector. Many other refinements are successfully applied, such as the consideration of a finite viscosity, and finite electrical and thermal conductivities.

Equations (5.12) can also be replaced by a different set of equations in which, for instance, equations (5.12a) and (5.12b) are replaced by four equations, corresponding to the first and second momenta of the general Boltzmann equation, written successively for the electrons and the positive ions. Equation (5.12d) must also be modified in a similar manner. In such conditions, the so-called electrostatic instabilities, related to the electrostatic types of propagation described in Chapter 4 can be predicted in addition to the MHD instabilities.

The consideration of an anisotropic pressure leads to further refinements of the model. It is understandable that a great effort is being devoted to improving the theory of waves in plasmas, since few waves have been studied experimentally or are even detectable. Many types of waves hardly known as such from the experimental point of view, can become important when the instability aspect of propagation in bounded plasmas is introduced.

5.5 THE BOLTZMANN EQUATION APPROACH[20, 21]

The Boltzmann equation approach to the problem of plasma instabilities is much more rigorous than the two preceding ones. Not only does it take into account phenomena of charge separation similar to the electrostatic waves, but also it considers phenomena related to the gyration of the particles in the magnetic field, especially if the radius of gyration cannot be regarded as infinitely small with respect to the other parameters involved. Such effects can be of significant importance in some cases. Often, the ideal quiescent fluid of the MHD approximation is stable in a given configuration for zero resistivity (frozen magnetic lines of force) but becomes unstable for a finite resistivity. However, the influence of a finite radius of gyration counteracts partially or totally this instability, because the same particle would travel in different regions, and can attain a kind of average stability, even if it crosses briefly regions of theoretically gross instability. Also, the longitudinal component of the motion along the magnetic lines of force can have a stabilising effect by averaging some influence over regions of instability combined with regions of stability.

Unfortunately, the formalism of the Boltzmann equation approach is complicated, except in a small number of exceptional cases. Rosenbluth and Rostoker's method, for instance, is one of the simplest. In their treatment[38] they consider very hot plasmas, in which electrons have large speeds, hence large gyration radii. On the other hand, the complication created by the assumption of a hot plasma is partially compensated by the fact that the probability of collision in a fully ionised plasma decreases drastically with increasing electronic velocity; hence, when the temperature is sufficiently high, a collisionless picture of the plasma becomes sufficiently valid and Vlasov's equation is then used instead of Boltzmann's. Let i be an index which is successively replaced by $(-)$ for the electronic component and by $(+)$ for the ionic component. Thus, the two Vlasov's equations can be written in one:

$$\frac{\partial f_i}{\partial t}+(\boldsymbol{v}_i\,.\,\nabla)f_i+(\boldsymbol{F}_i\,.\,\nabla_v)f_i = 0 \tag{5.22}$$

The acceleration $\boldsymbol{a}_i$ is given by $\boldsymbol{a}_i = q_i(\boldsymbol{E}+\boldsymbol{v}_i\times\boldsymbol{B})/M_i$. The fields are related to the charge density ϱ_e and the current density $\boldsymbol{J}$ by Maxwell's relations for a negligible displacement current:

$$\nabla\,.\,\boldsymbol{E} = \frac{e}{\varepsilon_0}(n^+-n^-) = \varrho_e/\varepsilon_0 \tag{5.23}$$

$$\nabla\times\boldsymbol{B} = \mu_0\boldsymbol{J}$$

Those quantities are in turn related to the Boltzmann distribution functions by the following relations, written for smoothed charged distributions:

$$\begin{aligned}\varrho_e &= \sum_i q_i \iiint f_i\, \mathrm{d}v_i\\ \boldsymbol{J} &= \sum_i q_i \iiint f_i \boldsymbol{v}_i\, \mathrm{d}v_i\end{aligned} \tag{5.24}$$

For a two-component plasma the sums comprise only two terms, one for the electrons and the other for the positive ions.

For each particle species, the Boltzmann distribution function f_i and the acceleration $\boldsymbol{a}_i$ of the particle are decomposed into an unperturbed and a perturbed component in the same manner as for equations (5.14):

$$\begin{aligned}f_i &= f_{0i}+\tilde{f}_i\\ \boldsymbol{a}_i &= \boldsymbol{a}_{0i}+\tilde{\boldsymbol{a}}_i\end{aligned} \tag{5.25}$$

The unperturbed f_{0i}, $\boldsymbol{a}_{0i}$ quantities must satisfy Vlasov's equation (5.22):

$$(\boldsymbol{v}_{0i}\,.\,\nabla)f_{0i}+(\boldsymbol{a}_{0i}\,.\,\nabla_v)f_{0i} = 0 \tag{5.26}$$

assuming implicitly that the unperturbed state is a steady state. After introducing equations (5.25) into equation (5.22), neglecting second-order terms, and eliminating zeroth order terms by subtracting equation (5.26), one obtains the equation describing the time revolution of the perturbation:

$$\frac{\partial \tilde{f}_i}{\partial t}+(\boldsymbol{v}_{0i}\,.\,\nabla)\tilde{f}_i+(\boldsymbol{a}_{0i}\,.\,\nabla_v)\tilde{f}_i = -(\tilde{\boldsymbol{a}}_i\,.\,\nabla_v)f_{0i} \tag{5.27}$$

The unperturbed acceleration is:

$$\boldsymbol{a}_{0i} = \frac{q_i}{M_i}(\boldsymbol{E}_0 + \boldsymbol{v}_{0i} \times \boldsymbol{B}_0)$$

in which the fields can be calculated as functions of the f_{0i}'s by using equations (5.23) and (5.24) written for the steady state values. The accelerations $\boldsymbol{a}_{0i}$ of perturbation can be evaluated by using a set of equations similar to the above but written with perturbation values. This is a direct consequence of the linear character of the equations for the field and for the motion.

Consider now the evaluation of the $\tilde{f}_i$'s by using equation (5.27). For each $\tilde{f}_i$, the left-hand side of equation (5.27) can be regarded as a total time derivative of the relevant $\tilde{f}_i$ in the μ-space (see Section 3.2), assuming that the observer follows the relevant species of particles in its motion. Hence, equation (5.27) can be written as:

$$\left(\frac{\mathrm{d}\tilde{f}_i}{\mathrm{d}t}\right)_{\text{motion}} = -(\tilde{\boldsymbol{a}}_i . \nabla_v) f_{0i} \tag{5.28}$$

In this equation, the right-hand side cannot be regarded as known. One implicitly assumes that the problem has first to be solved for the unperturbed state; hence the f_{0i}'s are known functions of x, y, z, v_x, v_y, v_z, but the $\tilde{\ }_i$'s are calculable only in terms of the perturbations $\tilde{f}_i$'s. Consequently, equation (5.28) is actually a set of implicit equations with the f_i's still underwritten in the right-hand side. The next step consists of integrating equation (5.28) along the orbits. As a first approximation, all the drifts can be neglected, hence the orbit in the physical space is a circle, and the orbit in the μ-space is projected as a circle in both the physical and the hodographic space, the hodographic representation of a circle with constant angular velocity being also a circle. The result of such an integration can be expressed as:

$$\tilde{f}_i = \tilde{f}_{i0} - \int_{t_0}^{t} (\tilde{\boldsymbol{a}}_i . \nabla_v) f_{0i}\, \mathrm{d}t \tag{5.29}$$

An implicit equation such as equation (5.29) can be solved by using Laplace transforms. Only a brief account of the method used can be given here; for more details, see the original paper by Rosenbluth and Rostoker[38]. As in the MHD approximation or the wave approach, definite 'modes' of instability corresponding to eigenfunctions should first be assumed. The eigenfunctions are written as:

$$\tilde{f}_i = \tilde{f}_{i0}(\boldsymbol{r}, \boldsymbol{v}) \exp(\nu t)$$
$$\tilde{\boldsymbol{E}} = \tilde{\boldsymbol{E}}_0(\boldsymbol{r}) \exp(\nu t) \tag{5.30}$$
$$\tilde{\boldsymbol{B}} = \tilde{\boldsymbol{B}}_0(\boldsymbol{r}) \exp(\nu t)$$

Approximate values, for instance those at the centre of the orbit modified by first order corrections in the orbit radius, can be assumed, and the different coefficients 'fitted' together in order to satisfy equations (5.23) and (5.24) written for the perturbations. Hence, for each mode, equation (5.29) becomes an algebraic equation, giving the value of ν. Real and positive values of ν or its complex values with a positive real part correspond to growing modes, thus instabilities.

5.6 CLASSIFICATION OF PLASMA INSTABILITIES

Some physicists have devoted their efforts to a better understanding of the very notion of instability, the classification of instabilities, and the prediction of still unknown ones. Two questions remain at the heart of the problem. First, is the plasma instability an essential feature of any plasma configuration confined by a magnetic field or are there absolutely stable confined configurations? Second, is it possible that new types of instabilities will again be discovered or are there valid reasons to believe that the known list of instabilities is approximately complete?

For the first question, although theoretical research indicates that instability is not an essential feature of the confined plasma, few physicists still believe that absolutely stable confinements do exist. Speaking in terms of the parameter ν introduced in equations (5.30), it is generally believed that configurations can be achieved in which the most unstable modes will have ν's with very small positive real parts. The relevant instabilities would grow exponentially or quasi-harmonically with time constants not too short say with respect to the average lifetime of a fuel nucleus in a thermonuclear reactor, thus allowing the latter to 'burn' with a satisfactory efficiency before the confinement is destroyed.

To answer the second question, it is essential to understand what drives an instability. In principle, an instability results from the conversion of potential energy into kinetic energy. The energy principle illustrates this well, in the limited scope of the MHD approximation. To be more accurate, the problem is related to the thermodynamic concept of 'free energy'. A thermonuclear plasma, for instance, must have an enormous free energy per unit mass. Thermonuclear plasmas confined by magnetic fields are probably non-existent throughout nature. This does not preclude entirely the possibility of a stable confinement lasting for a sufficient length of time, at least for small perturbations. The free energy of a confined plasma can be resolved into five 'energy reservoirs', each reservoir being able to drive a relevant class of instabilities.

(1) The first reservoir arises from the *pressure gradient* and seems to be inevitable because the plasma has necessarily a finite extension in space. The relevant class of instabilities is designated as '*universal*'[40]. Under the influence of the pressure gradient, the particles tend to drift across the magnetic lines of force, but they are deflected differently owing to their sign, hence a current perpendicular to the pressure gradient and the magnetic field arises. The interaction of this current with the magnetic field is a confining force. However, theoretical investigations show that under certain circumstances, growing undulatory fluctuations can degenerate into significant instabilities, which have in many cases sufficiently long time constants. The knowledge of these modes of instability is still limited. However, it is likely that they cannot be avoided, although it is hoped to increase almost arbitrarily their time constants.

(2) Another reservoir results from the interaction of the plasma as a whole with its confining magnetic field. The plasma as a whole is diamagnetic, that is,

the orbital motion of the particles tends to generate magnetic fields opposed to the confining one. It is known from the electromagnetic field theory that a diamagnetic body tends to occupy regions of minimum magnetic field. Since the confining magnetic field must somehow tend towards zero at infinite distance, the plasma is always more stable outside the confining zone. The simplest example of this instability is the *flute* or ripple instability in which the surface separating the plasma from a vacuum becomes rippled. First, the situation of the 'hole on top of a mountain' seen in Figure 5.1c can be duplicated with magnetic fields to overcome this type of instability. 'Minimum-B' configurations have been obtained experimentally. When the particles drift significantly along heterogeneous lines of force, the 'minimum-B' condition ceases to be absolute; instead, the quantity ds/B, integrated over the path s of the gyration centre, must be minimum. This is in general easier to achieve than a strict 'minimum-B'. Second, since the drifting plasma tends to carry the magnetic lines of force with it, it is advisable to achieve configurations in which the magnetic lines of force are 'interwoven' so that the possibility of drifting becomes limited.

(3) The third reservoir results from anisotropic and non-Maxwellian velocity distributions. In statistical physics, it is demonstrated that the Maxwellian distribution of velocities corresponds to the maximum entropy and any decrease in the entropy leads to an increase in the free energy.

(4) A fourth reservoir, more or less related to the preceding one, results from directed beams of particles and is related to the beam–plasma interaction problem.

(5) The last reservoir is represented by the magnetic field energy. It is indeed more likely to drive instabilities if a significant part or the totality of the magnetic field results from currents carried by the plasma itself. This class of instabilities comprises the most catastrophic ones: the *kink* and the *sausage* instabilities.

Plasmas are extremely inventive in finding detailed ways to exploit what is thermodynamically permissible. In other words, once an energy reservoir exists, one can be reasonably sure that the relevant class of instabilities exists too. However, since the number of terms in the expression of the free energy is finite, the number of instability modes is necessarily finite too.

In the following the three most important instabilities will be treated in some detail; these are the ripple instability, the kink instability, and the sausage instability.

The ripple instability

Consider a tri-rectangular frame (Figure 5.2) where Oz is vertical, the gravitational field $\boldsymbol{g}$ points toward $-z$ and a homogeneous magnetic field is in the x-direction. A homogeneous plasma exists above the xy-plane and a vacuum below. Assume that the plasma pressure is negligible with respect to the magnetic pressure so that the magnetic field becomes constant. In equilibrium, the particles gyrate and drift horizontally; none of those motions affects the neutrality of the plasma, noting the fact that it is supported against gravity by the magnetic pressure.

This system can be studied as a heavy fluid supported by a lighter one (the magnetic field) in pressure equilibrium, and the macroscopic equations can be

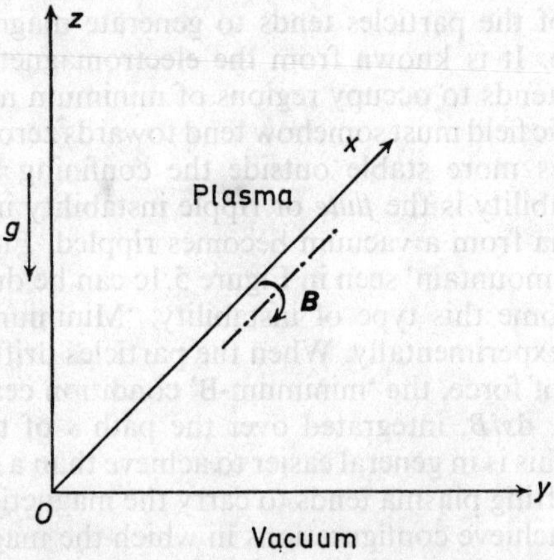

Fig. 5.2. Plane separating a plasma from vacuum before the occurrence of a ripple instability. The plasma is homogeneous and above the xy-plane. The vacuum is below

used for the particle guiding centres. If the unstable motion of the plasma is slow compared to the gyrofrequency, one can write for the equation of continuity:

$$\frac{\partial n}{\partial t}+\boldsymbol{v}.\nabla n+n\nabla.\boldsymbol{v} = 0 \tag{5.31}$$

where n is the number density and:

$$\boldsymbol{v} = \frac{\boldsymbol{g}\times\boldsymbol{i}}{\omega_c}+\frac{\boldsymbol{E}\times\boldsymbol{i}}{B}$$

with $\boldsymbol{g}$ as the gravitational field, $\boldsymbol{E}$ as the electric field, $\boldsymbol{B}$ as the magnetic field, $\boldsymbol{i}$ as unit vector in the x-direction and ω_c as the gyrofrequency. Now consider small density perturbations around equilibrium as shown by equations (5.14) $n = n_0(z)+\tilde{n}(y, z, t)$ with $n_0 \gg \tilde{n}$. After taking into account the relation $\nabla\times\boldsymbol{E} = 0$, and neglecting the second-order perturbations, equation (5.31) becomes:

$$\frac{\partial \tilde{n}}{\partial t}+\frac{g}{\omega_c}\frac{\partial \tilde{n}}{\partial t}-\frac{E_y}{B}\frac{\partial n_0}{\partial z} = 0 \tag{5.32}$$

The solution of this equation is of the form:

$$\tilde{n} = \tilde{n}_0 \exp{(\omega t+\mathrm{j}ky)} \tag{5.33}$$

and equation (5.32) becomes:

$$\left(\omega+\mathrm{j}\frac{kg}{\omega_c}\right)\tilde{n} = \frac{E_y}{B}\frac{\partial n_0}{\partial z}$$

If now the subscript *e* is used for electrons, and *i* for ions, the perturbations on the ionic and the electronic densities become respectively:

$$\tilde{n}_i = \frac{\dfrac{E_x}{B}\dfrac{\partial n_{0i}}{\partial z}}{\omega + \mathrm{j}\dfrac{kg}{\omega_{ci}}} \tag{5.34}$$

$$\tilde{n}_e = \frac{\dfrac{E_x}{B}\dfrac{\partial n_{0e}}{\partial z}}{\omega + \mathrm{j}\dfrac{kg}{\omega_{ce}}} \tag{5.35}$$

Assuming that the plasma is on the average neutral and that the phase velocity of the wave is much larger than g/ω_{ci}, equations (5.34) and (5.35) can be combined by Poisson's law to yield:

$$\frac{\mathrm{d}^2V}{\mathrm{d}z^2} - k^2\left[1 + \frac{1}{\omega^2}\left(\frac{M_i g}{\varepsilon_0 B^2}\frac{\partial n_0}{\partial z}\right)\right]V = 0 \tag{5.36}$$

where V is the electric potential and M_i is the mass of an ion. Equation (5.36) can be written in the form:

$$\frac{\mathrm{d}^2V}{\mathrm{d}z^2} - k^2\mathcal{A}V = 0$$

The potential should be bounded, that is, it cannot increase indefinitely with z, thus $\mathcal{A} \leqslant 0$ and for the fastest mode corresponding to a very small constant of propagation $\mathcal{A} = 0$, and consequently:

$$\omega^2 = -\frac{M_i g}{\varepsilon_0 B^2}\frac{\partial n_0}{\partial z} \tag{5.37}$$

If ω^2 is negative, the perturbation will be oscillating and the system will be stable, whereas if ω^2 is positive the amplitude of the perturbation will grow with time, resulting in an instability of the system. Equation (5.37) can also be written as:

$$\omega^2 = -\frac{g}{n_0}\frac{\partial n_0}{\partial z} \tag{5.38}$$

and the condition of stability becomes such that the gravity should be in the same direction as $\partial n_0/\partial z$. If g is in the opposite direction to $\partial n_0/\partial z$ the system becomes unstable. In other words, the ripple instability, which is also called the Taylor instability, arises from the drifts and the motion becomes unstable when the drift $\boldsymbol{E}\times\boldsymbol{B}$ becomes amplified. This instability is represented in Figure 5.3. Thus, for the plasma to be stable, the lines of force of the confining magnetic field should always have their centre of curvature outside the plasma.

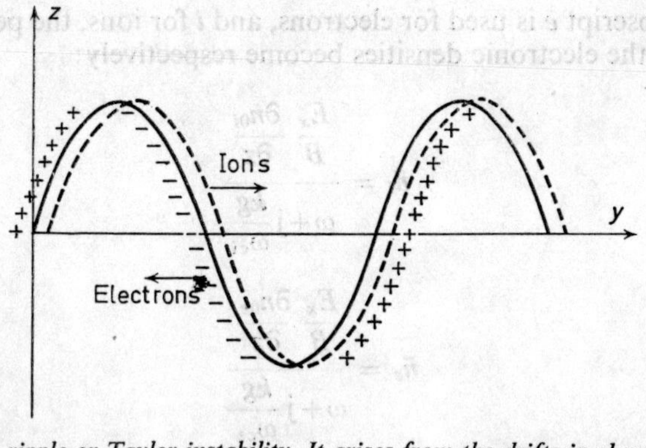

Fig. 5.3. The ripple or Taylor instability. It arises from the drifts in charged particles

The flute instability[29]

The ripple instability occurs also in configurations where the surface separating the plasma from the vacuum is not a plane. Such configurations are cylindrical e.g. pinch configuration and mirror machines (see Chapter 7). In the case of

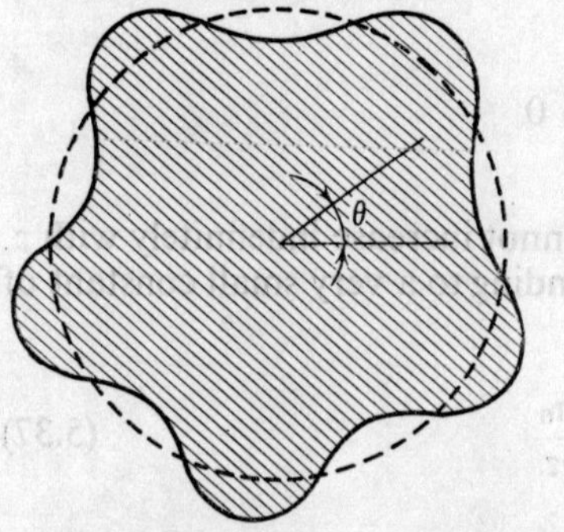

Fig. 5.4. The flute instability in a pinch configuration for $n_\theta = 5$. This figure is a cross-section of the plasma column

a pinch configuration, the ripple instability is called *flute* instability. Figure 5.4 illustrates the case of a flute instability in a pinch configuration for $n_\theta = 5$, where n_θ is given in equation (5.20).

The kink instability (Figure 5.5)

This is a ripple instability for $n_\theta = 1$. It can be assumed that the presence of the kink does not change significantly the plasma pressure at the interface. On the other hand, from a simple examination of Figure 5.5, it is clear that the magnetic lines of force are more concentrated on the concave portion of the kink and less on the convex portion, than in the case of a normal rectilinear column. Hence, there is a net unbalance, the result of which is to amplify the kink, until

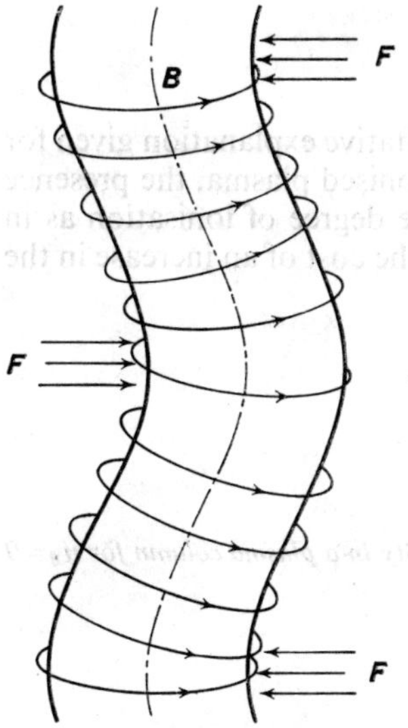

Fig. 5.5. The kink instability in a plasma column for $n_\theta = 1$

eventually the column is broken. The rupture can result in a strong inductive effect, causing the acceleration of privileged classes of ions; with the heavy isotopes of hydrogen, this can result in neutron emission from fusion reactions.

If g is replaced by v^2/r_c in equation (5.38), where r_c is the radius of curvature of the magnetic field lines, it can be seen that r_c should have the same sign as $\frac{1}{n_0}\frac{\partial n_0}{\partial z}$ for the system to be stable. That is, if $\frac{1}{n_0}\frac{\partial n_0}{\partial z}$ is positive, the system will be unstable for $r_c < 0$ and stable for $r_c > 0$, as noted before. It can also be shown that the condition of stability can be expressed as:

$$\int_1^2 (\mathrm{d}l/r_c r_0 B) > 0 \tag{5.39}$$

where $\mathrm{d}l$ is along the magnetic line of force such that $\mathrm{d}l \cong r_c \mathrm{d}\theta$ and r_0 is the radius of the column. From Maxwell's equation, Br^2 is constant and equation (5.39) becomes:

$$\int_1^2 (\mathrm{d}\theta/B^{3/2}) > 0 \tag{5.40}$$

where 1 and 2 are the turning points on the trajectory. Neither the condition $r_c > 0$ nor equation (5.40) are satisfied in a plasma column. Therefore, the kink instability cannot be avoided in this configuration unless some new method of stabilisation is introduced into the system (see Chapter 7).

The sausage instability (Figure 5.6)

This is a ripple instability for $n_\theta = 0$; the same qualitative explanation given for the kink instability holds here. In an already fully ionised plasma, the presence of a constriction cannot result in an increase of the degree of ionisation as in an ordinary discharge; the current can only flow at the cost of an increase in the

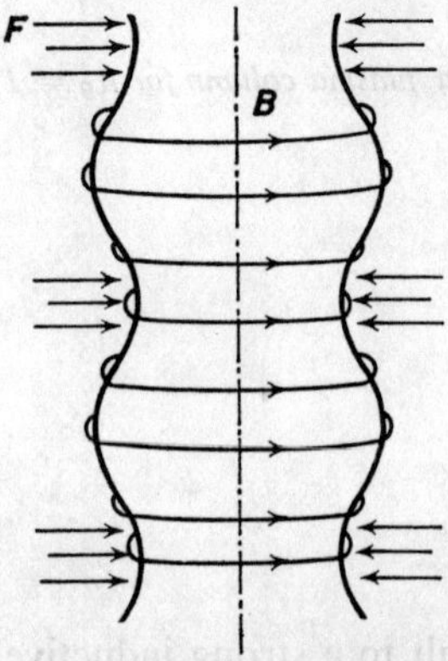

Fig. 5.6. The sausage instability in a plasma column for $n_\theta = 0$

drift velocity. Therefore, the plasma pressure is fairly uniform, whereas the magnetic pressure is greater where the discharge is already more constricted, and smaller where it is less constricted than the average. Hence, any irregularity

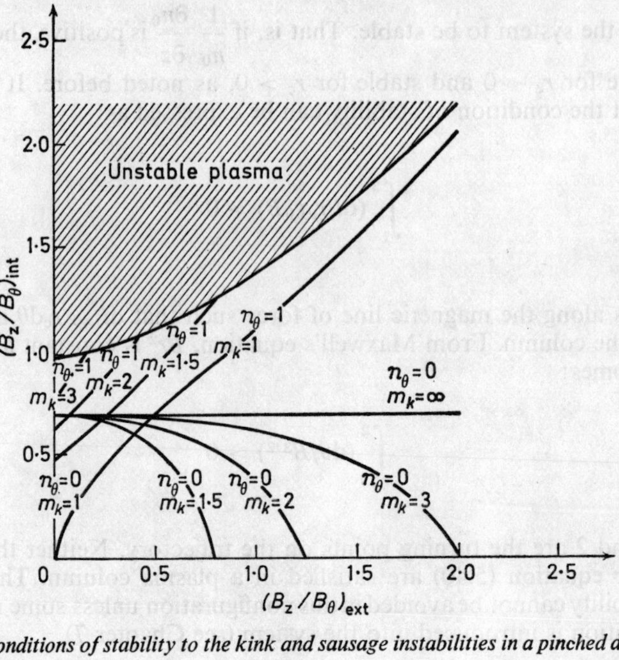

Fig. 5.7. Conditions of stability to the kink and sausage instabilities in a pinched discharge[44]

in the column diameter tends to be amplified. Again, the eventual state is a rupture of the column, with acceleration of privileged classes of particles, and in heavy hydrogen isotopes, neutron emission by a non-thermonuclear process.

The problem of stability of the types $n_\theta = 0$ and $n_\theta = 1$ has been studied by many authors[31, 43]. Tayler[44], for instance, used a normal mode method to study the stability of a pinch geometry surrounded by a conducting wall, and where the annular plasma sheath is immersed in an internal and an external magnetic field. A pinched plasma can remain only if the external pressure is equal to (or larger) than the internal pressure, thus $(B_\theta^2 + B_{\text{zext}}^2) \geqslant B_{\text{zint}}^2$. The results of Tayler's study are reported in Figure 5.7 for both $n_\theta = 0$ and $n_\theta = 1$ and for the pinch ratio $m_k = r_0/r$ as parameters, r_0 being the radius of the discharge. The stability area is limited by the curve $(B_\theta^2 + B_{\text{zext}}^2) = B_{\text{zint}}^2$ and the curve given by n_θ and m_k. This area decreases with m_k increasing, and thus, for decreasing values of the axial external magnetic field. For $m_k = 5$, stability will be possible only if there is no axial external magnetic field.

5.7 INSTABILITY AND TURBULENCE[2, 4, 28]

Turbulence results from instability when a flow changes its state from laminar to turbulent[13, 14]. This problem was studied in hydrodynamics first by Reynolds[32, 36]. To study the transition of a fluid (or a plasma) from the laminar to the turbulent state it is convenient to investigate a parameter that increases with the instability. In hydrodynamics, the Reynolds number is such a parameter.

The occurrence of turbulent flow is a result of the viscosity of the fluid. Reynolds used an apparatus similar to that shown in Figure 5.8 to demonstrate

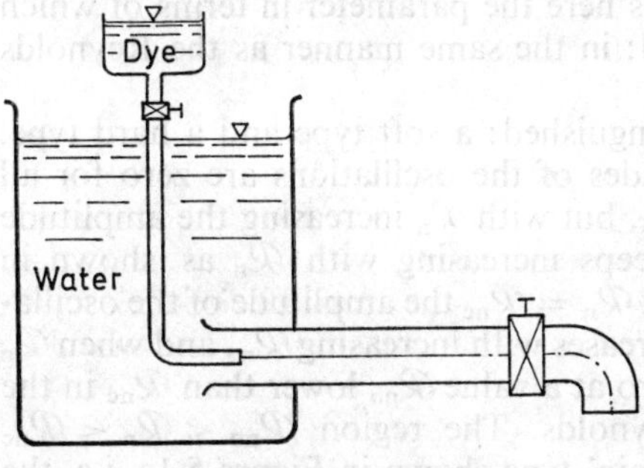

Fig. 5.8. Reynolds' apparatus demonstrating turbulent and laminar flows

that at low velocities a thin filament of dye issuing from the orifice was maintained intact in the pipe along its axis. However, when greater velocities were attained, reaching a critical value v_c, the filament broke up and diffused in the flowing water of the pipe. Reynolds found that this critical velocity was a function of the degree of quiescence of the water and increased with the increasing quiescence. In the case where the dye filament was not mixing with the water, Reynolds deduced that this was a laminar flow regime; and when the filament was mixing, the regime was termed turbulent. When the flow is turbulent, it is necessary to decrease its velocity to a *lower critical value* to make it laminar, the *upper critical velocity* being the value of the velocity making the transition from laminar to turbulent. Reynolds' findings were generalised by the introduction of a dimensionless term $\mathcal{R}_n$ called the Reynolds number and defined as $\bar{v}d\varrho/v$ where $\bar{v}$ is the mean velocity of the fluid in the pipe, d is the diameter of the pipe, ϱ and v are respectively the density and the viscosity of the fluid flowing in the pipe. This number, $\mathcal{R}_n$, determines the upper and lower critical velocities for all fluids flowing in all sizes and shapes of pipes.

Turbulence has been studied in plasma flows ever since 1949 when Bohm and his coworkers[8] showed that random oscillations of the electric field created by an instability enhanced the diffusion in a plasma flow. Therefore, plasma turbulence comes about as a result of the excitation of a large number of collective degrees

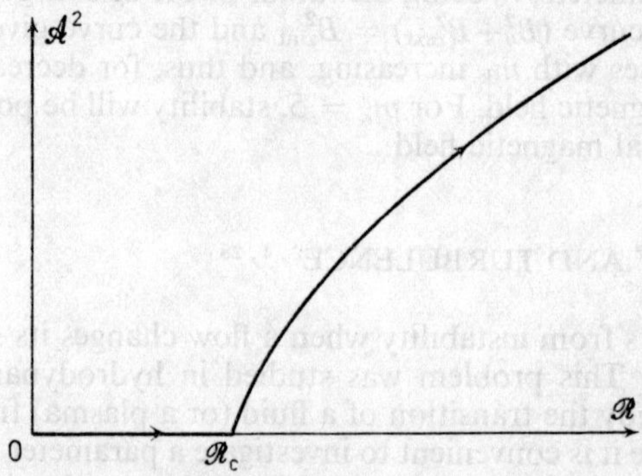

Fig. 5.9. Soft excitation[28]. The amplitude $\mathcal{A}^2$ of the oscillation is zero for all values of $\mathcal{R}$ lower than $\mathcal{R}_c$. The amplitude starts increasing at $\mathcal{R} = \mathcal{R}_c$

of freedom. The magnetic field becomes here the parameter in terms of which the turbulence of the plasma is studied; in the same manner as the Reynolds number for a hydrodynamic flow.

Two types of excitation can be distinguished: a soft type and a hard type. In the soft excitation type the amplitudes of the oscillations are zero for all values lower than the critical value $\mathcal{R}_{nc}$, but with $\mathcal{R}_n$ increasing the amplitude starts to increase at $\mathcal{R}_n = \mathcal{R}_{nc}$ and keeps increasing with $\mathcal{R}_n$ as shown in Figure 5.9. In the hard excitation type at $\mathcal{R}_n = \mathcal{R}_{nc}$ the amplitude of the oscillation jumps to a finite value and then increases with increasing $\mathcal{R}_n$, and when $\mathcal{R}_n$ is decreased, the amplitude drops to zero at a value $\mathcal{R}_{no}$ lower than $\mathcal{R}_{nc}$ in the same manner as was noticed by Reynolds. The region $\mathcal{R}_{no} < \mathcal{R}_n < \mathcal{R}_{nc}$ corresponds to the 'hole on the mountain' type shown in Figure 5.1c, i.e. the system is stable for small perturbations but becomes unstable as the perturbation becomes larger than a critical value. This excitation is represented in Figure 5.10. It was shown by Landau[32] that the transition from the laminar state to the turbulent state is a result of an instability as well. Landau considered the soft excitation of a system to be comprised of an infinite number of degrees of freedom. For $\mathcal{R}_n < \mathcal{R}_{nc}$ there were no degrees of freedom in excess, but as $\mathcal{R}_n > \mathcal{R}_{nc}$, new modes are excited at the critical values $\mathcal{R}_{n1}, \mathcal{R}_{n2}, \ldots$ representing additional degrees of freedom to the system; further and further degrees of freedom are thus generated until the turbulent state is reached. When $\mathcal{R}_n$ is not much larger than $\mathcal{R}_{n1}$, the amplitude of the turbulence can be studied by a

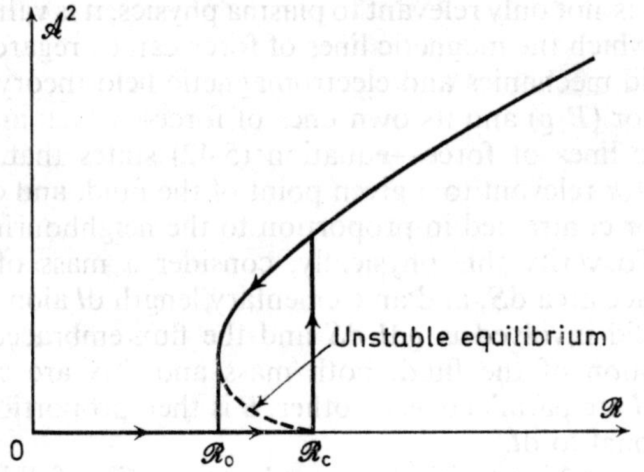

Fig. 5.10. Hard excitation[28]. The system is stable for small perturbations but becomes unstable as the perturbation becomes larger than the critical value $\mathcal{R}_c$

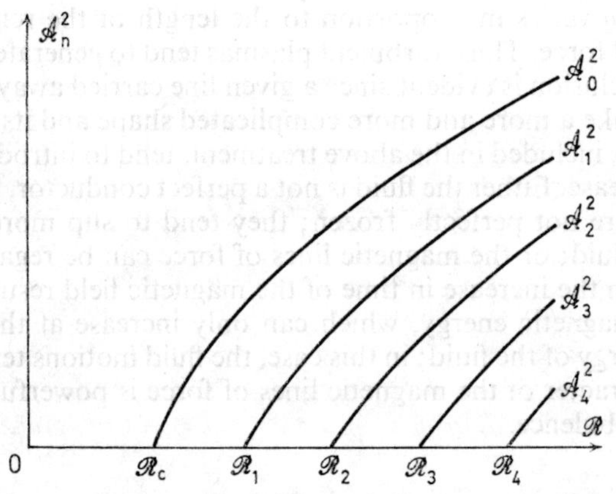

Fig. 5.11. Soft excitation of a system having a large number of degrees of freedom

quasi-linear method[41]. The excitation of a system with a large number of degrees of freedom is represented in Figure 5.11.

The variation of the magnetic induction with the plasma density can be obtained by considering equation (3.58) in the case of a negligible resistivity. It becomes:

$$\partial \boldsymbol{B}/\partial t = \nabla \times (\boldsymbol{v}_d \times \boldsymbol{B}) \tag{5.41}$$

expressing that the magnetic lines of force are 'frozen' in the plasma. Taking into consideration equations (2.15) and (3.35), equation (5.41) becomes after expanding the curl of the cross product:

$$\mathrm{D}\boldsymbol{B}/\mathrm{D}t = (\boldsymbol{B}.\nabla)\boldsymbol{v}_d - \boldsymbol{B}(\nabla.\boldsymbol{v}_d) = (\boldsymbol{B}.\nabla)\boldsymbol{v}_d + (\boldsymbol{B}/\varrho)(\mathrm{D}\varrho/\mathrm{D}t)$$

where $\mathrm{D}/\mathrm{D}t = \partial/\partial t + \boldsymbol{v}_d.\nabla$ as usual. After some elementary rearrangements this equation becomes:

$$\frac{\mathrm{D}}{\mathrm{D}t}\left(\frac{\boldsymbol{B}}{\varrho}\right) = \left[\left(\frac{\boldsymbol{B}}{\varrho}\right).\nabla\right]\boldsymbol{v}_d \tag{5.42}$$

Equation (5.42) is not only relevant to plasma physics, it is valid for any conducting medium in which the magnetic lines of force can be regarded as frozen, and is known in fluid mechanics and electromagnetic field theory. Considering the field of the vector ($\boldsymbol{B}/\varrho$) and its own lines of force—which are in fact identical to the magnetic lines of force—equation (5.42) states that, during the fluid motion, the vector relevant to a given point of the fluid, and carried away with it, is stretched or contracted in proportion to the neighbouring segment of the line of force. To verify this physically, consider a mass of fluid carrying a contour of surface area $\mathrm{d}\boldsymbol{S}$, and an elementary length $\mathrm{d}\boldsymbol{l}$ along the line of force. The mass of fluid involved is $\varrho \mathrm{d}\boldsymbol{l}.\mathrm{d}\boldsymbol{S}$ and the flux embraced by $\mathrm{d}\boldsymbol{S}$ is $\boldsymbol{B}.\mathrm{d}\boldsymbol{S}$. During the motion of the fluid, both mass and flux are conservative since vectors $\boldsymbol{B}$ and $\mathrm{d}\boldsymbol{l}$ are parallel to each other. $\boldsymbol{B}$ is then proportional to $\varrho \mathrm{d}\boldsymbol{l}$, hence, $\boldsymbol{B}/\varrho$ is proportional to $\mathrm{d}\boldsymbol{l}$.

This theorem, combined with the preceding one, gives full information about the time evolution of the magnetic field if the plasma motions are known. The magnetic lines of force are carried away with the fluid, and the local value of $\boldsymbol{B}$ is such that $\boldsymbol{B}/\varrho$ varies in proportion to the length of the relevant element of magnetic line of force. Thus, turbulent plasmas tend to generate higher magnetic fields. This conclusion is evident since a given line carried away with a turbulent fluid tends to take a more and more complicated shape and its length increases. Two effects, not included in the above treatment, tend to introduce some saturation in this increase. Either the fluid is not a perfect conductor, and the magnetic lines of force are not perfectly frozen; they tend to slip more and more with respect to the fluid; or the magnetic lines of force can be regarded as perfectly frozen, but then the increase in time of the magnetic field results in an increase of the electromagnetic energy, which can only increase at the expense of the mechanical energy of the fluid; in this case, the fluid motions tend to slow down. The frozen character of the magnetic lines of force is powerful enough to limit severely the turbulence.

5.8 QUASI-LINEAR APPROXIMATION OF TURBULENCE[24]

The system behaves with increasing $\mathcal{R}_\mathrm{n}$ in different ways depending on its characteristics. The transition from a laminar to a turbulent flow might be rapid in the case where a large number of interacting modes are excited. These higher modes are sometimes only harmonics of the fundamental first excited mode. In this case, a degree of freedom has been excited, represented by a non-linear excitation of finite amplitude. The excited modes may also interact only weakly with each other resulting in a *weak turbulence* of the flow.

A fully ionised plasma in a toroidal magnetic field is convectively unstable because of its diamagnetism. The same can be said of a weakly ionised plasma in a strong magnetic field. This example of magnetic field-induced laminar convection has been treated briefly in Section 2.6 where Figure 2.10 shows that for $H > H_c$ the ionisation frequency increased instead of decreasing with increasing magnetic field. This problem has been studied theoretically by Timofeev[47] and observed experimentally by Golant[23].

Another example of magnetic field-induced convection occurs in a positive column in the presence of a homogeneous magnetic field, the positive column (of a discharge) being the seat of a longitudinal current. Hoh[25] studied the convection of such a weakly ionised plasma by strong magnetic fields. His results are summarised in Figure 5.12 which represents the relationship between the longitudinal electric field $\boldsymbol{E}$ and the magnetic field strength $\boldsymbol{H}$ in the positive column of a helium discharge. It is observed that for low values of $\boldsymbol{H}$, $\boldsymbol{E}$ decreased monotonically with $\boldsymbol{H}$ increasing, obeying the classical result in which the diffusion coefficient will decrease, thus $\boldsymbol{E}$ decreases with $\boldsymbol{B}$ increasing as seen in Section 2.6. As soon as $\boldsymbol{H}$ reaches a critical value $\boldsymbol{H}_c$, $\boldsymbol{E}$ starts increasing with

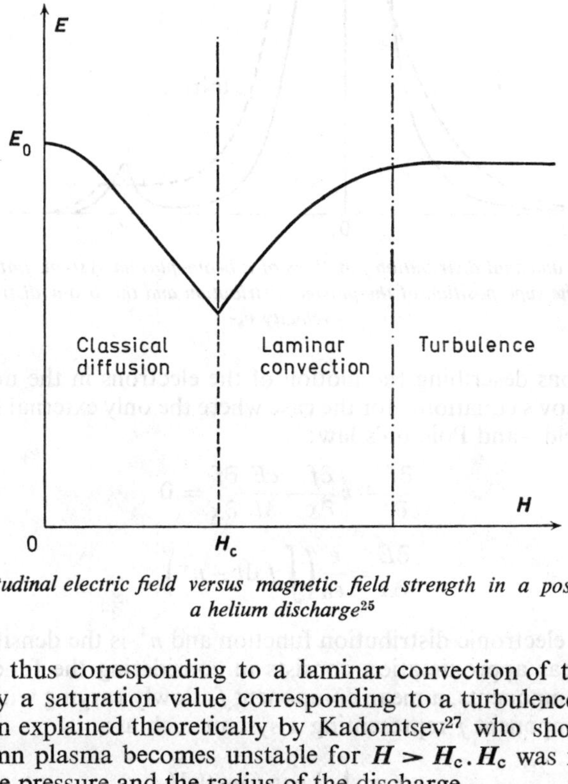

Fig. 5.12. Longitudinal electric field versus magnetic field strength in a positive column of a helium discharge[25]

$\boldsymbol{H}$ increasing, thus corresponding to a laminar convection of the plasma. It reaches finally a saturation value corresponding to a turbulence region. This effect has been explained theoretically by Kadomtsev[27] who showed that the positive column plasma becomes unstable for $\boldsymbol{H} > \boldsymbol{H}_c$. $\boldsymbol{H}_c$ was found to be a function of the pressure and the radius of the discharge.

Consider now the case of a collisionless plasma in which Langmuir oscillations are excited by an electron beam. This problem has been studied by Drummond and Pines[16, 17] by applying the quasi-linear approximation. The distribution of the electronic velocities of the beam–plasma system is composed initially of the superposition of the plasma distribution and the beam distribution of mean velocity v_b (Figure 5.13). This system is unstable since Langmuir waves with phase velocities such that $df/dv > 0$ will increase with time until the entire region where $df/dv > 0$ is covered by such waves. If the thermal electron density of the plasma is much larger than the particle density of the beam then the growth factor α of the waves will be much smaller than the wave frequency ω. In this case the interaction between the different modes of the wave can be neglected, leading to the application of a quasi-linear approximation.

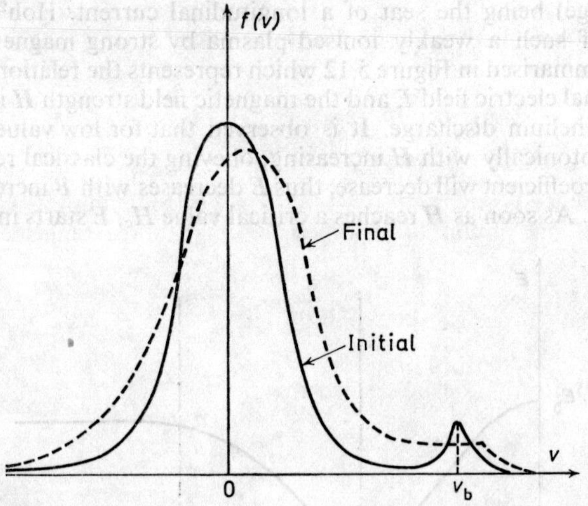

Fig. 5.13. Initial and final distribution functions of a beam–plasma system. Initially, the system is composed of the superposition of the plasma distribution and the beam distribution of mean velocity v_b

The equations describing the motion of the electrons in the unidimensional space are Vlasov's equation—for the case where the only external force is due to the electric field—and Poisson's law:

$$\frac{\partial f}{\partial t}+v\frac{\partial f}{\partial x}-\frac{eE}{M}\frac{\partial f}{\partial x}=0$$
$$\frac{\partial E}{\partial x}=\frac{e}{\varepsilon_0}\left(\int f\,\mathrm{d}v-n^+\right) \tag{5.43}$$

where f is the electronic distribution function and n^+ is the density of the ions. The quasi-linear approximation consists in considering the function f as the sum of two components, a mean component f_0 slowly varying with time and an oscillating component $\tilde{f}$ representing a system such as:

$$\tilde{f}=\int f_k \exp\left(\mathrm{j}\omega_k t\right)\mathrm{d}k \tag{5.44}$$

where the subscript k represents the characteristics of the kth frequency mode. Assuming that the system is in a large but finite space and that it satisfies periodic boundary conditions, one can linearise and Fourier-analyse equations (5.43) and obtain:

$$(\partial f_k/\partial t)+\mathrm{j}kvf_k-(e/M)E_k(\partial f_0/\partial v)=0 \tag{5.45a}$$

$$(\partial f_0/\partial t)=(e/M)\int E_k(\partial f_k/\partial v)\,\mathrm{d}k \tag{5.45b}$$

$$\mathrm{j}kE_k=(e/\varepsilon_0)\int f_k\,\mathrm{d}k \tag{5.45c}$$

it being noted that $(\partial f_0/\partial x) = 0$. To solve these equations, one assumes a normal mode solution of the form given by equation (5.44). This leads to a value of f_k such that:

$$f_k = (e/M)\frac{E_k}{\mathrm{j}(\omega_0 - kv) - \alpha_k}(\partial f_0/\partial v)$$

where the frequency is such that $\omega_k = \omega_0 + \mathrm{j}\alpha_k$. Introducing this value of f_k in equation (5.45b), one obtains after some mathematical rearrangements:

$$\frac{\partial f_0}{\partial t} = \frac{\partial}{\partial v}\left(\mathcal{D}_\mathrm{v}\frac{\partial f_0}{\partial v}\right) \tag{5.46}$$

where $\mathcal{D}_\mathrm{v}$ is a diffusion constant given by

$$\mathcal{D}_\mathrm{v} = \frac{e^2}{M^2}\int \frac{E_k^2(t)\,\mathrm{d}k}{(\omega_k - kv)^2 + \alpha_k^2} \tag{5.47}$$

When f_0 varies only slowly with time, one can write for E_k^2:

$$\mathrm{d}E_k^2/\mathrm{d}t = 2\alpha_k E_k^2 \tag{5.48}$$

Consider now the evolution of unstable oscillations in the plasma by examining first the resonance electrons group, that is, for $\alpha_k \ll \omega_k$.

In this case, the denominator of the complex value of equation (5.47) becomes:

$$\frac{1}{(\omega_0 - kv) + \mathrm{j}\alpha_k} = \frac{P}{(\omega_k - kv)} - \mathrm{j}\pi\delta(\omega_k - kv)$$

where $\delta(\omega_k - kv)$ is a Dirac delta function. The term $P/(\omega_k - kv)$ represents the principal part which cancels out in equation (5.45a). Thus, equation (5.47) can be written as:

$$\mathcal{D}_\mathrm{v} = \frac{\pi e^2}{M^2}\int \delta(\omega_k - kv)E_k^2\,\mathrm{d}k \tag{5.49}$$

This coefficient is equal to zero for all the values of α_k equal to or smaller than zero. Due to the diffusion, the distribution function will be flattened as shown in Figure 5.13 establishing a spectrum of supra-thermal oscillations. Landau[15] has shown that:

$$\alpha_k = \frac{\pi\omega_0^3}{2k^2 n}(\partial f_0/\partial v)_{v=\omega/k} \tag{5.50}$$

for Langmuir oscillations. Using equations (5.46), (5.48) and (5.49), one obtains for the spectrum:

$$f + \frac{\partial}{\partial v}\left(\frac{e^2}{M^2}\frac{nE_v^2}{\varepsilon_0 v^3}\right) = \text{constant}$$

which yields for the amplitude of the oscillation in the steady state.

$$E_v^2(t=\infty) = \frac{M^2\omega_0 v^3}{e^2 n}\int_{v_1}^{v_2}[f_0(t=\infty)-f_0(t=0)]\,dv$$

$f_0(t=\infty)$ is constant and v_1 is the velocity at which the plateau of the curve is reached and is determined by the condition that the total number of resonance particles should remain constant.

A second case is the non-resonant electrons for which $(\omega_k-kv)^2 \gg \alpha_k^2$. Their diffusion coefficient $\mathcal{D}_v$ is given by equation (5.47) after neglecting α_k^2 in the denominator of the integrals. Taking f_0 approximately equal to $f_0(t=0)$ on the right-hand side of equation (5.46), and considering both equations (5.48) and (5.46), one obtains:

$$f_0(t=\infty)-f_0(t=0) = \frac{e^2}{2M^2}\int E_k^2(t=\infty)\frac{\partial}{\partial v}\left[\frac{1}{(\omega-kv)^2}\frac{\partial f}{\partial v}\right]dk$$

and for $t=\infty$ the distribution function will reach the form shown by a dotted curve in Figure 5.13. It can be shown that the entire momentum lost by the beam and half the energy are gained by the thermal electrons, whereas the electric field gained the remaining half energy lost by the beam.

5.9 WEAK TURBULENCE[28]

The states described in Section 5.8 are so weakly excited that they can hardly be considered as turbulent states; non-linear interactions between oscillations were neglected although they are most important in the definition of turbulence.

The criterion of a weak turbulence is a weak interaction between the waves, in other words when the growth rate α of perturbation becomes much smaller than the angular frequency ω. First, consider the case where the resonance interaction between oscillations and particles is negligible, such as in the case of a cold plasma. Such a model can be represented by the MHD equations; the latter are similar in form, and are based on the Boltzmann equation given by equation (3.20). The term $\partial f/\partial t)_{coll}$ in the equation represents here collisions between electrons and waves. Taking the Fourier transform of equation (3.20), one finds in general:

$$(\omega-\omega_k-j\alpha_k)\mathcal{C}_{k\omega} = \int \mathcal{B}_{k\omega,\,k'\omega'}\,\mathcal{C}_{k'\omega'}\mathcal{C}_{k-k',\,\omega-\omega'}\,dk'\,d\omega' \qquad (5.51)$$

in which the term of the left-hand side represents the linear part of the equation and $\mathcal{C}_{k\omega}$ represents the Fourier transform of f or in general of any real function. The term on the right-hand side is the Fourier transform of the quadratic terms represented by the collision term describing the interaction between the different harmonics of the wave. The matrix element of this interaction is represented by the expression $\mathcal{B}_{k\omega,\,k'\omega'}$. In conditions near the equilibrium state, equation (5.51) becomes at zero approximation:

$$(\omega-\omega_k)\mathcal{C}_{k\omega}^{(0)} = 0 \qquad (5.52)$$

After multiplying equation (5.52) by $\mathscr{C}^{(0)*}_{k'\omega'}$ and taking the average with respect to the random phases of the various oscillations, one obtains when the oscillations are stationary and homogeneous in space:

$$\langle \mathscr{C}_{k\omega}\mathscr{C}^{*}_{k'\omega'}\rangle_0 = f_k\delta(\omega-\omega_k)\,\delta(\boldsymbol{k}-\boldsymbol{k}')\,\delta(\omega-\omega')$$

where f_k is the spectral distribution of the electric field. If now on the right-hand side of equation (5.51), $\mathscr{C}_{k'\omega'}$ and $\mathscr{C}_{k-k',\,\omega-\omega'}$ are replaced respectively by $\mathscr{C}^{(0)}_{k'\omega'}$ and $\mathscr{C}^{(0)}_{k-k',\,\omega-\omega'}$, a non-linear inducing oscillation $\mathscr{C}^{(1)}_{k\omega}$ will appear. The appearance of this oscillation does not lead necessarily to a damping and in the growth factor α_k, a term $-\nu$ representing the damping due to higher approximations is usually introduced. Equation (5.51) becomes for the first approximation:

$$\mathscr{C}^{(1)}_{k\omega} = (\omega-\omega_k+\mathrm{j}\nu)^{-1}\int \mathscr{B}_{k\omega,\,k'\omega'}\mathscr{C}^{(0)}_{k'\omega'}\mathscr{C}^{(0)}_{k-k',\,\omega-\omega'}\,\mathrm{d}\boldsymbol{k}'\,\mathrm{d}\omega' \tag{5.53}$$

To determine the magnitude of the damping factor, equation (5.51) should be multiplied by $\mathscr{C}^{*}_{k'\omega'}$ and the result averaged with respect to the random phases of the oscillations. The right-hand side of this equation goes to zero in the zeroth order approximation; in the first order approximation, it becomes, however:

$$(\omega-\omega_k-\mathrm{j}\alpha_k)f_{k\omega} = f_{k\omega}\int \mathscr{B}_{k\omega,\,k'\omega'}\frac{\mathscr{B}_{k''\omega'',\,k\omega}+\mathscr{B}_{k''\omega'',\,-k'-\omega'}}{\omega''-\omega_{k''}+\mathrm{j}\nu}f_{k'\omega'}\,\mathrm{d}\boldsymbol{k}'\,\mathrm{d}\omega'$$

$$+f_{k\omega}\int \mathscr{B}_{k\omega,\,k'\omega'}\frac{\mathscr{B}_{k'\omega',\,k\omega}+\mathscr{B}_{k'\omega',\,-k''-\omega''}}{\omega'-\omega_{k'}+\mathrm{j}\nu}f_{k''\omega''}\,\mathrm{d}\boldsymbol{k}'\,\mathrm{d}\omega'$$

$$+(\omega-\omega_k-\mathrm{j}\nu)^{-1}\int \mathscr{B}_{k\omega,\,k'\omega'}(\mathscr{B}^{*}_{k\omega,\,k'\omega'}+\mathscr{B}^{*}_{k\omega,\,k''\omega''})f_{k'\omega'}f_{k''\omega''}\,\mathrm{d}\boldsymbol{k}'\,\mathrm{d}\omega' \tag{5.54}$$

with $\boldsymbol{k}''=\boldsymbol{k}-\boldsymbol{k}'$ and $\omega''=\omega-\omega'$. Under the condition $\alpha \ll \omega$, the real parts of the first and second terms of the right-hand side of equation (5.54) can be neglected. The terms proportional to $f_{k\omega}$ can now be combined with the left-hand side to yield an overall growth factor α_k; in the limit of the first approximation, ν can be replaced by $-\alpha_k$ in the third term of the right-hand side of the equation. Thus, equation (5.54) becomes symmetrical and such that:

$$[(\omega-\omega_k)^2+\alpha_k^2]f_{k\omega} = \frac{1}{2}\int |\mathscr{B}^{t}_{k\omega,\,k'\omega'}|^2 f_{k'\omega'}f_{k''\omega''}\,\mathrm{d}\boldsymbol{k}'\,\mathrm{d}\omega' \tag{5.55}$$

with $\mathscr{B}^{t}_{k\omega,\,k'\omega'} = \mathscr{B}_{k\omega,\,k'\omega'}+\mathscr{B}_{k\omega,\,k''\omega''}$. The kinetic equation for sustained oscillations is obtained by replacing $f_{k\omega}$ in equation (5.55) by $f_k\delta(\omega-\omega_k)$ and using the approximation,

$$\mathrm{Im}\,(\omega'-\omega_{k'}+\mathrm{j}\nu)^{-1} = -\pi\delta(\omega'-\omega_{k'})$$

one obtains finally for a small growth rate $\tilde{\alpha}_k$:

$$\tilde{\alpha}_k f_k = \alpha_k f_k - \frac{\pi}{2}\int [2\mathscr{B}^{t}_{kk'}\mathscr{B}^{t}_{k'k}f_k f_{k''} - |\mathscr{B}^{t}_{k\omega,\,k'\omega'}|^2 f_{k'}f_{k''}]\,\delta(\omega_k-\omega_{k'}-\omega_{k''})\,\mathrm{d}\boldsymbol{k}' \tag{5.56}$$

For a transparent medium the quantity $\mathcal{B}^t_{k\omega,\,k'\omega'}$ is real. If a cubic term is added to equation (5.51), similar to the existing quadratic one, a new term

$$\int \mathcal{A}_{k\omega,\,k'\omega'} f_{k\omega} f_{k'\omega'}\, \mathrm{d}\boldsymbol{k}'\, \mathrm{d}\omega'$$

would appear in the right-hand side of equation (5.54), with:

$$\mathcal{A}_{k\omega,\,k'\omega'} = \mathcal{A}_{k\omega,\,k'\omega',\,-k'-\omega'} + \mathcal{A}_{k\omega,\,k\omega,\,k'\omega'} + \mathcal{A}_{k\omega,\,k'\omega',\,k\omega}$$

Equation (5.56) is a steady state equation for the spectrum of oscillation where the growth due to instability compensates exactly the damping of the wave resulting from the non-linear interactions.

The conservation of energy of the wave can be described by an equation similar to equation (3.48). For a homogeneous plasma, one has:

$$(\partial\mathcal{E}/\partial t) + \nabla . Q = 0$$

where $\mathcal{E}$ is the energy of the kth wave and Q is the flux of energy. Introducing the spectral function f_k of the electric field, the above equation becomes:

$$(\partial f_k/\partial t) + \boldsymbol{v}_k . (\partial f_k/\partial \boldsymbol{r}) = 0$$

where $\boldsymbol{v}_k$ is the group velocity of the kth wave. If α_k is its growth rate, then a term $2\alpha_k f_k$ should be introduced in the equation; thus:

$$(\partial f_k/\partial t) + \boldsymbol{v}_k . (\partial f_k/\partial \boldsymbol{r}) = 2\alpha_k f_k$$

To consider a weakly inhomogeneous plasma, the variation of the spectral function with respect to $\boldsymbol{k}$ should be taken into account. This leads to a new term $(\partial\omega_k/\partial\boldsymbol{r}).(\partial f_k/\partial\boldsymbol{k})$ and the kinetic equation for the waves becomes:

$$[(\partial f_k/\partial t) + \boldsymbol{v}_k . (\partial f_k/\partial \boldsymbol{r}) - (\partial\omega_k/\partial\boldsymbol{r}).(\partial f_k/\partial\boldsymbol{k}) = 2\alpha_k f_k$$

Equation (5.56) is a steady state equation for the spectrum of oscillation where the growth due to instability is exactly compensated by the damping of the wave resulting from the non-linear interactions. To take into account these non-linear effects, α_k should simply be replaced by $\tilde{\alpha}_k$ given by equation (5.56) in the kinetic equation of the waves. This leads to the following basic kinetic equation describing the weakly turbulent systems:

$$\frac{\partial f_k}{\partial t} + \boldsymbol{v}_k \cdot \frac{\partial f_k}{\partial \boldsymbol{r}} - \frac{\partial \omega_k}{\partial \boldsymbol{r}} \cdot \frac{\mathrm{d} f_k}{\mathrm{d}\boldsymbol{k}} = 2\alpha_k f_k - \pi \int [2\mathcal{B}^t_{kk'}\mathcal{B}^t_{k'k} f_k f_{k''} - |\mathcal{B}^t_{kk'}|^2 f_{k'} f_{k''}]$$
$$\delta(\omega_k - \omega_{k'} - \omega_{k''})\, \mathrm{d}\boldsymbol{k}'$$

Note that for a homogeneous medium $\partial\omega_k/\partial\boldsymbol{r} = 0$. Note also that the conservation of energy leads to $\omega_k = \omega_{k'} + \omega_{k''}$. This condition limits greatly the possible area of interaction in the $\boldsymbol{k}$-space. Here the dispersion relation $\omega = \omega(\boldsymbol{k})$ leads to three possible wave processes: the decay of the $\boldsymbol{k}$-wave into $\boldsymbol{k}'$ or $\boldsymbol{k}''$ and the degeneration of the waves $\boldsymbol{k}'$ and $\boldsymbol{k}''$ into one single wave. This dispersion relation is usually divided into one degenerate and one non-degenerate relation. As an example, Figure 5.14 shows a non-degenerate relation in an isotropic medium in which the phase velocity v_p decreases with $\boldsymbol{k}$ and a degenerate relation in the medium where v_p increases with $\boldsymbol{k}$.

The kinetic equation for waves has been applied to plasma physics by many workers. As an example of degenerate interaction, Oraevskii and Sagdeev[35] studied the interaction between the Langmuir waves and the ion-sound waves. It was found[12] that Langmuir oscillations must first excite ion oscillations before they can interact with one another. With their increasing amplitude, Langmuir

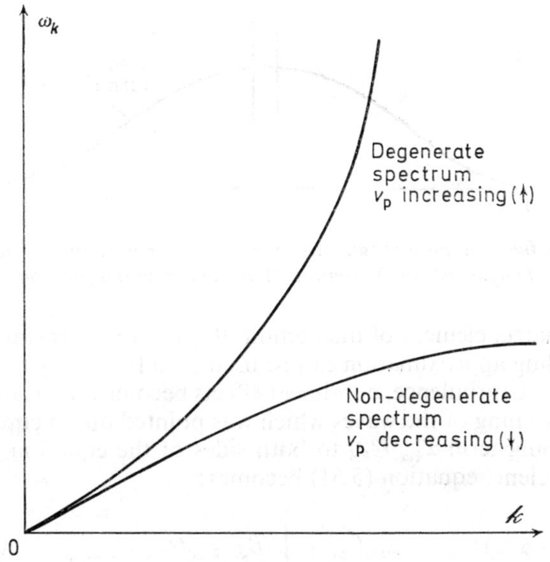

Fig. 5.14. Examples of degenerate and non-degenerate spectra in an isotropic medium[28]*. For the degenerate spectrum the phase velocity increases with k, whereas it decreases for the non-degenerate spectrum*

waves become scattered at the ion inhomogeneities. Galeev and Oraevskii[22] studied oscillations of an ideally-conducting plasma in a homogeneous magnetic field. The result was the interaction of the Alfvén waves with the magnetoacoustic waves, which is another example of a degenerate mode. The effect of the thermal motion of particles was considered by many authors[46], with special attention given to the interaction between the waves and the particles as seen in Chapter 4.

5.10 STRONG TURBULENCE[19, 45]

With increasing turbulence, the frequency spectrum of the oscillations broadens as a result of an increase in the interaction between the various waves. When the turbulence becomes very strong the spectrum of $f_{k\omega}$ with respect to the frequency can no longer be considered as a δ-function as shown in Figure 5.15. However,

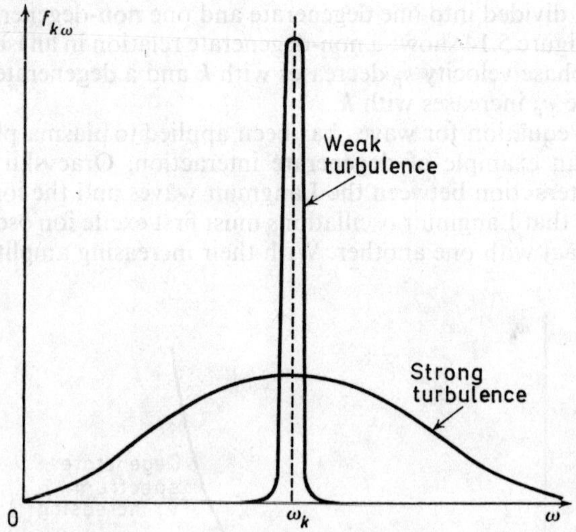

Fig. 5.15. Spectral functions for weak and strong turbulences. For a strong turbulence the spectral function cannot be considered any longer as a δ-function

as long as the matrix element of interaction $\mathcal{B}_{k\omega,\, k'\omega'}$ remains smaller than unity the weak coupling approximation can be used even for strong turbulence.

With increasing turbulence, non-linear effects become important. One of these effects is the damping of the waves which was pointed out in equation (5.51) by adding a damping term $\alpha^*_{k\omega}\, \mathcal{C}_{k\omega}$ to both sides of the equation, $\alpha^*_{k\omega}$ being the damping coefficient; equation (5.51) becomes:

$$(\omega-\omega_k+\alpha^*_{k\omega})\,\mathcal{C}_{k\omega} = \alpha^*_{k\omega}\mathcal{C}_{k\omega}+\int \mathcal{B}_{k\omega,k'\omega'}\mathcal{C}_{k'\omega'}\mathcal{C}_{k-k',\omega-\omega'}\,\mathrm{d}\boldsymbol{k}'\,\mathrm{d}\omega' \quad (5.57)$$

In this equation the term $\mathrm{j}\alpha_k$ has been included in ω_k, which is considered a complex quantity, and the damping term is now removed from the right-hand side of the equation. In this side only the effect of the interaction between the different modes remains, and this effect is small as long as $\mathcal{B}$ remains smaller than unity. Assume now that:

$$\mathcal{C}_{k\omega} = \mathcal{C}^{(0)}_{k\omega}+\mathcal{C}^{(1)}_{k\omega} \quad (5.58)$$

with the first approximation term much smaller than the zero approximation term. Substituting equation (5.33) in equation (5.57) and considering only the non-linear terms, one obtains:

$$\mathcal{C}^{(1)}_{k\omega} = \frac{\int \mathcal{B}_{k\omega,\,k'\omega'}\mathcal{C}^{(0)}_{k'\omega'}\mathcal{C}^{(0)}_{k''\omega''}\,\mathrm{d}\boldsymbol{k}'\,\mathrm{d}\omega'}{(\omega-\omega_k+\alpha^*_{k\omega})}$$

an equation similar to equation (5.53). Multiplying equation (5.57) by $\mathcal{C}^*_{k\omega}$, averaging the product over the random phases of the various oscillations, and taking into consideration equation (5.58), one obtains a relation similar to equation (5.54). If in this equation $\alpha^*_{k\omega}$ is defined to eliminate the terms proportional to $f_{k\omega}$, one obtains a system of two equations:

$$f_{k\omega} = \frac{\frac{1}{2}\int |\mathcal{B}^t_{k\omega,\,k'\omega'}|^2 f_{k'\omega'} f_{k''\omega''}\,\mathrm{d}\boldsymbol{k}'\,\mathrm{d}\omega'}{|\omega-\omega_k+\alpha^*_{k\omega}|^2} \tag{5.59}$$

$$\alpha^*_{k\omega} = \int \frac{\mathcal{B}^t_{k\omega,\,k''\omega''}\mathcal{B}^t_{k''\omega'',\,k\omega,\,k'\omega'}\,\mathrm{d}\boldsymbol{k}'\,\mathrm{d}\omega'}{(\omega''-\omega_{k''}+\alpha^*_{k''\omega''})} \tag{5.60}$$

where the different terms are as defined in equation (5.55). To eliminate $\alpha^*_{k\omega}$ from equations (5.59) and (5.60) put $t_{k\omega} = (\omega-\omega_k+\alpha^*_{k\omega})^{-1}$ and $t^0_{k\omega} = (\omega-\omega_k)^{-1}$; then obtain:

$$f_{k\omega} = \frac{1}{2}|t_{k\omega}|^2 \int |\mathcal{B}^t_{k\omega,\,k'\omega'}|^2 f_{k'\omega'} f_{k''\omega''}\,\mathrm{d}\boldsymbol{k}'\,\mathrm{d}\omega'$$

$$t_{k\omega} = t^0_{k\omega} - t^0_{k\omega}t_{k\omega}\int t_{k''\omega''}\mathcal{B}^t_{k\omega,\,k''\omega''}\mathcal{B}^t_{k''\omega'',\,k\omega}f_{k'\omega'}\,\mathrm{d}\boldsymbol{k}'\,\mathrm{d}\omega'$$

$t_{k\omega}$ represents a Green function describing the turbulent response of the medium to a 'driving force' $\boldsymbol{F}_{k\omega}$ whereas $t^0_{k\omega}$ is the zero approximation Green function.

As an example consider Langmuir oscillations in a collisionless plasma. In this case Poisson's law and Vlasov's equation become respectively:

$$V_{k\omega} = -\frac{e}{\varepsilon_0 k^2}\int \boldsymbol{F}_{k\omega}\,.\,\mathrm{d}\boldsymbol{v} \tag{5.61}$$

$$(\omega-\boldsymbol{k}.\boldsymbol{v})\boldsymbol{F}_{k\omega} = \frac{e}{M}V_{k\omega}\boldsymbol{k}\frac{\partial \boldsymbol{F}}{\partial \boldsymbol{v}} + \frac{e}{M}\frac{\partial}{\partial \boldsymbol{v}}\int \boldsymbol{k}'\cdot(V_{k'\omega'}\boldsymbol{F}_{k''\omega''} - \langle V_{k'\omega'}\boldsymbol{F}_{k''\omega''}\rangle)\,\mathrm{d}\boldsymbol{k}'\,\mathrm{d}\omega' \tag{5.62}$$

where $V_{k\omega}$ is the electric potential. Taking the zero and first approximations one obtains:

$$\begin{aligned} \boldsymbol{F}_{k\omega} &= \boldsymbol{F}^{(0)}_{k\omega} + \boldsymbol{F}^{(1)}_{k\omega} \\ V_{k\omega} &= V^{(0)}_{k\omega} + V^{(1)}_{k\omega} \end{aligned} \tag{5.63}$$

Introducing equations (5.63) into equations (5.61) and (5.62), one obtains for the first approximation:

$$V^{(1)}_{k\omega} = -\frac{e}{\varepsilon_0 \boldsymbol{k}^2}\int \boldsymbol{F}^{(1)}_{k\omega}.\mathrm{d}\boldsymbol{v}$$

$$(\omega-\boldsymbol{k}.\boldsymbol{v})\boldsymbol{F}^{(1)}_{k\omega} - \frac{e}{M}V^{(1)}_{k\omega}\boldsymbol{k}\frac{\partial \boldsymbol{F}}{\partial \boldsymbol{v}} = \frac{e}{M}\frac{\partial}{\partial \boldsymbol{v}}\int \boldsymbol{k}'V^{(0)}_{k'\omega'}\boldsymbol{F}^{(0)}_{k''\omega''}.\mathrm{d}\boldsymbol{k}'\,\mathrm{d}\omega'$$

These equations are the basis of the turbulent response study, leading to three equations with three unknown values. This system was studied for the case of a plasma in a strong magnetic field by Mikhailovskii[34]. The result is a weak coupling system of equations for a strong turbulence. This argument considers only resonant interactions between the waves. To consider both resonance and adiabatic wave interactions assume that, in equation (5.51), α_k/ω_k is small and that ω_k increases monotonically with increasing $\boldsymbol{k}$. The lifetime of the waves $\boldsymbol{k}$, ω, is then equal to approximately $1/\alpha$ and the space of its region is about $\omega/\alpha\boldsymbol{k}$. When this wave disappears it is replaced immediately by another wave which is independent from it and which results from the beat interaction. The turbulent motion of a continuous plasma can thus be considered as a succession of wave packets. The lifetime of these packets is very large for weak turbulence and can be considered as Fourier components instead of wave packets as seen before. Therefore, in a region of wave number $\boldsymbol{k}$, there will be oscillations of frequency ω and much slower oscillations of frequency $\omega' \ll \alpha_k$; these oscillations can be considered permanent as compared to oscillations $\boldsymbol{k}$, ω. Thus, a strong correlation will occur between the wave packets that describe the wave $\boldsymbol{k}$, ω and the Fourier components that describe the wave $\boldsymbol{k}'$, ω' leading to an adiabatic variation of $\boldsymbol{k}$ and ω. Since the Fourier components $\mathcal{C}_{k\omega}$ and $\mathcal{C}_{k''\omega''}$ describe the same wave packet the interaction between the fast wave and the slow wave cannot be considered here as the resonant input to the fast wave from the neighbouring wave $\boldsymbol{k}'' = \boldsymbol{k}-\boldsymbol{k}'$, $\omega'' = \omega-\omega'$.

For large values of α/ω, that is, for a strong turbulence, a distinction should be made between resonant and adiabatic interactions, and a wave-packet representation should replace the Fourier one. For each wave packet $\boldsymbol{k}$, ω, the region of integration of equation (5.51) can be divided into three parts: (1) the central region for $\boldsymbol{k}' \cong \boldsymbol{k}$, $\omega' = \omega$; (2) the short wave region ($\alpha \ll \omega$); and (3) the long wave region. Equation (5.51) can be used for the Fourier components of the short wave region. Calculations similar to the ones above would lead in the case of an improved weak coupling between the waves to the spectral relation:

$$f_k \, \mathrm{d}f = \boldsymbol{k}^{-5/3} \, \mathrm{d}\boldsymbol{k} \tag{5.64}$$

This problem of turbulence in plasma is becoming more and more important. For more details, Kadomtsev's monograph should be consulted[28].

REFERENCES AND BIBLIOGRAPHY

1. ABRAHAM, E. E., CRAWFORD, L. W. and MILLS, D. M., 'Observation of a "Hose" Instability of an Electron Beam in a Plasma', *J. appl. Phys.*, **38**, No. 2, 911 (1967).
2. ABRAMOVICH, G. N., *Theory of Turbulent Jets*, M.I.T. Press, Cambridge, Mass. (1963).
3. BALESCU, R., 'Irreversible Process in an Ionized Plasma', *Physics Fluids*, **3**, 52 (1960).
4. BATCHELOR, G. K., 'The Spontaneous Magnetic Field in a Conducting Liquid in Turbulent Motion', *Proc. R. Soc.*, **A201**, 405 (1950).
5. BERNSTEIN, I. B., GREENE, J. M. and KRUSKAL, M. D., 'Exact Non-Linear Plasma Oscillations', *Phys. Rev.*, **108**, 546 (1957).
6. BERNSTEIN, I. B., FRIEMAN, E. A., KRUSKAL, M. D. and KULSRUD, R. M., 'An Energy Principle for Hydromagnetic Stability Problems', *Proc. R. Soc.*, **A244**, 17 (1958).
7. BERNSTEIN, I. B. and KULSRUD, R. M., 'Ion Wave Instabilities', *Physics Fluids*, **3**, 937 (1960).
8. BOHM, D. *et al.*, *The Characteristics of Electrical Discharges in Magnetic Fields* (edited by GUTHRIE and WAKERLING) USAEC, New York (1949).

9. BORODIN, A. V., GAVRIN, P. P., KOVAN, I. A., PATRUSHEV, B. I., NEDOSEEV, S. L., RUSANOV, V. D. and KAMENETSKII, D. A., 'Magneto-Acoustic Oscillations and the Instability of an Induction Pinch', *Zh. éksp. teor. Fiz.*, **41**, No. 8, 317 (1961).
10. BUNEMAN, O., 'Maintenance of Equilibrium by Instabilities', *J. nucl. Energy*, Part C, **2**, 119 (1961).
11. BUNEMAN, O., LEVY, R. H. and LINSON, L. M., 'The Stability of Crossed-Field Electron Beams', *J. appl. Phys.*, **37**, 3203 (1966).
12. CAMAC, M. *et al.*, 'Shock Waves in Collision-Free Plasmas', *Nucl. Fusion*, Suppl., Part 2, 423 (1962).
13. CHANDRASEKHAR, S., 'Problems of Stability in Hydrodynamics and Hydromagnetics', *Mon. Not. R. astr. Soc.*, **113**, 667 (1953).
14. CHANDRASEKHAR, S., 'Hydromagnetic Turbulence', *Proc. R. Soc.*, **233A**, 322 (1955).
15. DAWSON, J., 'A One-Dimensional Plasma Model', *Physics Fluids*, **5**, 445 (1962).
16. DRUMMOND, W. E. and PINES, D., 'Nonlinear Stability of Plasma Oscillations', *Nucl. Fusion*, Suppl., Part 3, 1049 (1962).
17. DRUMMOND, W. E. and PINES, D., 'Nonlinear Plasma Oscillations', *Ann. Phys.*, **28**, 478 (1964).
18. DUPREE, T. H., 'Plasma Turbulence', *Advances in Plasma Physics* (edited by A. SIMON), Interscience, New York (1968).
19. EDWARDS, S. F., 'Strong Turbulence', *Proc. 2nd Orsay Summer Institute*, Gordon and Breach, New York, 57 (1969).
20. FRIEMAN, E., BODNER, S. and RUTHERFORD, P., 'Some New Results on the Quasi-Linear Theory of Plasma Instabilities', *Physics Fluids*, **6**, No. 9, 1298 (1963).
21. FRIEMAN, E. and RUTHERFORD, P., 'Kinetic Theory of a Weakly Unstable Plasma', *Ann. Phys.*, **28**, 134 (1964).
22. GALEEV, A. A. and ORAEVSKII, V. N., 'The Stability of Alfvén Waves', *Doklady Akad. Nauk SSSR*, **147**, 71 (1962).
23. GOLANT, V. E., 'Diffusion of Charged Particles in a Plasma in a Magnetic Field', *Soviet Phys. Usp.*, **74**, 161 (1963).
24. HINZE, J. O., *Turbulence*, McGraw-Hill, New York (1959).
25. HOH, F. C., 'Low Temperature Diffusion in a Magnetic Field', *Rev. Mod. Phys.*, **34**, 267 (1962).
26. JEFFREY, A. and TANIUTI, T., *MHD Stability and Thermonuclear Containment*, Academic Press, New York (1966).
27. KADOMTSEV, B. B., 'On the Convective Instability of a Plasma Filament', *Zh. éksp. teor. Fiz.*, **37**, 1096 (1959).
28. KADOMTSEV, B. B., *Plasma Turbulence*, Academic Press, New York (1965).
29. KADOMTSEV, B. B. and ROGUTSE, P. O., 'Flute Instability of a Plasma in Toroidal Geometry', *Dokl. Akad. Nauk, SSSR*, **170**, No. 10, 811 (1966).
30. KRALL, N. A., 'Drift Instabilities', *Advances in Plasma Physics* (edited by A. SIMON), Interscience, New York (1968).
31. KRUSKAL, M. D. and SCHWARZSCHILD, M., 'Some Instabilities of a Completely Ionized Gas', *Proc. R. Soc.*, **A233**, 348 (1954).
32. LANDAU, L. D. and LIFSHITZ, E. M., *Mechanics of Continuous Media*, Gostekhizdat, Moscow, U.S.S.R.
33. LENARD, A., 'On the Bogoliubov Kinetic Equation for a Spatially Homogeneous Plasma', *Ann. Phys.*, **3**, 390 (1960).
34. MIKHAILOVSKII, A. B., *Problems in Plasma Theory* (edited by M. A. LEONTOV), Vol. 3, Atomizdat (1963).
35. ORAEVSKII, V. N. and SAGDEEV, R. Z., 'Stability of Steady-State Longitudinal Plasma Oscillations', *Zhur. tekh. Fiz.*, **32**, 1291 (1962).
36. REYNOLDS, O., 'An Experimental Investigation of the Circumstances which Determine whether the Motion of Water Shall be Direct or Sinuous and of the Low Resistance in Parallel Channels', *Phil. Trans. R. Soc.*, **174**, Part III, 935 (1883).

37. ROSENBLUTH, M. N., 'Stability of the Pinch', *USAEC Rept.*, LA-2030 (1956).
38. ROSENBLUTH, M. N. and ROSTOKER, N., 'Theoretical Structure of Plasma Equations' *Proc. 2nd U.N. Conf. on Peaceful Uses of Atomic Energy*, **31,** 144 (1958).
39. SAGDEEV, R. Z. and GALEEV, A. A., *Nonlinear Plasma Theory*, Benjamin, New York (1969).
40. SATO, N. *et al.*, 'Electron Heating Effects and Universal Stability in Cesium Plasma', *Phys. Lett.*, **24A**, 293 (1967).
41. STUART, J. T., 'The Stability of Viscous Flow Between Parallel Planes in the Presence of a Coplanar Field', *Proc. R. Soc.*, **A221,** 189 (1954).
42. SUYDAM, B. R., 'Stability of a Linear Pinch', *Proc. of the 2nd U.N. Conf. on Peaceful Uses of Atomic Energy*, **31,** 354 (1958).
43. TAYLER, R. J., 'Hydromagnetic Instabilities', *Proc. phys. Soc.* **B70,** 31 (1957).
44. TAYLER, R. J., 'The Influence of an Axial Magnetic Field on the Stability of a Constricted Gas Discharge', *Proc. phys. Soc.*, **B70,** 1049 (1957).
45. TCHEN, C. M., 'Cascade Mechanism in the Wave Mixing Processes of Turbulence', *Proc. of the 2nd Orsay Summer Institute*, Gordon and Breach, New York, 197 (1969).
46. THOMPSON, W. B. and HUBBARD, J., 'Long-Range Forces and Diffusion Coefficients of a Plasma', *Rev. Mod. Phys.*, **32,** 714 (1960).
47. TIMOFEEV, A. V., 'Convection of a Weakly Ionised Plasma in a Toroidal Magnetic Field', *Zhur. tekh. Fiz.*, **33,** 776 (1963).
48. TSYTOVICH, V. N., *Nonlinear Effects in Plasma*, Plenum Press, New York (1970).

6 Plasma diagnostic methods

6.1 INTRODUCTION

Plasma diagnostics is one of the most important fields of modern experimental physics. Numerous diagnostic methods have been proposed and most of them successfully tried. However, there is an urgent need for development of still further diagnostic methods to meet all the required conditions. For instance, in the field of thermonuclear fusion, the introduction of any kind of solid material inside the heart of the plasma will be strongly prohibited.

The two parameters of greatest interest in plasma diagnostics are the temperature and the particle density. In high speed, high temperature plasma flows, parameters such as the drift velocity and the electrical conductivity become important. In general, the transport coefficients of mobility and diffusion are of interest in diagnostic technique investigations. This is also true for the parameters describing the processes of generation and annihilation of a plasma.

The three conventional methods of plasma diagnostics are: (1) spectroscopic methods, (2) use of probes, and (3) use of waves. Spectroscopic methods have been important as ways of investigating hot gases and high-pressure plasmas. Two important techniques can be recognised here; these are the method using atomic line spectra for studying high temperature plasmas and the line reversal method for measuring the electronic excitation temperature. Another method involves molecular spectra for measuring rotational temperatures as well as vibrational and vibration–rotational temperatures.

One of the first techniques for measuring plasma properties was the use of electrostatic probes. This method was developed in the early twenties by Mott-Smith and Langmuir[35] and is called the Langmuir probe method. Electrostatic probes may simply consist of an insulated wire inserted into the plasma with the extremity exposed. The probe is connected to a d.c. power supply capable of biasing it. The current collected by the probe contains almost all the required information about the plasma to be investigated. Often the presence of the probe has no sensible disturbing effect in the plasma. However, this is not always the case, especially when strong magnetic fields are present, where the probe has a tendency to perturb (exhaust) a long channel parallel to the magnetic line of force. In this case, the probe current will depend not only on the plasma parameters, but also on the way the plasma is generated and maintained.

More recently, microwave techniques have become the most widely used for diagnostics of a plasma. Heald and Wharton[23] listed several methods for measuring the plasma electron density applicable for a density range between 10^{14} and 10^{20} particles/m^3. Two methods are of greatest importance: (1) microwave diagnostics by resonant cavity, and (2) microwave diagnostics by transmission lines. The first method is based on the fact that the dielectric constant of a plasma is different from that of a vacuum, the difference being related to the plasma parameters, and mainly to its electron density. The second method is based on the fact that when an electromagnetic wave travels through a plasma, it undergoes two major modifications: (1) a phase shift, due to the different index of refraction, and (2) an attenuation due to the energy loss in collisions; both are related to the parameters of the plasma.

The use of lasers for diagnostics is presently being investigated. The laser delivers[17] an intense pulse of monochromatic and perfectly coherent light. When such a beam falls upon a very dense plasma, in the range of 10^{17} carriers/cm^3, a certain amount of light is diffracted. The monochromatic character of this diffracted light is somewhat altered by the Doppler effect, so that the width of the diffracted line becomes proportional to the square root of the temperature. The diffracted light is generally observed by means of a photomultiplier (for the intensity) and a spectrometer (for the Doppler broadening).

In general, a diagnostic method is a way of expressing the quantity to be measured as an easily observable physical quantity, by the use of a suitable device. A host of methods other than those listed above do exist and some of them are mentioned in the following sections.

6.2 OPTICAL METHODS[3]

Little information about the plasma can be obtained by visual observation or *normal photography;* however, such methods are not entirely devoid of interest if properly used, taking into account their own limitations. Most plasmas are relatively transparent in and near the visible part of the spectrum, so that the emission comes from the bulk of the plasma, and not just from the surface. But the light emitted by the plasma itself generally corresponds to an exceptional behaviour if it consists of the characteristic lines of the gas; 20–30 absorptions and re-emissions (radiation trapping) can occur between the initial release of an energy quantum and its eventual finding of a way out. Consequently, microphotometering the intensity of the light emitted by a plasma can only give a vague idea of the relative density of this plasma from point to point. The method is thus generally confined to limited investigations, such as the observation of dark regions or sheaths for which it offers a convenient method of making quantitative measurements.

If the plasma is going through an exactly recurring cycle, the optical observation is meaningful only if aided by some kind of *stroboscope*. At industrial frequencies, a widely used type of stroboscope is the 'razorblade' one shown in Figure 6.1(a). A large disc, about half a metre in diameter, is rotated by means of a synchronous motor. A hole is pierced near the edge of the disc, and a narrow radial slit is shaped by gluing two razor blades close together. If the plasma is cylindrical, the slit is oriented perpendicularly to its axis and the plasma is observed by means of an optical refractor [Figure 6.1(b)]. The image obtained in this way, we should point out, is not an instantaneous picture of the discharge as the time evolution of the discharge combines itself with the sweeping by the slit. For instance, assume that a discharge strikes instantaneously while the slit crosses the field of observation, then the areas swept by the slit before the instant of striking will appear dark, while those swept after this instant will appear luminous. Thus, some care must be exercised in interpreting

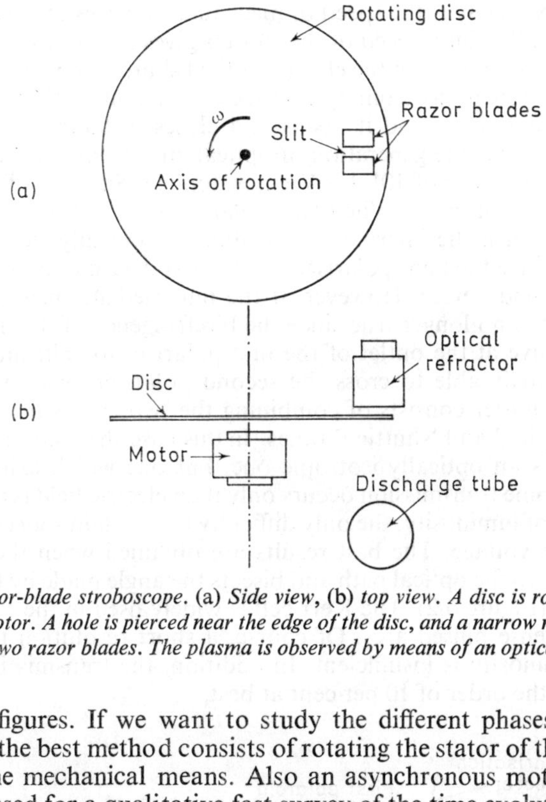

Fig. 6.1. The razor-blade stroboscope. (a) *Side view,* (b) *top view. A disc is rotated by means of a synchronous motor. A hole is pierced near the edge of the disc, and a narrow radial slit is shaped by gluing two razor blades. The plasma is observed by means of an optical refractor*

the observed figures. If we want to study the different phases of a periodic phenomenon, the best method consists of rotating the stator of the synchronous motor by some mechanical means. Also an asynchronous motor with a slow slip could be used for a qualitative fast survey of the time evolution of the phenomenon: if the slip is sufficiently small, the apparent evolution is slowed down by several hundred times.

If a plasma phenomenon is periodic and if a mirror is rotating nearby in synchronism with it, the figure viewed in any direction perpendicular to the rotation axis is a representation of the plasma in some given state. The time evolution can be observed, either by rotating the means of observation around the mirror, by rotating once more the stator of the motor driving the mirror with an endless screw device, or by using a low-slip asynchronous motor. In practice the axis of rotation is generally either parallel or perpendicular to some axis of the plasma. If the latter is filamentary (e.g. spark) no screening is necessary. But in general, one will observe only the portion of the plasma visible through a slit parallel to the said axis.

If the plasma is very luminous, there is no real need for periodicity. In such a case photography must be used; the method is called the '*Streak Camera*'. It is not advisable, if otherwise possible, to let the phenomenon occur at random; it is more interesting to trigger it by the rotation of the mirror, when the latter passes through the correct angular position. In some cases, such as in the observation of lightning, this triggering is completely out of the question, and obtaining a good picture is then purely a matter of chance and patience.

Ultra high-speed photography with mechanical shutters suffers from serious shortcomings. Not only is the speed of mechanical devices insufficient for most applications, but the finite speed of the shutter gives rise to distortions.

This explains the success of the electro-optical shutter known as the *Kerr Cell* which is free from such shortcomings. Besides, its resolution time can be lowered to the nanosecond range and it opens and closes instantaneously. The Kerr effect in a dielectric is the generation of optical anisotropic properties when an electric field of the order of 10^4–10^5 V/cm is applied. Such an effect occurs, for instance, with nitrobenzene. When two polarising devices are set at 90° from each other and when the intermediate medium is optically isotropic, no light is transmitted, since the light polarised by the first device is entirely inadequate to cross the second device. However, if the intermediate medium is optically anisotropic, this is no longer true since the birefringence of the medium causes the polarised wave at the outlet of the first polariser to split into two waves, each of them partly able to cross the second polariser. The principle of the electro-optical shutter consists of combining the two effects in order to attain very short 'opening' and 'shutting' times. In this case the medium between the two polarisers is an optically isotropic one, but one which exhibits the Kerr effect wherein some transmission occurs only if an electric field is applied. Apart from problems of luminosity, the only difficulty is to obtain short enough pulses with the proper voltage. The best results are obtained when the electric field is perpendicular to the optical path and bisects the angle made by the two planes of polarisation (Figure 6.2). The Kerr cell is widely used in the study of sparks and of very intense pulsed arcs. Of course, a short resolution time is useless if the total luminosity is insufficient. In addition, the transmission of the cell itself is only of the order of 10 per cent at best.

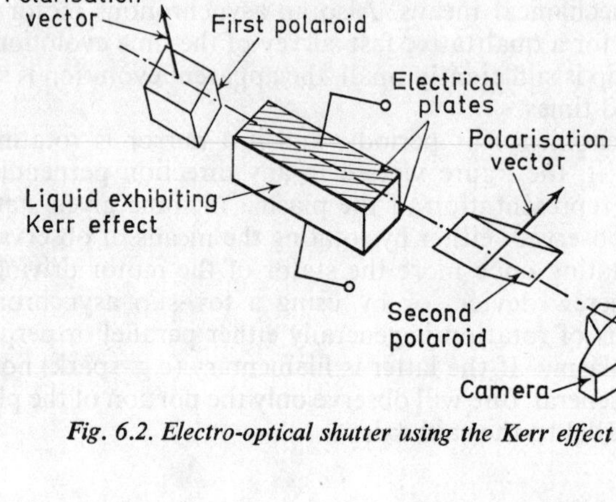

Fig. 6.2. Electro-optical shutter using the Kerr effect

Recently, new methods have been sought for attaining approximately the same result as the Kerr cell, avoiding if possible the important attenuation of light, by using a special type of camera with an image converter biased by an electronic fast gate. The plasma is examined by an optical system similar to that of an ordinary camera except that the photographic plate is replaced by a photocell which acts as the first element of an image converter. When the image of the plasma is focused on this photocell, electrons are emitted on the opposite side in proportion to the local light intensity. The electrons are then focused by electronic lenses analogous to those used in electronic microscopy techniques. The result is an electronic image of the photocathode, hence an optical image of the plasma, on a fluorescent screen similar to that in a cathode-ray tube.

So far, the only interest in the gated image converter is to permit the observation of weakly luminous plasmas, since the electronic image converter in the middle can introduce a significant amount of light amplification. Applications of this simplified technique have been made in several fields of science, for example, in astronomy. But if the source is luminous enough, a stroboscopic device can be added in the following manner: somewhere in the electronic section, a grid is introduced; it can be polarised negatively enough to repel all the electrons coming from the photocathode. This polarisation should be permanent, except that, during a brief instant, a short positive pulse is applied so that temporarily the camera as a whole is operating as described above. Resolution times in the nanosecond range have been attained in practice, and the main limiting effect seems to be the amount of light available from the plasma.

In practice, not one but three positive pulses are successively applied; in synchronism with them, an electric field perpendicular to the axis of the image converter is applied by increasing steps so that three corresponding images occur, not in the same location on the fluorescent screen but in three different locations. Thus, three successive images of the plasma, corresponding to three different instants of its life, can be obtained on the same photographic plate. Built-in electronic circuits can effect both operations automatically. The time interval between successive pictures can, in principle, be adjusted at will and in practice can be lowered to the same order of magnitude as the resolution time. The major advantages of this type of camera over a camera with a Kerr cell are (1) that several pictures can be taken, corresponding to *successive* states of the *same* event, and (2) some light amplification can be introduced by the image converter section.

If a complete picture of the plasma is not required, and can be replaced by a unique signal indicating the passage at a given location of a strong luminous front (e.g. shock wave), it is generally advisable, for economical reasons, to replace the camera by a set of *photomultipliers*. Assume, for instance, that the coefficient γ_t of electron multiplication is equal to three, and consider an electron falling on a first multiplying electrode (dynode); it will give rise to the emission of three electrons, which can be accelerated toward a second dynode. With twelve dynodes the amplification is 3^{11}, which is of the order of 10^5. Further amplification can be achieved by ordinary electronic tubes, the grid of the first one being polarised by the voltage drop induced in a resistor by the current of the last dynode. In order to use the same device as a photomultiplier the first dynode must be preceded by a photocell. The latter can be excited by the light emitted by a convenient type of 'phosphor' which can be a crystal, a plastic material, or even a liquid in a transparent container. The phosphorescence is excited by any kind of radiation coming from the plasma, including visible light, ultra-violet light, and X-rays. If convenient 'windows' are provided, corpuscular radiations can be detected, even neutrons when the phosphor is 'loaded' with a boron compound, thus converting the neutron into an alpha particle.

6.3 SPECTROSCOPIC METHODS[16]

All the spectroscopic optical methods use an optical spectrograph having a sufficient resolving power. A picture of the plasma spectrum is taken by means of this device. Only exceptionally good pictures can be used at all and the exposure time must be chosen carefully. Also, the exposure of the photographic film must be calibrated with care.

The next operation is the microphotometry of the density of the film by interposition of the latter between a lamp and a photocell, with a narrow slit delimiting the illuminated part of the film, the movement being a slow motion perpendicular to the slit. Thus a graph is obtained, giving an accurate reading of the light intensity versus the wavelength. Alternatively, such a graph is obtained directly by scanning the spectrum itself by means of a photocell. If the spectrum recorded is continuous as in the carbon arc, information can be obtained about the gas temperature, provided the contribution from the electrodes can be eliminated since it is an important parasitic part. But in general, the spectrum is made of definite spectral lines, and information can be obtained by analysing and comparing the lines. In certain types of low-luminosity plasmas, such as those obtained in thermonuclear experiments, impurities are purposely introduced into the gas in order to increase the luminosity and to assure that well-known lines appear.

It is possible, from a comparison of the spectral line intensities at different wavelengths, to determine the temperature of the exciting agent, which is in general the electronic component of the plasma. Each line corresponds to a certain intensity, obtained by integration of the curve on the photometric recording, provided no distortion or saturation occurs by improper choice of the exposure time. This line can thus be translated into a number of photons per unit time, hence a number of exciting collisions per unit time. Since the number of exciting collisions is a function of the temperature of the exciting agent, for each spectral line, a comparison of the relative intensities gives the value of this parameter. In principle, the shortest wavelengths require the highest temperatures. If a significant amount of step-excitation occurs, the method may become very intricate.

Two important effects used in spectroscopic measurements are the *Zeeman* and the *Stark* effects. The Zeeman effect is a modification appearing in the spectrum under the influence of a magnetic field. The Stark effect is usually the splitting of the lines in the spectrum in the presence of an electric field. In both cases the local value of the field can be calculated from the modification of the spectrum. However, a region of homogeneous field must be chosen, otherwise the final result may be a broadening rather than a splitting and the interpretation is different. For instance, in strongly or fully ionised plasmas, most of the atoms or ions emitting the radiation are in relatively large electric fields, owing to the microcomponent due to the charge carriers themselves. When the density of the carrier pair is increased, each emitter is, on the average, situated in a greater electric field at the instant of the emission. This gives rise to a special kind of Stark effect, different from that observed in a homogeneous electric field, and called Stark broadening of the spectral line instead of a splitting. This broadening evidently results from the superposition of many different splittings in the highly heterogeneous and time-dependent microfield. Both the ions and the electrons of the plasma play a role, the former in a quasi-static way, and the latter in the manner of impacts. The half-width of the spectral line is a function of the carrier pair density. This function is linear under certain conditions but may sometimes acquire an important quadratic term. It can serve as a means of measuring the plasma density, either by careful analytical treatment or after suitable calibration[19].

Another important phenomenon is the Doppler effect, which is a wavelength modification under the influence of a velocity component along the line of sight. If the distance between the source and the observer decreases, then the wavelength appears shorter, whereas when the light source recedes from the observer, there is an apparent increase in its wavelength. If the plasma approaches or recedes as a whole, there is a general displacement of the spectral lines, and the quantity measured is the drift velocity. Such a measurement is possible only under exceptional conditions. In general, the Doppler effect is associated with the random velocity rather than with the drift velocity. The result is once more a broadening of the spectral line. However, this Doppler broadening is frequently negligible with respect to the other causes of broadening. For the Doppler broadening to become significant, high temperatures are required as in thermonuclear studies. Alternatively, low pressures tend to emphasise Doppler broadening by reducing all the other causes.

The Doppler frequency shift $\Delta\nu = \nu - \nu_0$ resulting from the relative velocity v of a radiating atom with respect to the observer is given by the well-known relation $(\Delta\nu/\nu) = (v/c)$ where c is the velocity of light in vacuum. If a velocity distribution is involved, the line acquires a profile, which for a hot, quiescent plasma, is perfectly symmetric and approximately Gaussian. When the emitting atoms of the plasma have a Maxwellian distribution, the shape is exactly Gaussian and of the form:

$$f_D(\nu) = (\pi^{1/2}/\delta\nu_D)\exp[-(\Delta\nu/\delta\nu_D)^2]$$

The Doppler half-width in frequency is therefore:

$$(\delta\nu_D/\nu_D) = \frac{\ln(2)}{c}(2kT/M)^{1/2} \tag{6.1}$$

where M is the mass of the emitting species. The temperature measured in this way is that of the component emitting the light, which is in fact some definite and perhaps insignificant kind of excited atoms, molecules or ions. The conclusions that can be drawn concerning the temperature of major components such as the positive ions depend upon the circumstances. In thermonuclear experiments for instance, where the method has been most in use, the excited component is, in fact, a purposely or accidentally introduced rare impurity. Owing to the order of magnitude of the temperature, it is ionised several times, so that the spectrum observed originates in excited states of this multiply-ionised ion. In such a case, the temperature measured is essentially that of the ionic component of the plasma.

We should finally mention two other broadening mechanisms especially efficient in high-pressure arcs: the resonance broadening and the Van der Waals broadening. The first one is a broadening mechanism in which the frequency of a spectral line is influenced by the presence in the vicinity of neutral atoms in the lower level of the transition. This mechanism is especially conspicuous for resonant lines and is proportional to the pressure (approximately 1 Å/torr for the sodium D line). The Van der Waals broadening is analogous to, but characteristic of, gas mixtures (e.g. argon–mercury discharges). It is two to three orders of magnitude less efficient than the resonance broadening.

6.4 ELECTROSTATIC AND MAGNETIC PROBES[10]

The simplest electrostatic probe is an auxiliary electrode inserted into the plasma and having dimensions much smaller than those of the main electrodes or the radius of the tube. This probe is called *floating* when no attempt is made to polarise it by an external voltage source. It was generally thought in early experiments that such a floating probe, like any kind of probe introduced into a solid or liquid conductor, would assume a potential equal to that of the point where it is inserted. Since the major contribution of Mott-Smith and Langmuir[35], this expectation is known to be false. The reason is that, in most plasmas, the average velocity of the electrons is significantly higher than that of the positive ions. Even if this were not true, it would require an improbable coincidence for the average velocities of both components to be equal. Assume further that the probe is initially at the space potential of the surrounding areas of the plasma. In such a case, it is hit per unit time by a larger number of electrons than ions so that the voltage of the probe will tend to decrease because of the accumulation of an excess of electrons. This creates a voltage drop between the probe and the surrounding plasma, the effect of which is to accelerate the positive ions and to brake the electrons. This process will go on until, eventually, the average velocity of both components becomes equal when they hit the probe. Since both densities are equal in the surrounding plasma, the number of hits has thus become equal for both components, hence the resulting current is zero. The probe has reached its so-called 'floating potential' which can be measured by a voltmeter, provided the latter draws a negligible amount of current from the plasma. The floating potential is in practice negative with respect to the space potential. In most practical cases, however, the difference is below 10 V. Consequently, the method of the floating probe can be used in two cases (1) in an unrestricted manner, if 10 V is a negligible quantity with respect to the quantities to be measured, or (2) for compressions only, if the difference between the floating potential and the space potential is expected to be almost constant so as to compare space potentials among themselves, for instance in order to deduce the electric fields.

The electric probe is called a *Langmuir probe* if it is connected to an external source of electric bias. The Langmuir probe characteristic is the curve obtained by plotting the probe current versus the probe voltage. The probe current is the result of the combination of ionic and electronic components. When the probe is negatively biased, it collects more positive ions than electrons so that as the bias increases the current eventually becomes a pure positive ion component. When the probe is biased positively with respect to the floating potential, it collects more electrons than positive ions and eventually, as the bias is increased, the electron component also becomes pure. In most cases, the ionic component is several hundred times smaller than the electronic one, so that the separation can be effected by approximate methods such as linear extrapolation of the curve obtained for very negative potentials.

If the probe potential is equal to the space potential and if no reflection or secondary emission occurs, then the probe current density is given by equation (1.51). If a retarding potential is applied, the current density at the probe tends to be reduced because part of the carriers are repelled or deflected away from the probe. The fractional decrease can be calculated from a statistical study of the orbital motions of the carriers in the probe sheath[35]. If the velocity distribution is isotropic, the fractional decrease for a given voltage is independent of the probe shape and depends only on the velocity distribution. If the latter is Maxwellian, then from Boltzmann statistics:

$$J_p = (n\bar{v}/4) \exp [-(eV_p/kT)] \tag{6.2}$$

Hence, a straightforward method of calculating T is obtained. If the logarithm of I_p, where $I_p = J_p S_p$ is plotted versus the probe voltage V_p a straight line is obtained, the slope of which is $1/T$ times a universal constant. In addition, if T is known, $\bar{v}$ can be calculated, hence the value of I_p at the space potential is directly related to n and enables one to calculate it easily. In general, the anisotropy introduced by the drift velocity is insufficient to render the above theory invalid[35].

If the probe voltage is accelerating, the current density is generally increased with respect to the value attained at the space potential. However, the fractional increase is less than the value given by an extrapolation of the preceding law. Consider, for instance, a Maxwell distribution. The effect of a retarding potential on such a distribution is to give an identical distribution of velocities; the number of particles involved is, however, reduced because of the many reflections and deflections. This is no longer the case for an accelerating potential since particles having an energy less than that corresponding to this potential will in any case be missing from the distribution.

In the case of an ideal plane probe, the latter collects, at the space potential, all the carriers available; the application of an accelerating potential can accelerate the individual carriers but it has no influence on their number so that the probe current will not increase beyond the value corresponding to the space potential. Hence, the latter corresponds to a well defined kink on the probe characteristic (Figure 6.3). This ideal case is never attained in practice. The reason is that, the application of an accelerating potential will deflect carriers toward the probe, hence, the current density collected will be greater than

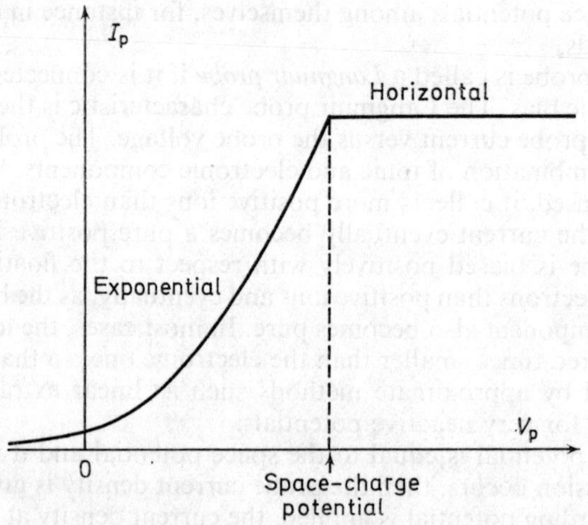

Fig. 6.3. Characteristic of an ideal Langmuir probe

at the space potential. Mott-Smith and Langmuir[35] have considered two extreme cases, the actual circumstances always lying somewhere in between.

If the plasma density is small enough, the Debye length is large, and the sheath dimensions can be large compared to those of the probe, provided that the probe voltage differs from the space potential at least by two or three times kT/e. In such a case, the current density at the probe is entirely governed by the orbital motion of the carriers around the probe, and the result turns out to be a constant for the ideal plane probe, a parabola with a horizontal axis for a cylindrical probe, and a linear function, tangent to that below the space potential for the spherical probe. The slope of a T_p versus V_p curve in the spherical probe and in a I_p^2 versus V_p curve in a cylindrical probe can be used as a substitute method for determining n[25].

If the plasma density is sufficiently high, the probe sheath thickness can no longer be regarded as large with respect to the probe dimensions, and the probe current is essentially limited by the sheath area, which imposes an upper limit to the current which can be drawn from the plasma. This upper limit current is $n\bar{v}S_s/4$, where S_s is the area of the sheath. If S_s is of the same order as S_p, most of the particles penetrating through the sheath are effectively collected. The principle involved in calculating S_s is that the space charge accumulated in the sheath must be such that the probe electric field does not extend beyond the limits of the sheath. When the Debye length becomes negligible with respect to the probe dimensions, $S_s \cong S_p$, so that there is no opportunity for an increase in probe current above the space potential, as was the case in an idealised plane probe. Figure 6.4 summarises the situation for spherical and cylindrical probes.

In practice, the sheath of a plane probe without guard rings undergoes deformations because of edge effects*, and the probe current for accelerating poten-

* Edge effects are theoretically present with other probe shapes, but usually of a lesser order of magnitude.

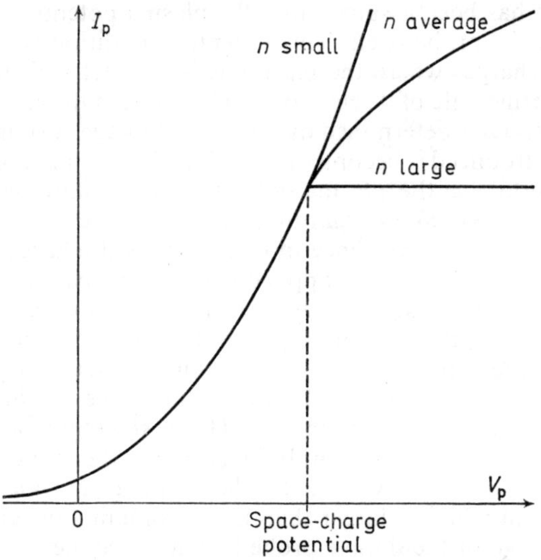

Fig. 6.4. Ideal characteristics of spherical and cylindrical Langmuir probes for several values of the plasma density

tials is essentially governed by these edge effects. An approximate theory has been developed by Mott-Smith and Langmuir[35], and practical graphic methods are presented by Hoyaux[25]. Figure 6.5 shows actual plane probe characteristics compared with the ideal characteristic. In this case, the space potential is not exactly given by the kink but has to be determined by a reconstruction of the ideal plane-probe characteristic from the actual one[25].

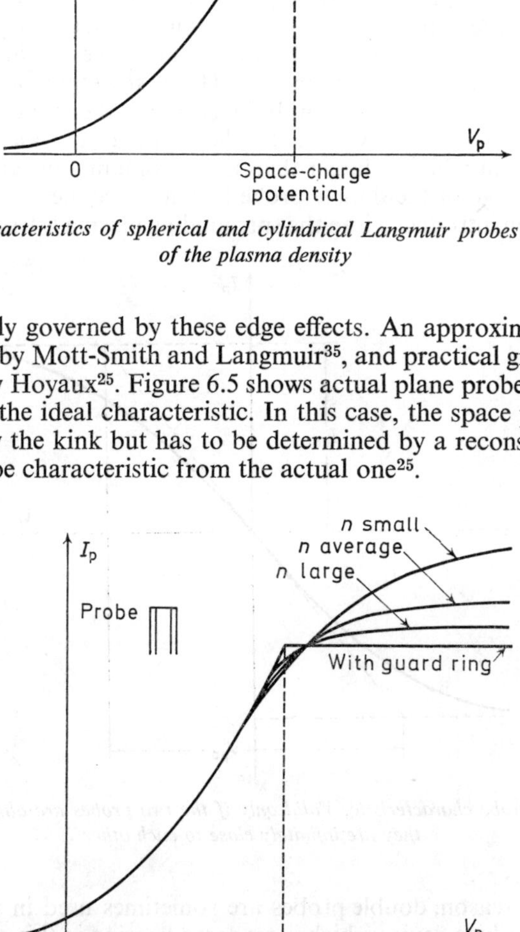

Fig. 6.5. Experimental characteristics of Langmuir probes and the effect of the addition of guard rings

Up to now it has been assumed that the plasma potential is independent of the presence of the probe itself. This potential condition is generally the case in ordinary discharges where the plasma is in contact with the outside world through the intermediate of its electrodes. However, in the absence of the probe, the plasma potential is determined by unknown leakage currents, and if a Langmuir probe is introduced it becomes by itself the best contact of the plasma with the external world, i.e. the plasma potential is essentially such that the probe is at the floating potential, disregarding its bias voltage.

To overcome this inconvenience in electrodeless discharges, *double probes*[28] have been suggested. The bias is applied between the two probes, and the sum of the probe currents is essentially zero. If the probes are identical, the double-probe characteristic can be deduced graphically from the single probe characteristic. One of the probes is always collecting an excess of positive ions, i.e. the total current cannot exceed the maximum value attained for the ionic branch of the single-probe characteristic. This is sometimes an advantage since it frequently happens that single probes melt under the influence of the current that they collect. It has, however, the shortcoming that all of the electronic part of the current which is larger than the maximum ionic current is ignored. Another advantage of the double probe is that, since the total current is zero, the perturbation introduced into the surrounding plasma is less than for a single

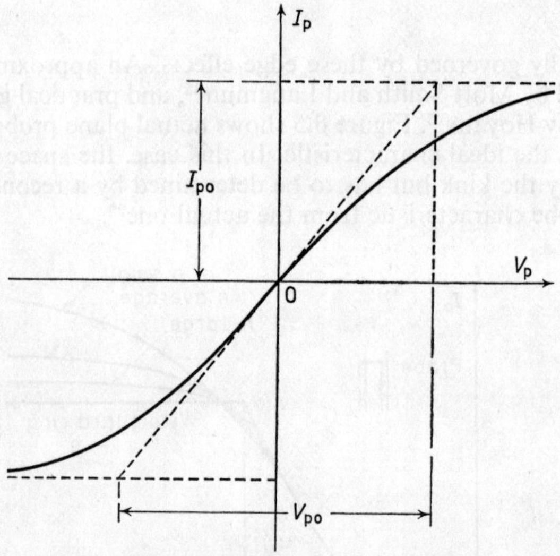

Fig. 6.6. Double-probe characteristic. Valid only if the two probes are almost identical and if they are infinitely close to each other

probe. For this reason, double probes are sometimes used in non-floating discharges; it is the bias device which is rendered floating in this case.

The theory of the double probe method is similar to that of the Langmuir probe. Consider the double-probe current versus voltage characteristic of Figure 6.6. This characteristic is slightly idealised, and strictly valid only if the two probes are almost identical, and infinitely close to each other. In the idealised case, the current versus voltage characteristic has two horizontal asymptotes, corresponding to the random current of the *ionic* component, whereas the slope at the origin is related to the *electronic* temperature, leading to:

$$T^- = eV_p/2k \tag{6.3}$$

where V_p is the space charge potential; also:

$$n^+ = \frac{I_p}{Se}\sqrt{\left(\frac{2\pi M^+}{kT^+}\right)} \tag{6.4}$$

I_p being the asymptotic value of the probe current. Equation (6.4) lacks accuracy, since there is no way of attaining T^+ except by an independent diagnostic means. In many types of discharges, T^+ is very close to the absolute temperature of the neutral component, but this does not seem to give good results. The probe collects only the ions accelerated toward it by ambipolar diffusion. For these, a temperature of some $0{\cdot}5\,T^-$ [T^- given by equation (6.3)] gives the best results. Besides, the asymptotes are generally not horizontal, and I_p has to be taken rather arbitrarily by analogy with sheath effects in Langmuir probes.

If the plasma is in a steady state, the probe voltage can be displayed horizontally on the screen of an oscilloscope, with the probe current, or a voltage proportional to it, displayed vertically. If the plasma is variable, a complicated figure, arising from the superposition of all the probe characteristics occurring at the same location for the different states of the plasma, would appear. If the discharge is periodic, or comprises, for instance, recurrent bursts occurring in a quasi-periodical manner, it is possible to select short positive pulses at a given arbitrary phase of the phenomenon, to introduce them at the input of a delay unit, in order to reproduce them arbitrarily delayed. The control grid of the oscilloscope is kept sufficiently negative at all times, save for this short arbitrarily delayed positive pulse. This has the result of wiping off all the probe characteristics, except that corresponding to a given, arbitrary state of the plasma. The time evolution can be followed by modifying the delay introduced by the delay unit[31].

The presence of a *magnetic field* permeating the plasma will alter significantly the shape of the probe characteristic. In 1936, Spivak and Reichrudel[43] studied the collection of electrons by cylindrical probes in a weak magnetic field. They found that the Langmuir orbital theory was valid only for large Debye sheaths. Thus, orbital motions control the collection of electrons in weak magnetic fields over a larger range of possible potential distributions. Later, Bickerton and Von Engel[7] studied electron collection by a plane parallel to a magnetic field for values of the Larmor radius equal to r_h, much larger or much smaller than r_h, where r_h is defined as:

$$r_h = (kT_2/4\pi n_0)^{1/2}/2q_1$$

where subscript (1) is related to the collected particles and subscript (2) is related to Maxwellian particles in the plasma; n_0 is the plasma density at an infinite distance away from the plane. Bickerton[7] assumed that electron motion is prescribed by perpendicular diffusion and mobility. If collisions during the last Larmor orbit are considered, Bickerton[7] reports that:

$$J = \frac{n_1 \bar{v}}{4}\left\{\frac{8+\omega_c^2\tau^2[1-\exp(-2\pi/\omega_c\tau)]}{2(4+\omega_c^2\tau^2)}\right\}$$

where ω_c is the electron gyrofrequency, τ is the mean collision time, and n_1 is the electronic density near the plane of the probe.

Although electric probes have been widely used in the study of electrical discharges, they are of little help in thermonuclear plasmas because they melt and they perturb the plasma. Most of the sounding work in thermonuclear experiments has been performed with magnetic probes, which are also best adapted to the study of fast transients. There are two different types of magnetic probes, the Rogowsky coil and the miniature internal probe.

The *Rogowsky coil* is a solenoid having a large length-to-diameter ratio and employing flexible materials for both the wire and the core. The coil is usually outside the vessel enclosing the plasma. Each loop of the Rogowsky coil picks up a signal proportional to the time derivative of the flux enclosed, hence of the local induction if the loop is small enough. Since the coil is long with respect to its diameter, the overall signal is an integral of the local values performed along a path which is symbolised by the curvilinear axis of the coil.

If the magnetic field is in a steady state, a signal can be picked up by moving the Rogowsky coil from a remote location where the field is negligible to a location well inside the configuration to be studied. An ordinary fluxmeter is generally sufficient to perform the integration mechanically. This method is not generally used in plasma physics. Another method is applicable in transient magnetic fields such as the measurement of the instantaneous value of the current in the linear pinch. In this method, the coil is fixed and the signal induced resolves this time from the local variations of the field. In fast transients, a fluxmeter is generally insufficient to perform the integration, and conventional RC circuits are preferred.

The *Rogowsky girdle*[12] consists of two sets of coils. One of the coils is wound around the entire plasma region to be investigated whereas the other encloses only a small current flow. The currents not enclosed, such as those going to the walls, are represented by a difference in induced voltage. The current profiles can be obtained by segmenting the coils. Also, high frequency fluctuations such as those associated with plasma instabilities can be measured by using a Rogowsky girdle of low inductance coils. In this manner perturbations of frequencies varying from 10 kHz to 10 MHz can be studied.

Local values of the magnetic field can be measured by means of devices similar in principle to the Rogowsky coil. These devices consist of linear inductive probes, small enough to reduce the perturbation introduced. A considerable use has been made of miniature magnetic probes, in the form of coils a few millimetres in diameter and in length, protected by pyrex or quartz thimbles and electrically shielded. During a given measurement, the probe is fixed so that the signal is proportional to the time derivative of the local magnetic field[11]. The integration is performed by a classical RC circuit and the result displayed on the screen of an oscilloscope. Frequently, separate probes are used to measure the three components of the field in a given area of the plasma.

In some experiments, sufficient accuracy has been attained to provide a direct experimental confirmation of Maxwell's equations. In other cases, the latter are used to calculate unknown parameters such as the current density. Linear arrays are used across current channels to measure hydromagnetic instabilities and current density contours in dense plasmas.

Magnetic probes are sensitive to stray fields because of their small resulting signals. They are also sensitive to temperature variations especially in the case of Hall probes.

In the steady state, the field can be measured if the probe is rotating around a line orthogonal to its axis. This considerably complicates the design of the probe as the coil is less than one centimetre in diameter and in length. Only limited application has been made of this method in plasma physics.

6.5 CAVITY PERTURBATION METHOD

Since the advent of radar, cheap microwave devices have been made widely available, and this has made possible considerable advances in microwave diagnostic techniques. Ordinary plasmas are almost entirely transparent to most electromagnetic waves. However, the situation is entirely different in the neighbourhood of the four plasma resonances: electron electrostatic resonance, ion electrostatic resonance, electron cyclotron resonance, and ion cyclotron resonance. In practice, ion resonances are generally in the range of metre waves, whereas electron resonances occur generally in the centimetre wave range. The latter is more interesting than the former, since directed propagation, intersecting only a narrow channel in the plasma, can take place. In the neighbourhood of its resonance, an optically thin plasma can easily become completely or almost completely opaque. At the same time it begins scattering, reflecting or emitting a non-negligible power. It is advisable to compare the phenomena both near and far from the resonances.

The incidence of a microwave on a plasma generally gives rise to the following phenomena: (1) absorption; (2) diffusion, reflection and secondary emission; (3) phase lag, caused by the difference in refraction index; (4) modification of the polarisation and birefringence, among other phenomena. All these parameters are easily related to the characteristic parameters of the plasma. One of the most straightforward is the refractive index measured from the phase lag, which has a direct relationship with the plasma density. Energy absorption, diffusion, reflection and occasional secondary emission are related to the plasma density and to the frequency of collision which depends upon the electron temperature. This is perhaps the most accurate method for determining this parameter when the Langmuir probe cannot be used. The polarisation and birefringence phenomena are related to the magnetic field and can help to measure it inside the plasma. In practice, the inhomogeneity of the plasma can be a source of considerable complication. The absorption, phase lag, etc., are generally integrated over the entire path of the microwave through the plasma. Several methods can be used to overcome this shortcoming. One of them consists in combining the microwave measurement with another diagnostic method, which can be purely qualitative, such as ordinary photography. The latter gives, for instance, a good picture of the relative density of the plasma, and the measurement can be converted into a quantitative one if an integrated value along a given path is determined by a microwave technique. A second method consists in exploring several different paths in the plasma in order to increase the amount of data.

Microwave measurements can be carried out by the use of a resonant cavity. In this case, the plasma is placed in a cylindrical cavity operating in the TM_{010} mode. The TM_{010} mode is characterised by the fact that the unperturbed component of the electric field is uniform along the length of the cavity and parallel to it. This TM_{010} mode should be made the lowest cavity resonance mode in the design of the cavity. This method can be used to measure the average plasma density in any cross-section of a discharge tube by inserting this tube into the cavity as shown in Figure 6.7. The plasma acts through its refractive

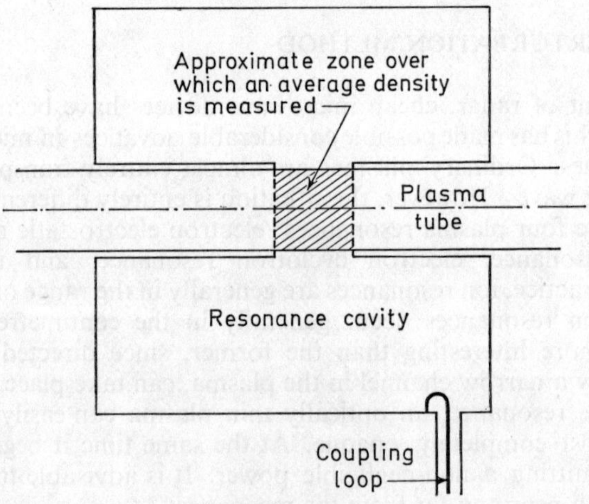

Fig. 6.7. Resonance cavity diagnostic method. The plasma is placed in a cylindrical cavity operating in the TM_{010} mode

index. Measurements of the resonant frequency of the cavity with and without plasma indicate that the latter introduces a detuning $\Delta\omega_0/\omega_0$ where ω_0 is the mode frequency. The amount of detuning permits an evaluation of the plasma density.

The frequency shift $\Delta\omega_0$ is given by Slater's perturbation theorem[42], or in magnitude:

$$\Delta\omega_0 = \frac{1}{4\mathcal{E}} \int_{\mathcal{V}_p} \sigma E^2 \, d\mathcal{V} \tag{6.5}$$

where $\mathcal{V}_p$ is the volume of the plasma inserted into the cavity, $\mathcal{E}$ is the average stored energy in the TM_{010} mode, and $\boldsymbol{E}$ is the unperturbed component of the electric field. The conductivity σ of a lossless plasma is given by $(\varepsilon_0\omega_p^2/\omega_0)$, where ε_0 is the permittivity of free space, and ω_p is the characteristic plasma frequency. The amount of stored energy $\mathcal{E}$ is $\varepsilon_0 \int_{\mathcal{V}_c} E^2 \, d\mathcal{V}$ where $\mathcal{V}_c$ is the volume of the cavity. Replacing $\mathcal{E}$ and σ by their values in equation (6.5), one finds readily for a uniform plasma:

$$(\Delta\omega_0/\omega_0) = (\omega_p/2\omega_0)^2 \frac{\int_{\mathcal{V}_p} E^2 \, d\mathcal{V}}{\int_{\mathcal{V}_c} E^2 \, d\mathcal{V}} \tag{6.6}$$

After integrating equation (6.6) over the cylindrical configuration, one finds:

$$(\Delta\omega_0/\omega_0) = 1{\cdot}85 m_\alpha^2 (\omega_p/\omega_0)^2 \, [J_0^2(2{\cdot}405 m_\alpha) + J_1^2(2{\cdot}405 m_\alpha)] \tag{6.7}$$

with $m_\alpha = (r_p/r_c)$, r_p being the radius of the plasma, and r_c the radius of the entire cavity. J_0 and J_1 are Bessel functions. The result is a simplification of the more exact expression obtained by Buchsbaum *et al.*[8]. The characteristic plasma frequency can thus be obtained from equation (6.7) and consequently the electronic density n can be deduced.

The quality factor of a resonance cavity is defined as the ratio of the time-average energy stored to the energy loss per cycle of oscillation. This factor is reduced under the effect of collisions in the plasma. An expression relating the quality factor to the collision frequency ν_c can be derived, thus permitting measurement of ν_c by the resonant cavity method.

6.6 MICROWAVE PROPAGATION TECHNIQUES[21]

Electromagnetic waves are dispersed in a plasma medium. This property is often used to measure the plasma density and consequently the charge density. Mathematically, the behaviour of charged particles in a free space can be described by two sets of equations; the Maxwell curl equations given by equations (3.4) and (3.8) with $\boldsymbol{D} = \varepsilon_0 \boldsymbol{E}$, $\boldsymbol{B} = \mu_0 \boldsymbol{H}$, and $\boldsymbol{J} = \sigma \boldsymbol{E}$, and the equation of motion of the particles (e.g. electrons):

$$\boldsymbol{E} = \frac{M}{e}\frac{d\boldsymbol{v}}{dt} + \frac{\nu_c M}{e}\boldsymbol{v} \tag{6.8}$$

where ν_c is the collision frequency and $\boldsymbol{v}$ is the electron velocity. It is assumed here that there is no applied magnetic field. If it is assumed that the electron is initially at rest and that the perturbations on the electric field quantities vary as $\exp(j\omega t)$, equation (6.8) can be written:

$$\boldsymbol{E} = \frac{M}{e}(\nu_c + j\omega)\boldsymbol{v} \tag{6.9}$$

The conduction current density $\boldsymbol{J}$ is:

$$\boldsymbol{J} = ne\boldsymbol{v} \tag{6.10}$$

Thus, if $\boldsymbol{v}$ is replaced by its value from equation (6.10) in equation (6.9), one finds:

$$\boldsymbol{J} = \frac{\omega_p^2}{\nu_c + j\omega}\,\varepsilon_0 \boldsymbol{E} \tag{6.11}$$

with ω_p being the characteristic plasma frequency given by equation (1.13). Introducing equation (6.9) into equation (3.8), one finds:

$$\nabla \times \boldsymbol{H} = \left(\frac{\omega_p^2}{\nu_c + j\omega} + j\omega\right)\varepsilon_0 \boldsymbol{E} \tag{6.12}$$

In the absence of an applied magnetic field, the plasma modifies the permittivity of the medium but keeps its permeability unchanged. From equation (6.12) the apparent permittivity of the plasma can be deduced, assuming that the overall current is a displacement current; it is:

$$\varepsilon_p = \varepsilon_0\left(1 - \frac{\omega_p^2}{\omega(\omega - j\nu_c)}\right) \tag{6.13}$$

If the effect of ions is not negligible, similar equations for these can be deduced.

The equation of an electromagnetic wave propagating in a magnetic field free plasma is:

$$\nabla^2 \boldsymbol{E} - (\omega_p/c)^2 \boldsymbol{E} = \frac{1}{c^2}\frac{\partial^2 \boldsymbol{E}}{\partial t^2} \tag{6.14}$$

where c is the velocity of light in vacuum. Taking a one-dimensional approach in which the electric field is in the y-direction and propagates along the x-axis, equation (6.14) becomes:

$$\frac{\partial^2 E_y}{\partial x^2} - (\omega_p/c)^2 E_y = \frac{1}{c^2}\frac{\partial^2 E_y}{\partial t^2} \tag{6.15}$$

The solution of which is:

$$E_y = E_0 \exp\left[j(kx - \omega t)\right] \tag{6.16}$$

where k is the constant of propagation of the electromagnetic wave. Introducing equation (6.16) into equation (6.15) one obtains:

$$k^2 + (\omega_p/c)^2 = (\omega/c)^2 \tag{6.17}$$

Equation (6.17) is the dispersion equation of the electromagnetic wave from which the value of k can be deduced:

$$k = (\omega/c)[1 - (\omega_p/\omega)^2]^{1/2} \tag{6.18}$$

Therefore, a wave of angular frequency ω will propagate for $\omega > \omega_p$. This wave will degenerate into a decreasing exponent for $\omega < \omega_p$. The phase velocity v_p of the propagating wave is defined as (ω/k), and taking into account equation (6.18), one obtains:

$$v_p = c[1 - (\omega_p/\omega)^2]^{-1/2}$$

The dielectric constant ε of the plasma is defined as $(c/v_p)^2$, thus:

$$\varepsilon = 1-(\omega_p/\omega)^2 \tag{6.19}$$

If collisions are considered, an equation identical to equation (6.13) can be deduced from equation (6.19) for the permittivity of the plasma. The propagation of the electromagnetic wave will thus determine the value of ω_p and consequently the value of n.

To use this method for plasma diagnostics, a commercial coaxial standing-wave detector can be used. The central conductor of the detector could be removed and replaced by the plasma tube as suggested by Burke and Crawford[9]. The plasma will thus be inductive for $\omega < \omega_p$ and transmission will occur as a result of the capacitance existing between the plasma and the outer conductor of the detector. The value L_p of the plasma inductance for $\omega \ll \omega_p$ will be:

$$L_p = \frac{l}{\varepsilon_0 \omega_p^2 S_p} \tag{6.20}$$

where S_p is the cross-section area of the plasma and l is the length of the line. The capacitance C_p of the plasma for a cylindrical geometry will be:

$$C_p = \frac{2\pi\varepsilon_0\varepsilon_g l}{\ln(r_b/r_p)+\varepsilon_g \ln(r_c/r_b)} \tag{6.21}$$

where ε_g is the relative permittivity of the glass walls confining the plasma discharge, r_p is the radius of the plasma, r_b is the outer radius of the plasma tube, and r_c is the radius of the outer conductor of the wave detector. The phase velocity of the wave can be deduced as a function of L_p and C_p, it is $l(L_pC_p)^{-1/2}$, and taking into account equations (6.20) and (6.21), one finds:

$$v_p = \frac{\omega_p r_p}{\sqrt{(2\varepsilon_g)}}[\ln(r_b/r_p)+\varepsilon_g \ln(r_c/r_b)]^{1/2} \tag{6.22}$$

An adequate slotted line could be added to the wave detector to observe the standing waves. A microwave probe is used to detect the waves and to locate the nodes, thus yielding the value of ω_p and consequently the value of n. Figure 6.8 illustrates the basic circuit of the transmission line diagnostic method as reported by Burke and Crawford[9].

When a magnetic field is applied to a plasma, the effect of the propagation of a plane polarised electromagnetic wave is to produce electric field and current density components perpendicular to their initial directions. The plane of polarisation of the wave is consequently rotated. The resulting circularly polarised electromagnetic wave propagating in the magnetic field direction will then have a complex effective dielectric constant equal to:

$$\bar{\varepsilon} = 1-\frac{(\omega_p/\omega)^2[1\pm(\omega_c/\omega)]+j(\nu_c/\omega)}{[1\pm(\omega_c/\omega)]^2+(\nu_c/\omega)^2} \tag{6.23}$$

where ω_c is the electron cyclotron frequency and where the minus or plus signs depend on whether the wave is right-or left-hand circularly polarised.

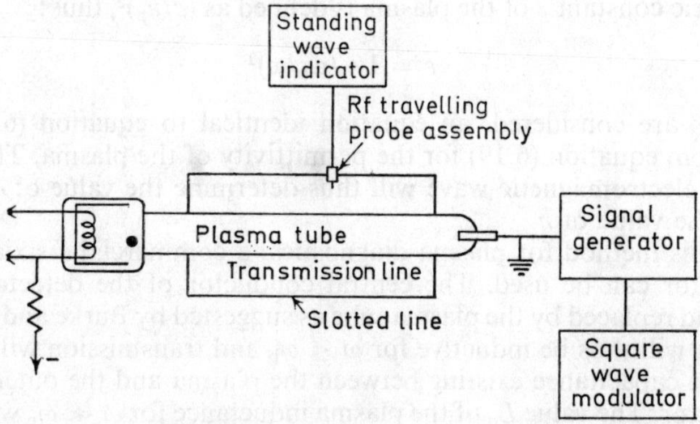

Fig. 6.8. Basic circuit for transmission line measurements. The central conductor of a commercial coaxial detector is removed and replaced by the plasma tube. A slotted line is added to observe the standing waves that are detected by a microwave probe[9]

6.7 DIPOLE RESONANCE METHODS

The dipole resonance method is based on quasi-static analysis and yields a resonance only when the plasma is present. The analysis is based on Laplace's equation. It is valid for frequencies smaller than the frequency at which the free space wavelength becomes comparable to the dimensions of the exciting electrode.

Consider a dielectric rod in free space subjected to a transverse electric field (Figure 6.9). It can be shown that the relationship between the electric field E_r inside the rod and the field E_0 far from the rod is:

$$E_r = \frac{2E_0}{1+\varepsilon_r} \tag{6.24}$$

ε_r being the relative permittivity of the rod. For a plasma column in an alternating electric field ε_r is given by equation (6.19), and equation (6.24) becomes:

$$E_r/E_0 = \frac{1}{1-\frac{1}{2}(\omega_p/\omega)^2} \tag{6.25}$$

In this case a resonance would occur at $\omega = \omega_p/\sqrt{2}$. However, the plasma column cannot be assimilated to a dielectric rod. Around the plasma is a glass tube whose walls have a certain thickness. An idealised cylindrical stripline system is assumed to surround the overall glass tube and is separated from it by an air layer. Applying Laplace's equation in cylindrical co-ordinates and assuming that it is valid for all the considered regions, i.e. plasma, glass walls, air layer, one finds an equation similar to equation (6.25), such as:

$$E_r/E_0 = \frac{1}{1-(1/m_p)(\omega_p/\omega)^2}$$

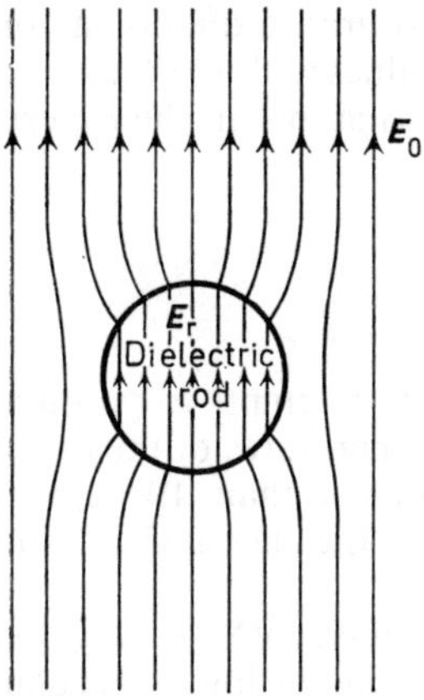

Fig. 6.9. Dielectric rod in an electric fied

where m_p is a dimensionless quantity which is a function of the geometry of the 'plasma tube–metal tube' system and of the dielectric properties of the glass tube enclosing the plasma. In this case the resonance frequency would be $\omega_p/\sqrt{m_p}$. The value of m_p is:

$$m_p = 1+\varepsilon_g\left\{\frac{1-[(1-\mathcal{A})/(1+\mathcal{A})]\,(r_a/r_b)^2}{1+[(1-\mathcal{A})/(1+\mathcal{A})]\,(r_a/r_b)^2}\right\} \tag{6.26}$$

where:

$$\mathcal{A} = \frac{1}{\varepsilon_g}\,\frac{1+(r_b/r_c)^2}{1-(r_b/r_c)^2}$$

and r_a, r_b and r_c are the radii of the internal wall of the glass tube, its external wall, and the stripline metal tube respectively. For $r_a = r_b$ and $r_c = \infty$, one finds that $\mathcal{A} = 1/\varepsilon_g$ and consequently $m_p = 2$. Computed numerical values of m_p are given by Crawford *et al.*[13]. To excite non-degenerate higher order resonances, the plasma column can be moved off centre in the stripline system and special types of electrodes should be used.

If the effect of collisions is considered, the value of the apparent dielectric constant is then given by equation (6.13); equation (6.25) becomes:

$$E_r/E_0 = \frac{1}{1-\dfrac{1}{m_p}\,\dfrac{\omega_p^2}{\omega(\omega-j\nu_c)}} \tag{6.27}$$

Values of m_p would be different from those given by equation (6.26) in the case where a non-zero electron temperature is considered[15].

Practically, the determination of the dipole resonant frequency is carried out over a range of current. The value of the plasma density is thus obtained directly from the resonance frequencies once the values of m_p are known. It is possible to display the resonance on an oscilloscope by modulating the discharge current. Often, resonances of lower amplitude are observed at higher currents[14].

By this same method a crude value of the collision frequency of the plasma can be found. Indeed, the power absorbed by the resonance is due to collisions. If tuning is used to decouple the detector from the signal source, a quality factor measurement can be obtained from the current trace. Values within a factor of 2 from the theoretical value, ω/ν_c, were obtained in this manner in the experiments reported by Burke and Crawford[9].

6.8 SHOCK TUBE MEASUREMENTS[24]

The development of shock tubes was a result of the many attempts to produce high temperature plasmas. Indeed, with shock tubes, temperatures of the order of 10^6K have been reached for plasma densities no smaller than 10^{16} cm^{-3}. Two important types of shock tubes can be distinguished: the chemical and the magnetic types.

The *chemical shock tube* consists of a long tube separated into two regions by a diaphragm. One region of the tube is filled with a gas of low molecular weight whereas the other region is filled with a gas of higher molecular weight. The low molecular weight gas is at a high pressure, whereas the higher molecular one is at a low pressure. When the diaphragm bursts, the gas at high pressure moves with a high speed and forms a shock wave in the low pressure gas. Thus the low molecular weight gas creates a very strong shock in the high molecular weight gas, since its Mach number in the high molecular weight gas is equal to several units. The shock will create a high temperature leading to the full ionisation of the gas. However, temperatures that can be reached by chemical shock tubes are limited by the escape velocity of the driver gas. To achieve higher temperatures, the driver gas can be replaced by a magnetic field created by the discharge of a capacitor bank.

This leads to the idea of a *magnetic shock tube*. Several geometrical configurations have been used for magnetically driven shock tubes. One that is frequently used approximates a one-dimensional flow in the coaxial geometry. In this geometry the driving condenser bank is discharged between two coaxial metal cylinders. This discharge produces a radial current sheet which in turn creates an azimuthal magnetic field. This magnetic field drives the current sheet in the annular spacing between the two coaxial cylinders. If the electrical conductivity of the sheet is high enough, the sheet will act as a piston and will generate a shock wave in front of it.

Kemp and Petschek[29] studied the flow in the shock tube by assuming that the current sheet acts as a solid piston that drives a shock wave in front of it at a certain distance filled with a shock-heated gas. Thus, Kemp and Petschek could treat separately the shock wave and the current sheet that can be assimilated with an expansion wave. The Alfvén Mach number $\mathcal{M}_A$ upstream of an ordinary gas-dynamic shock is:

$$\mathcal{M}_{A1} = (v_s/v_A) \geqslant [(\gamma+1)/(\gamma-1)]^{1/2}$$

where v_s is the shock speed and v_A is the Alfvén speed given by $B_z/\sqrt{(\mu_0 \varrho)}$, ϱ being the density of the medium and B_z the component of the magnetic field in the z-direction; γ is the ratio of specific heats and the subscript (1) denotes conditions upstream of the shock. For $1 \leqslant \mathcal{M}_{A1} \leqslant [(\gamma+1)/(\gamma-1)]^{1/2}$ the shock may become a magnetohydrodynamic switch-on shock and a tangential magnetic field $B_{\theta 2}$ will be created behind the shock. For both cases of an ordinary gas-dynamic shock and an MHD switch-on shock the conductivity is assumed infinite on both sides of the shock. Thus, there will be no electric field tangential to the shock in shock-stationary co-ordinates. The axial field interacts with the sheet current, introducing a tangential force $J_r B_z$ that produces an azimuthal acceleration of the fluid in the current sheet. By spreading through the gas, the current sheet reduces this acceleration and the axial field increases in intensity thus pushing the current sheet toward the shock front until it reaches it for:

$$\mathscr{M}_{A1} = [(\gamma+1)/(\gamma-1)]^{1/2} \tag{6.28}$$

The sheet will move away from the shock when $\mathscr{M}_{A1}$ is smaller than the value of equation (6.28) and part of its current will flow in the shock front.

In the snowplough model, the current sheet is assumed to sweep the gas and accelerate it to the sheet velocity. This is a limiting case of the Kemp–Petschek model for ϱ_2/ϱ_1 going to infinity, the subscript (2) here referring to conditions behind the shock. For this model the front speed is given by $\varrho_1 v_s^2 = B_\theta^2/2\mu_0$, where B_θ is the driving field in the θ-direction.

Shock tubes have been used for measuring plasma conductivity. One of the first experimental devices was the shock tube used by Lin *et al.*[32] shown in Figure 6.10. An argon plasma was driven into the tube through a d.c. magnetic field formed by a coil. The radial component $\boldsymbol{B}_r$ of the field interacts with the plasma velocity $\boldsymbol{v}$ to yield an azimuthal electric field $\boldsymbol{E}_\theta = \boldsymbol{v} \times \boldsymbol{B}_r$. The value of

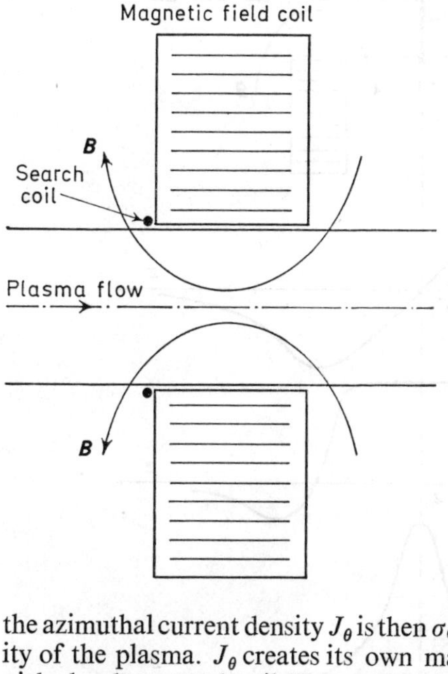

Fig. 6.10. Magnetic field configuration in a shock tube with a search coil[32]

the azimuthal current density J_θ is then $\sigma_0 v B_r$, where σ_0 is the electrical conductivity of the plasma. J_θ creates its own magnetic field, the flux of which can be picked up by a search coil. The response of this search coil is then compared with its response to an aluminium plug of known conductivity, velocity and dimensions. The comparison of the two responses yields the *a priori* unknown electrical conductivity σ_0 of the plasma.

Lin's method has been improved by Pain[36] by using high values of a pulsed magnetic field and by comparing its effect on the plasma with the effect of a low value d.c. field acting on the same plasma. The configuration of the applied magnetic field was a cusp formed by two coils in opposition, with a search coil in the centre where the magnetic field is zero (Figure 6.11). The field coils were pulsed over a time period long in comparison with the flow time of the plasma between the coils. In other words, the field imposed on the plasma was independent of time. Prior to running any tests on the plasma, discs of known conductivity were inserted within the zone defined by the field coils. The field coils were pulsed with the discs at various positions and the resulting measurements were integrated from oscilloscope traces to determine the value of the magnetic field. The function $\partial B_z/\partial z$ as given in Figure 6.11(c) was derived from Figure 6.11(b) and similarly $\mathcal{B}m_B(z)\,\partial B_z/\partial z$ as given in Figure 6.11(d) was derived from

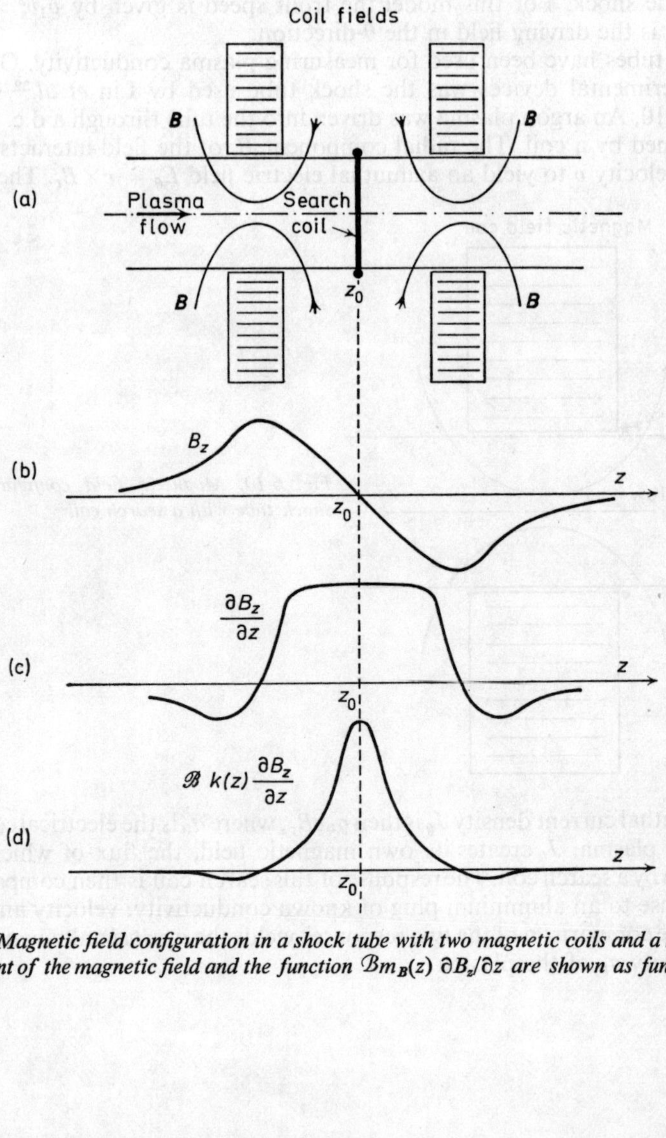

Fig. 6.11. Magnetic field configuration in a shock tube with two magnetic coils and a search coil. The gradient of the magnetic field and the function $\mathcal{B}m_B(z)\,\partial B_z/\partial z$ are shown as functions of z

Figure 6.11(c). The factor $\mathcal{B}$ is a constant and $m_B(z)$ represents the fraction of the field that intersects the search coil when the test disc is at a distance z from the search coil. The expression relating the calibrated data to the voltage induced by a plasma pulse moving through the coil is:

$$\int V\,\mathrm{d}t = \int \mathcal{B} m_B(z) \frac{\partial B_z}{\partial z} v\sigma_0\,\mathrm{d}z \qquad (6.29)$$

where V is the induced voltage from the plasma pulse. The left-hand side of equation (6.29) can be found from oscilloscope measurements when a plasma pulse is passed through the system. The conductivity can then be found since the right-hand side can be integrated by increments of Δz.

6.9 LASER TECHNIQUES[45]*

The conventional diagnostic techniques deal with two types of plasmas. The first type is the low density ($2\times10^4 < n < 10^{13}$ cm^{-3}), low temperature ($T \cong 10^3$K) plasma, such as that of the positive column or the ionosphere. Diagnostic methods here take advantage of the long wavelength microwave probing in low density media, or current flow near the metallic probes. The second is the high density ($n > 10^{15}$ cm^{-3}), high temperature ($T > 10^5$K) plasma, such as that of the theta-pinch discharges or lightning strokes. In this case, diagnostic tech niques use the Stark broadening of spectral lines, optical interferometry, or plasma refractivity.

Therefore, there exists a region ranging between 10^{13} and 10^{15} cm^{-3} in electron density and between 10^3 and 10^5K in electron temperature which cannot be adequately probed with the conventional diagnostic techniques. This range can be covered by four laser diagnostic methods: laser feedback interferometry, laser heterodyning, and laser perturbation techniques. These methods are based upon three different effects occurring during the interaction between matter and light: the phase shift of light propagating through matter, the resonance absorption and stimulated emission of light by excited species, and the scattering of light by particles.

The *laser feedback interferometer* is made of a conventional gas laser and an external mirror system as illustrated in Figure 6.12. It was first developed into a sensitive metrological instrument[34] capable of measuring distances with a relative error no larger than 10^{-9}. In plasma diagnostics, laser interferometry has been further developed[18] to measure changes in plasma refractive index corresponding to electron densities of the order of 10^{12} cm^{-3} by measuring changes in optical lengths.

In Figure 6.12, a conventional He–Ne gas laser oscillates simultaneously on the 6328 Å ($3s_2 \rightarrow 2p_4$) and 3·39 μm ($3s_2 \rightarrow 3p_4$) transitions in neon. These two laser transitions share the $3s_2$ state in neutral neon. Consequently, oscillation on the 3·39 μm transition decreases the population inversion available for oscillacion at 6328 Å. Therefore, one can observe directly, upon the red transition, thanges in the infra-red laser output, except for a 180° phase change. The inter-

* To understand this section, it is advisable to first read Chapter 10.

13

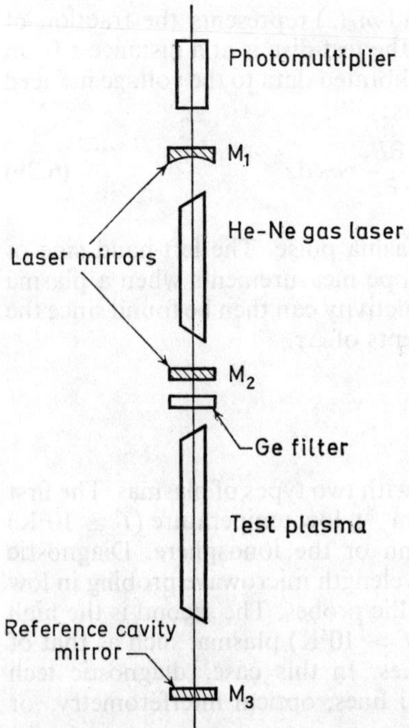

Fig. 6.12. Schematic diagram of a laser feedback interferometer using a flat reflector M_3 in the reference cavity[45]

ferometric probing can thus be performed at 3·39 μm to use the increased sensitivity of the wavelength, but the probing signal can be sampled at 6328 Å with a visible multiplier. The 3·39 μm output beam of the He–Ne laser makes a double passage through the test plasma by reflection from an external mirror M_3. The beam re-enters then the laser cavity $M_1 : M_2$ to interact with the laser discharge. The Ge filter is used to filter the 6328 Å beam.

As it crosses the test plasma, the probing beam suffers a phase shift:

$$\phi = \frac{el\lambda}{4\pi M\varepsilon_0 c^2} \tag{6.30}$$

where l is the length of the path in the plasma and λ is the wavelength. If the total phase shift experienced after two crossings of the plasma equals an integral number of probing wavelengths, then the beam re-entering the laser cavity will reinforce the 3·39 μm laser oscillations. If the total phase shift is equal to 180° then a partial cancellation of laser fields results. Therefore, the laser output through mirror M_1 is a function of the optical length of the reference cavity $M_2 : M_3$, and the 3·39 μm laser output displays a 'fringing' amplitude modulation as shown in Figure 6.13, which can be obtained from a fast response photomultiplier. The time history of successive $\lambda/2$ changes in reference cavity optical length can be recorded by counting the number of fringes produced after starting the discharge. The variation with time of electron density can then be obtained by using equation (6.30).

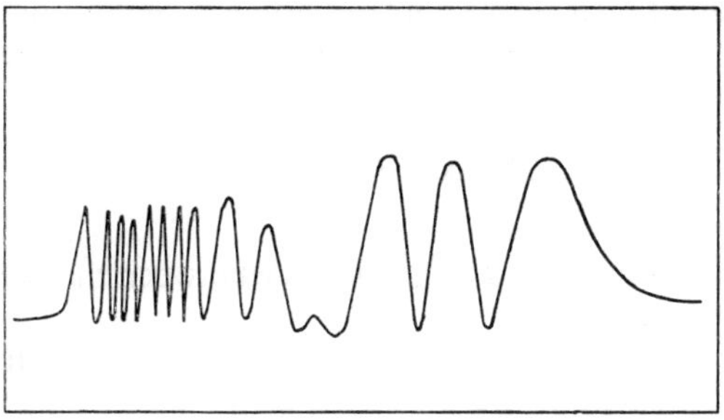

Fig. 6.13. Interferometric 'fringing' of the laser output due to a time-varying plasma contained within the external reference cavity[45]

The *laser heterodyne* system[33] was developed as a result of the search for increased sensitivity in plasma refractive index measurements. Electron densities several orders of magnitude lower than those measured by laser feedback interferometers can be measured by this method. Instruments using such a method are sensitive enough to measure metastable densities.

In this method, the signal to be studied is mixed in a non-linear device with an oscillator signal of comparable frequency. The difference frequency is then

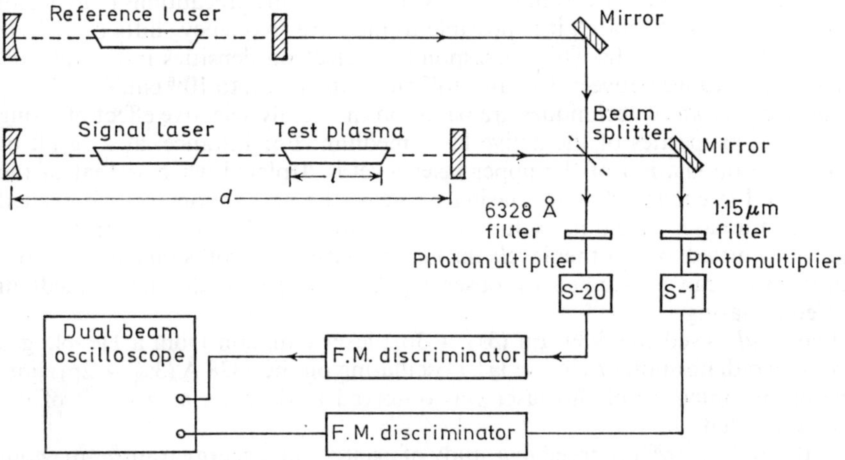

Fig. 6.14. Schematic diagram of a wavelength He–Ne laser heterodyne interferometer[45]

extracted from the mixing products. The resulting heat frequency signal has a much lower frequency than the two initial signals. It can be processed conveniently with a great reduction in the signal-to-noise ratio.

As an example, the laser heterodyne interferometer is shown in Figure 6.14. It is an optical version of the heterodyne system and consists of two independent lasers: a signal laser containing the test plasma within its resonators and a reference laser serving as a fixed frequency oscillator. The signal laser cavity possesses longitudinal resonances corresponding to half an integral number of the wavelengths between the mirrors; thus the resonance frequencies are:

$$f_{\mathrm{tk}} = n_{\mathrm{k}} \frac{c}{2} \frac{1}{d} \tag{6.31}$$

where n_k is an integral number, d is the distance between the mirrors, and c is the velocity of light in vacuum.

When a plasma of length l and refractive index ν_{R} is placed in the laser cavity, the new resonant frequencies are:

$$f^*_{\mathrm{tk}} = n_{\mathrm{k}} \frac{c}{2} \frac{1}{(d-l)+l_{\mathrm{n}}} \tag{6.32}$$

The plasma-induced frequency shift Δf is equal to $(f^*_{\mathrm{tk}} - f_{\mathrm{tk}})$. Thus using equations (6.31) and (6.32), one obtains after some rearrangements:

$$\Delta f = \frac{e^2 n^- \lambda}{8\pi^2 M \varepsilon_0 c d} \tag{6.33}$$

The frequency shift is then directly proportional to the plasma electron density n^- and the probing laser wavelength λ. By using extreme precautions to avoid all possible sources of 'noise', it is possible to measure frequency shifts of 20 kHz or even down to 1 kHz. This corresponds to electron densities in the difficult intermediate range between 10^{13} and 10^{15} cm^{-3} and down to 10^{10} cm^{-3}.

Laser perturbation techniques are based on the highly selective effect of lasing on certain properties of the active laser medium. For instance, as a result of lasing, the population of the upper laser level is depleted whereas that of the lower level is enhanced. Thus, lasing perturbs the populations of two excited levels in the laser medium, excluding all the other levels. This perturbation is thus transmitted to the other levels through radiative and collisional interaction. This transfer can be studied by observing the changes in the active medium caused by lasing.

Bell *et al.*[5] used the 3·39 μm ($3s_2 \rightarrow 3p_4$) laser emission from a He–Ne gas laser to irradiate another He–Ne laser oscillating on the 6328 Å ($3s_2 \rightarrow 2p_4$) line. The output intensity of this laser was observed to decrease as a result of the 3·39 μm radiation.

Parks and Javan[37] observed in a study of excited state energy transfer in neon, that changes in spontaneous emission from the $2s_2$ level due to lasing on the 1·15 μm ($2s_2 \rightarrow 2p_4$) transition are also present on lines originating from the neighbouring $2s_3$ level. Since these two levels are so close to each other and are not optically connected, this was attributed to collisions between the atoms of the two excited states. Using this method Parks and Javan measured the ratio of Einstein coefficients for the $2s_2$ and $2s_3$ neon levels. They also determined the collisional transfer cross-sections of excitation between these two levels by varying the neutral neon pressure in the test cell. They demonstrated that laser perturbation techniques could be used to measure basic atomic constants such as lifetimes and cross-sections.

Scattering of laser light from plasma particles has been one of the basic effects used in plasma diagnostic methods. The technique based on this method consists of measuring the quantity of incident radiation scattered into a certain direction by the test plasma, and determining the spectral profile of the scattered signal. This signal is a function of the incident radiation, the scattering angle and the plasma itself. To characterise it, Salpeter[41] determined a parameter α_L such as:

$$\alpha_L = \frac{\lambda}{r_D} \frac{1}{\sin(\theta/2)}$$

where λ is the probing optical wavelength, r_D is the Debye screening length defined by equation (1.4), and θ is the angle through which the incident light is scattered.

If $\alpha_L \ll 1$, the scattered signal starts behaving in a non-collective manner, each electron of the plasma becomes an individual scattering centre. The amount of scattered radiation is then a function of the number of electrons and the Thomson scattering cross-section*. The experimental device can, then, be calibrated against the Rayleigh scattering† from the neutral component to obtain the electron density from a measurement of the Thomson scattered radiation. If the electronic distribution of velocities is Maxwellian, then the intensity of the scattered light becomes a Gaussian function of the wavelength. This would lead to the value of the electronic temperature. In general, the spectral profile of the scattered light can lead to the distribution of electronic velocities.

If $\alpha_L \gg 1$, the behaviour of the scattered signal becomes fully collective. Its profile would consist of a central narrow peak, Doppler-broadened at the ion temperature T^+ with two minor peaks symmetrically displaced to either side of the narrow peak by the electronic plasma frequency ω_p. This leads to the value of both T^+ and n^-. Using this method electron densities of the order of 10^{17} electrons/cm^3 and temperatures of about 200 eV have been measured[39].

In conclusion, laser diagnostic methods have permitted the measurement of electron densities in ranges where conventional methods do not operate properly. The increased sensitivity has made it possible to do without the restrictive assumption of local thermodynamic equilibrium.

6.10 MISCELLANEOUS DIAGNOSTIC METHODS[22, 27]

One of the most important methods of measuring the plasma density is by the use of the microwave *interferometer*[30]. This device consists of a transmitter and a receiver between which two paths are provided; one of them includes the explored plasma and the other is a known path (sometimes a plasma itself) introducing calibrated attenuation and phase lag. A situation is created

* Natural frequencies of the scattering particles are small compared to frequency of wave.
† Natural frequencies of the scattering particles are large compared to frequency of wave.

in which the waves transmitted by both paths arrive at the receiver with the same amplitude and the same phase lag; they are opposed to each other in the detector unit (see Figure 6.15). For most of the laboratory plasmas of interest, the cut-off wavelength falls in the centimetre or millimetre range. The densest plasmas achieved so far cannot be explored by this method because their cut-off wavelength is below the millimetre range. The cut-off is generally noted in practice by the disappearance of some interference pattern on the receiver side. If the frequency could be varied, this would be an accurate method for measuring plasma densities. Approximately the same result can be attained in transient plasmas, where, for a given frequency, propagation takes place as long

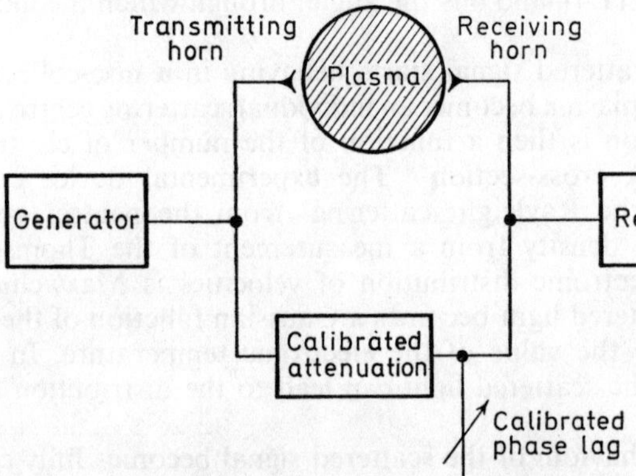

Figure 6.15. The principle of a microwave interferometer

as the plasma density is below the critical value corresponding to the characteristic plasma frequency and disappears at this value. For densities below the cut-off value n_c, the phase difference between the propagation through the plasma and through an equivalent thickness of vacuum is:

$$\Delta\phi = \frac{\omega}{c}\left[d - \int_0^d \left(1 - \frac{n}{n_c}\right)^{1/2} \mathrm{d}z\right] \tag{6.34}$$

in which d is the plasma thickness. The integral takes into account the fact that n is generally not constant along the path. For $n = 0$, $\Delta\phi = 0$. In general, the corresponding signal is then opposed to that of the second path so that the resultant is zero. Assume now that n increases with time. An interference pattern appears. Each time the phase lag varies by 2π, compensation is reached, neglecting the wave absorption through the plasma. Between zero and the cut-off value, there is thus a certain number of fringes of interference. According to equation (6.34) they correspond to regularly spaced values of n. Above cut-off, only the comparison path transmits a signal, which remains constant, its value being generally 1/4 of that at the maximum in the interference pattern, because quadratic detectors are used. The overall figure assumes the shape of Figure 6.16, in which five fringes, corresponding respectively to 20, 40, 60, 80 and 100 per cent of the cut-off value, have been assumed for the sake of argument. The time evolution of n is easily deduced from the interference pattern as indicated, the only operation required being to count the number of fringes between $n = 0$ and $n = n_c$. Standard equipment exists for the wavelengths of 8, 4 and 2 mm. The power emitted is in the milliwatt range. Echo sounding of the ionosphere is also based on the cut-off principle, but owing to the lower plasma density, the relevant wavelength is much larger.

Both in astrophysics and in thermonuclear experiments, it often happens that the plasma is a conspicuous *emitter of microwaves*. These microwaves arise from two different processes. The most frequent one is made of free-free tran-

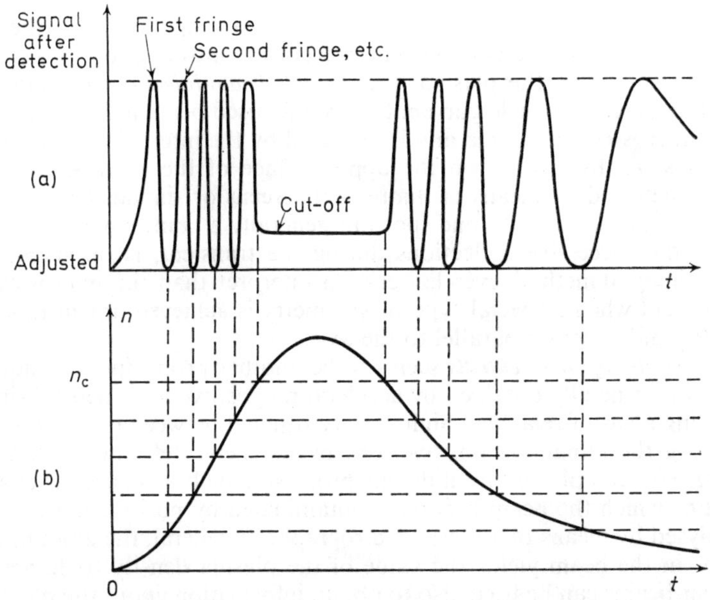

Fig. 6.16. Interference pattern (a) *and time evolution* (b) *of the plasma particle density in the case of diagnostics by microwave interferometer*

sitions in the electron component of the plasma; this results in a continuum, the intensity of which is related to the electron temperature, but not according to the Stefan–Boltzmann law. However, the relationship between these two parameters is beginning to be sufficiently well known for this kind of emission to become a powerful means of diagnostics whenever the emitted power is large enough. Another way by which a plasma can emit microwaves is by bound-bound transitions similar to those considered in the case of visible light, but with a much longer wavelength. The characteristic wavelength of 21 cm (the so-called hydrogen 'song'), which is a forbidden transition, is the best known example in astrophysics. Other gases like ammonia have well known wavelengths in the same range.

Some types of plasma physics devices considered most promising to the thermonuclear field are especially suited for the emission of *particle beams*, which can be analysed by several methods. In the case of the mirror machine, the beams in question are simply parts of the plasma which find their way through 'holes' in the magnetic confinement field. It is often interesting to separate the positive and negative components by electrostatic means, then to analyse either or both. Among the most frequent types of analysis are: (1) time-of-flight measurements, if the particle beam is emitted in bursts; (2) velocity spectrum measurements by magnetic deflection of the beam, if the mass is known; and (3) mass spectrographic measurements, for instance to determine the exact composition of the positive-ion component of the plasma, if it is a mixture of several types of ions.

In thermonuclear experiments, where the plasma temperature and the extreme delicacy of the magnetic confinement prevent the introduction of any kind of solid material inside the plasma, the *particle-beam-injection* technique seems to be promising. A particle gun emits a well-focused beam towards the plasma. The beam crosses the plasma and is deflected by the internal electric and magnetic fields. At the outlet, on the opposite face of the plasma, the beam is generally detected by means of fluorescent screens or similar devices. In transient phenomena, not only one spot but generally a continuous line, resulting from all the successive deflections during the transient, is observed. This is again an integral method, even less easy to interpret than the microwave technique, except when a special type of symmetry is achieved, for instance, axial symmetry and injection parallel to the axis.

The *injection of fast neutrals* seems to be another promising technique. Fast neutrals are generally obtained by accelerating positive ions, neutralising them by crossing a gas of relatively high density situated between two thin foils, and eliminating the remaining positive ions by magnetic deflection. The neutrals interfere with the plasma by different processes, such as charge exchange, as a result of which the beam becomes contaminated by positive ions. The latter are analysed by means of a mass spectrograph. In general, the amount of positive ions in the beam yields the value of the plasma density. It happens that particular beams can be used also to obtain information about the plasma temperature. This is the case when the neutrals are such that no charge exchange takes place to any significant extent. The main source of ionisation in the neutral beam is then electron collision, and since the electron speeds still largely exceed those of the neutrals of the beam, they are governed by the electron temperature.

In very dense plasmas with strong magnetic fields, the *propagation* and *attenuation* properties of *Alfvèn waves* might become of practical use in diagnostics. More generally, any kind of wave propagation, if suitably instrumented, can provide useful information. Similarly, any kind of *radiation* emitted spontaneously or under stimulation by the plasma is also likely to become a diagnostic tool. Such is the case for *X-rays* emitted by plasmas in thermonuclear xeperiments. Up to now, this method does not seem promising since the emission is not intense enough, and besides, parasitic sources of X-rays, relevant to plasma instabilities, tend to mask the phenomenon. In the future, this will become one of the most intense radiations of the thermonuclear reactor, so that it is unlikely that it would not be used as a diagnostic tool.

A true thermonuclear plasma would be a notable *neutron* emitter, and the study of the latter should provide useful information about the phenomena occurring in the plasma itself. But the parasitic phenomena are too obvious for the method to have been of any use to date. However, this situation is likely to change in the future since the neutron emission is increased much more rapidly than the bremsstrahlung or the cyclotron radiation as a function of temperature.

REFERENCES AND BIBLIOGRAPHY

1. AISENBERG, S., 'Multiple Probe Measurements in High Frequency Plasmas', *23rd Ann. Conf. Phys. Electronics*, MIT, Cambridge, Mass. (1963).
2. ALLEN, T. K., 'A Spectroscopic Study of Plasma Electron Oscillations', *Proc. phys. Soc.*, **68A,** 676 (1955).
3. ANDREWS, C. L., *Optics of the Electromagnetic Spectrum*, Prentice-Hall, Englewood Cliffs, N.J. (1960).
4. BARANGER, M. and MOZER, B., 'Light as a Plasma Probe', *Phys. Rev.*, **123,** 25 (1961).
5. BELL, B. E., BLOOM, A. and REMPEL, R. C., 'On Irradiation of an He–Ne Laser by Laser Emission from another He–Ne Gas Laser', *Proc. 3rd International Congress on Quantum Electronics*, Columbia University Press, Vol. 2, 1347 (1964).
6. BERNSTEIN, I. B. and RABINOWITZ, I. N., 'Theory of Electrostatic Probes in Low Density Plasmas', *Physics Fluids*, **2,** 112 (1959).
7. BICKERTON, R. J. and VON ENGEL, A., 'The Positive Column in a Longitudinal Magnetic Field', *Proc. phys. Soc.*, **B69,** 468 (1946).
8. BUCHSBAUM, S. J., MOWER, L. and BROWN, S. C., 'Interaction Between Cold Plasmas and Guided Electromagnetic Waves', *Physics Fluids*, **3,** 806 (1960).
9. BURKE, B. E. and CRAWFORD, F. W., 'Some Diagnostic Experiments to Accompany a Plasma Physics Course', *Microwave Lab. Report* No. 1170, Stanford University, Stanford, Calif., May (1964).
10. CHEN, F. F., 'Electrostatic Probes', *Plasma Diagnostic Studies* (edited by R. HUDDLESTONE and S. LEONARD), Academic Press, New York (1965).
11. COLGATE, S., FERGUSON, J. and FURTH, H., 'A Toroidal Stabilized Pinch', *2nd U.N. Conf. Peaceful Uses of Atomic Energy*, Geneva, Switzerland (1958).
12. COOPER, J., 'On the High Frequency Response of a Rogowsky Coil', *Plasma Phys., J. Nucl. Energy C.*, **5,** 285 (1963).
13. CRAWFORD, F. W., KINO, G. S. and CANNARA, A. B., 'Dipole Resonances of a Plasma in a Magnetic Field', *J. appl. Phys.*, **34,** 3168 (1963).
14. CRAWFORD, F. W., 'Resonances of a Cylindrical Plasma Column', *Microwave Lab. Report* No. 1045, Stanford University, Stanford, Calif. (1963).
15. CRAWFORD, F. W., 'Internal Resonances of a Discharge Column', *J. appl. Phys.*, **35,** No. 5, 1365 (1964).
16. DICKERMAN, P. J. (Ed.), *Optical Spectroscopic Measurements of High Temperatures*, University of Chicago Press (1961).

17. DOUGALL, A. A., 'Optical Maser Probing Theory for Magneto-Plasma Diagnostics', *4th Symp. Engng Aspects of MHD*, Berkeley, Calif. (1963).
18. GERARDO, J. B., VERDEYEN, J. T. and CUSINOW, M. A., 'High Frequence Laser Interferometer', *J. Appl. Phys.*, **36**, 2146 (1965).
19. GLASSTONE, S. and LOVBERG, R. H., *Controlled Thermonuclear Reactions*, Van Nostrand, Princeton, N.J. (1960).
20. GOLOVIN, I. N. *et al.*, 'Stable Plasma Column in Longitudinal Magnetic Field', *2nd U.N. Conf. Peaceful Uses of Atomic Energy*, Geneva, Switzerland (1958).
21. GOULD, R. W., in *Plasma Dynamics* (edited by CLAUSER) Addison-Wesley, Reading, Mass. (1960).
22. HARDING, G., *et al.*, 'Diagnostic Techniques Used in Controlled Thermonuclear Research at Harwell', *2nd Q.N. Conf. Peaceful Uses of Atomic Energy*, Geneva, Switzerland, Vol. 32, 365 (1958).
23. HEALD, M. A. and WHARTON, C. B., *Plasma Diagnostics with Microwaves*, Wiley, New York (1965).
24. HEYWOOD, J. B., 'Experiments in a Magnetically Driven Shock Tube with an Axial Magnetic Field', *Physics Fluids*, **9**, No. 6, 1150 (1966).
25. HOYAUX, M. F., *Contribution a l'Étude de la Théorie des Sondes dans les Décharges Electriques*, Publications de l'AIM-37-38 (Belgium) (1946–7).
26. HOYAUX, M. F., 'Improvements on a Device for Plotting Electrostatic Probe Characteristics in Variable Plasmas', *IEE Trans. on Instrumentation and Measurement*, 77, March (1967).
27. HUDDLESTONE, R. and LEONARD, S. L. (Eds.), *Plasma Diagnostic Techniques*, Academic Press, New York (1965).
28. JOHNSON, E. O. and MALTER, L., 'A Floating Double Probe Method for Measuring the Dynamic Characteristics of a Low Pressure Discharge', *Phys. Rev.*, **80**, 58 (1950).
29. KEMP, N. H. and PETSCHEK, H. E., 'Theory of the Flow in the Magnetic Annular Shock Tube', *Physics Fluids*, **2**, 599 (1959).
30. KLEIN, A., 'Some Results Using Optical Interferometry for Plasma Diagnostics', *Physics Fluids*, **6**, 310 (1963).
31. LEDRUS, R., HOYAUX, M. F. *et al.*, 'Le Plasmographe Synchronisé', *Revue gén. Elect.*, **66**, 513 (1957).
32. LIN, S. C., RESLER, E. A. and KANTROWITZ, A., 'Electrical Conductivity of Highly Ionized Argon Produced by Shock Waves', *J. appl. Phys.*, **26**, 95 (1955).
33. LIU, C. S., CHERRINGTON, B. E. and VERDEYEN, J. T., 'Laser Heterodyne Interferometer', *Bull. Am. phys. Soc.*, **13**, 266 (1968).
34. MINKOWITZ, S., 'Laser Feedback Interferometer as Metrological Instrument', *Opt. Spect.*, 64, May/June (1968).
35. MOTT-SMITH, H. and LANGMUIR, I., 'The Theory of Probes in Gaseous Discharges', *Phys. Rev.*, **28**, 727 (1926).
36. PAIN, H. J., 'Diagnostic Techniques', *Magnetohydrodynamic Generation of Electric Power* (edited by R. A. COOMBE), Reinhold, New York (1964).
37. PARKS, J. H. and JAVAN, A., 'On Excited State Energy in Neon', *Phys. Rev.*, **139**, A 1351 (1965).
38. PEACOCK, G. R. and LANZA, G., 'On Spectroscopic Information from a Plasma', *2nd Symp. Engng Aspects of MHD*, Philadelphia, Pa., March (1961).
39. RAMSDEN, S. A. and DAVIES, W. E., 'Radiation Scattered from the Plasma Produced by a Focused Laser Beam', *Phys. Rev. Lett.*, **16**, 303 (1966).
40. RAYLEIGH (Lord), *The Becquerel Rays and the Properties of Radium*, Arnold, London (1906).
41. SALPETER, E. E., 'Electron Density Fluctuations in a Plasma', *Phys. Rev.*, **120**, 1528 (1960).
42. SLATER, J. C., *Microwave Electronics*, Van Nostrand, Princeton, N.J. (1950).
43. SPIVACK, G. V. and REICHRUDEL, E. M., 'Collection of Electrons by Cylindrical Probes in a Weak Magnetic Field', *Phys. Z. Sowj. Un.*, **9**, 655 (1936).

44. STERN, O. and DACUS, E., 'Piezoelectric Probe for Plasma Research', *Rev. scient. Instrum.*, **32,** 140 (1961).

45. WEAVER, L. A., *An Introduction to Laser Techniques Applicable to Plasma Diagnostics*, Res. Report 68-ICI-VALAM-R2, Westinghouse Res. Labs, Sept. 1968.

7 Thermonuclear fusion

7.1 INTRODUCTION

It has long been suspected that energy could be released from the nuclear fusion of the light elements, i.e. from an interaction in which two light nuclei combine to form a heavier one. It was also known that ordinary helium of mass 4 possesses an extraordinary stability, hence, that reactions having this nuclide as an end product are especially exo-energetic. In 1938, it was discovered that the conversion of ordinary hydrogen into ordinary helium is responsible for the energy production in almost all the stars of the universe.

In the laboratory, fusion reactions on a small scale have been performed for more than three decades by means of particle accelerators. However, under no known condition, except for a special one that will be described later, can a particle accelerator be made to generate more energy than it consumes.

It was previously thought that if an energy-producing fusion reaction were to be achieved in a laboratory-scale experiment, it would be in a very hot plasma; in order to prevent excessive loss of energy toward the walls of an enclosing vessel, this plasma had to be confined magnetically in a space narrower than that available in the vacuum vessel. The vacuum vessel would be lined with a *magnetic bottle*, which is equivalent to a thermal insulator much more efficient than the material ones, and unlikely to be destroyed on contact with the hot plasma.

In *thermonuclear plasma*, a significant fusion energy release can be expected. Such a plasma can be studied theoretically by using a model in which inelastic collisions between ion pairs may lead to the formation of heavier ions with an energy release. The rate of this energy release can be calculated once the collision cross-section is known. Here, both reacting bodies are moving with comparable velocities, instead of one being relatively at rest[35].

The rate of a thermonuclear reaction increases rapidly with the ionic temperature, and becomes significant for a temperature of the order of 10^7K. This temperature corresponds to an average kinetic energy of the order of 10^4 eV. It should be remarked, however, that most of the actual fusion reactions involve particles, far in the tail of the Maxwell distribution, i.e. having energies about one order of magnitude above the average.

At such temperatures, three orders of magnitude above those corresponding to the ionisation potentials of most of the elements, the plasma is *fully ionised.* In this respect, the thermonuclear experiment is more complicated than the future thermonuclear *reactor*, since this property of full ionisation has not yet been achieved in the present research.

The temperature referred to above is the *ionic* temperature. It does not have to be equal to the electron temperature; in fact, electron temperature is often significantly lower than the ionic temperature, i.e. a situation reversed with respect to most of the electrical-discharge plasmas. This time, the ions gain energy from the thermonuclear reaction and can only transmit it to the electrons by the inefficient elastic collisions. Besides, the electrons of a thermonuclear plasma lose their energy in several processes. Again, in the typical experiment in which the thermonuclear reaction condition is approached but not achieved, the situation may be different. As long as the onset of a sustained thermonuclear reaction is not attained, the electrons may receive more energy from, say, an electric field than the ions, as they do in ordinary discharges.

Another important parameter in the thermonuclear plasma is the pressure. This parameter should be as high as possible in order to increase the reaction rate per unit volume. However, if the plasma is to be magnetically confined, the maximum equivalent magnetic pressure available is limited, to values of the same order of magnitude as the atmospheric pressure. Owing to the high temperature, however, this value corresponds to approximately the same density of particles as in the positive column of a high-current mercury-vapour rectifier.

Many problems should be first solved before even considering any power production from a would-be fusion reactor. First, it is necessary to find ways of confining a stable plasma of deuterium at sufficiently high density and energy. In other words, acceptable solutions should be found to such problems as containing the plasma and stabilising it, as well as injecting the fuel in the confinement and recovering the escaping part of it. Later, it would become necessary to enclose the entire plasma region in a lasting physical container, to breed tritium to fuel the fusion reaction, and to recover thermal energy from the products of the fusion reaction.

7.2 CONDITIONS FOR THERMONUCLEAR FUSION[3]

Hydrogen isotopes, deuterium ($^2D_1$), and tritium ($^3T_1$), are the most practical fusion fuels. The frequency of occurrence of a fusion reaction is measured in terms of the fusion cross-section, the said cross-section being defined as a probability of interaction. Indeed, the probability per unit time that a projectile nucleus undergoes a fusion is nQv where n is the density of the target particles in the plasma, v is the velocity of the projectile nucleus, and Q is the area presented by a single target nucleus to the incoming projectile. This area is by definition the fusion cross-section. The most important reactions[40] are listed in *Table 7.1*, where n indicates a neutron.

Reactions (1) and (2) are almost equally probable. The total energy released by Reaction (1) is equal to (2·45+0·82) = 3·27 MeV, whereas Reaction (2) releases (3·02+1·01) = 4·03 MeV. The tritium ($^3T_1$), produced by Reaction (2) fuses almost immediately with a deuterium to form a neutron and a helium-4

Table 7.1 THERMONUCLEAR REACTIONS

Reaction number	*Reactions*	*Fusion cross-sections* (cm^2)
1	$^2D_1 + {}^2D_1 \rightarrow {}^3He_2(0{\cdot}82\ MeV) + n(2{\cdot}45\ MeV)$	10^{-26} at 60 keV*
2	$^2D_1 + {}^2D_1 \rightarrow {}^3T_1(1{\cdot}01\ MeV) + {}^1H_1(3{\cdot}02\ MeV)$	10^{-26} at 60 keV
3	$^2D_1 + {}^3T_1 \rightarrow {}^4He_2(3{\cdot}5\ MeV) + n(14{\cdot}10\ MeV)$	2×10^{-24} at 60 keV
4	$^2D_1 + {}^3He_2 \rightarrow {}^4He_2(3{\cdot}6\ MeV) + {}^1H_1(14{\cdot}7\ MeV)$	2×10^{-27} at 60 keV
5	$^3T_1 + {}^3T_1 \rightarrow {}^4He_2(3{\cdot}8\ MeV) + 2n(7{\cdot}6\ MeV)$	$\sim 10^{-26}$ at 60 keV
6	$^3He_2 + {}^3He_2 \rightarrow {}^4He_2(4{\cdot}3\ MeV) + 2\,{}^1H_1(8{\cdot}5\ MeV)$	very small at 60 keV
7	$^2D_1 + {}^6Li_3 \rightarrow 2\,{}^4He_2 + 22{\cdot}4\ MeV$	$\sim 2 \times 10^{-27}$ at 150 keV
8	$^1H_1 + {}^6Li_3 \rightarrow {}^3He_2 + {}^4He_2 + 17{\cdot}4\ MeV$	$\sim 10^{-28}$ at 300 keV
9	$^6Li_3 + n \rightarrow {}^4He_2 + {}^3T_1 + 4{\cdot}6\ MeV$	Nuclear fission
10	$^7Li_2 + {}^2D_1 \rightarrow {}^8Be_4 + n + 15{\cdot}0\ MeV$	$\sim 10^{-27}$ at 200 keV

* Energy of the deuterium atom.

nucleus as in Reaction (3). Reaction (3) has a large cross-section and releases 17·6 MeV. In view of this reaction, it is, however, necessary to produce tritium independently from Reaction (2). This can be achieved by using lithium-6, as shown by Reaction (9). To fulfil the need for a primary source of neutrons in view of this reaction, use can be made of a nuclear fission reactor.

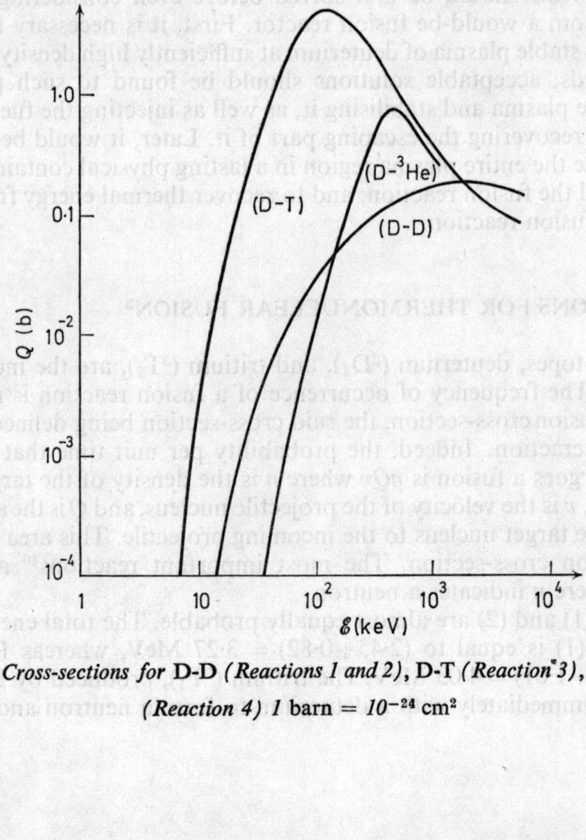

Fig. 7.1. Cross-sections for D-D *(Reactions 1 and 2)*, D-T *(Reaction 3), and* D-^{3}He *(Reaction 4) 1* barn = *10^{-24}* cm^2

Figure 7.1 shows the variations of the different cross-sections with temperature for Reactions (1), (2), (3) and (4). At low energies, the cross-section is extremely small; it is given by a Gamow factor:

$$Q = (\mathcal{A}/\mathcal{E}) \exp(-\mathcal{B}/\sqrt{\mathcal{E}}) \tag{7.1}$$

where $\mathcal{A}$ and $\mathcal{B}$ are constant values and $\mathcal{E}$ is the energy. This cross-section increases rapidly with energy until a maximum value is reached and then decreases for higher energies.

The rate $\dot{\mathcal{V}}$ of thermonuclear energy release as a function of the ionic temperature can be given by the following practical expression:[1]

$$\dot{\mathcal{V}} = \mathcal{A}_1 n_1 n_2 (\mathcal{B}_1/T_i)^{2/3} \exp[-(\mathcal{B}_1/T_i)^{1/3}] \tag{7.2}$$

where $\dot{\mathcal{V}}$ is given in cm³/s, $\mathcal{A}_1$ and $\mathcal{B}_1$ are constants particular to a given reaction, n_1 and n_2 are the concentrations in cm^{-3} of relevant species of ions, and T_i is the ionic temperature in eV. Constants $\mathcal{A}_1$ and $\mathcal{B}_1$ are:

$\mathcal{A}_1 = 75\times10^{-22}$ cm³/s, $\mathcal{B}_1 = 6{\cdot}602$ MeV for Reactions (1) and (2), and
$\mathcal{A}_1 = 11\times10^{-16}$ cm³/s, $\mathcal{B}_1 = 15{\cdot}763$ MeV for Reaction (3).

Figure 7.2 shows the temperature variations of $\dot{\mathcal{V}}$ for deuterium–deuterium, deuterium–tritium, and deuterium–helium-3 reactions. The most important

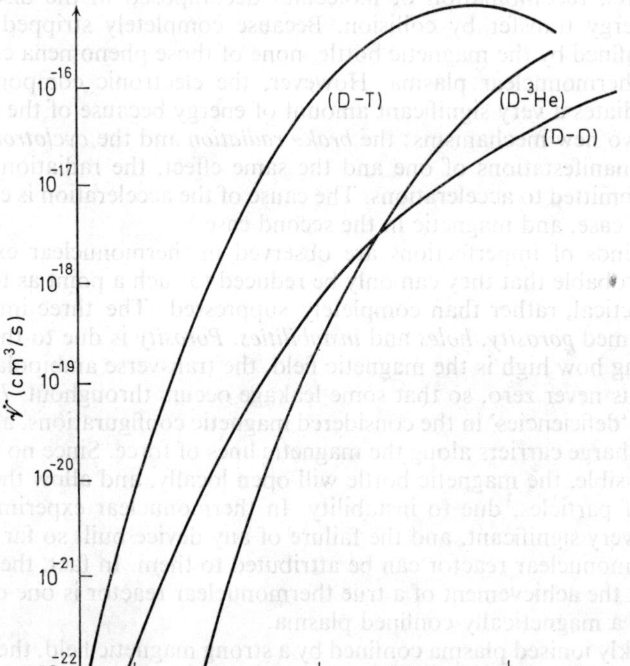

Fig. 7.2. Reaction rates for D-D *(Reactions 1 and 2),* D-T *(Reaction 3), and* D-³He *(Reaction 4)*

problem becomes that of obtaining conditions which lead to self-sustaining fusion reactions. These conditions imply notably the attainment of a critical fuel mass and the attainment of a minimum temperature of ignition.

The first condition is related to the plasma pressure. Since the elements are created above their ionisation potentials, the plasma becomes fully ionised. The pressure at the start of the experiment should not be smaller than 10^{-3} mm Hg. This pressure will correspond to a particle density of the order of 10^{14} cm^{-3} at the existing high temperatures.

The '*ignition*' *temperature* is the temperature at which the energy production equals the energy loss. Consider a thermonuclear plasma ideally confined by a magnetic bottle. For the fusion reactions to be sustained, a delicate equilibrium between the energy generated by the fusion reactions and the energy leaving the plasma through the magnetic bottle should exist.

Some types of fusion reactions generate *neutrons*. It is clear that neutrons cannot be confined by the magnetic bottle, since their charge is zero and their intrinsic magnetic moment much too small to have any significant effect. Only the kinetic energy of charged fragments contributes in any significant manner in maintaining the temperature of the ionic gas.

The mechanisms by which an ordinary plasma yields energy to the confining walls or the external world are: radiation of excited atoms, recombination of carrier pairs, recombination of molecules decomposed in the discharge and kinetic energy transfer by collision. Because completely stripped nuclei are ideally confined by the magnetic bottle, none of those phenomena can occur in an ideal thermonuclear plasma. However, the electronic component of the plasma radiates a very significant amount of energy because of the cumulative effect of two new mechanisms: the *brake radiation* and the *cyclotron radiation*. Both are manifestations of one and the same effect, the radiation of electric charges submitted to accelerations. The cause of the acceleration is electrostatic in the first case, and magnetic in the second case.

Three kinds of imperfections are observed in thermonuclear experiments, and it is probable that they can only be reduced to such a point as to make the device practical, rather than completely suppressed. The three imperfections may be termed *porosity*, *holes* and *instabilities*. *Porosity* is due to the fact that, disregarding how high is the magnetic field, the transverse ambipolar diffusion coefficient is never zero, so that some leakage occurs throughout. *Holes* result from local 'deficiencies' in the considered magnetic configurations, allowing the escape of charge carriers along the magnetic lines of force. Since no true steady state is possible, the magnetic bottle will open locally, and allow the escape of a bunch of particles, due to instability. In thermonuclear experiments, these losses are very significant, and the failure of any device built so far to become a true thermonuclear reactor can be attributed to them. In fact, the significant problem in the achievement of a true thermonuclear reactor is one of attaining stability in a magnetically-confined plasma.

In a weakly ionised plasma confined by a strong magnetic field, the transverse mobility and the diffusion coefficient are inversely proportional to the square of the magnetic induction. In the fully ionised plasma, the porosity is zero if the conductivity is infinite. In a pure magnetic confinement ($E = 0$) the drift velocity $\boldsymbol{v}$ is essentially perpendicular to both $\boldsymbol{B}$ and ∇p_r. Hence, there is no motion of ionised matter along ∇p_r, and the confinement is perfect. This perfection disappears, however, when the conductivity is finite. In such a case,

$$\Delta v_\perp \sim -\frac{\sigma}{B^2} \nabla p_r \tag{7.3}$$

where $\Delta v_\perp$ is an additional term to $v_\perp$ due to the pressure gradient. The diffusion depends upon the pressure gradient and since σ is, in a fully ionised plasma, independent of the pressure, an increase in the carrier density increases similarly the number of scattering centres, by contradistinction with the weakly ionised gas. Like the resistivity of the fully ionised plasma, the porosity is the result of collisions between dissimilar charge carriers. Collisions between like particles produce no such effects at least in the first order of approximation.

In thermonuclear experiments, the conditions from which equation (7.3) is deduced are never satisfied. Bohm's relation

$$\Delta v_\perp \sim -\frac{\sigma}{16B} \nabla p_r \tag{7.4}$$

was somewhat of a guess. However, the $1/B$ variation seems to be well confirmed by experiments. For practical values of the parameters, the leakage given by equation (7.4) is much more significant than that given by equation (7.3). It is generally agreed that the difference between equations (7.3) and (7.4) can be attributed to the intervention of *plasma turbulence*. According to this theory, equation (7.4) is not a constant drift, but the average value of a strong fluctua-

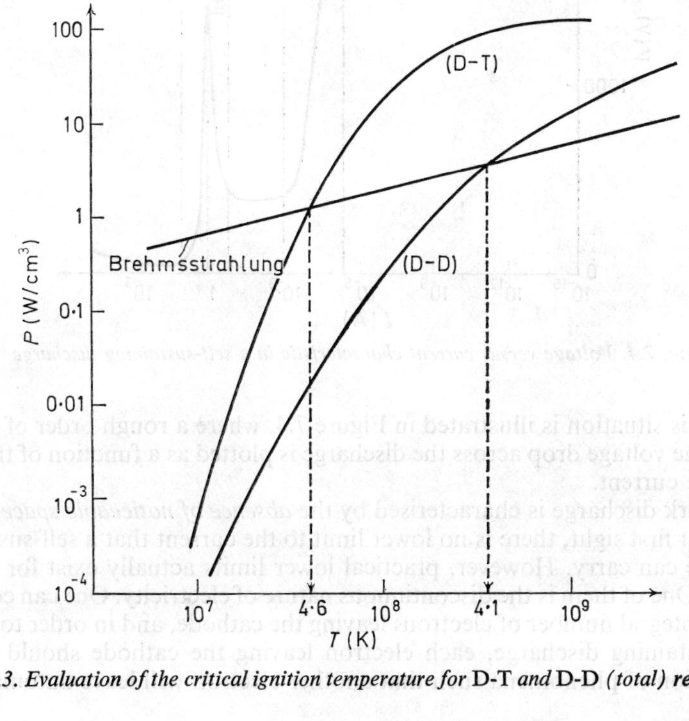

Fig. 7.3. Evaluation of the critical ignition temperature for D-T *and* D-D *(total) reactions*

tion. In fact, the situation is intermediate between the pure cross-field diffusion described by equation (7.3) and the plasma instabilities.

For the expected fusion 'reactors', the loss by brake radiation is the most significant. If only this loss is considered, the ignition temperature would be 46×10^6K for the deuterium–tritium reaction and 410×10^6K for the deuterium–deuterium reaction. This is illustrated in Figure 7.3.

7.3 PLASMA DISCHARGES[21]

The inclusion of a section on plasma discharges in a chapter on thermonuclear fusion might seem out of place. However, it is felt that a good understanding of discharges might help to grasp ideas which were at the origin of thermonuclear research. A *self-sustaining discharge* between electrodes without sharp tips or edges belongs always to one of the three major types: dark discharge, glow discharge, arc discharge, or to a transition region between two of those three

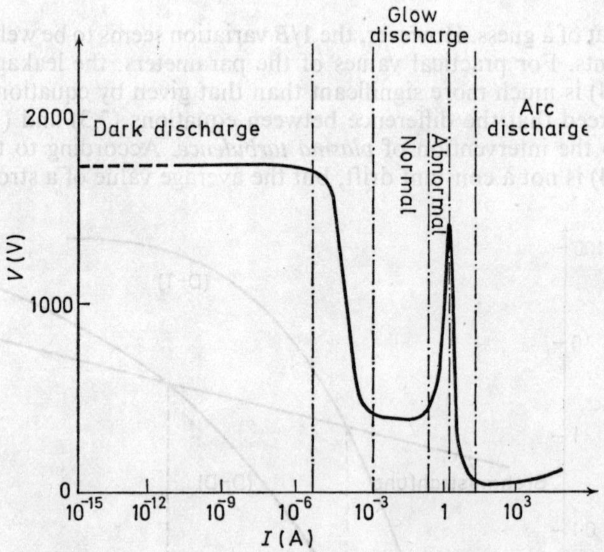

Fig. 7.4. Voltage versus current characteristic in a self-sustaining discharge

types. This situation is illustrated in Figure 7.4, where a rough order of magnitude of the voltage drop across the discharge is plotted as a function of the total discharge current.

The dark discharge is characterised by the *absence of noticeable space-charge effects.* At first sight, there is no lower limit to the current that a self-sustaining discharge can carry. However, practical lower limits actually exist for several reasons. One of them is the discontinuous nature of electricity. One can consider only an integral number of electrons leaving the cathode, and in order to obtain a self-sustaining discharge, each electron leaving the cathode should trigger a succession of phenomena such that another electron will leave automatically the cathode, either as a result of its heating by the discharge itself, or as a result of positive ion, photon, or metastable bombardment, or any other process. There is an obvious lower limit when only *one* primary electron is at work at a time. Such a phenomenon is subjected to *fluctuations* and this gives another lower limit situated much higher in the range of currents and corresponding to the requirement that, in the course of the fluctuations of ionisation, the latter never goes to zero, which would prevent striking back. There is a third and generally more stringent limit, because there is always an external ionising agent, such as cosmic rays or parasitic radioactivity, and the discharge cannot drop below a certain current without becoming non-self-sustaining, and governed essentially by this external agent. The existence of an upper limit is obvious. Regardless of the mode by which the ionisation is generated, there is, from a certain threshold upward a space charge effect which introduces distortions with respect to the uniform field postulated in the dark discharges. This is the onset of the transition toward the glow discharges.

Consider now *one* electron emitted by the cathode. This electron will move toward the anode and ionise the gas. The number of ionisation processes will be $[\exp(\lambda_t d)-1]$, where d is the distance between the anode and the cathode, and λ_t is Townsend first coefficient. As a result of the passage of this electron, the condition of self-sustainment is that the number of created electrons equals the number of disappearing ones as expressed in equation (2.75).

The mechanism of space separation is independent of the type of discharge. Thus, if the current is increased in a dark discharge, the field distribution is no longer that of a capacitor of the same shape. The field in the neighbourhood of both electrodes tends to exceed the 'capacitor' value, while a reduction is attained in between to compensate. The field increase in the neighbourhood of the cathode is by far the most significant. The mechanism of electron extraction is considerably enhanced because the ionic velocity rises, during the last free path before colliding with the cathode, considerably above its average value. This tends to increase the discharge current, and the increase would be catastrophic, if a limiting resistor were not introduced in series with the discharge.

As the transition toward the *glow-discharge* takes place, three subdivisions can be introduced in the discharge: cathodic region, anodic region, and an intermediate region which is the positive column. The voltage drops of those individual regions carry corresponding names: cathodic fall, anodic fall, and column fall. The end of the transition is marked by a new phenomenon: the *stabilisation of the cathodic fall*, which stops decreasing as a function of the current and stays constant for a while. This phenomenon is called a glow-discharge.

In a range of currents which, for normal discharge tubes a few centimetres in diameter, lies between 10^{-4} and 10^{-3} A, the cathodic fall is nearly constant for a given discharge and is dependent on the nature of the gas and the surface material of the cathode[5]. It ranges between 100 and 500 V. Since the anodic fall and the column fall are relatively unimportant, the constancy of the cathodic fall is reflected in the total fall, and this, over a current range of at least one order of magnitude. Such a constancy indicates that a steady electron emission is at work at the cathode. This is further emphasised by the luminous appearance in the neighbourhood of the cathode. There is a characteristic glowing layer, the thickness of which depends only upon pressure, and the surface area of which is proportional to the discharge current. The stabilisation of the cathodic fall ceases when the area of this layer tends to exceed the total area available. This layer is not the only one, and most of the glow discharges are characterised by luminous phenomena. Figure 7.5 is a negative picture of a typical glow discharge. Although some regions can be too thin to be visible to the unaided eye, there are always eight regions in sucession from the cathode to the anode: (1) the Aston dark space; (2) the cathodic glow; (3) the Hittorf dark space; (4) the

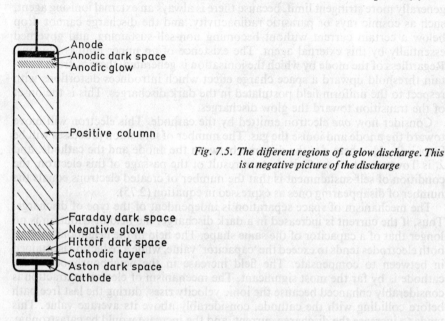

Fig. 7.5. The different regions of a glow discharge. This is a negative picture of the discharge

negative glow; (5) the Faraday dark space; (6) the positive column; (7) the anodic glow, and (8) the anodic dark space.

The *colour* of the luminous regions is characteristic of the gas in which the discharge is established. In general, the three main luminous regions have different colours[43]. They help to identify the gas. Some gases, like hydrogen and nitrogen, display a different set of colours if the walls are conductors or insulators. This is due to molecular decomposition and recombination, the first one always occurs in diatomic gases, but the second one is much more pronounced if the walls conduct electricity.

The density of the plasma in glow discharges is in a range where Langmuir probe measurements can easily be performed. In addition to this diagnostic means, other methods, such as the use of the Stark effect or the beam deflection method to determine local values of the electric field, have contributed in giving a complete picture of this type of discharge. By and large, there is an excess of negative charge near the anode, an excess of positive charge in the cathodic region while the positive column is neutral in spite of its name. In the different graphs of Figure 7.6, the field intensities are derived both from the space derivative of potential measurements and from direct measurements, whereas the space charges are all determined from space derivatives of field intensities.

A glow discharge in which the current density is above a critical value J_m, has a higher cathodic fall, and is called 'abnormal'. In ordinary cases, stable abnormal discharges can be observed over one to two orders of magnitude in current. Then, suddenly, the discharge degenerates into an arc, with a significant decrease in cathodic fall and increase in current. The cathodic fall reaches a maximum at a current between one or two orders of magnitude above the upper limit of the normal glow and then decreases over a comparable range. Around the peak, the intense bombardment of the cathode by an increased number of positive ions of increased energy must have triggered a new mechanism to produce electrons.

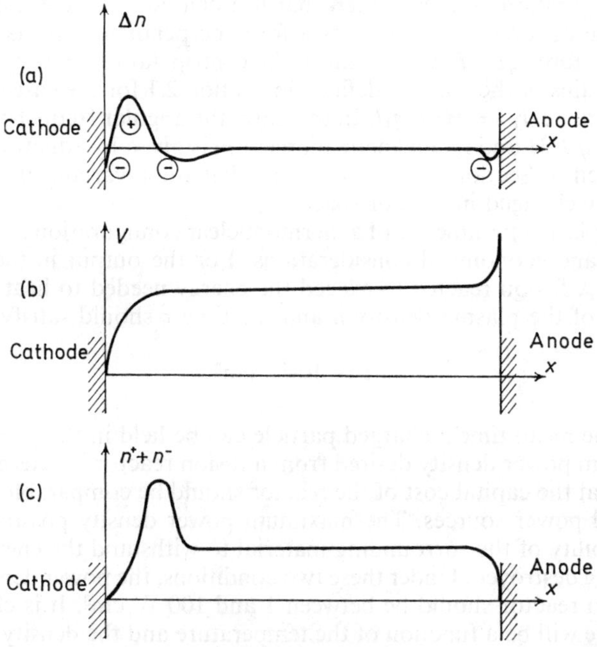

Fig. 7.6. Space-charge (a), *space potential* (b) *and average carrier density* (c) *as a function of space in a glow discharge*

This mechanism differs from one type of arc to another, but is always characterised by the appearance of a bright spot on the cathodic surface, designated as the 'cathodic spot'. The early appearance of the cathodic spot is difficult to observe due to its instability.

There is no unique definition for the arc, except that the current is in the range of amperes and upward, that the cathodic fall is at least one order of magnitude below that of the glow discharge and that the cathodic mechanism involves a cathodic spot. Such a definition includes actual discharges involving very different physical processes and a complete theory for all the types of arc discharges still remains a remote objective. It seems, however, that the best subdivision to introduce among arcs is between *high pressure arcs* and *low pressure arcs*, the critical pressure being about one order of magnitude below atmospheric pressure.

7.4 PLASMA CONFINEMENT[24]

The problem of confinement of the thermonuclear plasma for a time long enough to extract energy has been one on which most of the research has concentrated. The thermonuclear plasma is at such a high temperature that it becomes fully ionised and consequently perfectly conducting. Such a conducting medium can be confined by magnetic fields, independently of any physical walls. Each electron and each ion of tritium and deuterium can be made to move along pre-established magnetic lines of force and prevented from escaping outside a certain configuration. We have seen that if a charged particle tries to cross a magnetic line of force, it experiences a force perpendicular to its direction of motion. This force $q\boldsymbol{v}\times\boldsymbol{B}$ must balance the centripetal force $Mv_\perp^2/r_0$, where r_0 is the gyroradius of the particle defined in Section 2.1 for the case of a uniform magnetic field, i.e. $r_0 = Mv_\perp/qB$. In this case, the angular frequency of the particle is $\omega_c = qB/M$ and its magnetic moment $\boldsymbol{m}$ is given by equation (2.11). We have also seen in Section 2.2 that $\boldsymbol{m}$ is an adiabatic invariant, if the magnetic field is slowly changed in time or space.

The main plasma parameters of a thermonuclear configuration are determined by physical and economical considerations. For the output in thermonuclear energy from a fusion reactor to exceed the energy needed to heat the plasma, the product of the plasma density n and the time τ should satisfy the *Lawson criterion:*

$$n\tau > 10^{14}\ \mathrm{s/cm^3} \tag{7.5}$$

where τ is the mean time a charged particle can be held in the plasma volume. The minimum power density desired from a fusion reactor is determined by the condition that the capital cost of the reactor should be comparable with that of conventional power sources. The maximum power density possible is limited by the capability of the surrounding material to withstand the energy released, without being destroyed. Under these two conditions, the power density released by the fusion reactor should be between 1 and 100 W/cm³. It is clear that the power release will be a function of the temperature and the density of the confined plasma. Under the circumstances, it is possible to consider three regimes of confinement; these are:

(1) high density, low temperature, short lifetime plasmas
($n = 10^{16}$–10^{19} cm^{-3}, $T = 10^8$K, $\tau = 100$–10^4 μs)
(2) low density, low temperature, long lifetime plasmas
($n = 3\times10^{14}$ cm^{-3}, $T = 10^8$K, $\tau = 1$ s)
(3) low density, high temperature, long lifetime plasmas
($n = 10^{14}$ cm^{-3}, $T = 10^9$K, $\tau = 1$ s)

To each one of these regimes corresponds a different concept of magnetic confinement. For instance, in open-ended configurations, operation is possible in regimes (1) or (2) depending on the rate at which particles are lost by collisions. In closed-ended configurations such as toroids, the charged particles can theoretically remain trapped for ever at the highest temperature and thus operation in regime (3) is possible.

Whatever is the form of the magnetic confinement, the plasma can always escape from the configuration. This escape is due to three different processes: (1) gross instabilities; (2) microinstabilities; and (3) *classical diffusion*. The first two processes have been studied in Chapter 5, whereas process (3) has been mentioned in Chapter 2. The latter can be explained as follows: due to Coulomb collisions, particles can move from one magnetic line of force to another and eventually escape altogether the confinement. The classical diffusion coefficient $\mathcal{D}_{\perp}$ perpendicular to the magnetic lines of force is:

$$\mathcal{D}_{\perp} \sim \frac{n}{T^{1/2}B^2} \tag{7.6}$$

This equation corresponds to equation (7.3). The classical collisional diffusion is due to the interaction of a particle with the electric fields of all the other particles. The classical diffusion time τ can be deduced from equation (7.6). If x is a characteristic distance, one would have $\tau = x^2/\mathcal{D}_{\perp}$ or:

$$\tau \sim x^2B^2T^{1/2}/n \tag{7.7}$$

Nevertheless due to microinstabilities, diffusion is enhanced following the empirical Bohm equation [equation (7.4)]. From equation (7.4), one finds for the diffusion perpendicular to the magnetic lines of force:

$$\mathcal{D}_{\perp\mathrm{B}} = \frac{kT_{\mathrm{e}}}{16eB} \tag{7.8}$$

from which Bohm diffusion time τ_{B} is deduced; it is:

$$\tau_{\mathrm{B}} = \frac{16ex^2B}{kT_{\mathrm{e}}} \tag{7.9}$$

The ratio β of the plasma pressure to the magnetic pressure is often used to compare several plasma confinements. This ratio can be quite small (about 10 per cent) for some toroidal confinements. It is sometimes as high as 40 per cent in some theta pinch experiments.

Some non-magnetic means of confinement have been suggested. One of these is the use of radio-frequency. Since the radiation is reflected by a plasma the characteristic frequency of which is larger than the electromagnetic plasma frequency, the plasma would be confined when the 'radiation pressure' equals the plasma pressure.

Even electric fields have been proposed for confining a plasma without the use of magnetic fields. The charge separation can be avoided by the use of a potential well for the ions and a non-equilibrium system for the electrons.

7.5 OPEN THERMONUCLEAR SYSTEMS

The pinch effect[8]

The expression 'magnetic pinch' is used to designate an extreme case of a magnetically self-constricted electrical discharge. In an effort to establish as high a current as possible in the discharge before the instabilities set in, investigators have been led to use larger and larger $\mathrm{d}I/\mathrm{d}t$ values, until this has become the relevant parameter instead of I. When $\mathrm{d}I/\mathrm{d}t$ is very large, the model is essentially dynamic, leading to the assumption of a zero penetration depth of the 'snow-plough' model.

Consider a pinched column of fully ionised plasma, the resistivity of which is negligible. Assume also that the current density is purely confined to the surface and that the magnetic field inside the plasma column is zero. At the boundary of the column, there is a discontinuity both in the plasma density and in the magnetic field. For a magnetic confinement, the magnetic pressure is, as seen in Section 3.9:

$$p_{\mathrm{r}} = \frac{B^2}{2\mu_0} \tag{7.10}$$

Since p_{r} is given by equation (1.58), and B is given by equation (3.1) for a cylindrical column of radius r, then:

$$nkT = \frac{\mu_0 I^2}{8\pi^2 r^2}$$

or, if S is the cross-sectional area of the column:

$$(nS)kT = \frac{\mu_0}{8\pi} I^2 \tag{7.11}$$

In a fully ionised ideally confined gas (nS) can be considered as an invariant; hence, the column tends to contract or expand according to whether or not I is larger or smaller than $(8\pi nSkT/\mu_0)^{1/2}$. If I is larger than the equilibrium value, the only way to reach a further equilibrium is through an increase of T. Hence a rupture of equilibrium [equation (7.11)] may lead to an increase in temperature independently of an actual Joule heating of the discharge. This heating is negligible in a simplified model with infinite conductivity.

In a rapidly contracting column,

$$p_{\mathrm{r}} + \frac{\mathrm{d}}{\mathrm{d}t}\left(\frac{M_{\mathrm{a}}}{S} v\right) = nkT \tag{7.12}$$

where M_a/S is the inward accelerated mass, and v is the absolute value of $\mathrm{d}r/\mathrm{d}t$. In pinch experiments, the velocity at which the 'magnetic wall' moves is so large with respect to the sound velocity that the inner parts of the plasma are entirely unperturbed by the compression, whereas all the material swept by the 'magnetic wall' tends to accumulate on a very thin sheath along the wall. Eventually, the 'static' pressure becomes negligible with respect to the 'dynamic' one; thus:

$$\frac{d}{dt}\left(\frac{M_a}{S}v\right) = \frac{\mu_0 I^2}{8\pi^2 r^2} \tag{7.13}$$

To evaluate (M_a/S) consider Figure 7.7. If n_0 is the initial electron density of the column of initial radius r_0, all the material initially situated between the cylinders of radii r and r_0 becomes, at the relevant time t, placed in a sheath of negligible thickness just below the radius r, and constitutes the accelerated mass.

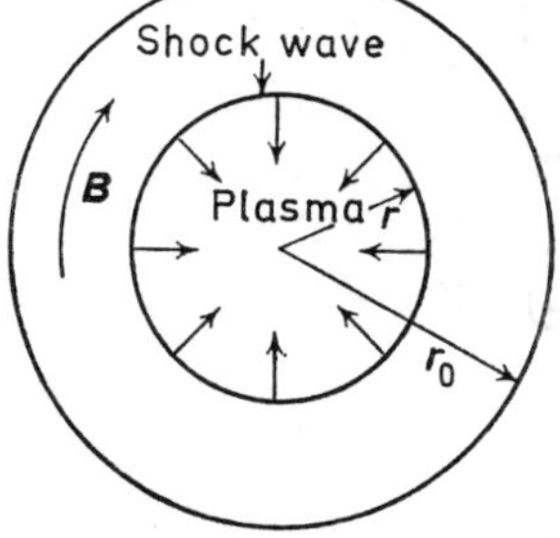

Fig. 7.7. Cross-section of a pinch column in its dynamic state

Since:

$$\frac{M_a}{S} = \frac{n_0 M}{2r}(r_0^2 - r^2) \tag{7.14}$$

where M is the electronic mass, then:

$$\frac{d}{dt}\left[\frac{n_0 M(r_0^2-r^2)}{2r}\frac{dr}{dt}\right] = -\frac{\mu_0 I^2}{8\pi^2 r^2} \tag{7.15}$$

The time variation of I has to be introduced in order to integrate equation (7.15). For instance, a constant dI/dt can be postulated as a first approximation.

The phenomenon occurring in the snow-plough model is a shock wave. It is characteristic of walls moving with velocities greater than the speed of sound. The zero thickness model is an oversimplified one; the true shock-wave has a finite thickness and a well defined structure. The snow-plough model can be refined by introducing this actual structure. Besides, the time variation of I can be computed by introducing the inductance of the plasma column as a variable impedance in the circuit.

Linear pinch experiments[9]

Linear pinch experiments are performed by discharging, into short-circuit around the discharge tube, a large bank of inductanceless capacitors, connected to a d.c. source of several tens of thousands of volts through a buffering resistor

(Figure 7.8). The discharge is triggered through a fast, low inductance switch, which can take the form of an air gap with variable pressure, or an externally triggered ignitron. Once the discharge is triggered, the phenomenon develops in a few microseconds. It is best followed by fast-shutter cameras and magnetic probes. The major phenomenon is the development of a radially collapsing plasma column in which a shock wave is produced. The column contracts down to a very small radius; then, the shock-wave is reflected and the column starts

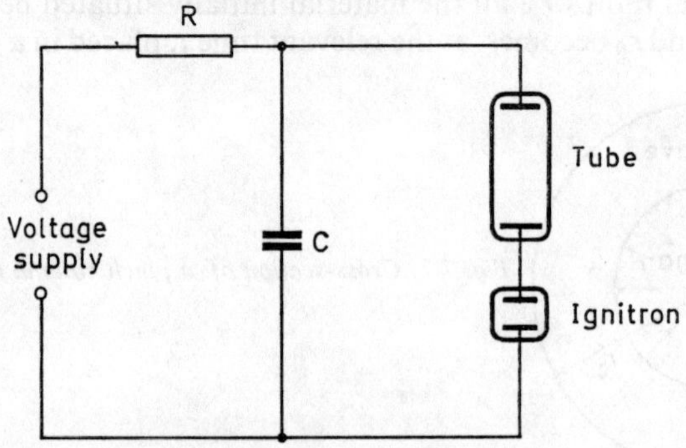

Fig. 7.8. Linear pinch experiment

expanding. However, the current has now attained a sufficiently high value, so that a new contraction is imposed. The same phenomenon occurs two or three times before the kink and sausage instabilities develop. The rupture of the column is accompanied by X-ray emission, and in some gases, neutron emission. Neither are observed while the column is reasonably stable.

Columbus II of Los Alamos Laboratories is one of the early linear pinch devices. The total energy stored by the discharge exceeded 10^5 J. The time spent to the first current peak of about 8×10^5 A was equal to 2·2 μs, whereas the value of dI/dt was 10^{12} A/s. The discharge produced 10^8 neutrons/pulse and the period of each pulse lasted for about 1·5 μs.

The stability duration of the linear pinch could be increased by two to three decimal orders of magnitude by using a longitudinal magnetic field and a coaxial return conductor. When the plasma becomes sufficiently conducting, it is almost

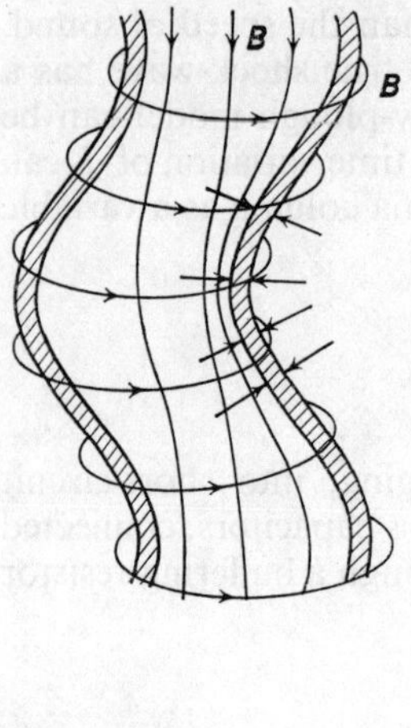

Fig. 7.9. Stabilised linear pinch. Resistance against kink instability

perfectly trapped inside the magnetic configuration. When the shock wave develops, it tends to carry the magnetic field with it, slowing down the column contraction. Eventually, a kind of steady state is reached, in which the discharge becomes hollow, its thickness being small with respect to its outer diameter. The main phenomenon is an equilibrium between two magnetic pressures, the pressure of the self-field from outside, and the pressure of the trapped field from inside. The plasma pressure has become almost irrelevant (Figures 7.9 and 7.10). The presence of this trapped magnetic field and the hollow arc structure elimi-

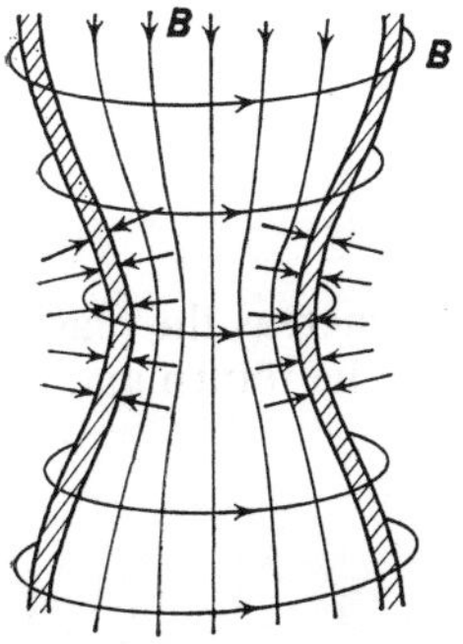

Fig. 7.10. Stabilised linear pinch. Resistance against sausage instability

nate the kink and the sausage instabilities. The local increase in the external pressure is counteracted almost immediately by an increase in the internal pressure, and the structure tends to stabilise after some oscillation. The role of the conducting wall is to contribute to the removal of overall lateral movements of the column (Figure 7.11).

A cylindrical or toroidal structure with a hollow plasma and the return conductor inside, would be stable with respect to the flute, kink and sausage instabilities. Considerable attention is thus being devoted to the so-called 'hard-core'

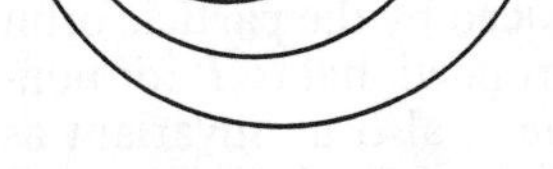

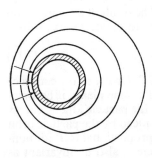

Fig. 7.11. The contribution of the conducting wall in removing the lateral movements of a pinch column

pinch, shown in Figure 7.12. The snow-plough model has been successfully applied to the outward motion of the inner edge of the plasma which appears to be much more stable than an inward moving front. A longitudinal magnetic field is generally used in order to improve the confinement conditions at the external limit of the plasma. A less extreme situation is achieved in the *triax pinch*, where the return current is parted between a hard core and an outer sleeve. In one coaxial triax pinch experiment the total energy stored was $0{\cdot}9\times10^4$

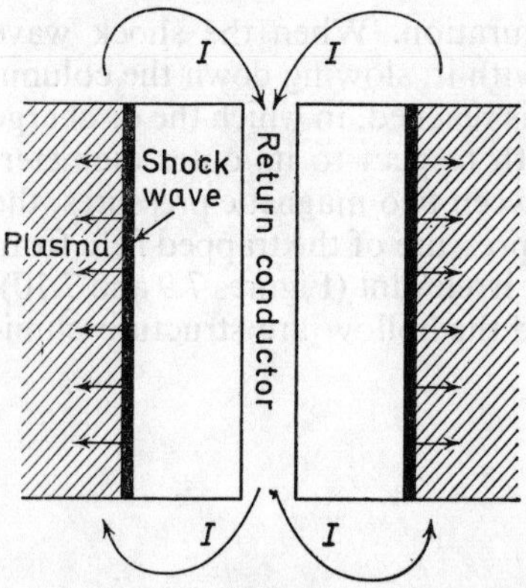

Fig. 7.12. The hard-core pinch

J and the time rise to the first current peak lasted for 3 μs whereas the plasma had a 15 eV energy. In the *inverse pinch*[2] device the roles of the return and the plasma conductor are completely reversed.

The mirror machines[44]

The adiabatic invariance of the magnetic moment in a space-dependent magnetic field leads to a magnetic field configuration known as a 'magnetic mirror'. Consider particles spinning in the *xy*-plane, the magnetic field being a function of *z*, and assume a symmetry around the *z*-axis. Equation (2.15) relevant to the magnetic field divergence can be written for a cylindrical configuration:

$$\frac{1}{r}\frac{\partial}{\partial r}(rB_r)+\frac{\partial B_z}{\partial z}=0 \qquad (7.16)$$

Assuming that B_z is much larger than B_r, equation (7.16) yields equation (2.14), after integration in the neighbourhood of the *z*-axis. The result of a combination of this radial component with the orbital motion is to create an axial force given by:

$$F_{\parallel} = -\boldsymbol{m}\partial B/\partial z \qquad (7.17)$$

where $\boldsymbol{m}$ is the magnetic moment defined in Section 2.2. Owing to the existence of this axial force, the particle orbit cannot stay in the same plane; it becomes shifted in the axial direction. If B is constant in time, the particle kinetic energy is invariant as stated in equation (2.17). The flux embraced by the particle orbit is given by $\pi r^2 B$. But since $\pi r^2 \omega$ is invariant and ω is proportional to B for non-relativistic motions, the flux embraced by the particle is also an invariant as shown by Figure 7.13. If the particle possesses an initial axial velocity towards the region of increasing field, it is retarded by the axial force given by equation (7.17). At the same time, its axial kinetic energy is changed into orbital kinetic energy, according to equation (2.17). If the field gradient is strong enough or extends sufficiently far, the particle may eventually be reflected at a point where the axial component of speed goes to zero, hence the name of '*magnetic mirror*'. The condition of reflection is that the axial component of speed is imaginary where the magnetic field is maximum. If $\sin\theta$ is equal to $v_\perp/v$, as v is invariant,

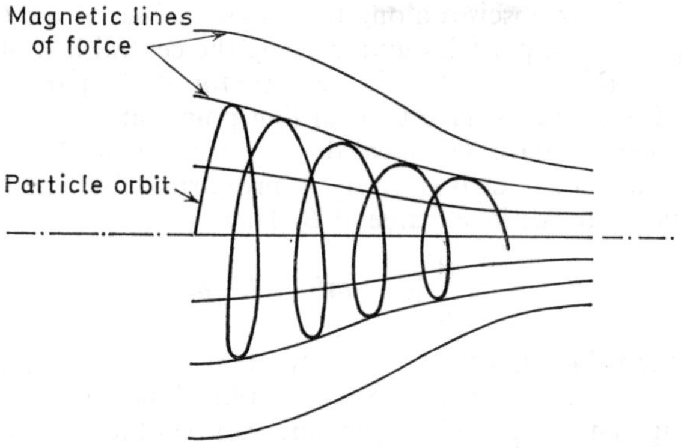

Fig. 7.13. The orbit of a particle in a mirror machine. The orbit describes a tube of force

a way to express the reflection condition is to state that sin θ calculated for the maximum value of B must be greater than one. But sin θ is also proportional to the square root of B so that the condition for reflection is:

$$\sin\theta \geqslant (B/B_{\max})^{1/2} \tag{7.18}$$

as stated in Section 2.2.

A *mirror machine* is made of two coaxial magnetic mirrors. If two similar mirrors are placed some distance apart along the same axis, the particles having a sufficient magnetic moment can be made to oscillate between the two as in a potential well. The retarding force exerted on a particle approaching a magnetic mirror is given by equation (7.17). If the magnetic moment is too small, the retarding force will be insufficient to reverse the axial motion and the particle will be lost. Between the collisions, $\boldsymbol{m}$ is invariant, but the collision can modify this and any particle not having leaked during a given free path can as well leak during the next one, if it suffers in the meantime a collision reducing the magnetic moment.

From equation (7.18), a 'loss cone' in the hodographic space can be defined. Any particle having the extreme of its velocity vector inside the loss cone will cross the mirror and be lost. Any particle having the extreme of its velocity vector outside the loss cone will be trapped. Thus, the velocity distribution is necessarily anisotropic. If the magnetic moment is invariant and if the plasma is collisionless, the situation is relatively simple: any particle doomed to be lost will be lost during the very first instants; any particle trapped remains trapped 'forever'.

Consider a collection of trapped particles in a steady state all of which have the same magnetic moment $\boldsymbol{m}$ and the same kinetic energy $\mathcal{E}$, and assume that the mirror machine is perfectly symmetrical with respect to its axis and to its median plane, in which there are n_0 particles per unit volume, n_0 being a function of the distance r to the axis. Considering one given magnetic tube of force and neglecting the azimuthal drift, the motion of the individual particle will be an oscillation between the two mirrors. Because of equations (2.11) and (2.17):

$$v_{\parallel} = \left[\frac{2}{M}(\mathcal{E} - mB)\right]^{1/2} \tag{7.19}$$

where $v_{\parallel}$ is $(2\mathcal{E}_{\parallel}/M^{1/2})$.

Let s be a curvilinear abscissa along the magnetic line of force, with $s = 0$ at the median plane. The particles gyrate along the central line of the relevant tube of force; the radius of gyration is the same for all the particles and varies inversely as B. The particles have two turning points at, say, $\pm s_m$ for which equation (7.19) goes to zero. The local density of the particles in the interval can easily be calculated, since it is inversely proportional to the local value of $v_{||}$, and proportional to the local value of B. Therefore,

$$n(s) = n_0 \frac{B}{B_0} [(\mathcal{E} - mB_0)/(\mathcal{E} - mB)]^{1/2} \tag{7.20}$$

in which B_0 is the value of B for $s = 0$ in the relevant tube of force. $n(s)$ goes to infinity for $s = s_m$; this refers to the oversimplified model and has no special physical meaning. The values of the pressure can be obtained similarly; there exists only one degree of freedom longitudinally; hence $\mathcal{E}_{||} = kT_{||}/2$ and since $p_{r_{||}} = nkT_{||}$, one obtains:

$$p_{r_{||}} = 2n_0 \frac{B}{B_0} [(\mathcal{E} - mB_0)/(\mathcal{E} - mB)]^{1/2} \mathcal{E}_{||} \tag{7.21}$$

Transversely, there are two degrees of freedom, then $\mathcal{E}_\perp = kT_\perp$, and therefore:

$$p_{r\perp} = n_0 m \frac{B^2}{B_0} [(\mathcal{E} - mB_0)/(\mathcal{E} - mB)]^{1/2} \tag{7.22}$$

By taking the derivative of $p_{r_{||}}/B$ with respect to s from equation (7.21), the following equation will be satisfied:

$$B \frac{\partial}{\partial s} (p_{r_{||}}/B) + (p_{r\perp}/B) \frac{\partial B}{\partial s} = 0 \tag{7.23}$$

This relation is general enough to be valid for all anisotropic plasmas.

If a significant increase in temperature is to be achieved in a mirror machine, either by shock-wave or adiabatic compression, the final volume must be much smaller than the initial one. This can be done by a *multistage compression* which involves a series of coaxial coils energised according to a well predetermined programme. The plasma is compressed in a first mirror stage. At the end of this first stage, the two mirrors are adjusted so they are unequal, so that the plasma goes through the hole in the weaker of the two and becomes trapped in a second mirror machine, of correspondingly smaller dimensions, located behind it. The same process is triggered in this second mirror machine, and so on.

In many mirror machines, the plasma is not generated inside the machine, but originates from a plasma gun. One can make use of the end holes (mirrors) as injection apertures. The particle spirals around a flux tube and passes through the magnetic mirror. Trapping can be achieved if the particle suffers at least one collision in between and increases its magnetic moment, so that the reflection condition is attained, or if the second mirror is stronger than the first one. However, in this latter case, the particle will come out through the inlet aperture if the plasma is in the steady state. In a non-steady state, this can be avoided if both mirrors have increasing magnetic fields. Then, the reflection condition will be attained even if the magnetic moment does not change in the meantime.

Consider now, the injection of a blob of plasma instead of a single particle. The plasma pressure can push away the magnetic nozzle at the entrance, and let it close behind it. If the blob remained unaltered while crossing the machine, the same phenomenon could occur to allow the blob to come out of the magnetic bottle; but this is unlikely, and at least some of the particles remain trapped. The proportion of trapped particles can be increased if two plasma guns are fired simultaneously at both ends of the mirror machine; then, the two blobs collide in the centre, and the number of individual particle collisions can be sufficient to enable most of them to remain trapped.

Beam injection has also been considered, but unless there is a variation in the magnetic field or a collision resulting in a variation of the magnetic moment, the beam will eventually emerge without being trapped. However, owing to the limited possibilities of particle injection in the form of beams, neither the first nor the second method is sufficiently powerful. A radically different approach has been used instead. In several machines, molecular ions of deuterium are created in an ion source, accelerated toward the confined volume, and decomposed inside the latter into a deuteron and a neutral deuterium atom. The latter escapes immediately from the confinement; but the former, having half the mass and approximately half the kinetic energy of the original ion, gyrates in a circle half the size and is generally trapped.

The Ogra machine built by Soviet scientists is an example of a mirror machine. This machine has a total length of 20 m, whereas the distance between the mirrors is 12 m. Molecular ions are injected at an energy of 200 keV through a magnetically shielded channel. In Ogra, a plasma density of 10^7 ions/cm^3 is achieved in a neutral gas density of 10^9 molecules/cm^3. The plasma is formed by dissociation when the beam of molecular hydrogen ions collides with the gas present. The temperature of the electrons in the configuration varies between 1 and 3 keV and their total path between the mirrors reaches the kilometre. The other characteristics of Ogra as well as other mirror machines are listed in *Table 7.2*.

Table 7.2 MIRROR MACHINES

Device	*Location*	$\mathcal{E}$ (keV)	B (T)	*Length* (m)	r_p (m)	T_i (K)	T_e (K)	β	n (cm^{-3})	τ (s)	$n\tau$ (s/cm^3)
Joffe's	Moscow		0·5		0·2	2×10^7	10^5	3×10^{-5}	10^{11}	10^{-4}	10^7
Post's	U.S.A.		1·0		0·03	10^7	2×10^7	7×10^{-2}	10^{13}	3×10^{-3}	3×10^{10}
Coens-gen's	U.S.A.		1·8		0·05	4×10^7	10^6	10×10^{-2}	10^{14}	9×10^{-5}	9×10^9
Ogra	Moscow	160	0·5	12	0·6	9×10^8	2×10^7	3×10^{-6}	10^7	10^{-3}	10^4
Ogranok	Moscow	20	0·3	1·6	0·25	10^8	3×10^5	3×10^{-6}	7×10^7	5×10^{-5}	$3{\cdot}5\times10^4$
DCX-1	Oak Ridge	600	1·0	0·8							3×10^7
DCX-2	Oak Ridge	600	1·7	4·0							
Alice I	Livermore	20	5·0	0·47							
Phoenix	Alder-maston	30	5·0	0·32							
MM II	France	200	1·2	0·40							
HX-0	Japan	600	0·25	0·70							
2X	Livermore		1·3			9×10^7	2×10^6		5×10^{13}	$0{\cdot}8\times10^{-3}$	4×10^{10}

The theta pinch[28]

The theta pinch is an electrodeless cylindrical discharge, the principle of which is that of a transformer. The primary coil is a heavy one-turn coil reinforced at both ends to make a mirror machine. The current is in the θ-direction, but such as to annul the magnetic field inside the plasma.

The experiment is performed with a very fast pulse, arising from the discharge of a low inductance capacitor and, apart from the fact that the current is now in the θ-direction and the magnetic field in the z-direction, the snow-plough model applies if the velocity of the magnetic wall exceeds that of sound, which is usually the case. The plasma heating is the result of a shock wave, but if the velocity of the magnetic wall is below that of sound, which is rather exceptional in this type of experiment, the plasma is compressed adiabatically. Temperatures at least one order of magnitude above those of the ordinary pinch have been attained. The conditions of stability are, in general, significantly improved.

Scylla 4 of Los Alamos Laboratories was developed in 1963. The primary capacitor bank feeding Scylla 4 delivered energy at the rate of $1{\cdot}5\times10^8$ W/s with a 10^7 A peak current. The secondary bank discharged at 2×10^7 A current to create a 22·5 T magnetic field at 2000 atmospheres. The power supply used (Zeus) had a storage capacity of 6×10^6 J. The capacitors can be charged from the commercial lines in minutes and discharge in few microseconds releasing a huge amount of energy. *Table 7.3* shows the characteristics of some theta-pinch devices. These devices are all of the high β type ($\beta = 1$).

Table 7.3 THETA-PINCH DEVICES

Device	*Location*	$\mathscr{E}_c$ (kJ)	B (T)	l_c (m)	T_i (K)	T_e (K)	n (cm^{-3})	τ (µs)	$n\tau$ (s/cm^3)
Aldermaston	Aldermaston	23	6·0	0·22	3×10^6			4·0	
Aldermaston	Aldermaston	45	8·6	0·21	10^7	3×10^6		2·5	
Scylla 1	Los Alamos	31	5·0	0·11	10^7	$4{\cdot}5\times10^6$	8×10^{16}	1·3	10^{11}
Scylla 2	Los Alamos	93	10·4	0·24	10^7		6×10^{16}	2·0	$1{\cdot}2\times10^{11}$
Scylla 3	Los Alamos	180	14·5	0·27				2·5	
Scylla 4	Los Alamos	$1{\cdot}5\times10^5$	22·5		6×10^7	5×10^{18}	5×10^{13}	4·0	2×10^{11}
Pharos	NRL	714	8·0	1·80	4×10^6	$3{\cdot}5\times10^6$	4×10^{17}	9·0	$3{\cdot}6\times10^{12}$
NRL 1	NRL	285	8·0	0·70	2×10^6	$2{\cdot}5\times10^7$	5×10^{16}	5·0	$2{\cdot}5\times10^{11}$
Rome 1	Rome	2	0·8	0·30				8·0	
Rome 2	Rome	15	1·0	0·35				1·0	
Jülich	Jülich	5	1·5	0·15	10^7	2×10^6		0·8	
USSR	USSR		1·0	0·55					
Betatron	Culham			8·00			2×10^{16}	19·5	4×10^{11}
Scyllac	Los Alamos			12·00					

The homopolar machine

In the homopolar machine, the plasma originates from one or two plasma guns and is injected axially through the holes of the confinement. The first tendency is for the formation of a dense plasma column in a roughly cylindrical zone along the axis of the mirror machine. This plasma column tends to diffuse by collisions towards the outer parts of the confinement zone. Between this highly ionised column and a cylindrical metallic anode coaxial with the mirror machine a voltage is applied, the plasma column playing the role of the cathode. Thus, an electric field, pointing radially toward the axis, is superimposed upon the magnetic field of the mirror machine. This induces a tendency for the particles to rotate, the linear velocity being E/B. The direction of the electric field is such that the confinement of the positive ions is improved up to a certain point. One should note that the drag on the plasma as a result of the presence of the neutral gas is responsible for the significant heating in this system.

The homopolar device of Livermore had a tube bore of 25 cm and a tube length of 10 cm. The magnetic field strength of the device was 1·3 T and the plasma density amounted to 10^{16} ions/cm^3. An important incidental use of such a device is as a fast discharge capacitor: dielectric constants as high as 10^7 and currents of 300 kA at a rate of 5×10^{11} A/s have been produced. The characteristic time was 10 μs, thus $n\tau = 10^{11}$ s/cm^3.

The cusped mirrors

The cusped geometry[17] corresponds to the case in which the two mirrors have opposite currents and a configuration in which the centre of curvature of the magnetic 'wall' is everywhere outside the plasma. Such a configuration is handicapped by the possibility of leakage all around the median plane (Figure 7.14).

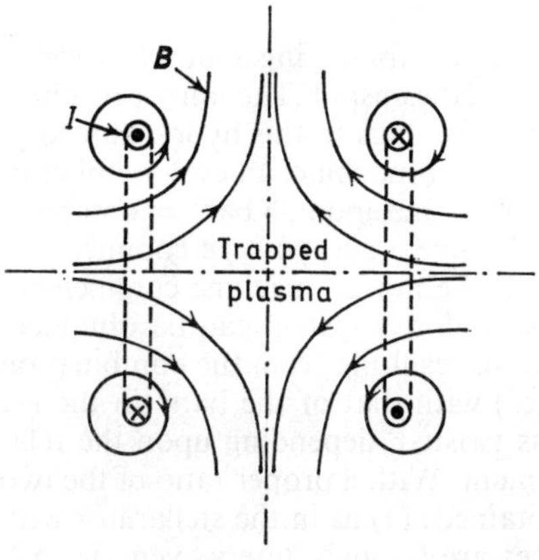

Fig. 7.14. Cross-section representation of the magnetic lines of force in the cusped geometry

The rate of leakage tends to decrease when a sufficiently high temperature is attained. The advantage of this configuration is two-fold. First, the hydromagnetic instabilities are theoretically non-existent. Second, the magnetic field inside the confined plasma is essentially zero. Therefore the cyclotron radiation is probably much lower than in any other configuration. Also the conditions of injection are significantly improved with respect to the ordinary mirror machine. The fact that the inner field is essentially zero frees the injected particles from the condition that they necessarily travel along the magnetic lines of force. This can be exploited by injecting the plasma blob obliquely through a hole in the geometry. Instead of being directed toward another hole, it is directed toward a plain magnetic wall. However, experiments in cusps are handicapped by the severe loss during the heating period.

The same principle is applied in the *picket fence* reactor formed by placing magnetic field coils around the plasma column. The coils are alternately opposed to each other and closure is obtained by biasing the end coils.This configuration, although stable, is extremely lossy. However, by creating an internal current configuration, it is possible to plug the leaks of the reactor. This led to the

Table 7.4 CUSPED-FIELD REACTORS

Experimeter	*Location*	B (T)	T_i (K)	T_e (K)	n (cm^{-3})	τ (μs)	$n\tau$ (s/cm^3)
Hagerman	Los Alamos	0·04	10^6		10^{13}	10	10^9
Nasedkin	USSR	0·6	4×10^5		10^{13}	100	10^8
Allen	Harwell	5·00	10^5	$1{\cdot}7\times10^5$	10^{16}	10	10^{11}

concept of the *caulked picket fence* reactor. The characteristics of three cusped-field reactors are reported in *Table 7.4* from the 1958 Genova Conference. It is worth noting that the cusped-field reactor does not seem promi sing as a future reactor.

Mirror machine with multipole

The best confinements seem to be those which combine ingeniously various features of closed confinements with features of the cuspid .The mirror machine with a superposed multipole-conductor drive belongs to this hybrid category, and seems to be promising. To an ordinary mirror machine, an even number of conducting bars (4 bars = quadrupole; 6 bars = hexapole; 8 bars = octupole) parallel to the axis is added. The bars would give a cusped-type confinement. If this type of confinement is combined with the mirror machine confinement with proper relative intensities of currents, a closed system can be obtained, but with a complicated magnetic field pattern, resulting from the combination of the field from the coils (in the r-z planes) with that of the bars (in the r-θ planes). The combined field is more or less twisted, depending upon the relative importance of the two types of confinement. With a proper ratio of the two fields, the two following advantages are obtained: (1) as in the stellarator with extra windings, the magnetic lines of forces are strongly interwoven, and (2) the field achieves a so-called 'minimum B' configuration. The minimum is along the axis for the quadrupole (for the hexapole and the octupole, there is a series of minima symmetrically disposed around the axis) but the axis itself corresponds to a slight maximum in 'the hole on top of the mountain' type.

The Ixion device[4]

The Ixion device is a combination of a mirror machine and a homopolar device. First a cloud of ionised deuterium gas is injected along the axis of a discharge chamber creating a central plasma electrode. Then a transverse electric field is applied between this electrode and the outer wall behaving as an anode. The combination of this transverse field and the axial magnetic field creates an energetic plasma rotating around the axis of the discharge.

Initial ionisation is either by pinch effect or by a tangential electric field. The plasma is further accelerated by a travelling magnetic wave. In Ixion III of 24 cm tube bore and 86 cm tube length, velocities of 5×10^4 m/s have been achieved. The magnetic field used has a 1·0 T strength, the plasma density was 10^{14} ions/cm^3 and the characteristic time, 1 ms, thus $n\tau = 10^{11}$ s/cm^3. It seems that this device would be more suitable for propulsion problems.

The magnetron device

This device consists of a central metallic anode in a long solenoidal magnetic field between two magnetic mirrors and a conducting outer wall. A radial electric field is applied between the anode and the outer wall. An anode plasma sheath is thus formed due to the low pressure of the discharge, in which energetic electrons rotate around the anode, ionising the gas. The ions formed are then accelerated in magnetron-like orbits out of the sheath. These ions will be diverted by the axial magnetic field and become confined.

7.6 CLOSED THERMONUCLEAR SYSTEMS

Toroidal pinch

The linear confinement leaves the plasma in contact with the electrodes at both ends of the column. Thus, a considerable amount of work has been performed to replace the linear pinch by an equivalent electrodeless discharge (Figure 7.15). This device is similar to a transformer in which the secondary is the discharge itself and the primary is a single turn coil made of a conducting paint

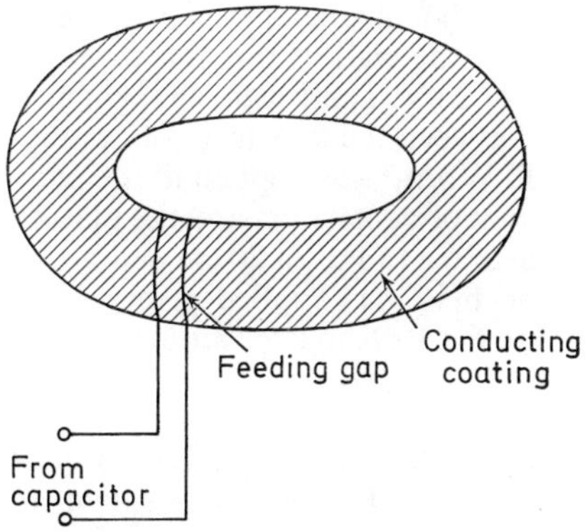

Fig. 7.15. The toroidal pinch

on the outside surface of the toroidal vessel, with a 'feeding gap'. The single turn is connected to the capacitor bank, and some initial ionisation is provided by an auxiliary discharge or by a small amount of radioactive paint.

It is possible to achieve equivalent toroidal configurations, such as the *Levitron* which appears as a promising approach toward the ultimate thermonuclear confinement. The levitron is a toroid with a hard core. In the absence of a discharge the hard core rests on a number of vertical fingers, which are mounted on pneumatic devices designed for the fastest possible downward motion, immediately before the discharge is started. This motion is much faster than that due to gravity, hence, the inner torus stays as if suspended by inertia. Once the discharge is started, the mutual influence of the discharge current, the current in the inner torus, and that in the outer sleeve results in a certain amount of 'magnetic levitation', which retards the descent significantly. After the discharge is over, the fingers take charge again; usually the downward motion is weak (1 cm) although not negligible. The discharge itself can have the properties of a hard-core or of a triax pinch; this depends upon the amount of current induced in the inner torus before the gas ionisation is triggered. As in the case of its linear equivalents, the levitron is devoid of any trace of flute instabilities related to finite electric conductivity, finite thermal conductivity, and finite gyration radii. The chances of finding adequate ways to suppress or significantly reduce such instabilities are not negligible.

The early experimental toroidal pinches were of small size. They were developed in the United States and in many countries of Europe. One of the first American devices, the Perhapsatron, was built in Los Alamos Laboratories. Perhapsatron S-4 is a small iron-core device which has a 7 cm minor diameter and a 35 cm major diameter. The discharge tube was made of quartz. An aluminium outer torus surrounding the discharge tube acted as a single turn primary and there were two iron cores. The apparatus generated a 320 kA peak pinch current in 12 μs. 10^7 neutrons have been generated for each pulse, although they were associated with instabilities. A revised model, Perhapsatron S-5, has been developed with some better characteristics. Similar devices have been built in France (Saclay Torus), United Kingdom (Sceptre), Sweden (Uppsala Torus), and the USSR (Moscow Torus). Their characteristics are reported in *Table 7.5*.

A much more sophisticated device, Zeta, has been developed in the United Kingdom. The torus of Zeta was made of one piece of aluminium metal, 2·5 cm thick. Its minor diameter was 1 m whereas its major diameter was as large as 3·7 m. A second segmented torus lies just inside the wall of the first to avoid any breaking down at the induced loop voltage of 2 kV. There were 48 segments, insulated from each other. The annular space between the two tori was evacuated. A stabilising axial induction of about 0·016 Wb/m^2 was applied to the plasma. The initial gas pressure was at about 10^{-4} mm of deuterium, and the plasma current reached about 10^5 A. The plasma pinched to a radius of 20 cm and broke up completely in 3000 μs. An rf preionisation field was applied prior to the pinch, helping to increase the ion temperature to 6×10^6K in later models. Diagnostic techniques such as magnetic probes and spectroscopic methods have been widely used in Zeta.

The most promising of all the thermonuclear reactors to date (1970) is of the Tokamak[14] series developed in the Soviet Union. This device is shown in

Table 7.5 TOROIDAL PINCHES

Device	*Location*	*Torus bore* (cm)	*Torus diameter* (cm)	I (A)	T_e (K)	T_i (K)	τ (μs)	n (cm^{-3})	$n\tau$ (s/cm^3)
Uppsala Torus	Sweden	28	130	10^6					
Saclay Torus	France	8	78	5×10^4			100	10^{16}	10^{12}
Alpha	USA	100	320	2×10^5					
Gamma Pinch	USA	10	60	3×10^5	5×10^5			10^{16}	
Sceptre	UK	30	110	2×10^5	10^5	3×10^6	100	10^{15}	10^{11}
Moscow Torus	USSR	48	125	2×10^5	3×10^5		20	4×10^{14}	$0{\cdot}8\times10^{10}$
Perhapsatron S-4	USA	14	70	3×10^5	10^5	10^6	10	5×10^{15}	$0{\cdot}5\times10^{11}$
Zeta	UK	100	370	10^5	5×10^5	6×10^6	10^3	10^{14}	10^{11}
Tokamak T-3A	USSR	35	300	$1{\cdot}2\times10^5$	10^7	5×10^6	3×10^4	5×10^{13}	$1{\cdot}5\times10^{11}$

Figure 7.16. This is a toroidal pinch with a very strong toroidal magnetic field B_ϕ generated by the current I_θ. A poloidal field results from a large current induced in the plasma by an iron core. Equilibrium is reached when this field presses against an outer heavy copper shell. The toroidal field is the main field of the device. In the present model (Tokamak T-3A) it has a strength of 3·5 T but might be increased to 5 T in the near future. A large transformer of 1·3

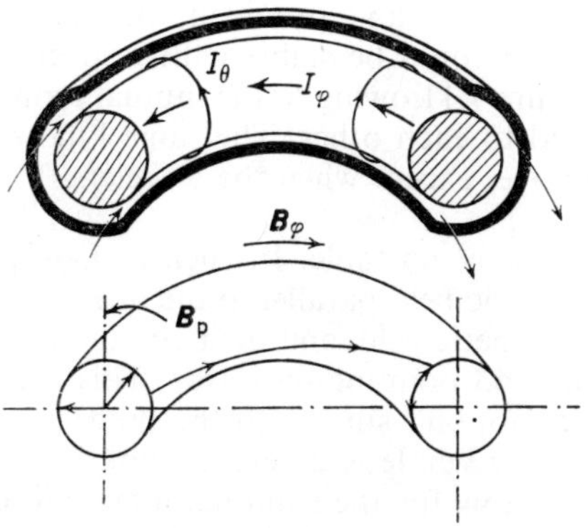

Fig. 7.16. Configuration of the Tokamak fusion experiment

V s produces plasma currents as high as $1{\cdot}2\times10^5$ A. The other characteristics of the device are: minor diameter 35 cm, major diameter 3 m, ion temperature 5×10^6K, plasma density 5×10^{13} cm^{-3}, and confinement time 30 ms or about 80 Bohm times. These values are about the closest to the future minimum fusion power reactor. For this reactor, at least 10 T field strength would be needed, that is about 3 times the present value, the current should be 10^6 A, about 10 times the present value, the ion temperature should be 20 times larger, the plasma density 4 times larger and the confinement time 50 times larger. The dimensions of such a reactor might be as high as 2 m for the plasma diameter and 10 m for the torus diameter. These are formidable values presenting a serious engineering challenge. Indeed a 10 T field will create a pressure on the I_θ coils as high as 41 400 kN/m^2. Nevertheless, it was shown that the confinement time in the Tokamak would increase with the radius r_p of the plasma, the plasma density n, and the ionic temperature T_i, as:

$$\tau \sim r_p^2 n T_i^{3/2} \tag{7.24}$$

Since n and T_i increase with the size of the device and the toroidal magnetic field, it becomes apparent that the minimum fusion reactor should be only 3 times larger than Tokamak T-3A. The characteristics of this device are shown in *Table 7.5* to be compared with the other toroidal pinches.

The stellarator

The stellarator[38] was initially conceived as a toroidal-like steady-state confinement. The simplest form is that of a system of coils which generates a magnetic field in a toroidal region of space. By contradistinction with the toroidal pinch the magnetic lines of force are here 'parallels' of the toroid instead of 'medians'. The magnetic confinement can be understood with a one-particle model: this particle will start spiralling around the magnetic lines of force and stay permanently at the same distance from the walls. However, as soon as the plasma pressure becomes significant, the magnetic field inside the plasma column becomes smaller than outside. Such a difference between the magnetic field inside and outside can only be maintained by electric currents inside the plasma. If a cylindrical configuration could be achieved, and provided the plasma column has a stability of its own, the system could be stable with respect to lateral displacements similar to those of Figure 7.11 owing to the mutual repulsion of the two currents which are opposed to each other. The same kind of stability can be assumed for kink and sausage, except when the wavelength is too short.

Pure toroidal stellarators are electrostatically unstable. In such configurations, the charge separation generates an electric field parallel to the linear axis of the toroid, hence at right angles to the magnetic field, and their combination results in an $\boldsymbol{E}\times\boldsymbol{B}$ drift; this drift is radial and pointing outwards. This is an outstanding example of plasma 'collective motion', similar to the flute instability; again, the behaviour of the plasma as a whole is different from that of individual particles, when the density is too low for their mutual interactions to be significant. In a toroidal stellarator, this drift would lead all but a few remaining particles to collide with the wall of the toroid in a time very short in comparison with the requirements of a practical experiment.

Particle motions perpendicular to the magnetic lines of force are strongly hampered with respect to those parallel to them. However, in the toroidal stellarator, once the charges have separated from each other, the same individual magnetic line of force is affected all along its length by the same electric charge density. This would not be the case if the magnetic lines of force were twisted in some way, so that the same one would alternately be in a positive and a negative excess charge. Then, owing to the high mobility of particles along the magnetic lines of force, charge recombination would occur at a satisfactory rate and prevent the $\boldsymbol{E} \times \boldsymbol{B}$ drift. Two ways have been proposed to impose such a torsion upon the magnetic lines of force. The first one is purely geometric, and has only been moderately successful; it consists in replacing the toroid by a '*figure of eight*'. The second one is purely electrical, preserving the toroidal configuration. It consists in adding auxiliary conductors wound on helices of very large pitch to the conductors belonging to the main coils and wound on helices of very small pitch. Neighbouring conductors would carry currents of opposite directions as shown in Figure 7.17. The pitch of a line of force depends upon its distance from the cylinder axis. This proves to be an important feature

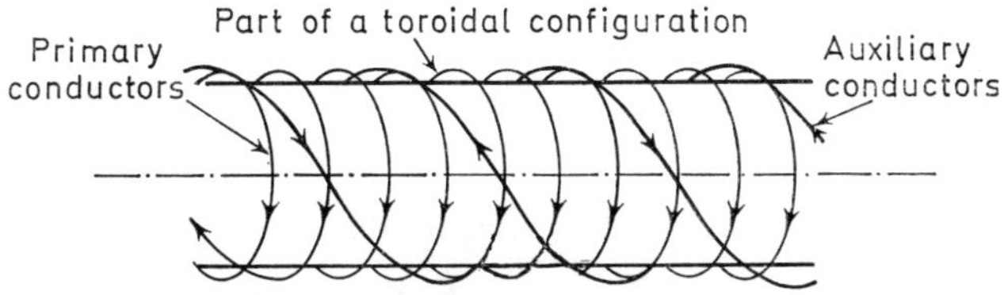

Fig. 7.17. Electrical method to impose a torsion upon the magnetic lines of force in a stellator

in preventing the flute instbility. Since the magnetic lines of force are interwoven, they acquire more rigidity.

The plasma generation and the first phase of the heating are similar to those used in the toroidal pinch, i.e. an electrodeless electrical discharge is initiated and maintained by passing a current in one or several coils inductively coupled with the discharge. Plasma heating is successful in this way until temperatures of about 10^6K are reached; then, it fails for several reasons. One of them is that the plasma has by then acquired a conductivity greater than that of copper, and that Joule heating has become negligible. Even more significant is the occurrence of an important instability, best described in the hodographic space.

In a normal plasma, the distribution of the charge carriers in the hodographic space is triply Gaussian. If a drift velocity is superimposed, the centre of weight in the hodographic space is slightly displaced with respect to the origin of co-ordinates. Now, if a strong electric field is applied, the distribution is fundamentally altered, and eventually tends to resolve into two bulks, owing to the fact that strongly accelerated electrons have, in all the gases, a lower probability of collision. This instability is known under the name of '*runaway electrons*'[11]. The runaway electrons are still confined magnetically, but in the hodographic space, their motion away from the bulk of the plasma is unstable.

Although a completely adequate theory of the runaway electrons can only be understood in terms of the general Boltzmann equation, the elementary concept can be grasped by means of a simplified orbital model. For a strongly ionised plasma, the scattering cross-section Q is essentially that of the interaction in a Coulomb field, which predicts a variation proportional to the inverse square of the kinetic energy, hence, as v^{-4} [equation (1.43)]. In the phenomenon of runaway electrons, the essential part of the total velocity v is the drift velocity v_d. Consider, now, in a one-dimensional model, an electron starting at $x = 0$ with a zero velocity. As long as no interaction intervenes, this electron is accelerated by a constant electric field E directed along the x-axis, with an acceleration eE/M. Its kinetic energy at a distance x from the origin is eEx, corresponding to an instantaneous drift velocity:

$$v_d = (2eEx/M)^{1/2} \tag{7.25}$$

The probability that this electron will be significantly deflected between x and $x+dx$ is proportional to the density n^+ of the positive ions, to dx and to the cross-section given by equation (1.43); hence:

$$f(x)\,dx) = \frac{K_2 n^+ \,dx}{E^2 x^2} \tag{7.26}$$

where K_2 is a suitable constant coefficient. If n^- is the number of electrons having survived up to the distance x, we can write:

$$\frac{dn^-}{n^-} = -\frac{K_2 n^+ \,dn}{E^2 x^2} \tag{7.27}$$

The integration of equation (7.27) cannot be performed from $x = 0$ because this is a special value. Let n_0^- be the number of electrons at distance x_0, then:

$$\ln (n^-/n_0^-) = \frac{K_2 n^+}{E^2}\left(\frac{1}{x} - \frac{1}{x_0}\right) \tag{7.28}$$

Even for $x = \infty$, n^- is still different from zero. Thus, a rapidly decreasing cross-section such as in equation (1.43) leads to a situation in which a finite fraction of the electrons have an infinite free path. Such electrons are accelerated indefinitely. Their fraction is given by:

$$n_\infty^-/n_0^- = \exp(-K_2 n^+/E^2 x_0) \tag{7.29}$$

The meaning of x_0 can be interpreted easily, since on the average, the electron does not start with zero velocity; it has a kinetic energy of the order of kT^-; hence $x_0 = kT^-/eE$, and introducing this equation into equation (7.29) we find:

$$n_\infty^-/n_0^- = \exp(-K_3 n^+/ET^-) \tag{7.30}$$

where K_3 is another suitable constant coefficient. Equation (7.30) indicates that the runaway phenomenon is favoured by an increase in the electric field and the electron temperature and a decrease in the plasma density.

There are several reasons for which the above theory cannot be entirely relied upon. First, equation (1.43) is only approximate and fails at low speeds, for which, even in strongly ionised media, a continuous friction model is adequate, this leading to a cross-section increasing proportionally with v_d. This variation combined with equation (1.43) leads to the concept that the cross-section Q has a maximum for a given value of v_d. This further leads to the concept of a critical field E_c, below which the mobility phenomenon described previously is entirely valid, and the fraction of runaways is negligibly small with respect to equation (7.30). On the other hand, if a significant fraction of the electrons become runaways, they tend to constitute a beam, and the recourse to the general Boltzmann equation becomes necessary. The runaway phenomenon can be considered as an instability in the hodographic space. Physically, the runaway electrons are still confined but become more and more strangers to the plasma.

Another consideration severely limiting the possibilities of Joule heating is that the current intensity must stay, at all times, much below the values obtained in a toroidal pinch, otherwise kink and sausage instabilities would reappear: the maximum current is designated as the *Kruskal limit*.

Among the other means designed, magnetic pumping and cyclotron wave heating should be mentioned. *Magnetic pumping* can only be understood in terms of anisotropic pressure and temperature. If an extra coil carrying an alternating current of adequate frequency is added to the coils contributing to the magnetic confinement, the magnetic lines of force will, locally, expand or contract radially. If the plasma conductivity is sufficient, it will be carried with them, thus undergoing alternative compressions and expansions.

Consider a simplified model with a square wave. The magnetic field has initially the value B_1, and as the result of the square wave, it is increased to some value B_2. Assume, however, that the steepness of this increase is not such as to jeopardise the conservation of magnetic moment. Because of equation (2.11) and the invariance of the magnetic moment, the energy $\mathcal{E}_\perp$ is increased in proportion to the increase of B:

$$\mathcal{E}_{\perp_2} = \mathcal{E}_{\perp_1}(B_2/B_1) \tag{7.31}$$

whereas $\mathcal{E}_{\parallel}$ remains unchanged. Now, assume that B stays constant at the value B_2 for a time larger than the collision time, so that equipartition occurs between $\mathcal{E}_{\parallel}$ and $\mathcal{E}_\perp$. Since $\mathcal{E}_\perp$ has two degrees of freedom and $\mathcal{E}_{\parallel}$ has only one, this equipartition leads to an energy one-third closer to the value of $\mathcal{E}_\perp$ before equipartition, i.e. at a value:

$$\mathcal{E}_3 = \tfrac{1}{3}(\mathcal{E}_{\parallel} + 2\mathcal{E}_{\perp_2}) \tag{7.32}$$

Now the expansion occurs, corresponding to a change in B from B_2 to B_1. As a result, $\mathcal{E}_\perp$ varies from $\mathcal{E}_{\perp_3}$ to some value $\mathcal{E}_{\perp_4}$ given by:

$$\mathcal{E}_{\perp_4} = \mathcal{E}_{\perp_3}(B_1/B_2) \tag{7.33}$$

or

$$\mathcal{E}_{\perp_4} = \tfrac{1}{3}\mathcal{E}_{\parallel}(B_1/B_2) + \tfrac{2}{3}\mathcal{E}_{\perp_1} \tag{7.34}$$

A new period of energy equipartition begins leading to an average value $\mathcal{E}_5$ given by:

$$\mathcal{E}_5 = \tfrac{1}{3}(2\mathcal{E}_{\perp_4} + \mathcal{E}_{\perp_3}) \tag{7.35}$$

or

$$\mathcal{E}_5 = \tfrac{2}{9}\mathcal{E}_{\parallel}(B_1/B_2) + \tfrac{4}{9}\mathcal{E}_{\perp_1} + \tfrac{1}{9}\mathcal{E}_{\parallel} + \tfrac{2}{9}\mathcal{E}_{\perp_1}(B_2/B_1) \tag{7.36}$$

In the case of a steady cycle, all the values of $\mathcal{E}$ with odd indices are in equipartition. Thus, if the cycle had actually continued for some time, $\mathcal{E}_{\perp_1}$ and

$\mathscr{E}_{\|1}$ would have had a common value $\mathscr{E}_1$. Hence, equation (7.30) can be written for a steady cycle:

$$\mathscr{E}^5 = (\mathscr{E}_1/9)\,[5+2(B_1/B_2)+2(B_2/B_1) \tag{7.37}$$

Let $B_2 = B_1(1+\delta_B)$, then after some approximations:

$$\mathscr{E}_5 \cong \mathscr{E}_1(1+\tfrac{2}{9}\delta_B^2) \tag{7.38}$$

The heating results from a second-order term resulting from a delay in the heat transfer between the perpendicular and the parallel components. This results in a hysteresis in the variation of the perpendicular temperature with respect to the magnetic pressure. As a result, the power developed in successive compressions and expansions is no longer zero. The efficiency of the magnetic pumping is increased if some resonance is sought with the ionic cyclotron motion or with any other characteristic time related to the ionic motion.

Electromagnetic waves have a cut-off at or near the plasma electrostatic frequency. But at very low frequencies, below the cyclotron frequency, propagation becomes possible again in the form of Alfvén waves. The name *cyclotron waves* is used to designate modified Alfvén waves propagating at frequencies immediately below the cyclotronic resonance. They are characterised by particle motions obeying the laws of the quasi-cyclotronic motion. As a result, the particles involved can acquire important motions which become more or less randomised by the collisions. In order to enhance the effect as much as possible, it is advisable to come as close to the actual resonance as possible. This is achieved by means of the so-called 'magnetic beach', which is a region of slowly decreasing magnetic field. The phenomenon can be visualised as follows: whereas the frequency of the wave is a constant imposed by the transmitter, the frequency of the cyclotron resonance decreases slowly as the wave proceeds along the magnetic lines of force; hence, the wave and the relevant particle motions come closer and closer to the ideal situation. It is understandable that true cyclotronic waves are strongly damped, since the large orbits of the individual particles can give rise to many collisions involving much randomisation of energy. This is the reason for which the wave has necessarily to be injected into the plasma at some frequency close to, but slightly below the true resonance. Both ion and electron cyclotron heating have been used successfully in practice.

The stellarator is interesting in that it seems to bear the promise of a much better confinement than the pinch. On the other hand, plasma heating appears as a major problem, and the results so far are limited to lower temperatures than in the pinch. Another problem is the so-called 'pump-out phenomenon consisting in a disappearance of the plasma after some time, thus terminating the experiment. It has been suggested that at least two mechanisms operate in conjunction; they comprise on one hand a charge transfer phenomeon, and on the other hand, a 'universal instability'. The future of the stellarator is entirely dependent upon the solution of such problems.

The most important device in this category, the C-Stellarator, has been developed at Allis-Chalmers with the collaboration of Princeton[37]. It is a refined model of a series of devices including B_1, B_2, and B_3-Stellarators. The characteristics of this device along with two other are reported in *Table 7.6*.

Table 7.6 STELLATOR DEVICES

Device	Location	B (T)	p_r (mm Hg)	$\frac{8\pi p_r}{H^2}$	T_e (K)	T_i (K)	n (cm^{-3})	τ (s)	$n\tau$ (s/cm^3)
Proto-Cleo	UK			10^{-5}	6×10^4	10^6	10^{11}	5×10^{-3}	5×10^8
Wendelstein II	Germany			10^{-9}	2×10^3	2×10^3	2×10^9	1	3×10^9
C-Stellarator	USA	5·5	2×10^{-10}	10^{-4}	10^6	5×10^8	5×10^{12}	10^{-3}	5×10^9

The astron system

The astron, suggested by Christofolios[7], makes use of an interesting and original feature. Up to now, the confining magnetic fields were always generated either by currents carried by the plasma itself, or currents carried by external coils or conductors, or combinations of both. However, a third way to generate magnetic fields is theoretically possible, this is by the use of charged particle beams of sufficiently high intensity. The main difficulty appears at once; since the confining magnetic fields are of the order of Wb/m^2, any beam capable of modifying them significantly must have a tremendously high intensity. The only hope of ever reaching such an intensity is by using relativistic electrons. The magnetic mirror configuration is used in order to achieve an almost cylindrical layer of relativistic electrons on one of the magnetic surfaces of the machine. This layer

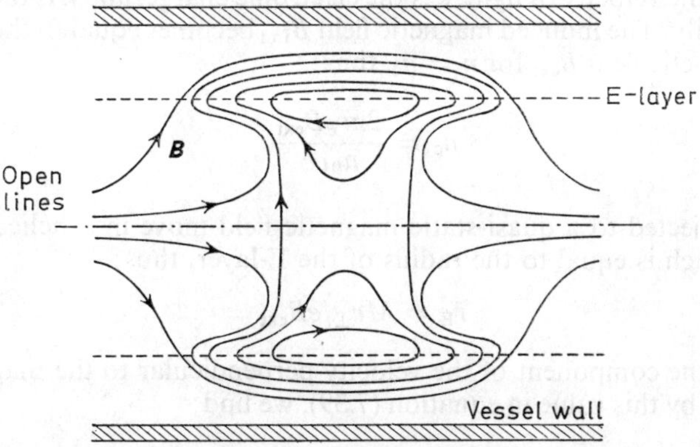

Fig. 7.18. Magnetic field configuration and E-layer in the astron geometry

generates an additional magnetic field such that a toroidal confinement is achieved (Figure 7.18). The confined plasma is heated by the relativistic electrons

The electron beam is injected at one end of an evacuated cylinder in the presence of an external axial magnetic field, with small bumps in the field near the cylinder ends. Under the effects of this field, electrons form a sheet of current circulating around the axis of the tube, the E-layer. This layer creates a magnetic field forming closed lines of force, thus confining a plasma of cold deuterium or deuterium–tritium gas. This gas collides with the electronic layer and is heated until it reaches temperatures sufficient for thermonuclear reactions to occur. One should note that in the astron system, the mirrors at both ends of the configuration mainly serve to contain the electrons in the E-layer but do not confine the plasma. The charged particles gyrate around the central axis while travelling axially from one mirror to another.

Since the electrons lose energy to the plasma in the E-layer, it is necessary to inject relativistic electrons continuously in order to maintain the plasma. The electrons should be introduced in such a way that they will disturb neither the electrons already injected nor the pressure of the layer. Electrons are injected at the top of the magnetic well and they proceed to its bottom by decreasing their axial energy. Their axial momentum then becomes small enough to confine them in the mirror field.

Once the electron bunch reaches the E-layer, it spreads in the axial direction within the mirror. Thus, the energy stored in the field of the electron bunch is converted into kinetic energy. The continuous injection of electron bunches increases the electron energy in the E-layer until the rate of escape from the fields limits this increase. The magnetic field inside the volume enclosed by the E-layer goes to zero when the electron density approaches a critical value n_c. To calculate n_c, consider the circulating sheet current per unit length H and the created induction B; thus if r_E is the radius of the E-layer one can write:

$$B = \frac{\mu_0 nec}{2\pi r_E}$$

in which c is the velocity of light, e is the electronic charge, and n is the electronic number density. The induced magnetic field B_{ind} becomes equal to the externally applied magnetic field B_{ext} for $n = n_c$, thus:

$$n_c = \frac{2\pi r_E B_{ext}}{\mu_0 ec} \tag{7.39}$$

Particles subjected to a quasi-static magnetic field move in a helical path, the radius of which is equal to the radius of the E-layer, thus:

$$r_E = Mv_\perp / eB_{ext} \tag{7.40}$$

where $v_\perp$ is the component of the velocity perpendicular to the magnetic field. Replacing r_E by this value in equation (7.39), we find:

$$n_c = (2\pi M / \mu_0 e^2 c) v_\perp \tag{7.41}$$

Thus n_c is directly proportional to $v_\perp$. Increasing the number of electrons beyond n_c causes an inversion in the magnetic lines of force forming the magnetic bottle around the E-layer.

Although the general energetic balance in the astron system seems favourable, the conditions for a stable confinement are not yet fully understood. The astron system seems to provide hopes for an acceptable thermonuclear confinement, although the final form of this approach is still hard to predict.

7.7 FUSION REACTORS[14]

Whatever is the final form of plasma confinement, it is already possible to have a conceptual view of the future core of the thermonuclear reactor. As shown in Figure 7.19, this reactor would consist of the following items: a vacuum wall; heat removal ducts; a neutron moderating and tritium breeding blanket[22]; a coil shield; a magnet and its mechanical support; a plasma injector; and a plasma removal device. This reactor would then be similar to the core of a fission breeder reactor.

Figure 7.19 can be a cross-section of the torus of Figure 7.16 representing the Tokamak reactor, or it can be the cross-section of any other reactor configuration. The centre of the cross-section is occupied by the plasma discharge

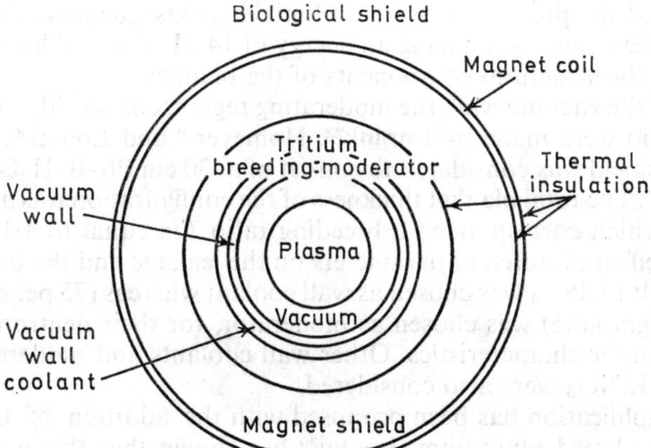

Fig. 7.19. Conceptual view of a steady-state D-T *fusion reactor*

which might have a radius of about 2 m. The plasma is surrounded by the vacuum wall through which the 14 MeV neutrons should pass from the plasma to the neutron moderating blanket. Depending on the coolant chosen for the walls and on their material, the wall temperature would range between 500 and 1000°C. The neutrons would lose their energy in the moderating blanket and become absorbed by lithium; whereas tritium would be bred. From nuclear cross-sections considerations, the moderating blanket might have a thickness of about 1 m. The coil shield surrounding the breeding region would also be 1 m thick. Outside the shield would lie 8 m thick superconducting magnets able to support pressures at least equal to 41 400 kN/m^2. The overall would then be shielded biologically. It is clear, therefore, that the minimum diameter of the torus cross-section of a thermonuclear reactor would be about 24 m. Smaller nuclear fusion reactors would not be economically feasible.

Building magnetic field coils in the 10 T range is a relatively simple problem to solve, and it is in the grasp of the present technology using a superconducting material. For instance, the present cost of a wire of niobium–tantalum alloy embedded in a copper matrix is about $4 per metre carrying a current of 1000 A.

This amount is comparable to that of pure copper, if it were to do the same function without excess heating. The operating costs of the magnet are almost insignificant compared to the other costs. Large magnets have already been built in the Soviet Union for the Tokamak device reaching 3·5 T and soon they will reach 5 T. In the United States similar magnets, such as the 2 T magnet of Argonne National Laboratory used for a bubble chamber, have been built. There is, however, the problem of the stresses caused by these strong magnetic fields that should be solved before building such large magnets.

This brings us to the problems of the blanket assembly of the thermonuclear reactor. This is the complex structure in which all the functions of the reactor are located. Studies have shown that it should be possible to build an adequate functional blanket assembly. However, research and development studies should be conducted in many fields before even foreseeing the final form of this assembly. For instance, materials science should help us in finding the material that will withstand the present stresses and limit the consequences of the radiation damage. Indeed, neutrons having an energy of 14 MeV would have a detrimental effect on the structural components of the blanket.

Studies of the vacuum wall, the moderating region, coil shield, and the tritium breeding ratio were made by Impink[22], Homeyer[20] and Lontai[30]. In all cases, the magnet shield was considered to consist of a '30 cm Pb–B–H_2O, 20 cm LiH, and 6 cm Pb'. The total blanket thickness of the configuration was assumed to be 120 cm to which corresponded a breeding ratio T/n equal to 1·16. They also studied the effect of different parameters on the leakage and the breeding ratio. The fused salt Li_2BeF_4 was chosen as wall coolant whereas (75 per cent Li_2BeF_4, 25 per cent graphite) was chosen as moderator, for their neutron moderating and heat transfer characteristics. Other wall coolants and moderators such as $LiBeF_3$ and $LiNO_3$ were also considered.

Power amplification has been proposed with the addition of U-238 to the moderator to breed plutonium. Lontai[30] has shown that this will result in a power gain of about 2, if U-238 fast fission is used. This gain factor might be increased to a factor of 30, thus making economical some otherwise uneconomical systems.

Other problems dealt with the influence of magnetic fields on the flow, erosion, material transfer, and heat transfer of the coolant flowing across and along the magnetic lines of force. Also, ways should be found to extract and recover efficiently the tritium bred in the blanket.

7.8 ENERGY PRODUCTION[10]

Energy could be produced from a fusion reactor by using the heat produced thermonuclearly either in a conventional plant to produce electrical energy or directly. Direct production of electrical energy can be obtained by confining and compressing the plasma by an increasing magnetic field. The compression increases the particle density which in turn increases the power released. The simultaneous increase of the pressure causes the plasma to expand against the magnetic lines of force. The magnetic field varies and induces electric currents in pick-up circuits, thus generating electric power directly. This is possible due to the fact that more than two thirds of the power emitted in a deuterium plasma is emitted as the kinetic energy of charged particles[27]. Those charged particles are held by the confining magnetic field and it is their kinetic energy that could be converted directly into electrical energy. For the case of a thermonuclear device using deuterium and tritium, about 80 per cent of the energy released would be in the form of kinetic energy of the neutrons. In this case a neutron moderator could be used to stop those neutrons and the heat released could be extracted by a suitable coolant.

The total energy released by a fusion reaction is the kinetic energy of the products of the reaction, which may be neutrals or charge carriers. In a deuterium gas, a pair of deuterons can fuse with equal probability following one or the other of the two reactions:

(1) $D+D \rightarrow (^3He+0{\cdot}82\ MeV)+(n+2{\cdot}45\ MeV)$

(2) $D+D \rightarrow (T+1{\cdot}01\ MeV)+(H+3{\cdot}02\ MeV)$

In the presence of tritium, the following secondary reaction occurs with much higher probability:

(3) $D+T \rightarrow (^4He+3{\cdot}5\ MeV)+(n+14{\cdot}1\ MeV)$

The fusion power density is then:

$$P_f/\mathcal{V} = \tfrac{1}{2}n_D n_D \dot{\mathcal{V}}_1\mathcal{E}_1+\tfrac{1}{2}n_D n_D\dot{\mathcal{V}}_2\mathcal{E}_2+n_D n_T\dot{\mathcal{V}}_3\mathcal{E}_3 \qquad (7.42)$$

where $\mathcal{E}_1$, $\mathcal{E}_2$ and $\mathcal{E}_3$ are the energies released by reactions 1, 2 and 3 respectively and n_D and n_T are the number densities of fuel deuterium and fuel tritium respectively. $\dot{\mathcal{V}}_1$, $\dot{\mathcal{V}}_2$ and $\dot{\mathcal{V}}_3$ are the average products over the velocity space of the cross-section Q, and the relative velocity v between reacting nuclei for reactions 1, 2 and 3 respectively, and $\mathcal{V}$ is the volume of the plasma. $\dot{\mathcal{V}}_{12} = \dot{\mathcal{V}}_1+\dot{\mathcal{V}}_2$ and $\dot{\mathcal{V}}_1$ are represented in Figure 7.2 as functions of the temperature.

Tritons are very short-lived compared to deuterons. Equilibrium is reached when the rate of consumption of tritons equals the rate of their formation; thus when

$$n_T = \tfrac{1}{2}n_D(\dot{\mathcal{V}}_2/\dot{\mathcal{V}}_3) \qquad (7.43)$$

Eliminating n_T from equation (7.42) leads to:

$$P_f/\mathcal{V} = \tfrac{1}{2}n_D^2\dot{\mathcal{V}}_{12}\bar{\mathcal{E}} \qquad (7.44)$$

after assuming that $\dot{\mathcal{V}}_1 = \dot{\mathcal{V}}_2$ and

$$\bar{\mathcal{E}} = \tfrac{1}{2}(\mathcal{E}_1+\mathcal{E}_2+\mathcal{E}_3) \qquad (7.45)$$

The losses in a fusion reactor result from the several imperfections mentioned in Section 7.2. If a deuterium gas is assumed, the two important losses of bremsstrahlung and cyclotron radiation can be given by:

$$P_b/\mathcal{V} = \tfrac{1}{2}\mathcal{C}_\beta n_D^2 T^{1/2} \qquad (7.46)$$
$$P_c/\mathcal{V} = \tfrac{1}{2}\mathcal{C}_\gamma n_D^2 B^2 T \qquad (7.47)$$

and the efficiency of the reactor becomes:

$$\eta = \frac{P_f - \mathcal{C}_\alpha P_f - P_b - P_c}{P_f} \tag{7.48}$$

where $\mathcal{C}_\alpha$ is the fraction of total fusion in the form of kinetic energy of the neutrons. Replacing P_b and P_c by their values, one finds[26]:

$$\eta = 1 - \mathcal{C}_\alpha \frac{\mathcal{C}_\beta n_D T^{1/2} + \mathcal{C}_\gamma B^2 T}{n_D \mathcal{V}_{12} \bar{\mathcal{E}}} \tag{7.49}$$

The energies produced by the three important reactions are plotted as functions of temperature in Figure 7.3 along with bremsstrahlung.

7.9 THERMONUCLEAR POWER PLANTS[13]

An outline of a nuclear power plant is shown in Figure 7.20. This scheme represents a 5000 MW(t) mirror machine the vacuum wall of which having a 10 m diameter and 20 m length. The vacuum wall is surrounded internally by a 60 cm thick lithium layer acting as a reflector, and externally by a 40 cm thick

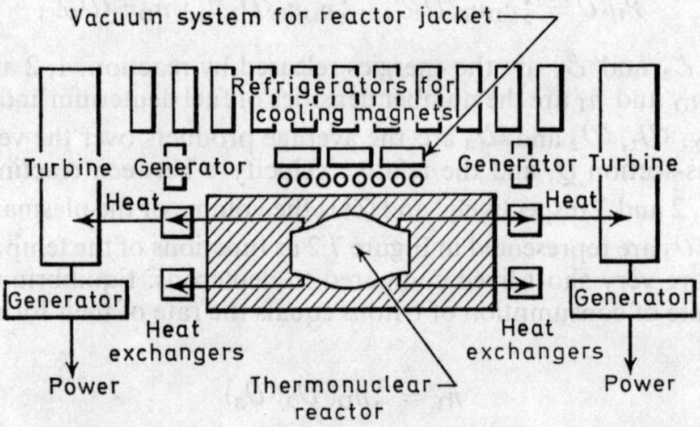

Fig. 7.20. Outline scheme of a thermonuclear power station

graphite layer. The reflector region is enclosed by niobium shells to provide passage for the lithium flow, thus allowing the absorption of about 99·5 per cent of the heat absorbed by the reflectors (T = 1000°C). It is surrounded first by a 2 cm thick multifoil reflective insulation and then by a 76 cm thick shield of borated water followed by 17·2 cm of lead to prevent neutrons and gamma particles from reaching the superconducting magnet. The latter is cooled cryogenically and has a power of 1 kW. The coils of the magnet, mounted just outside the shield, are insulated from it thermally by a region 4 cm thick containing a liquid nitrogen-cooled buffer plate. The magnet is held in place by titanium hoops. The entire system is surrounded again by another 4 cm layer of thermal insulation.

In the system of Figure 7.20, the thermal energy produced in the reflector region is converted into electricity by using a high temperature thermodynamic cycle. To do this a binary vapour cycle is used by boiling potassium at 1000°C, expanding it through a turbine and then condensing it at 600°C. The heat thus generated is used to produce steam at 570°C and 27 600 kN/cm^2 to be expanded in a conventional steam turbine. If the condenser is at 27°C, the thermal efficiency of the entire system would be as high as 58 per cent. To produce electricity, a pair of 1400 MW(e) Rankine-cycle systems operate in parallel with the thermonuclear reactor acting as the heat source. The potassium boilers installed at each end of the reactor are heated by the lithium circulating in the reflector cooling circuit.

It should be noted here, that the thermonuclear plant just described is of the minimum size possible. Future thermonuclear power reactors would certainly reach the 20 000 MW(t) range, if not more. Open-ended systems might be significantly smaller than the closed-ended systems.

The feasibility of a thermonuclear power plant will finally be in terms of the cost per thermal kilowatt produced. This cost is equal to the ratio of the cost per unit length of the vacuum wall, the moderator, the magnet and its support, the plasma injectors, and the pumps over the net output per unit length. It is clear that the cost per unit power will decrease with increase of confinement time. Indeed, a too short confinement time will increase the need for recycling and raise the injection and pumping capital costs. It also reduces the net power output by requiring more input power. Capital costs will, of course, depend on the desired strength of the magnetic field and on the fractions of the injected D–T fuel burned in the reactor, and lost in the exhaust. It is usually believed that for a thermonuclear reactor to become economically feasible as a source of a large amount of electricity, the cost of the thermal kilowatt produced should be lowered to about $20. Then, it seems safe to predict the year A.D. 2000 as the earliest date for any large thermonuclear fusion system to operate commercially.

The above analysis applies to a steady-state thermonuclear system. There is also the possibility that thermonuclear power evolves into pulsed systems. However, all the studies have been made for steady-state systems to avoid introducing the time parameter into the already unsolvable equations.

7.10 FUTURE TRENDS[24]

None of the present fusion reactors can be judged satisfactory as yet. Practically, stability is a problem still waiting for adequate solutions, and even the theoretical work on instability is far from final. Bremsstrahlung and 'runaway' electrons seem to pose formidable problems for energy production. It is, however, a fact that the great amount of work done in this field ever since the early 1950s has led to a good understanding of the relevant types of plasmas. The fusion research programme has grown tremendously in the US and in Europe (Britain, USSR, France, Sweden, Germany, Netherlands) ever since the beginning. This is proof enough that scientists working in this field strongly believe that there is no fundamental barrier to the solution of all the problems.

The most advanced thermonuclear scheme is the Russian Tokamak T-3A machine. The magnetic field used in this machine will soon be increased from

3·5 T to 5 T leading to Tokamak T-4; T-6 and T-7 are already under construction. The Tokamak machine is planned to approach scientific feasibility by 1974 with the T-10 machine. This machine will have a plasma radius about twice that of T-3A and will operate at 6 T.

An American Tokamak, called Ormak, has just been constructed at Oak Ridge. It is smaller than T-3A, but it will be expanded to this size soon. At Princeton, work on stellarators has been discontinued. Work will, however, be continued on the levitated superconducting spherator and the floating multipole device. The *spherator* is a toroidal sheared-field device. The C-Stellarator will be converted into a Tokamak device operating at 5 T but having a smaller plasma radius and a smaller iron-core than the T-3A. Plans are being made for the construction of two new machines: the advanced Tokamak (AT) and the proto-large Tokamak (Proto LT), followed by the large Tokamak (LT). The latter, the closest to proving scientific feasibility, is planned for 1973. It will reach an ion temperature of 6×10^{7}K at densities of $1{\cdot}2\times10^{14}$ cm^{-3} and confinement times of 0·25 s, ($n\tau = 3\times10^{13}$ s/cm^3). At Livermore, the mirror-machine programme is being continued. Schemes are being studied for direct conversion of the end loss energy into electricity. The 2X-machine is being modified to push towards ignition by the year 1973. A new mirror-machine is also being studied; this is the *baseball*-machine, the geometry of its coil is shaped like a baseball seam to produce a deep magnetic well. Work on Alice will also be continued with Alice II. The astron programme is also developing. At Culham (UK) research on toroidal pinch systems is planned, based on the success of the toroidal Zeta pinch.

It is therefore a fact that the gap between conditions obtained experimentally and those needed by a fusion reactor is being gradually closed. What will be the final form of the fusion reactor cannot yet be foreseen accurately. However, a clear demonstration that a thermonuclear plasma can be confined magnetically is almost at hand.

A new field of plasma physics was born with the studies of thermonuclear fusion. This is the physics of ultra-high temperature plasmas. The understanding of the interaction of these plasmas with electromagnetic fields will lead to many scientific and technological applications related to fusion power. One of these applications is the direct exploitation of the fusion reaction products as propellants for rocket propulsion in space[31]. The problems of a fusion rocket motor are similar to those of a thermonuclear power plant: practical methods should be found to magnetically confine the plasma in a stable configuration and heat it up to the reaction temperatures. Further, acceptable thrust-to-weight-ratios should be taken into consideration. Because of the tremendous amount of energy produced, these ratios are extremely large for thermonuclear fusion. If this propulsion method succeeds, manned interplanetary vehicle communications will become as simple as the present intercontinental communications.

REFERENCES AND BIBLIOGRAPHY

1. ALLIS, W. P., *Nuclear Fusion*, Van Nostrand, Princeton, N.J. (1960).
2. ANDERSON, O. A., FURTH, H. P., STONE, J. M. and WRIGHT, R. E., *Physics Fluids*, **1**, 489 (1958).

3. ARTSIMOVICH, L. A., *Controlled Thermonuclear Reactions*, Gordon and Breach, New York (1962).
4. BOYER, K., HAMMEL, J. E., LONGMIRE, C. L., NAGLE, D., RIBE, F. L. and RIESENFELD, W. B., 'Theoretical and Experimental Discussion of Ixion, a Possible Thermonuclear Device', *2nd U.N. Conf. Peaceful Uses of Atomic Energy*, Geneva, Paper P/2383 (1958).
5. BROWN, S. C., *Basic Data in Plasma Physics*, MIT Press, Cambridge, Mass. (1959).
6. CARRUTHERS, R., DAVENPORT, P. A. and MITCHELL, J. T. D., *The Economic Generation of Power from Thermonuclear Fusion*, U.K. Atomic Energy Authority Research Group, Rpt. CLM-R8S (1967).
7. CHRISTOFOLIOS, N. C., 'Astron Thermonuclear Reactor', *2nd U.N. Conf. Peaceful Uses of Atomic Energy*, Geneva, Paper P/2446 (1958).
8. COLE, G. H., 'The Pinch Effect', *Sci. Prog., Lond.*, **47,** 437 (1959).
9. COLGATE, S. A., FERGUSON, J. and FURTH, H., 'A Toroidal Stabilized Pinch', *2nd U.N. Conf. Peaceful Uses of Atomic Energy*, Geneva, Vol. 32, 129 (1958).
10. DOUGAL, A. A., 'Problems and Progress in Control of Thermonuclear Fusion for Electrical Power Production', Oklahoma State University, *Proc. 2nd Ann. Energy Conv. and Storage Conf.*, Oct. 12 (1964).
11. DREICER, H., 'Electron and Ion Runaway in a Fully Ionized Gas', *Phys. Rev.*, **117,** 329 (1960).
12. FOWLER, T. K. and POST, R. F., 'Progress Toward Fusion Power', *Scient. Am.*, **215,** No. 6, 21 (1966).
13. FRAAS, A. P., 'A Potassium-Steam Binary Vapor Cycle for a Molten-Salt Reactor Power Plant', *J. Engng Pwr*, Series **A, 88,** No. 4, 355, Oct. (1966).
14. FRAAS, A. P. and ROSE, D. J., 'Fusion Reactors as Means of Meeting Total Energy Requirements', *ASME Winter Meeting*, Los Angeles, Nov. 16–20 (1969).
15. FURTH, H. P., 'Minimum-Average β for Toruses', *Advances in Plasma Physics* (edited by A. SIMON), Interscience, New York (1968).
16. GLASSTONE, S. and LOVBERG, R. H., *Controlled Thermonuclear Reactions*, Van Nostrand, Princeton, N.J. (1960).
17. GRAD, H., 'Plasma Trapping in Cusped Geometries', *Phys. Rev. Lett.*, **5,** No. 5, 222 (1960).
18. GUTHRIE, A. and WAKERLING, R. K. (Eds.), *The Characteristics of Electric Discharges*, McGraw-Hill, New York (1949).
19. HARRISON, E. R., 'Runaway and Suprathermal Particles', *Plasma Phys., J. Nucl. Energy C*, **1,** 105 (1960).
20. HOMEYER, W. G., *Thermal and Chemical Aspects of the Thermonuclear Blanket Problem*, Technical Rpt. No. 435, MIT, Electronics Lab., Cambridge, Mass. (1965).
21. HOYAUX, M. F., *Arc Physics*, Springer-Verlag, New York (1968).
22. IMPINK, A. J., Jr., *Neutron Economy in Fusion Reactor Blanket Assemblies*, Technical Rpt. No. 434, MIT, Electronics Lab., Cambridge, Mass. (1965).
23. JEFFREY, A. and TANIUTI, T., *MHD Stability and Thermonuclear Containment*, Academic Press, New York (1966).
24. KADOMTSEV, B. B. and STEFANOSVKY, A. M., 'A Survey of the 3rd International Conference on Plasma Physics and Controlled Nuclear Fusion Research', *Nucl. Fusion*, Special Supplement, **9** (1969).
25. KELLER, C. and SCHMIDT, D., *Industrial Closed-Cycle Gas Turbines for Conventional and Nuclear Fuel*, ASME Paper 67-GT-10, Gas Turbine Conf., Houston, Texas, March 5–9 (1967).
26. KETTANI, M. A., *Direct Energy Conversion*, Addison-Wesley, Chapter 4, Reading, Mass. (1970).
27. KURCHATOV I, V., 'Kurchatov on Thermonuclear Reactors' *Engineering, Lond .*, **185,** No. 4804, 431 (1958)
28. LITTLE, E. M., QUINN, W. E. and SAWYER, G. A., 'Plasma End Losses and Heating in the "Low Pressure" Regime of a Theta Pinch', *Physics Fluids*, **8,** No. 6, 1168 (1965).

29. LOEB, L. B., *Fundamental Processes of Electrical Discharges in Gases*, Wiley, New York (1939).
30. LONTAI, L. N., *Study of a Thermonuclear Reactor Blanket with Fissile Nuclides*, Technical Rept. No. 436, MIT, Electronics Lab., Cambridge, Mass. (1965).
31. MASLIN, S. H., 'Fusion for Space Propulsion' *Trans. IRE*, Vol MIL–3, No. 4, 52 (1959)
32. MOTLEY, R., LUSTIG, C. D. and SANDERS, S., 'Synchrotron Radiation from Runaway Electrons in the Stellarator', *Plasma Phys., J. Nucl. Energy C*, **3,** 17 (1961).
33. POST, R. F., 'Some Aspects of the Economics of Fusion Reactors', *2nd Symp. the Engng Aspects of MHD*, Philadelphia, Pa., March (1961).
34. POST, R. F., 'The Status of Mirror Research', *Advances in Plasma Physics* (edited by A. SIMON), Interscience, New York (1968).
35. ROSE, D. J. and CLARK, M., *Plasmas and Controlled Fusion*, MIT Press, Cambridge, Mass. (1961).
36. ROSE, D. J., 'Engineering Feasibility of Controlled Fusion, A Review', *Nucl. Fusion*, **9,** 183 (1969).
37. SCAG, D. T. and KILGOUR, A. E., 'The C-Stellarator System', *Allis-Chalmers elect. Rev.*, First Quarter (1962).
38. SPITZER, L., Jr., 'The Stellarator', *Scient. Am.*, **199,** 28 (1958).
39. TAYLOR, J. B., 'Plasma Containment and Stability Theory', *Proc. R. Soc.*, **A304,** 335 (1968).
40. TUCK, J. L., *Thermonuclear Reaction Rates*, LAMS-1640, March (1954).
41. TUCK, J. L., 'Controlled Thermonuclear Research at Los Alamos', *Proc. 2nd U.N. Conf. Peaceful Uses of Atomic Energy*, Vol. 31, 3, Geneva (1958).
42. TUMA, D. T. and LICHTENBERG, A. J., 'Electron-Cyclotron Heating in a Mirror Machine' *Plasma Phys.*, **9,** 87 (1967).
43. WESTPHAL, W., *Elektrizitätsbewegung im Gases*, Springer-Verlag, Berlin (1926).
44. YAMATO, H., IIYOSHI, A., ROTHMAN, M. A., SINCLAIR, P. M. and YOSHIKAWA, S., 'Confinement of Energetic Plasma in Magnetic Mirrors in a Model C-Stellarator', *Physics Fluids*, **10,** No. 4, 756 (1967).

8 Magnetohydrodynamic applications

8.1 INTRODUCTION

Magnetohydrodynamics, MHD, is the science dealing with moving conducting fluids in electromagnetic fields. With reference to plasma physics, the word has often a more restrictive meaning, notably, that the plasma is regarded as a unique fluid in which charge separations are disregarded. Most of the equations of MHD appear therefore as limiting cases of the equations of plasma physics, when the number density of carrier pairs tends toward infinity, carrying the Debye screening length towards zero and the plasma frequency towards infinity.

One of the oldest applications of MHD is the direct generation of electricity. The effect upon which this application is based was first discovered by Faraday about one hundred and forty years ago[16]. Faraday noticed that an electric field was produced in a direction perpendicular to a magnetic field and to the velocity of mercury flowing perpendicularly to the field. Five years later, Faraday noticed that the Thames river itself would act as an 'MHD' generator due to the interaction of its salty and thus conducting water with the terrestrial geomagnetic field.

The possibility of using a plasma as working fluid for MHD power generation was seriously investigated in the period between the two wars. Then, the results were rather unconvincing. Since World War II, much attention has been devoted to means of converting heat into electricity without making use of the traditional heat engines and rotating machines. The advance in materials science and technology led to this renewal of interest. Also, the opportunities offered by new sources of energy, such as nuclear energy, have some appeal when combined with the new conversion schemes.

Plasma acceleration, and propulsion, is based on the same principles as MHD generation but in reverse. Plasma acceleration, however, is very different from charged particle acceleration, since, in most cases, a basically electric force would be entirely unoperative as such, the plasma being quasi-neutral.

Two forces may be operative in plasma acceleration or in direct energy conversion: a magnetic $\boldsymbol{J}\times\boldsymbol{B}$ force and a pressure gradient. Of the two, only the first one is specific to a plasma since the second one exists also in non-ionised gases, and is fully exploited in nozzle devices. In MHD generators and accelerators, the major role is thus played by magnetic forces; however, the pressure-gradient effect is sometimes far from being negligible.

Consider an MHD duct (in principle a pipe travelled by a conducting fluid and subjected to a magnetic field) that can be used either as a generator or as a propulsor[6]. Poynting's law yields the electrical power density for such a duct; it is:

$$P = -\boldsymbol{J}.\boldsymbol{E} \tag{8.1}$$

A loaded duct has a braking effect on the working plasma whereas in an accelerator, this effect will be reversed in an accelerating one. In general:

$$P' = -\boldsymbol{v}.(\boldsymbol{J}\times\boldsymbol{B}) \tag{8.2}$$

where P' is the braking power, that can be positive (for the generator) or negative (for the propulsor), $\boldsymbol{v}$ being the velocity of the working plasma. Considering the Hall effect in a duct where the working fluid flows in a direction perpendicular to the applied magnetic field, one can write:

$$\begin{aligned} \boldsymbol{E} &= \boldsymbol{i}E_x + \boldsymbol{j}E_y \\ \boldsymbol{J} &= \boldsymbol{i}J_x + \boldsymbol{j}J_y \end{aligned} \tag{8.3}$$

if $\boldsymbol{v} = \boldsymbol{i}v$ and $\boldsymbol{B} = \boldsymbol{k}B$ where $\boldsymbol{i}, \boldsymbol{j}$ and $\boldsymbol{k}$ are the unit vectors in the x-, y- and z-directions respectively. Equations (8.3) can be written in dimensionless form and complex notation. Taking the x-direction as the real axis and the y-direction as the imaginary one, equations (8.1) and (8.2) become respectively:

$$P = -\sigma v^2 B^2 (m_x \delta_x + m_y \delta_y) \tag{8.4}$$

$$P' = \sigma v^2 B^2 \delta_y \tag{8.5}$$

where $\delta_x = J_x/\sigma vB$, $\delta_y = J_y/\sigma vB$, $m_x = E_x/vB$ and $m_y = E_y/vB$. Ohm's law, with ion slip and relativistic effects neglected, can be written as:

$$\boldsymbol{J} = \sigma(\boldsymbol{E} + \boldsymbol{v}\times\boldsymbol{B}) \tag{8.6}$$

or in dimensionless notation $\delta_x = m_x$, and $\delta_y = (m_y - 1)$. Eliminating m_x and m_y from equations (8.4) and (8.5) and letting $P_{\max} = \sigma(vB)^2/4$, one obtains:

$$\delta_x^2 + (\delta_y + \tfrac{1}{2})^2 = \tfrac{1}{4}[1 - (P/P_{\max})] \tag{8.7}$$

$$(P'/P_{\max}) = -4\delta_y \tag{8.8}$$

Depending on the signs of P and P', the MHD duct would function as a generator, a pump, or a brake. The duct is a generator for $P > 0$ and $P' > 0$; it is a pump for $P < 0$ and $P' < 0$ and it is a brake for $P < 0$ and $P' > 0$. Figure 8.1 illustrates these three different types by the use of equations (8.7) and (8.8).

In spite of unusual technical difficulties, and owing to the existence of stringent demands in several fields of applications, work on the solutions to the problem of MHD generation and acceleration is now making very important progress. In the early 1960s the MHD effects have been considered in a host of

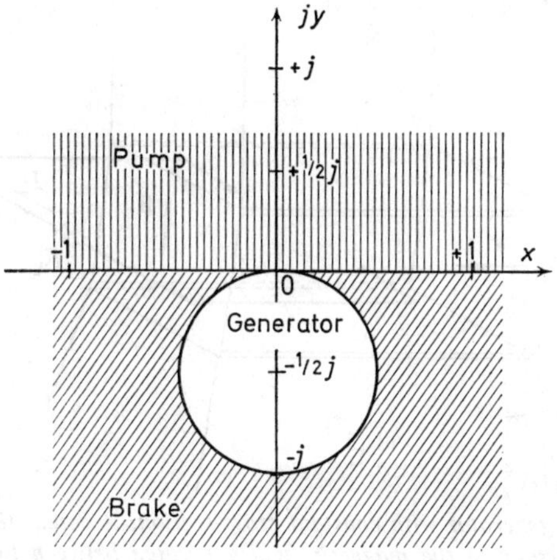

Fig. 8.1. Operating ranges of an MHD duct. ○, *generator;* ◉, *brake;* ◉, *pump*

industrial problems. When the viscosity of the plasma is not negligible, MHD lubrication can be of interest. When the thermal conductivity is of importance, the MHD effect can be used for thermal convection. These methods are still in the field of investigation and deserve further research effort.

8.2 MHD POWER GENERATION[8, 30, 40]

MHD power generation is based upon the Faraday effect which can be described as follows: an e.m.f. is induced in a circuit when the flux of induction through it is changed. Instead of using a solid metal, however, a flowing ionised gas is utilised in MHD generators. The MHD generators producing d.c. power can be divided into four categories: (1) linear generators; (2) vortex generators; (3) radial outflow generators; and (4) annular Hall generators.

Linear generators

A linear generator is shown in Figure 8.2. In this case, a conducting fluid medium flows in a direction perpendicular to the magnetic field generated between two poles of a magnet. Electrodes are placed in the direction parallel to both the magnetic field and the velocity of the flow. When these electrodes are connected through an external load an electric current will flow through the load as a result of the induced electric field. The direction of this field will depend on the way these electrodes are connected. Four important types of linear generators can be distinguished. These are: (1) continuous electrode generators ($E_{\parallel} = 0$); (2) parallel-connected segmented electrode generators ($J_{\parallel} = 0$); (3) Hall generators ($E_{\perp} = 0$); and (4) series-connected segmented electrode generators ($E_{\perp} \sim E_{\parallel}$). These connections are shown in Figure 8.3. The electric field and current density are referred to the direction of plasma flow, the electric

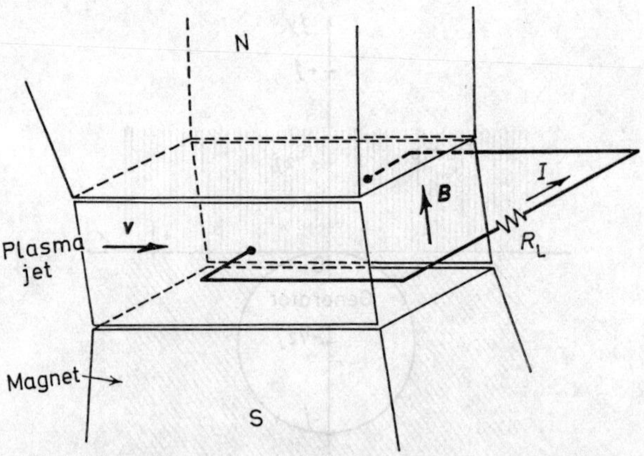

Fig. 8.2. Linear MHD generator with constant cross-section. A conducting fluid medium flows in a direction perpendicular to the magnetic field generated between two poles of a magnet. Electrodes are parallel to both magnetic field and fluid velocity

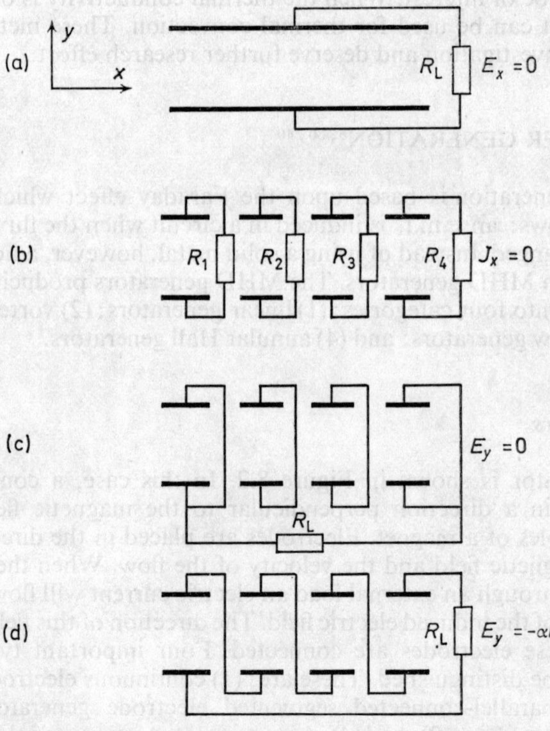

Fig. 8.3. Electrode configuration for various linear generator types. (a) *Continuous electrode generator;* (b) *parallel-connected segmented electrode generator;* (c) *Hall generator;* (d) *series-connected segmented electrode generator*

current component parallel to the plasma flow being called the Faraday current, whereas the component perpendicular to it is the Hall component.

The continuous electrode generator is practical only if the Hall component of the current is negligible, that is when the ratio (ω_c/ν_c) of the cyclotron frequency over the collision frequency is much smaller than unity. This will occur in dense plasmas in relatively low magnetic fields. Segmentation of the electrodes is introduced to avoid losing the Hall component of the current (generators 2). However, this will introduce an unpractical multiplicity of loads which can be avoided by connecting the loads to make one single load leading to the series-connected electrode generators. When the Faraday current becomes negligible as compared to the Hall current, the Faraday component can be short-circuited altogether leading to the Hall effect generator.

The analysis of a linear generator can be made by using the MHD equations of Chapter 3, applied to the special geometry considered. Usually a unidimensional model is treated in which each cross-section of the duct is assumed as homogeneous. This is a crude approximation that gives, however, a good idea about the operation of an MHD generator. Further, the plasma is assumed to be in its steady state, non-viscid and neutral. Under those assumptions the different MHD equations become: equation (3.41) of motion:

$$JB\,\mathrm{d}x + \varrho v\,\mathrm{d}v + \mathrm{d}p_r = 0 \tag{8.9}$$

Equation of conservation of energy:

$$\varrho\,\mathrm{d}(Q/M_p) = \varrho C_p\,\mathrm{d}T - \mathrm{d}p_r \tag{8.10}$$

(This result has been obtained after assuming that the plasma is adiabatic and incompressible. Q is the heat flux, M_p is the mass of the flow, and C_p is the heat capacity at constant pressure.) Equation of state is given by equation (1.58) or $p_r = \varrho \mathcal{R} T$ [where $\mathcal{R}$ is $(C_p - C_v)$ as stated in Section 1.10]. Equation (3.35) of continuity:

$$\varrho v = (M_p/S_p) = \text{constant} \tag{8.11}$$

where S_p is the cross-section of the MHD duct. To these equations, equation (8.6) should be added as well as the Maxwell curl equations yielding $E =$ constant and $\mathrm{d}B = \mu J\,\mathrm{d}x$. The voltage between the electrodes is $V = El_y$ where l_y is the interelectrode distance, and if l_z is the distance between the pole pieces of the magnet then $S_p = l_y l_z$. If J^2/σ is merely the Joule heating per unit volume, one may write:

$$\varrho v\,\mathrm{d}(Q/M_p) = (J^2/\sigma)\,\mathrm{d}x \tag{8.12}$$

There is a set of eight equations with the nine unknowns, S_p, p_r, T, ϱ, v, B, J, σ, and x. To solve this system a further assumption should be introduced. This leads to five major different approaches of analysis: (1) constant velocity analysis by taking v constant, (2) constant temperature analysis, by taking T constant, (3) constant pressure analysis, by taking p_r constant, (4) constant area analysis, by taking S_p constant, and (5) constant Mach number analysis, by taking (v^2/T) constant. However, the engineer should keep in mind the overall desired characteristic of a generator and not the 'solvability' of a problem. For the constant area generator, Thompson[43] has shown that, if the device is optimised both in final velocity and in the external load, the overall efficiency cannot be larger than $1/\gamma^2$, which is 36 per cent for a monatomic gas. The physical reason for such a limitation is easy to understand. If the plasma stream loses part of its velocity in a pipe of constant cross-section, its density must necessarily increase. If the flow is adiabatic, this requires an increase in pressure which gives rise to an increase in temperature. If the Joule heating is taken into account, an extra source of heating is introduced giving rise necessarily to a greater final pressure and a greater final temperature; hence, part of the fluid kinetic energy is converted into heat and cannot be converted into electrical energy. This limitation in overall theoretical efficiency is related to the assumption that the cross-section offered to the plasma flow is constant. It seems more reasonable to expand the flow as the energy is converted to electricity. In this case a constant temperature model can be taken, thus offering the possibility of converting the kinetic energy of the fluid, or a constant speed model offering the possibility of converting its thermal energy. Care, however, should be taken to avoid a drastic reduction in the plasma conductivity when the temperature falls.

Numerical example

Design data of an MHD generator using helium as working fluid has been reported by Gunson *et al.*[22]. The output of the generator was calculated to be 315·5 MW(e) at 11 550 V and 80 per cent efficiency. The inlet cross-section was 813×813 mm whereas the outlet cross-section was 813×1448 mm, the overall length of the generator being 27·43 m. At the input, the generator had the following characteristics: pressure = 4·425 atm, temperature = 1340°C, flow velocity = 1775 m/s, and Mach number = 0·75. At the output the velocity was the same whereas the other quantities changed in the following manner; pressure = 1·892 atm, temperature = 954°C, and Mach number = 0·86. The applied magnetic field did not exceed 10 Wb/m².

Other types of generators

In the vortex generators the working plasma is introduced tangentially into an outer cylinder and withdrawn in a spiral path along the surface of an inner coaxial cylinder. The magnetic field is applied in the axial direction perpendicularly to the flow, and the electrodes are the two coaxial cylinders. To permit a long magnetic interaction length, the diameter of the inner cylinder should be made much smaller than that of the outer cylinder, thus allowing the flow to make several revolutions before leaving the generator. This is a continuous electrode converter similar to the linear generator. It is therefore, basically a Faraday generator.

In the radial outflow generator, the gas is injected outwardly from the inner cylinder, the magnetic field being applied axially. This is a variation of the Hall generator in which the Faraday current is rather a source of loss.

The annular, Hall generator is similar to the one above. However, in this case the magnetic field is radial, being produced by a cylindrical permanent magnet. This generator requires only one annular electrode at each end of the geometry. It has, however, the disadvantage of needing a large wall area.

MHD materials for power generation[25]

Materials used in MHD generators can be separated into two types: materials used as working fluids and materials used in the walls and electrodes.

The working fluid is chosen for its high conductivity and its chemical stability as well as its price especially in the open-cycle systems. In such systems working fluids such as hydrogen, alcohol and fuel oil mixed with air and oxygen were considered. In closed cycle systems, more expensive materials can be used since they may be recovered; only the noble gases have been seriously taken into consideration because of their chemical stability. To increase the conductivity of these gases, they should be seeded with some metallic materials having low ionisation potentials, such as potassium and caesium. The conductivity of the plasma is increased drastically by increasing the amount of seed until an optimum seeding ratio of a few percent is reached. Figure 8.4 illustrates the conductivity of several seeded noble gases as function of temperature. The conductivity is also a function of the temperature and pressure in the gas. The optimum values are reached at a temperature of about 2000°C.

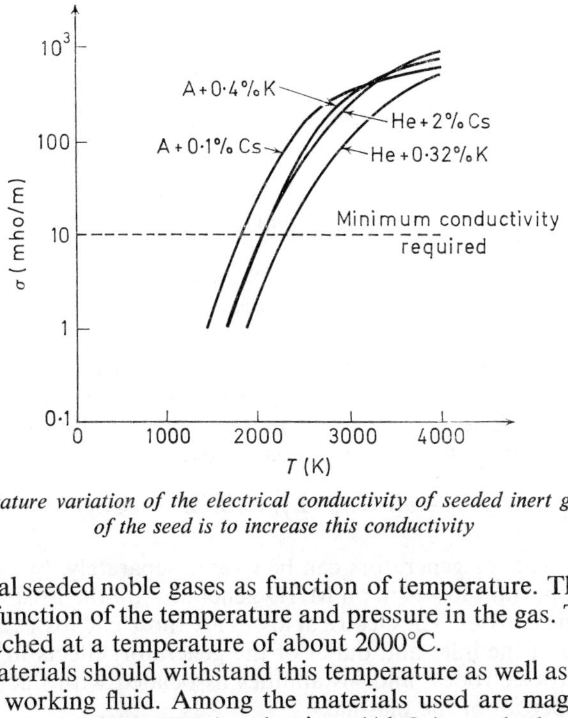

Fig. 8.4. Temperature variation of the electrical conductivity of seeded inert gases. The effect of the seed is to increase this conductivity

The wall materials should withstand this temperature as well as the corrosive action of the working fluid. Among the materials used are magnesia (MgO), strontium zirconate ($SrO.ZrO_2$), alumina (Al_2O_3), and thorium dioxide (ThO_2). Every one of these materials has some serious disadvantage. Ceramic oxides seem, however, to be rather attractive, although it is unlikely that materials having a lifetime of only a few weeks will be found in the near future. Another solution to the problem has been proposed; using a cooling system for the walls. This will necessitate materials of good thermal conductivity such as alumina. Again the problem of short-circuiting becomes acute and can be solved only through segmentation of the walls.

The material of the electrodes is required to be a good conductor at the high temperature of the fluid and to make good electrical contact with the working plasma. This is obtained by using materials having good thermionic emissivity such as tungsten, rhenium, tantalum and its carbonates, nitrates and borates.

Power output and losses

The power generated by an MHD converter is given by equation (8.1) which can be written in the following manner:

$$P = -\sigma(E_x J_x + E_y J_y) \tag{8.13}$$

when the components of $\boldsymbol{J}$ and $\boldsymbol{E}$ in the x and y directions are considered. If the effect of collisions is taken into account, Ohm's generalised law yields, in a scalar notation:

$$J_x = \mathcal{A}E_x - \mathcal{B}(E_y - vB) \tag{8.14}$$

$$J_y = \mathcal{A}(E_y - vB) + \mathcal{B}E_x \tag{8.15}$$

where

$$\mathcal{A} = \frac{[\nu_{\text{en}}^2 + \omega_{\text{i}}\omega_{\text{e}}(\nu_{\text{en}}/\nu_{\text{in}})]\sigma}{[\nu_{\text{en}} + \omega_{\text{i}}\omega_{\text{e}}(\nu_{\text{en}}/\nu_{\text{in}})]^2 + \omega_{\text{e}}^2}$$

$$\mathcal{B} = \frac{\sigma\omega_{\text{e}}\nu_{\text{en}}}{[\nu_{\text{en}} + \omega_{\text{i}}\omega_{\text{e}}(\nu_{\text{en}}/\nu_{\text{in}})]^2 + \omega_{\text{e}}^2}$$

with ω_{e} and ω_{i} being the cyclotron frequencies for electrons and ions respectively, ν_{en} is the collision frequency between electrons and neutrals whereas ν_{in} is the collision frequency between ions and neutrals. Introducing equations (8.14) and (8.15) into equation (8.13) one finds:

$$P = -\sigma[\mathcal{A}(E_x^2 + E_y^2) + vB(\mathcal{B}E_x - \mathcal{A}E_y)] \tag{8.16}$$

The different types of generators can be treated separately, by equation (8.16).

The two important losses in an MHD generator are the losses at the magnet and the electrical losses. The electrical losses are those end losses associated with eddy currents at the inlet and outlet of the generator, effects in the vicinity of the segmented electrodes, and instabilities associated with fluctuations in the electrical properties of the plasma.

MHD power generation systems

Four types of MHD generation systems can be distinguished. These are: (1) open cycle systems without recovery; (2) open cycle systems with recovery; (3) two working fluid systems[14]; and (4) closed cycle systems[1].

In the open cycle system, fuel is burned with the oxidiser to which seed is added without preheating. The oxidiser is usually oxygen and the fuel is a fossil fuel, for economical reasons. The working fluid is accelerated by a nozzle to the generator from which it is exhausted to the atmosphere. There is no recovery of the seed. Such a system may be of interest only for small-duration generators. An efficiency of 15 per cent has been reached with this system.

Oxygen is an expensive oxidiser, therefore the working gas should be compressed to allow the use of air as oxidiser which should be preheated to permit thermal ionisation. The seed should also be recovered. Thus, a regenerative cycle becomes necessary to recover the heat output of the MHD generator

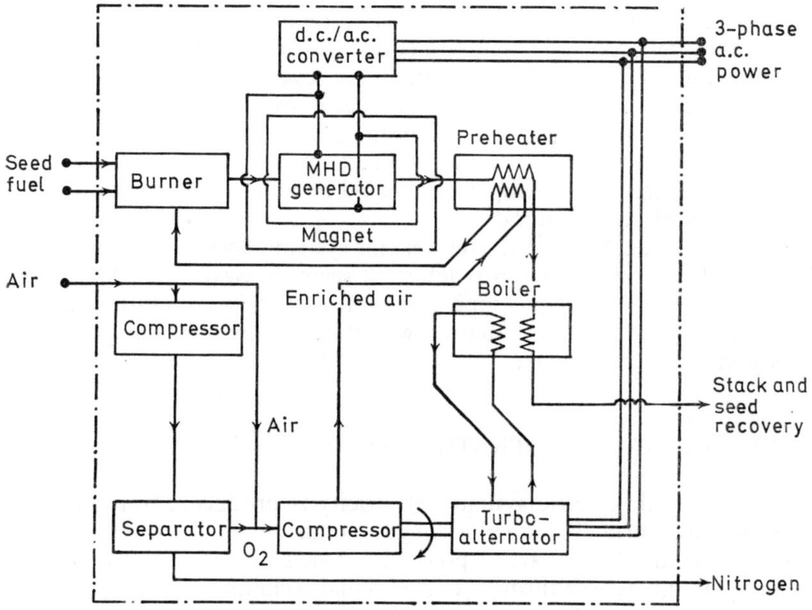

Fig. 8.5. MHD open-cycle system with recovery of the seed. Air is used as oxidiser

working, say, as a 'topper' for a steam turbine plant. The fuel usually used in such a system is coal and the ashes are removed by a cyclone burner (Figure 8.5). Efficiencies of about 56 per cent are predicted for this system.

In the MHD generator two types of energies are converted into electricity: the thermal energy and the mechanical energy of the working gas. If these two roles are separated by the use of two different working gases, operation at much lower temperatures would be possible thus avoiding many problems related to high temperature. Efficiencies of 50 per cent are predicted for the two working fluid system.

In the closed cycle system, the working plasma in a heat exchanger is expanded through a nozzle, passed through the generator and pumped back through the heat exchanger after being cooled. Thus, the working plasma is made in a closed cycle. The heat source could be a nuclear source supplied by a nuclear reactor or an in-pile heat exchanger.

Economic aspects[23]

A certain number of economic studies about MHD generator power plants are available. In these studies, the installation cost per kilowatt and the anticipated price per kilowatt-hour are not discouraging, but definitely lack accuracy. It seems that with only conventional techniques at stake, the final cost savings will be distinctly marginal. In particular, if the electrical power is generated in d.c. and eventually converted into a.c. either by rotating machines or electrical discharge devices, the beneficial margin if any will be entirely consumed in this process. However, some possibilities of breakthrough do exist. For instance, it is likely that supraconductivity can be used on an industrial scale. With tin–niobium alloy coils, cooled by liquid helium, the possibility of generating magnetic inductions of 10–15 Wb/m^2 in volumes of several cubic metres can be anticipated. Under such conditions, the electrical power consumed in the coil becomes negligible with respect to that consummed in the cryostat. The latter is about 5 per cent of the power consumed in a conventional coil of the same size. Also, the problem of high temperature could be alleviated by the use of some non-thermal ionisation method.

The success of an MHD-nuclear[41] reactor system depends on progress in the field of high temperature nuclear reactors. It seems however, that MHD generators would be successful as 'toppers' for conventional plants rather than main sources of power. The latter situation would be feasible with the development of thermonuclear technology.

8.3 MHD A.C. POWER GENERATION[33, 47]

In most large-scale electric systems, electricity is produced, transmitted, and processed in a.c. form. The simplest way to obtain a.c. power would seem to be by converting the d.c. electrical power produced by the MHD generator of Section 8.2 to a.c. power through conventional techniques. However, this new stage would undoubtedly complicate the overall system and make it more expensive. This led workers in the field to investigate some methods to produce a.c. electrical power directly.

The elimination of the conventional rectifier is, however, not the only advantage of the MHD a.c. power generator. At least two other advantages might be of interest. The first is the possibility of using electrodeless generators and thus avoiding all the problems related to the contact of solid materials with a hot and corrosive medium. The second is the possibility offered of having the overall resistivity of the gas governed by the mobility of the electrons, rather than that of the ions. Indeed, the electronic mobility is much larger than the ionic one and if an electron current is made to flow around a closed circuit much larger conductivities can be obtained in the working fluid.

One can distinguish two types of a.c. MHD generators: electrodeless generators; and generators with electrodes.

Electrodeless generators

In this category the working plasma is confined away from the walls by the magnetic field. It has less opportunity of parasitic cooling and erosion. In some cases, the coils are used to generate a d.c. magnetic field, and the effect of the interaction with the plasma is to perturb this d.c. field. In this case an a.c. current is generated in the d.c. coils and can be picked up by transformer-like devices, which leave the d.c. component unperturbed and transfer only the a.c. to their secondary coils. In other cases, the magnetic field is purely a.c. and is the result of the plasma interaction itself, combined with an electric resonant circuit. Such devices can only be understood, like the conventional triode oscillator if a state of oscillation is assumed to exist automatically at the resonant frequency.

In the first category the generator would be a variable velocity device. Consider a pipe surrounded by a coil and assume that a d.c. current exists in the coil and that the magnetic pressure corresponding to the d.c. component of the magnetic field is large with respect to the plasma pressure. If plasma blobs travel down the tube, one can assume that, around each of them, the magnetic lines of force are locally pushed farther away from the axis than their normal position in the absence of plasma, and this configuration travels down the tube with the plasma blob. This moving configuration of magnetic lines of force is thus cut by the conductors of the coil; hence, electric signals are generated in the turns at the beginning and the end of each blob, creating a travelling wave along the coil. By convenient inductive coupling, this energy can be transferred to an external circuit. Colgate[7] has suggested the possibility of using such a scheme with a single blob travelling back and forth, provided it receives heat at both ends. For instance, the plasma might contain some fissionable fuel and constitute temporarily a critical configuration under the influence of the pressure increase arising from the collision between the plasma and a dead end. Such critical configuration would be reached alternately at both ends of the tube. A better proposition would be to use the same method in a torus like configuration where the plasma blob would be travelling around between a heat source and a heat sink and where coils would be arranged to pick up the electric pulses from the two positions on the torus at equal distances to the heat-sink heat-source axis.

The second type of generator would be a variable magnetic field device. This is a wave-type generator since an alternating magnetic field, the phase of which varies along the MHD channel, corresponds to a travelling magnetic wave[7, 46]. In this device, the flowing plasma is decelerated by the variable induction and its kinetic energy is transferred into a.c. electrical energy in the field coils. The feasibility of this generator remains questionable. Indeed, the power generated in a unit duct volume is at most $\sigma v^2 B^2/8$. The reactive power necessary to alternate a unit volume of the field at a frequency ω is $\omega B^2/4\mu_0$. Thus the ratio of produced power over the power necessary to alternate the field is:

$$(P_{\mathrm{m}}/P_{\mathrm{B}}) = \frac{\mu_0}{4}\frac{\sigma v^2}{\omega} \tag{8.17}$$

This ratio would be encouraging only if the plasma has a very high conductivity and is travelling at a very large speed, whereas the field is working at a low frequency. Further, field-swinging devices are very expensive.

Generators with electrodes

These generators can be divided into three categories: (1) generators with varying gas velocity; (2) generators with varying gas conductivity; and (3) generators with varying electric field.

Varying the velocity over ducts of any appreciable length would itself prove difficult. It would also be necessarily accompanied by fluctuations in temperature and pressure.

If non-thermal ionisation is used, a pulsating ionisation source could be used to achieve a variable gas conductivity. The same result can be obtained by introducing charges of fuel and oxygen at discrete intervals to produce hot spots in the working gas. The current obtained would be a modulated d.c. current. This is a self-defeating result since one would not avoid the additional expense of d.c.–a.c. converters.

McCune[32] proposed the use of the Hall instability of Velikhov[44] to produce a.c. power in an MHD generator (Figure 8.6). In this generator there are a Hall circuit and a Faraday circuit. The potential difference in the Faraday circuit

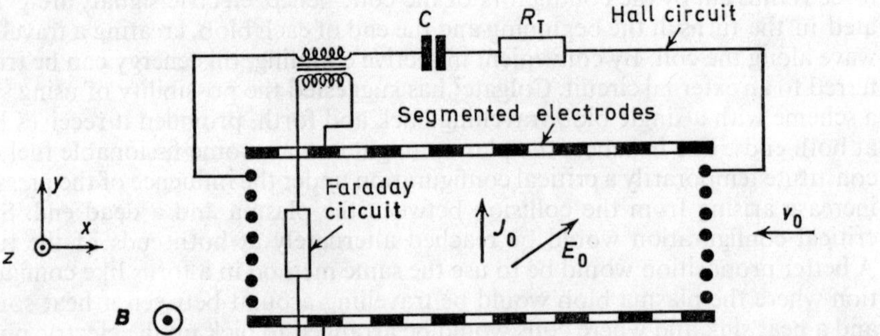

Fig. 8.6. Circuitry of an a.c. MHD generator. It is composed of a Hall circuit and a Faraday circuit. A Hall current is generated in the load connected to the electrodes of the Hall circuit. (From McCune[32] by courtesy of Pergamon Press)

is modulated uniformly along the duct at a frequency ω. The linear MHD duct is in a magnetic field perpendicular to the direction of the flow and parallel to the plane of the segmented electrodes. The d.c. component of power operates in the Faraday circuit at a large value of (ω_c/ν_c). Because of the modulation along the duct, a Hall current is generated in the axial direction and collected by the Hall electrodes; this current flows in the Hall circuit through the external load. The capacitance in the Hall circuit is chosen for matched conditions at the operating frequency. McCune concluded that although such an a.c. generator is feasible, the linear geometry investigated has a low power level. Other configurations might be usefully considered.

Finally one should note that it is theoretically possible to obtain an a.c. MHD generator by varying the applied magnetic field, in a configuration with electrodes.

In general, although a.c. MHD generators seem to give good promise, they are not economically competitive. They may become so, if and when a much higher plasma conductivity can be obtained.

8.4 MHD PROPULSION[26, 29]

The reverse or motor mode of the MHD effect can be used for MHD propulsion. In fact the same engine might work as a generator or as a propulsor, as seen in Section 8.1, depending whether the force on the working fluid is an accelerating force or a braking one. In such an MHD propulsor, the electrical energy would be converted directly into kinetic energy of a plasma body.

The interest in MHD propulsion grew first as a result of the U.S. space programme. The conventional space propulsors are chemical rockets that give thrust-to-weight ratios larger than unity. However, since the potential energy of a chemical reaction is limited, the specific impulses* of chemical rockets are limited to about 6 min. The specific impulse is raised by using nuclear rockets with a propellant of low molecular weight such as hydrogen. Thus in order that a propulsor is of interest in space applications it must have a large thrust-to-weight ratio, a light weight power supply and a specific impulse as large as possible. MHD propulsion offers the possibility of efficient operation[21] at high specific impulses reaching 5 h and more.

More recently the MHD effect has been considered for the propulsion of submarines. Indeed, if one realises that sea water is an ionic medium, one recognises the possibility of using MHD propulsion for ships. But since the surface of the sea is of a rather turbulent nature, the study is limited to using the MHD principle for propelling submarines. The advantage of this system is its lack of moving solid parts. This will result in silent operation. To a military submarine prowling the dephts, silence can often mean survival. The noise produced by the propeller of a conventional submarine is a very real disadvantage in avoiding detection.

Of all the methods proposed for MHD acceleration, one can distinguish three main categories: (1) crossed field accelerators; (2) accelerators using the Hall effect; and (3) electrodeless accelerators. These devices are also of interest in plasma injection techniques (e.g. for fusion devices), plasmoid generators and coating guns.

Crossed-field accelerators

The basic principle of crossed-field MHD propulsion is shown in Figure 8.7. The magnetic field B_a is constant and applied perpendicularly to the flow, while an electric field E_a is applied perpendicularly to both the magnetic field and the

* The specific impulse is the ratio of the rocket exhaust velocity to the gravitational acceleration at the surface of the earth.

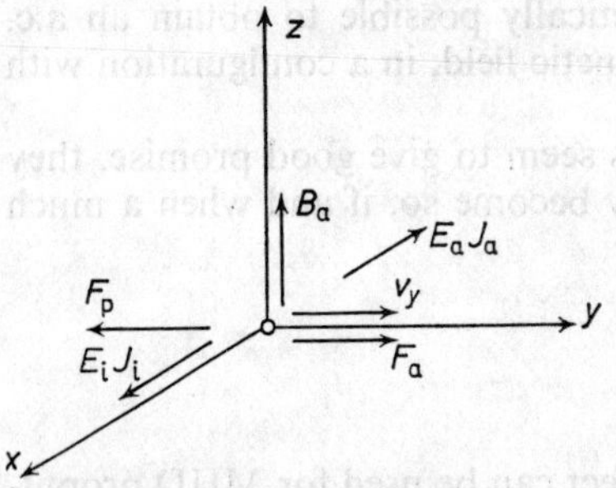

Fig. 8.7. Basic cross-field configuration for an MHD accelerator

direction of flow. The velocity of the working plasma interacts with the applied magnetic field to produce an induced electric field $\boldsymbol{E}_i = \boldsymbol{v}\times\boldsymbol{B}_a$. This gives rise to a force $\boldsymbol{F}_i = \sigma\boldsymbol{E}_i\times\boldsymbol{B}_a$ where $\sigma\boldsymbol{E}_i$ is the induced current density $\boldsymbol{J}_i$. However, the applied electric field induces in turn, a conduction current density $\boldsymbol{J}_a = \sigma\boldsymbol{E}_a$ which interacts with the applied magnetic field to create a force $\boldsymbol{F}_a = \sigma\boldsymbol{E}_a\times\boldsymbol{B}_a$. The total force in the y-direction will be therefore:

$$F_y = \sigma(E_a - vB_a)B_a \tag{8.18}$$

For the duct to act as a propulsor, it is necessary to have $E_a > vB_a$. This imposes basic restrictions on the system, since smaller exhaust velocities require smaller applied electric fields.

The total kinetic power transferred to the plasma is:

$$P_{kin} = I_y\, \mathrm{d}B(v_e - v_i) \tag{8.19}$$

where I_y is the total current in the y-direction, d is the distance across the channel, v_e is the exit velocity, and v_i is the input velocity. The power lost by Joule heating of the plasma is:

$$P_j = R_p I_y^2 t \tag{8.20}$$

where R_p is the resistance of the plasma, and t is time. The efficiency of the device is $\eta = P_{kin}/(P_{kin}+P_j)$. Introducing equations (8.19) and (8.20), one finds:

$$\eta = \frac{1}{1+\dfrac{1}{\sigma B^2}\dfrac{T_m}{\mathcal{V}(v_e - v_i)}}$$

where $\mathcal{V}$ is the volume of the channel and $T_m = I_y\,\mathrm{d}B$ is the mechanical torque. Ideally the output velocity is much larger than the input velocity, thus $v_e - v_i \cong v_e$. Since, for a given application, v_e will be specified, the figure of merit of the propulsor will be:

$$Z_p = \sigma B^2 \mathcal{V} \tag{8.21}$$

which should be as large as possible. The efficiency is therefore:

$$\eta = \frac{1}{1+\dfrac{T_m}{Z_p v_e}} \tag{8.22}$$

The electrical conductivity of a fully ionised plasma is given by Spitzer[38] as:

$$\sigma = 0{\cdot}0153 T_e^{3/2}/[\ln(\Lambda)] \qquad [1/\Omega\ \mathrm{m}]$$

where $\ln(\Lambda)$ is a slowly varying function of density, and T_e is the electron temperature. Higher conductivity implies higher temperature and thus the upper limit of the conductivity is eventually fixed by the electrode erosion and cooling problems. The fact that $(B^2\mathcal{V})$ should be large is a shortcoming in space applications since it implies a weight increase of the propulsor.

Experimental work on the crossed-field accelerator has been performed by Demetriades[11] who obtained a 44 per cent efficiency. It is felt that higher efficiencies could be reached with specific impulses in the 1500 to 4000 s range[39].

One of the inherent disadvantages of the cylindrical linear geometry with a constant cross-section is the loss of energy due to temperature gradients and thermal agitation of the plasma. If the plasma is permitted to expand in a nozzle, those losses can be greatly reduced. Indeed, this expansion has the effect of converting the Joule heat into kinetic energy.

Figure 8.8 shows the so-called 'source flow model' of a crossed-field device with an expansion nozzle. This model has been postulated by Sherman[37].

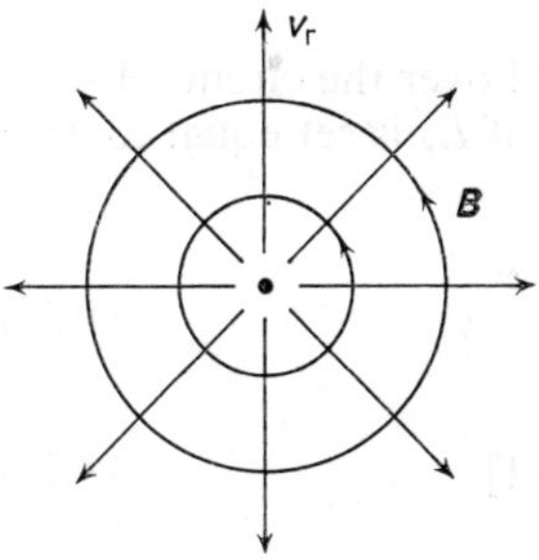

Fig. 8.8. Source flow model of an MHD cross-field propulsor

Calculations employing this model have indicated that efficiencies larger than 70 per cent are possible with an accelerator length of only 8 cm, producing specific impulses in the range of 2200 s.

Hall current accelerators

If m_F is defined as the ratio of the magnetic body force to the fluid inertial force, it will be:

$$m_F = \frac{\sigma_i B_i^2 d}{\varrho_i v_i} \tag{8.23}$$

at the inlet of the duct. It is obvious that m_F should be made as large as possible. In the crossed-field accelerator this is done by increasing B and decreasing ϱ_i.

The Hall current density is:

$$J_H = \sigma \frac{|\boldsymbol{J}\times\boldsymbol{B}|}{nq} = \frac{JvM_p}{qBd} m_F \tag{8.24}$$

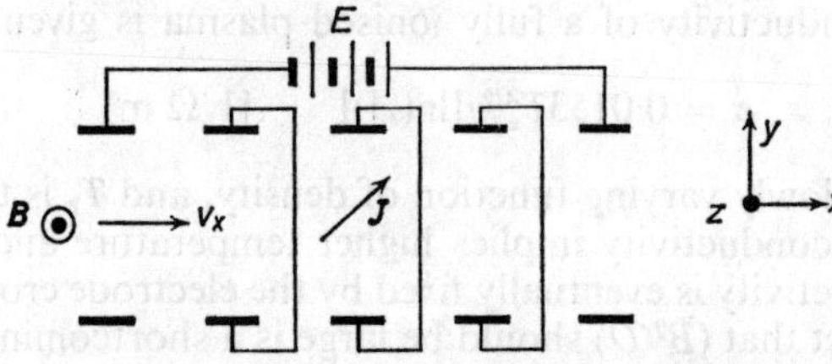

Fig. 8.9. The Hall current accelerator

Thus, increasing m_F results in an increased Hall current density. Hence, it would be desirable to utilise this current in an accelerator. Such an accelerator is shown in Figure 8.9 where $\sigma J_a B_z/nq$ is the current density in the y-direction, and this current interacts with the applied magnetic field to produce a force in the x-direction. Solving the generalised Ohm's law:

$$\boldsymbol{J} = \sigma\left(\boldsymbol{E}+\boldsymbol{v}\times\boldsymbol{B}-\frac{\boldsymbol{J}\times\boldsymbol{B}}{nq}\right)$$

leads to equations (8.14) and (8.15).

If $\mathcal{K}$ is defined as the ratio of the actual electric field over the circuit electric field, equations (8.14) and (8.15) become respectively, if E_y is set equal to zero and ω_i is considered negligible:

$$J_x = \frac{\sigma v_x B_z(\omega_e/\nu_e)}{1+(\omega_e/\nu_e)^2}(1+\mathcal{K}) \tag{8.25}$$

$$J_y = \frac{\sigma v_x B_z}{1+(\omega_e/\nu_e)^2}[(\omega_e/\nu_e)^2\mathcal{K}-1] \tag{8.26}$$

The force in the x-direction is $J_y B_z$, thus:

$$F_x = \frac{\sigma v_x B_z^2}{1+(\omega_e/\nu_e)^2}[(\omega_e/\nu_e)^2\mathcal{K}-1] \tag{8.27}$$

which implies that $\mathcal{K} > (\omega_e/\nu_e)^{-2}$ is necessary for the acceleration to occur. Defining the efficiency as $(F_x v_x/E_x J_x)$ one finds:

$$\eta = \frac{(\omega_e/\nu_e)^2\mathcal{K}-1}{(\omega_e/\nu_e)^2\mathcal{K}(1+\mathcal{K})} \tag{8.28}$$

The optimum value of $\mathcal{K}$ is found by setting $\mathrm{d}\eta/\mathrm{d}\mathcal{K} = 0$, it is:

$$\mathcal{K}_{\text{opt}} = \frac{1+\sqrt{[1+(\omega_e/\nu_e)^2]}}{(\omega_e/\nu_e)^2} \tag{8.29}$$

for which:

$$\eta_{\text{opt}} = \frac{(\omega_e/\nu_e)^2[1+(\omega_e/\nu_e)^2]^{1/2}}{\{1+[1+(\omega_e/\nu_e)^2]^{1/2}\}\{1+(\omega_e/\nu_e)^2+[1+(\omega_e/\nu_e)^2]^{1/2}\}} \tag{8.30}$$

For large values of (ω_e/ν_e) this becomes:

$$\eta_{\max} = \frac{1}{1+2(\nu_e/\omega_e)} \tag{8.31}$$

which indicates that very high efficiencies might be obtainable. It is therefore evident that, for large values of (ω_e/ν_e) the Hall accelerator is comparable to the crossed-field propulsor.

An improvement to the Hall accelerator is the application of an oblique magnetic field, making an angle ϕ with the y-direction. The electric field is then applied in the y-direction causing a current flow which interacts with the x-component of the applied magnetic field to produce a Hall current J_z. This current interacts with the y-component of the magnetic field to produce the accelerating force. For $\phi = 0$ no acceleration is obtained. An expansion nozzle can be added to the Hall accelerator to improve its efficiency. Patrick and Powers[34] have experimented with such a device using a cylindrical geometry and the basic field configuration of Figure 8.9. They found an efficiency of 25 per cent and a specific impulse of 600 s, using a 50° nozzle.

Pulsed accelerators[19, 20]

The devices discussed so far have been continuous-flow devices, which is not a disadvantage in itself. However, a limitation arises with erosion of the container walls due to the high temperatures. There is also a decrease in efficiency at such temperatures due to the heat transfer through the walls. One method of partially circumventing this problem is to operate on a pulsed basis.

The basic operating principle of pulsed devices is illustrated in Figure 8.10. When the switch in this figure is closed, the capacitor discharges initiating the current density $\boldsymbol{J}$ and a magnetic field encircles the discharge. The Lorentz force $\boldsymbol{J}\times\boldsymbol{B}$ blows the plasma away from the electrodes. Due to random energy losses

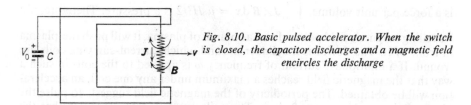

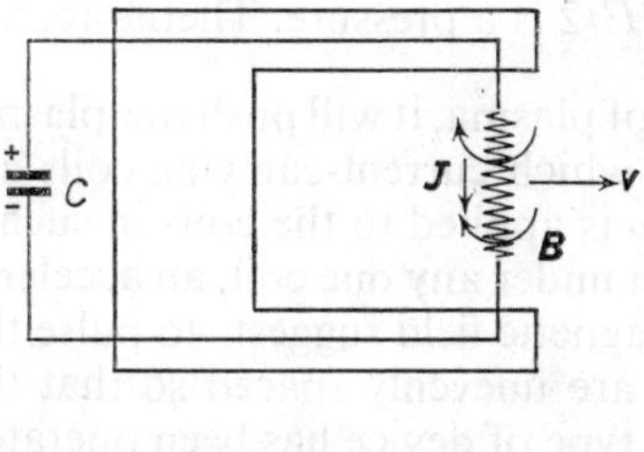

Fig. 8.10. Basic pulsed accelerator. When the switch is closed, the capacitor discharges and a magnetic field encircles the discharge

from lateral expansions, and short transit time from the electrode region, the simple device of Figure 8.10 is very inefficient[39].

Three modifications of the basic idea of Figure 8.10 are shown in Figure 8.11. Figure 8.11(a) shows a T-tube geometry[24], in which the arc is confined for a longer period of time. Efficiencies achieved by this device are of the order of 5–10 per cent. Figure 8.11(b) shows a plasma rail accelerator which has the advantage that the accelerating force acts on the plasma for a longer period of time, since the rails carry the current along the accelerating discharge. Only

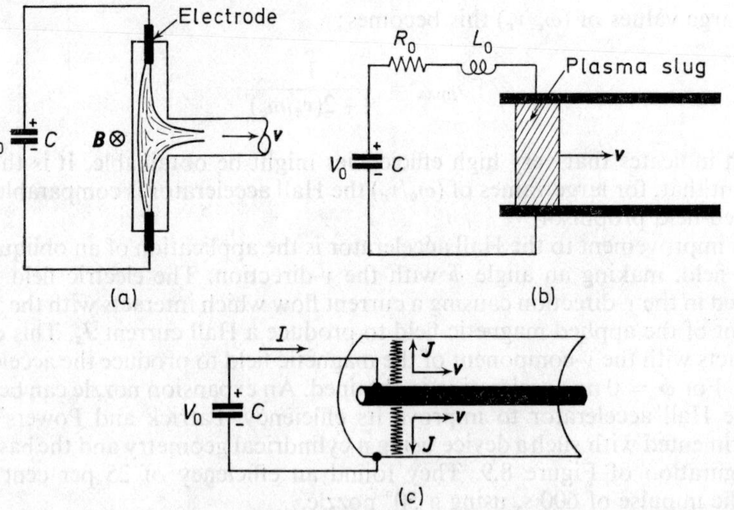

Fig. 8.11. Other types of plasma accelerators. (a) *The T-tube accelerator;* (b) *the plasma rail accelerator;* (c) *the coaxial plasma gun accelerator*

low efficiencies[39] have been achieved with this device; but it is felt that efficiencies as high as 70 per cent are possible with the specific impulse as high as 42 h[9]. Figure 8.11(c) shows a device which both confines the discharge and keeps the accelerating force acting on it. Known as the coaxial plasma gun[31], this device has achieved efficiencies as high as 50 per cent with specific impulses in the range of 5000–25 000 s.

Travelling wave tube

Although pulsing offers a possibility of reducing the wall erosion problems over extended periods of time, the electrode erosion problem still exists. Since $\boldsymbol{J}\times\boldsymbol{B}$ is a force per unit volume, $\int_{x_1}^{x_2} \boldsymbol{J}\times\boldsymbol{B}\,\mathrm{d}x = \mu_0 H^2/2$ is a pressure. Therefore, if a travelling magnetic field passes down a column of plasma, it will push the plasma in front of it. The device is a cylinder around which current-carrying coils are wound. If a multiphase current of frequency ω is applied to the coils in such a way that the magnetic field reaches a maximum under any one coil, an acceleration will be obtained. The periodicity of the magnetic field suggests to pulse the plasma into the duct every $2\pi/\omega$ s. The coils are unevenly spaced so that the transit time between the coils is constant. This type of device has been operated with a specific impulse around 2600 s, but with a low efficiency[28].

Other solutions to the wall problem have been considered. First, it is possible for the plasma to consist entirely or partially of electrode material. This, of course, has very limited applications in long term space flights unless a method of replacing the electrodes is found. Also, the pinch effect can be used to preserve the walls. The azimuthal current interacting with a weak axial magnetic field generates a force which is directed radially inward and pushes the plasma away from the walls of the propulsing cylinder.

Submarine propulsion systems[13, 18]

Most of the preceding devices are potentially suitable for injection in fusion devices. The problem of submarine propulsion is slightly different and warrants a brief description.

The basic principle is shown in Figure 8.12 where a current carrying conductor is placed axially along the hull of the submarine. As indicated by the cross-section, the conductor is surrounded by an electrode. From the field pattern

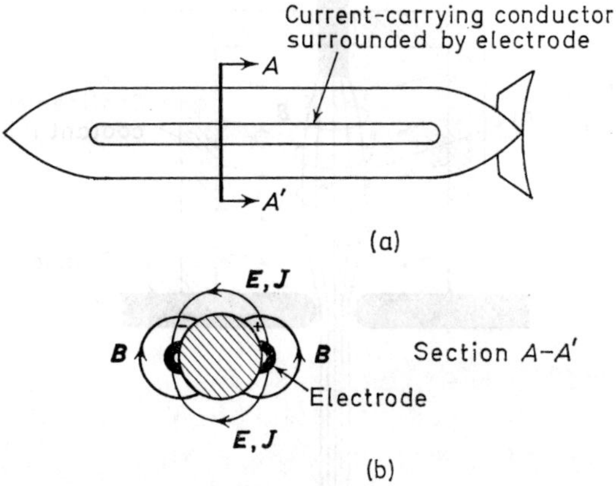

Fig. 8.12. Submarine propulsion. (a) *Side view;* (b) *cross-section with field pattern*

generated by the configuration it can be seen that a force $\boldsymbol{J} \times \boldsymbol{B}$ is produced pointing along the axis of the submarine and can be used for propulsion. Way[45] has shown that efficiencies higher than 80 per cent can be obtained at speeds in excess of 15 m/s (30 knots) and displacements around 10^5 t. Way employed 6 electrodes in his calculations instead of the two shown in Figure 8.12. A test model was designed and constructed during the spring of 1966. The model was of bipolar field arrangement, 3 m long with 4080 kg displacement. Tests were made and the design operating speed of 1·44 km/h was obtained.

Conclusions

The field of MHD propulsion is rich with capabilities; however, there is, in so far as space applications are concerned, one major drawback. All propulsors have thrust-to-weight ratios much smaller than unity; this means that they must be placed in earth stationary orbits by a chemical or nuclear rocket engine.

8.5 PLASMA TORCH AND ROCKETS

Basically, the plasma torch comprises an arc discharge between an anode and a cathode with a hole in the cathode. Permanent gas injection is provided so that part of the plasma is permanently propelled through the cathode hole. It would be a mistake, however, to consider that this expulsion is simply the result of the mechanical gas pressure, since at least part of the force actually results from the magnetic pressure of a self-confined positive column.

Water vapour is frequently used as a propulsive gas, and as wall and electrode coolant as well. It is introduced symmetrically all around the positive column of the arc through a porous wall (Figure 8.13). Under such conditions, wall heating is prevented by a very thin liquid layer all around the discharge on the wall. The

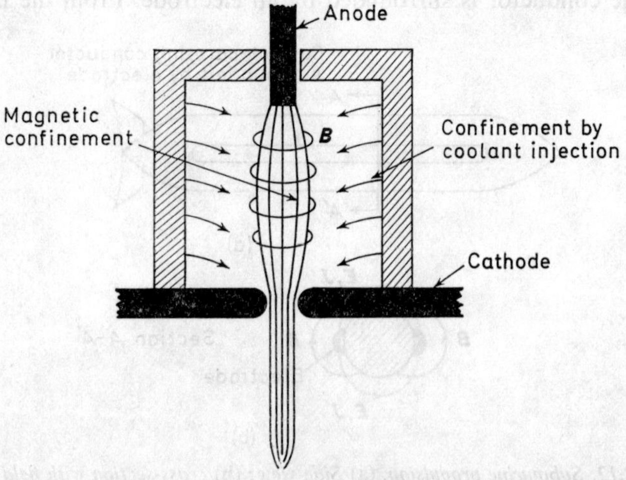

Fig. 8.13. The plasma torch. The plasma jet extending from the hole in the cathode is much hotter than ordinary flames

magnetic confinement contributes also the the thermal insulation of the wall from the hot plasma. Noble gases are also used and in such a case liquid coolant is required in ducts within the wall and the electrodes.

The plasma jet extending from the hole in the cathode is much hotter than ordinary flames and temperatures between 10 000 and 20 000°C are easily attained. Only part of the energy is transmitted as random kinetic energy of the plasma constituents. If the plasma jet is brought in contact with a solid surface favouring recombination, an important additional heat generation appears as a result of such recombination.

The plasma torch has already been used extensively as a cutting and welding device for metals in which high melting points render it difficult to work with ordinary flames or electric arcs. Other applications such as alloying in gaseous phase, purification by distillation, are also in a more or less advanced stage.

The operation of a plasma rocket requires an electric source. The use of the electric supply to accelerate charged particles would seem unattractive because important space charge effects limit severely the operation, unless particles of both signs are accelerated separately, but reunited as a plasma as soon as possible. Consequently, the plasma generation and acceleration appears as an attractive way of improving the performance of rockets beyond that attainable by chemical means.

Although the plasma velocity can be high, the possible mass flow is not expected to be very high, even if an improvement of several orders of magnitude can be obtained with respect to that presently available. Thus, the total thrust will probably always be low with respect to the thrust of a chemical rocket. Moreover, since a plasma device must operate in a fairly good vacuum, there is little chance that plasma rockets can ever operate correctly in the presence of a counter-pressure such as the atmospheric pressure at the surface of the earth for instance. However, there is an extended field of applications for rockets with a low thrust and operative only in the absence of an atmosphere, since the problem of liberating important masses from the gravitational field of the earth has already been solved successfully. Once the vehicle is in interplanetary space, due to the duration of the flights and to the small acceleration needed, satisfactory velocities can eventually be reached.

A number of devices capable of emitting plasma jets with significant mass flows and with velocities significantly higher than those of chemical rockets have been proposed. For instance, in a combustion chamber-like vessel, a light gas could be heated by a conventional arc. Since the energy involved in the ionisation process has to be recuperated, care must be taken to allow recombination of the plasma during the expansion in the nozzle. A possible device operating under such conditions is represented in Figure 8.14. The jet velocity might be of the order of 15 km/s.

The next step consists of taking advantage of the magnetic confinement to obtain an increase in temperature. The arc can either be entirely self-confined or provided with an extra stabilising longitudinal magnetic field. These devices could operate at high pressures and possibly in a relatively dense atmosphere.

Even though the lateral confinement of the discharge is good, the plasma is still in contact with the electrodes at both ends. Magnetic mirrors have been

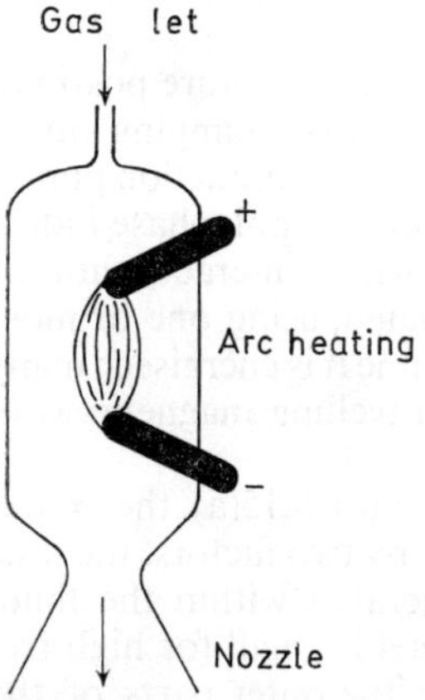

Fig. 8.14. Principle of an MHD rocket. The jet velocity might be of the order of 15 km/s

suggested as a possible means of introducing a temperature difference between the active part of the plasma and that in contact with the electrodes. Velocities in the range of 10–30 km/s might be eventually attained with such devices.

Since one of the major problems is electrode heating, electrodeless discharges appear to have attractive features and a device with a magnetically induced azimuthal current might be of interest. The magnetic confinement might be improved by a superposed d.c. magnetic field. Also, the principle of the conical theta pinch could be applied. Devices related to the rail propulsor or its improved version used for plasma injection in mirror machines are not entirely excluded. All these devices are theoretically capable of producing plasma velocities of 100 km/s and above.

Still higher velocities can be reached in at least two ways. The first one consists of accelerating an ion beam which emerges from an ion gun, provided the positive charge of this beam is neutralised as soon as possible by electron injection. Another method consists of using a thermonuclear combustion chamber. The latter might be a mirror machine with a strong asymmetry. The thermonuclear reaction in the confined zone would heat the plasma, and the particle 'loss' through the weaker of the two mirrors could be used as a propelling jet. Velocities of thousands of kilometres per second are theoretically attainable.

Finally, it is interesting to mention a very remote, but not unlikely possibility offered by the use of antimatter. Minute amounts of antimatter have been made, in the form of individual particles, with giant accelerators. When antimatter is brought in contact with matter, both dematerialise with a tremendous energy release equal to $\mathscr{E} = Mc^2$. If ponderable amounts of antimatter could be generated, any possibility of mechanical confinement would be excluded. However, magnetic confinement would offer attractive features, provided it is stabilised sufficiently. To create the jet, equal amounts of matter and antimatter would have to be injected into a magnetic confinement region, made to react, and the jet generated in the same way as in the magnetic mirror propulsor. The velocity would be an important fraction of that of light in a vacuum. Interstellar travel might be considered confidently on such a basis.

8.6 MHD PUMPS[2]

MHD induction pumps have various present applications and future possibilities. Their uses include pumping fluids for atomic reactors, pumping liquid metals for MHD lubrication and even in MHD generators. The induction pumps may work in several modes. One type of induction pump uses polyphase induction coils; these coils create a travelling magnetic wave which interacts with the MHD fluid. Pulse-operated pumps can also be constructed, using one or more coils. One coil will provide a single pulse pumping each time it is energised; many coils can be used and pulsed sequentially creating a travelling magnetic wave to move the fluid.

In the travelling magnetic field-type pump shown in Figure 8.15(a), the excitation frequency of the travelling magnetic field is limited by two factors, the time response of the MHD fluid and the skin effects generated within the fluid. Penetration of the magnetic field into the MHD channel is small for high frequencies and the electromagnetic fields will interact in the outer parts of the channel. At very high frequencies all the electromagnetic forces are concentrated in thin layers near the channel boundaries and the flow is controlled almost entirely by the mechanical pressure gradient. Although this extreme situation should be avoided, the frequency must, however, be high enough to provide the damping out of all harmonic components of the electromagnetic forces by the inertia of the fluid.

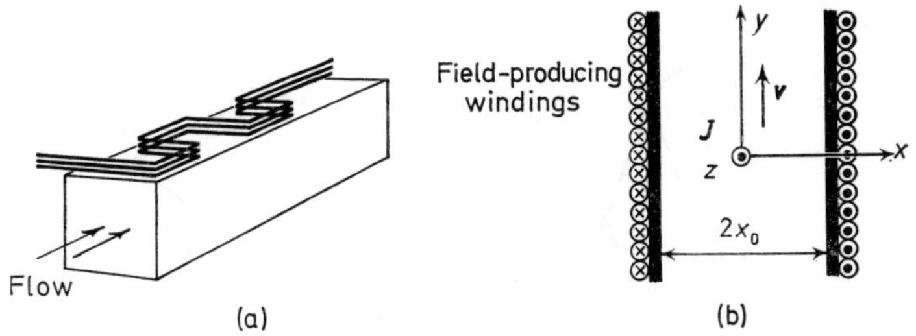

Fig. 8.15. The MHD induction. Travelling magnetic field type. (a) *Field-inducing multiphase coils;* (b) *cross-section of the pipe*

The analysis of such a pump has been reported by Hughes and Young[27]. In this analysis, a turbulent flow with a flat velocity profile is assumed. The skin friction between the fluid and the channel walls is neglected. Polyphase windings are used, supplying a magnetic field $H_0 \cos(\omega t - ky)$ in the y-direction. The motion of a slug of fluid in a channel is given by Maxwell's equations and Newton's law of motion. The combination of the two curl Maxwell's equations yields:

$$\nabla^2 \boldsymbol{H} + \mu_0 \sigma \nabla \times (\boldsymbol{v} \times \boldsymbol{H}) = \mu_0 \sigma (\partial \boldsymbol{H} / \partial t) \tag{8.32}$$

When using liquid metals in MHD induction pumps, time averages of the velocity and pressure can be used, since the relaxation time of liquid metals is much larger than the period of the lowest frequency of the travelling magnetic wave. Newton's law of motion can then be written:

$$\frac{\partial p_r}{\partial y} = \frac{\mu_0}{2x_0 t_p} \int_0^{t_p} \int_{-x_0}^{x_0} (\boldsymbol{J} \times \boldsymbol{H})_y \, dx \, dt \tag{8.33}$$

where $2x_0$ is the width of the pump and t_p is the electrical period. The excitation can be sinusoidal, such as:

$$\boldsymbol{H} = [\boldsymbol{i}H_x(x) + \boldsymbol{j}H_y(x)] \exp[j(\omega t - ky)] \tag{8.34}$$

Equations (8.32), (8.33) and (8.34) are basic to the solution of this problem. They lead to an equation giving the ratio of the driving pressure gradient to the maximum magnetic force as a function of the slip, $[1 - (v_y/v_p)]$, (v_p being the phase velocity of the wave), and the frequency. These results are shown in Figure 8.16. The maximum efficiency of the pump occurs when the depth of penetration is equal to half the channel width, the depth of penetration being defined as:

$$d_w = (\omega \mu_0 \sigma / 2)^{-1/2}$$

The ratio of pump width to skin depth varies from about 13·3 at 10 per cent slip to 5 at 70 per cent slip.

A solution to the energy conversion problems associated with a d.c. transmission of power and with the d.c. generation from MHD generators has been proposed. To convert d.c. currents and voltages to different levels or to convert

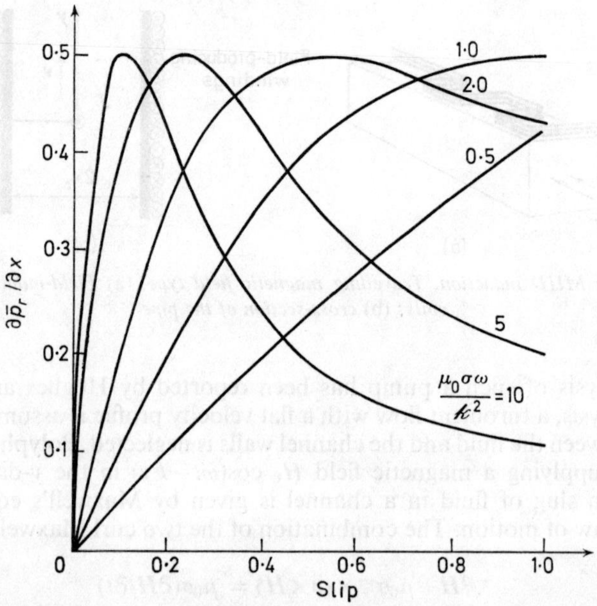

Fig. 8.16. The pressure gradient versus slip for a laminar flow at several values of $\mu_0\sigma\omega/k^2$

a.c. to d.c. or vice versa, an MHD equivalent of the electrical motor-generator can be set up. The d.c. power would be used to feed an MHD pump; the output of this pump would be fed to an MHD generator. To convert d.c. voltages and currents to different levels, electrode-type MHD pumps and generators could be used.

8.7 MHD AMPLIFICATION

The MHD amplifier is illustrated in Figure 8.17. The fluid is forced radially inward in a cylindrical space in which an axial magnetic field is generated by a set of coils. To prevent the fluid from spiralling and to keep it in the radial direction, radial vanes are used. An even fluid flow is permitted in the axial direction by the use of output and input plenums.

To solve this problem the effect of gravity should be added to equation (3.41) of motion. Assuming that the working fluid is viscous and in its steady state, equation (3.41) becomes then:

$$\varrho(\boldsymbol{v}\,.\,\nabla)\boldsymbol{v} = -\nabla p_{\mathrm{r}} + \boldsymbol{\Psi} + \boldsymbol{J}\times\boldsymbol{B} + \varrho_{\mathrm{e}}\boldsymbol{E} + \varrho\boldsymbol{F}_{\mathrm{g}} \tag{8.35}$$

where $\boldsymbol{F}_{\mathrm{g}}$ is the force of gravity. Further, because of the fluid neutrality the term $\varrho_{\mathrm{e}}\boldsymbol{E}$ is negligible. If the fluid is assumed incompressible the term $\varrho(\boldsymbol{v}\,.\,\nabla)\boldsymbol{v}$ will be also negligible. Under the assumption of a constant viscosity, $\boldsymbol{\Psi}$ becomes $\upsilon\nabla^2\boldsymbol{v}$. Gravitational forces may also be neglected since the velocity is mostly in

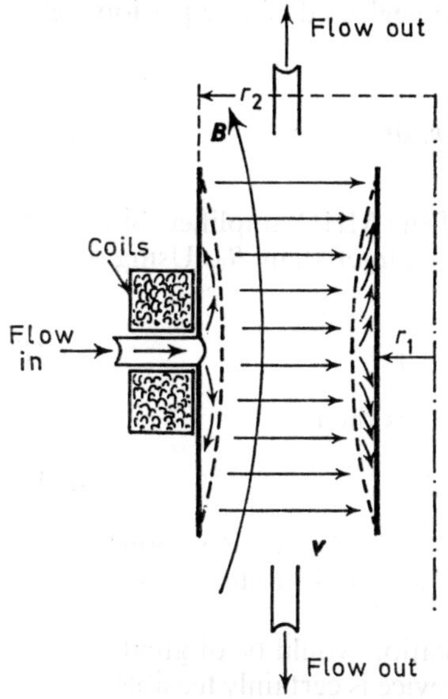

Fig. 8.17. MHD magnetic amplifier. The fluid is forced radially inwards in a cylindrical space in which an axial magnetic field is generated by a set of coils

the radial direction due to the presence of the vanes. Taking all those assumptions into consideration, equation (8.35) becomes:

$$\nabla p_r = v\nabla^2 \boldsymbol{v} + \boldsymbol{J}\times\boldsymbol{B} \tag{8.36}$$

Taking account of the cylindrical symmetry, assuming that the flow is exclusively in the radial direction, and that the coils are arranged such that over the area of concern $\boldsymbol{B} = \boldsymbol{a}_z B_0$, the induced current density becomes:

$$\boldsymbol{J} = \sigma(\boldsymbol{v}\times\boldsymbol{B}) = -\boldsymbol{a}_\theta \sigma v_r B_\theta \tag{8.37}$$

Consequently, equation (8.36) yields the two following scalar equations:

$$\frac{\partial p_r}{\partial r} = v\left(\frac{\partial^2 v_r}{\partial r^2} + \frac{1}{r}\frac{\partial v_r}{\partial r} + \frac{\partial^2 v_r}{\partial z^2} - \frac{v_r}{r^2}\right) - v_r B_0^2 \tag{8.38}$$

and $\partial p_r/\partial z = 0$. Thus p_r is function of r only. The variation of v_r in the z-direction can also be neglected, thus equation (8.38) becomes:

$$\frac{d^2 v_r}{dr^2} + \frac{1}{r}\frac{dv_r}{dr} - \left(\frac{B_0^2}{v} + \frac{1}{r^2}\right) v_r = \frac{1}{v}\frac{dp_r}{dr} \tag{8.39}$$

This is a Bessel differential equation, the solution of which would lead to the distribution of v_r. However, even the assumption of a constant pressure gradient would not lead to a straight solution of this equation. Since there are two unknowns: the velocity and the pressure, another equation is needed. This equation might be the equation of conservation of energy. However, interest is in the investigation of the feasibility of such a device. Consider Maxwell's equation

(3.4) in which the displacement current has been neglected. This equation yields after considering the above assumptions:

$$B_z = B_0 - \mu\sigma B_0 \int_{r_1}^{r_2} v_r \, dr \tag{8.40}$$

where r_1 and r_2 are the inner and outer radii of the MHD amplifier. Since v_r is in the negative r-direction, it is clear that B_z is larger than B_0. Using a limit technique, one can write:

$$B_z \leqslant B_0(1+\mu\sigma\Delta r v_{max}) \tag{8.41}$$

where $\Delta r = (r_2 - r_1)$. The factor of amplification is then:

$$B_z/B_0 \leqslant 1+\mu\sigma\Delta r v_{max} \tag{8.42}$$

One can see that for an amplification factor of 2 and for Δr equal to about 10 cm, and $\sigma = 100$ mho/m, the maximum velocity of the fluid should be larger than 10^5 m/s.

It can be concluded that good MHD amplification would be of great interest if somehow the fluid was made magnetic. The device is certainly feasible.

8.8 MHD LUBRICATION

In this application it is the viscosity of the fluid that is put into use. The forces of inertia are assumed negligible compared to the viscous, pressure, and electromagnetic forces. Liquid metals are used as lubricants under high temperature conditions but they are often less than satisfactory because of their relatively low viscosity. By using the MHD effect the liquid metal could be made more viscous in appearance, thus allowing it to support greater loads. Indeed, a body force may be created as a result of the interaction between the circulating currents induced by the motion of a conducting fluid through an applied magnetic field and this field itself. This body force will pump the liquid metal between the bearing surfaces enhancing the hydrodynamical lubricating effect. The body force created will be $\boldsymbol{J} \times \boldsymbol{B}$. The induced current $\boldsymbol{J}$ is given by equation (8.37); thus, approximately $|F| = \sigma v B^2$. It can be seen that for $v = 10$ m/s, $F = 10^8$ N/m³ for liquid sodium. Although this figure is rather low, the feasibility of such a device does not seem to be questionable; higher figures may be attained.

Hughes and Young[27] reported the analysis of several MHD lubricating devices. A hydrostatic thrust bearing is shown in Figure 8.18. Screen electrodes are arranged in the radial direction as shown in the figure. The lubricant is supplied under pressure in the plenum recess and flows radially outward to the atmosphere in an axially applied magnetic field. The induced current flows in the azimuthal direction and interacts with the axial magnetic field, thus retarding the radial flow of the fluid. Assuming that the electrodes are ideal conductors, the bearings as non-magnetic insulators, and the fluid as a viscous flow in its steady-state and of constant conductivity, Hughes and Young[27] solved the equation of motion. After using Maxwell's equation, they found that by supplying an electrical cur-

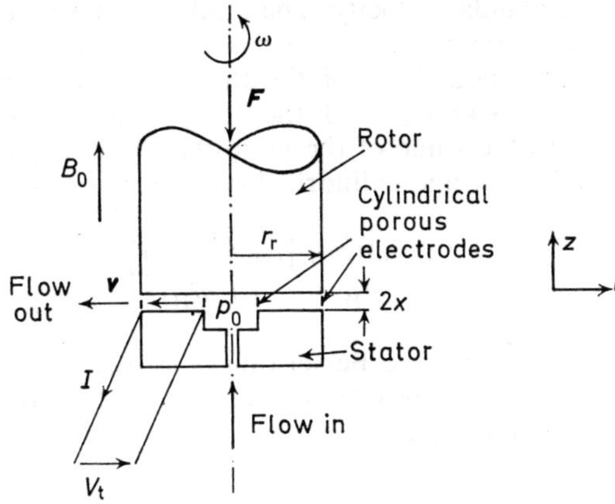

Fig. 8.18. The hydrostatic thrust bearing. Screen electrodes are arranged in the radial direction. The lubricant flows radially outwards in an axially applied magnetic field

rent from an external power source the frictional drag torque on the rotating bearing can be made equal to zero. This current is:

$$I' = -\pi\omega\sqrt{\sigma r^2} \coth(\mathcal{M}_H) \tag{8.43}$$

where ω is the frequency of rotation of the bearing, r is the radius of the rotor, and $\mathcal{M}_H$ is the Hartmann number defined as:

$$\mathcal{M}_H = xB_0\sqrt{(\sigma/v)} \tag{8.44}$$

$2x$ being the distance between the rotor and the stator of the hydrostatic thrust bearing.

A device similar to the hydrostatic thrust bearing is the MHD viscous coupler. The system serves as an electric control of torque-speed characteristics. It consists of a conducting incompressible fluid film separating two rotating insulated plates closely spaced in an axial magnetic field. A voltage applied externally between two concentric electrodes would control the torque-speed relationship. The two plates are rotating at different speeds.

Other problems studied by Hughes and Young deal with MHD inclined slider bearings with transverse or tangential magnetic fields. The externally applied magnetic field can give rise to body forces capable of modifying the pressure distribution in the lubricant film.

When a potential is applied between the rotor and the stator of a thrust bearing, an axial electric current is induced through the fluid. This current gives rise to a tangential magnetic field that creates a radial body force. This pinch effect increases the load capacity of the bearing.

The MHD effect can also be used as a basis of a viscometer. The viscometer uses a torque as experimental parameter. This torque is exerted by the shear stress of the tested fluid on the rotating walls of the viscometer. The shear stress varies with the viscosity of the tested liquid. An MHD viscometer would consist of two concentric cylinders, the inner one being stationary and the outer one rotating at a constant angular velocity. The fluid to be tested remains between the two cylinders, and an axial magnetic field is applied giving rise to an induced e.m.f. If the electrical conductivity of the fluid is known together with the velocity distribution, measurement of the e.m.f. induced between the two cylinders would lead to the value of the viscosity. Assuming a steady state, a laminar incompressible flow and cylindrical symmetry, one would find for the induced e.m.f.:

$$V_0 = \frac{(rB_0)(r\omega)}{\mathcal{M}_{\mathrm{H}}} \frac{[I_0(\mathcal{M}_{\mathrm{H}})-1]}{I_1(\mathcal{M}_{\mathrm{H}})} \tag{8.45}$$

where $I_0(\mathcal{M}_{\mathrm{H}})$ and $I_1(\mathcal{M}_{\mathrm{H}})$ are the Bessel functions and r is the radius of the viscometer. The MHD viscometer would be definitely of interest for fluids having high electrical conductivities.

8.9 MHD THERMAL CONVECTION[17, 41]

There are two basic types of MHD thermal convection problems: forced and free MHD convection. The heat exchange between a wall and a fluid when the flow is forced along the wall by some external means is called heat transfer by forced convection. The heat exchanged between a wall and a fluid when the fluid is set in motion by the temperature difference between the wall surface and the surrounding fluid is called heat transfer by free or natural convection. Obviously, the type of MHD thermal convection which predominates for a certain application would depend entirely on the problem at hand. For example, an analysis of the heat transfer from the surface of a re-entry vehicle would be primarily concerned with forced convection heat transfer, due to the high flow velocities existing a short distance above the surface. On the other hand, there are many applications where the flow velocities are almost zero such as in plasma containment problems or in MHD propulsion systems.

An important MHD natural convection problem is discussed by Hughes and Young[27]. It is an electrically conducting fluid contained between two vertical plates as shown in Figure 8.19. The channel is assumed sufficiently long to insure a fully developed flow and sufficiently deep to allow a one-dimensional approach. Under these assumptions, the basic equations of continuity, momentum and energy, reduce in cartesian co-ordinates, to: $\partial v_x/\partial x = 0$,

$$\nu_{\mathrm{k}} \frac{\mathrm{d}^2 v_x}{\mathrm{d}y^2} - \frac{J_z B_0}{\varrho} + g\beta(T-T_0) = 0 \tag{8.46}$$

$$\mathrm{d}p_{\mathrm{r}}/\mathrm{d}y = J_z B_x \tag{8.47}$$

$$\varkappa \frac{\mathrm{d}^2 T}{\mathrm{d}y^2} + \frac{J_z^2}{\sigma} + \varrho\nu_{\mathrm{k}}(\mathrm{d}v_x/\mathrm{d}y)^2 = 0 \tag{8.48}$$

where ν_{k} is the kinematic viscosity, $\varkappa$ is the heat conductivity, g is the gradient of gravity, and β is the expansivity:

$$\beta = -(1/\varrho)(\partial\varrho/\partial T)_{p_{\mathrm{r}}}$$

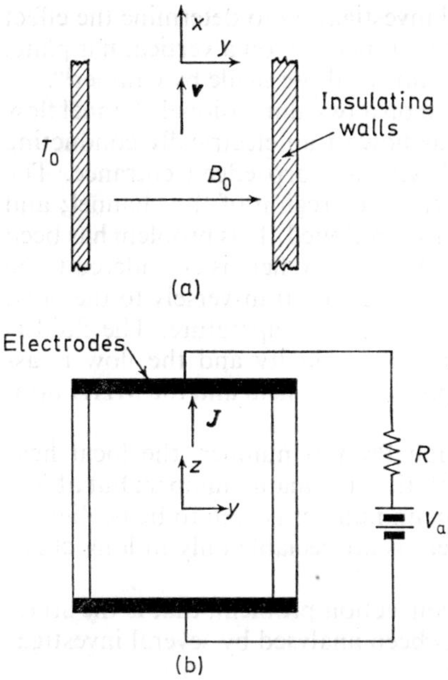

Fig. 8.19. Configuration for MHD thermal convection. (a) *Upper view;* (b) *side view*

It is assumed that the flow is laminar and that the difference in temperature is small compared to the ambient temperature T_0 at which the fluid density is ϱ_0. Equations (8.46) and (8.47) become, after the introduction of generalised Ohm's law [equation (8.6)],

$$\nu_k \frac{d^2 v_x}{dy^2} + g\beta(T - T_0) + \frac{\sigma}{\varrho}(E_z + v_x B_0)B_0 = 0 \tag{8.49}$$

$$\frac{dp_r}{dy} - \sigma(E_z + v_x B_0)B_x = 0 \tag{8.50}$$

From Maxwell's equations it can be seen that E_z is constant. Hughes and Young transformed those equations into dimensionless ones and solved them by using adequate boundary conditions, assuming a small buoyancy, and viscous as well as Joule heatings, thus allowing the use of a perturbation technique. It was found that the applied e.m.f. must be either zero or small enough to cause an electromagnetic body force that is not larger than the buoyant force. It was also found that the velocity was a linear function of the voltage drop across the electrodes when viscous and Joule heatings are neglected. This velocity increases also with the electrical conductivity of the fluid and becomes maximum for an infinite σ. The rate of heat transfer through the walls is seen to increase with increasing electrical conductivity and decreasing Hartmann number.

Similar results have been obtained by Cramer[10] who discussed the constant temperature parallel plate free convection MHD problem. Cramer assumed a two dimensional problem and neglected the Joule and viscous heating terms, with the further restriction of a fully developed flow.

Enery[15] has conducted an experimental investigation to determine the effect of a magnetic field upon the free convection of mercury on a vertical flat plate. The assumptions made by Enery were the same as those made by Cramer[10].

A simple MHD forced convection is based on a two dimensional channel flow model. Consider the incompressible viscous flow of an electrically conducting fluid through a two dimensional channel with a sharp edged entrance. The problem is twofold; first, the flow in the entrance region of the channel; and second, the flow downstream, where it is fully developed. This problem has been studied by Dhanak[12] under its first aspect. A magnetic field is considered to be applied uniformly along the length of the channel and transversely to the flow. The channel walls are assumed to be at a uniform temperature. The fluid is entering at a uniform temperature and a uniform velocity and the flow is assumed to be steady. There is no external applied electric field and the MHD approximation is made.

Dhanak's[12] analysis shows that at a high Prandtl number, the local heat transfer in the entrance region increases with the Hartmann number; but at low Prandtl number the influence of the Hartmann number is seen to be negligible. Furthermore, the effect of the magnetic field is appreciable only in long channels.

The second aspect of the forced MHD convection problem, that is the study of the flow in the channel downstream, has been analysed by several investigators. One such study is presented by Yen[48].

8.10 MHD STIRRING[13]

This problem is similar to the MHD pumping studied above. Indeed, in many industrial furnaces, travelling flux waves are used to stir a molten charge. Induction stirring can be achieved as shown in Figure 8.20. Excitation coils are arranged around the furnace in polyphase windings. They can thus induce a

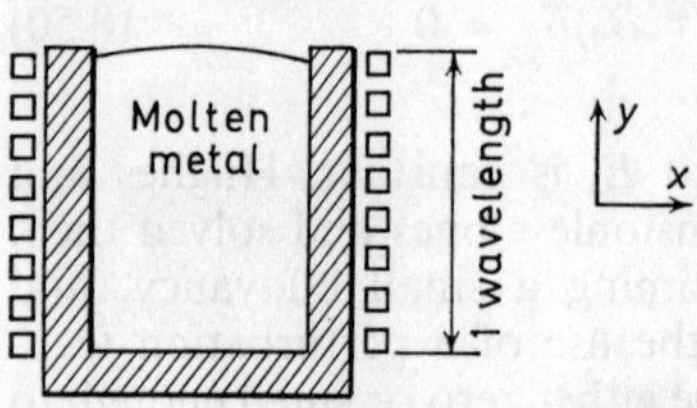

Fig. 8.20. MHD stirring. The current in the coils is such that $I = I_0 \cos(\omega_0 t - ky)$

magnetic-field-wave travelling in the axial direction. This magnetic field will interact with the axial component of the fluid velocity to induce a radial eddy current density. This in turn reacts with the applied magnetic field producing an axial stirring force along with a radial pinch.

Hughes and Young[27] studied this problem by assuming a model with parallel planes instead of the more realistic cylindrical model. They considered an infinite geometry. They neglected the thermomechanical coupling so that the motion becomes independent of the gravitational forces and the fluid density becomes independent of the temperature. In spite of the fact that the coupling between the fluid velocity and the magnetic field has been neglected, Hughes and Young's analysis shows that the induction stirring could be maximised by a proper choice of wavelength and frequency of the applied magnetic field.

It can be concluded that the MHD effects can be of interest in many industrial applications. The problems discussed in this chapter are not the only ones which have been considered. Many other ideas such as the use of boundary layers around re-entry vehicles to produce electricity, have been proposed. Some of these ideas might some day stimulate a much wider interest.

REFERENCES AND BIBLIOGRAPHY

1. BIENERT, W. B., YOUNG, W. H. and ZAVODNY, E. N., 'Electrical Output From Closed Loop MPD Experiment Using Auxiliary Ionization', *Advd Energy Conversion*, **6,** No. 1, 25 (1966).
2. BLAKE, L. R., 'Conduction and Induction Pumps for Liquid Metals', *Proc. Instn elect. Engrs*, **104A,** 49 (1957).
3. BLUE, E., *Torsional MHD Waves in the Presence of Finite Viscosity*, AF SOR TN-54-57 (1957).
4. BOOTH, L. A., 'Prospects for a 1000 Mw(e) Nuclear Reactor/MHD Power Plant', *6th Symp. Engng Aspects of MHD*, Pittsburgh, Pa., March (1965).
5. BREWER, G. R. *et al.*, 'Ionic and Plasma Propulsion for Space Vehicles', *Proc. IRE*, Special Issue on Plasma, Dec. (1961).
6. BÜRGEL, B., 'A Graphical Method for the Investigation of MHD Generators', *Brown Boveri Rev.*, **49,** No. 12, 493 (1962).
7. COLGATE, S. A. and AAMODT, R. L., 'Plasma Reactor Promises Direct Electric Power', *Nucleonics*, **15,** No. 8, 50 (1957).
8. COOMBE, R. A. (Ed.), *MHD Generation of Electrical Power*, Chapman and Hall, London (1964).
9. CORLISS, W. R., *Propulsion Systems for Space Flight*, McGraw-Hill, New York (1960).
10. CRAMER, K. R., 'Several MHD Free Convection Solutions', *J. Heat Transfer*, Series C, **85,** 35 (1963).
11. DEMETRIADES, S. T., 'Plasma Propulsion', *Astronautics*, March 7 (1962).
12. DHANAK, A. M., 'Heat Transfer in MHD Flow in an Entrance Section', *J. Heat Transfer*, Series C, **87,** 231 (1965).
13. DORAGH, R. A., *MHD Ship Propulsion Using Superconducting Magnets*, Society of Naval, Architects and Marine Engrs., Nov. (1963).
14. ELLIOTT, D. G., 'Two-Fluid MHD Cycles for Nuclear Electric Power Conversion', *ARS Jl*, **32,** 924 (1962).
15. ENERY, A. F., 'The Effect of Magnetic Field Upon the Free Convection of a Conducting Fluid', *J. Heat Transfer*, Series C, **85,** 119 (1963).
16. FARADAY, M., 'Experimental Researchs in Electricity' reprint from *Phil. Trans. R. Soc.* 2nd edn (1831–1836) R. and J. E. Taylor, London (1849).
17. FARN and HUGHES, *Combined Free and Forced Convection in MHD*, ASME Rept. 66-WA/HT-38 (1966).
18. FRIAUF, J. B., 'Electromagnetic Ship Propulsion', *J. Am. Soc. nav. Engrs*, Feb. (1961).
19. GLOERSEN, P., GOROWITZ, B. and PALM, W., 'Experimental Performance of a Pulsed Gas Entry Coaxial Plasma Accelerator', *ARS Jl*, **31,** 1158 (1961).
20. GLOERSEN, P., GOROWITZ, B., HOVIS, W. A. Jr. and THOMAS, R. B. Jr., 'An Investigation of the Properties of a Repetitively Fired Two-Stage Coaxial Plasma Engine', *3rd Annual Symp. Engng Aspects of MHD* (1963).
21. GOURDINE, M. C., 'Recent Advances in MHD Propulsion', *ARS Jl*, **31,** 1670 (1961).
22. GUNSON, W. E., SMITH, E. E., TSU, T. C. and WRIGHT, J. H., 'MHD Power Conversion', *Nucleonics*, **21,** No. 7, 43 (1963).

23. HAMILTON, S., 'MHD for Power Stations', *2nd Symp. the Engn Aspects of MHD*, Philadelphia, Pa., March (1961).
24. HARNED, B. W., 'Magnetic Effect in a T-Tube', *ARS Jl*, **30**, 656 (1950).
25. HEPWORTH, M. A. and ARTHUR, G., 'Ceramic Materials for MPD Power Generation', *1st Inter. Symp. Electrical Power Generation*, Newcastle upon Tyne, Sept. (1962).
26. HOGAN, W. T., 'Experiments with a Transient D.C. Crossed Field Accelerator at High Power Levels', *Proc. 3rd Annual Symp. Engng Aspects of MHD* (1963).
27. HUGHES, W. F. and YOUNG, F. J., *The Electrodynamics of Fluids*, Wiley, New York (1966).
28. JANES, G. S., DOTSON, J. and WILSON, T., 'Electrostatic Acceleration of Neutral Plasmas, Momentum Transfer through Magnetic Fields', *3rd Symp. Adv Propulsion Concepts*, Oct. 2–4 (1962).
29. KASH, S. W. (Ed.), *Plasma Acceleration*, Stanford University Press (1960).
30. KETTANI, M. A., *Direct Energy Conversion*, Addison-Wesley, Reading, Mass., Chapter 5 (1970).
31. MARSHALL, J. Jr., 'Performance of a Hydromagnetic Plasma Gun', *Physics Fluids*, **3**, 134 (1960).
32. MCCUNE, J. E., 'Linear Theory of an MHD Oscillator', *Advd Energy Conversion*, **5**, 221 (1965).
33. MESSERLE, H. K., 'The Travelling-Wave Plasma Conversion', *J. Fluid Mech.*, **19**, Part 4, 527 (1964).
34. PATRICK, R. M. and POWERS, W. E., 'Plasma Flow in a Magnetic Annular Arc Nozzle', *3rd Symp. Adv. Propulsion Concepts*, Cincinnati, Ohio, Oct. (1962).
35. RASHAD, A. R., 'Theoretical Studies on Travelling Wave Plasma Engines', *3rd Aerospace Sc. Meeting*, AIAA Paper No. 66–72, New York, Jan. (1966).
36. ROSA, R. J., 'Power Conversion', *Plasma Physics in Theory and Application* (edited by W. B. KUNKEL), McGraw-Hill, New York (1966).
37. SHERMAN, A., 'Theoretical Performance of a Crossed Field MHD Accelerator', *ARS Jl*, **32**, 414 (1962).
38. SPITZER, L., *Physics of Fully Ionized Gases*, Interscience, New York (1956).
39. SUTTON, G. W. and SHERMAN, A., *Engineering MHD*, McGraw-Hill, New York (1965).
40. SWIFT-HOOK, D. T., 'MHD Generation', *Direct Generation of Electricity* (edited by K. H. SPRING), Academic Press, New York (1965).
41. SZEWCZYK, A., 'Combined Forced and Free Convection Laminar Flow', *J. Heat Transfer*, **86**, Series C, 501 (1964).
42. TALAAT, M. E., 'MHD Electric Power Generators', *Advd Energy Conversion*, **1**, 19 (1961).
43. THOMPSON, W. B., *An Introduction to Plasma Physics*, Pergamon, New York (1962).
44. VELIKHOV, E. P., 'Hall Instability of Current Carrying Slightly Ionizing Plasmas', *1st Inter. Symp. MHD Power Generation*, Newcastle upon Tyne (1962).
45. WAY, S., *Electromagnetic Propulsion for Cargo Submarines*, Adv. Marine Vehicle Meeting (AJAA/SNAMR), Norfolk, Va., May 25 (1967).
46. WOODSON, H. H., 'MHD A. C. Power Generation', *Pacific Energy Conversion Conf.*, San Francisco, Aug. 12–16 (1962).
47. WOODSON, H. H., 'A. C. Power Generation with Transverse Current MHD Conduction Machines', *IEEE Trans., PAS*, **86**, 1066 (1965).
48. YEN, J. T., 'Effect of Wall Electrical Conductance on MHD Heat Transfer in a Channel', *J. Heat Transfer*, Series C, **85**, 371 (1963).

9 Thermionic applications

9.1 INTRODUCTION

Before the nineteenth century, it was already known that a negatively charged metallic body loses its charge more rapidly when heated. However, thermionic emission of electrons was not discovered until 1883 by Edison[11]. This effect which came to be known as the Edison effect can be described in the following: a current was observed to flow through a galvanometer connecting a plate to the positive end of a battery-heated filament, the space between the plate and the filament being a vacuum.

In 1904, Fleming[13] translated this effect into a thermionic diode rectifier. This led to the development of vacuum tubes and their use for amplification, microwave generation, and control. It is relevant to say that all present day electron tubes with heated filaments or cathodes are applications of thermionic emission. These tubes remained unique in their functions during the first half of this century, and until the appearance of solid-state devices.

In 1915, Schlichter[37] was the first to hint at the possibility of using the Edison effect to convert heat directly into electricity. However, because of the low efficiencies that were prevalent among early models of thermionic converters little development work was accomplished until the last decade. With a better understanding of the parameters which effect the converter efficiency, and improved materials, great increases have been made in the power outputs and efficiencies of recent thermionic converters.

More recently, radio-frequency oscillations in thermionic diodes have been considered as a means of directly generating radio-frequency energy. The possible use of thermionic tubes as amplifiers widens the range of application of thermionic devices for communication purposes.

Thermionic emission can be defined as the release of electrons from a hot metallic surface (emitter). The heat supplies energy to the valence electrons of the metal, allowing them to escape from the surface and travel away from it. In a diode those electrons are collected by a colder electrode (collector). An electric current is obtained if an external load is placed between the emitter and the collector, thus closing the circuit.

9.2 THERMIONIC EMISSION ANALYSIS[30]

The Bohr atomic theory suggests that an atom has a positively charged nucleus surrounded by a number of electrons that are established in definite states or orbits. The different orbits represent energy levels for the electrons. In a non-ionised atom the negative charge of these electrons is exactly balanced by the positive charge of the protons in the nucleus.

In metals, a great many atoms are forced very close together in a regular pattern called a crystal lattice. In this lattice, the forces of attraction between the atoms are very large. As a result, each of the original energy levels is split into a large number of levels. Consequently, the energy necessary to raise an electron from one level to another may become very small and the resulting energy levels may represent an almost continuous band.

As the temperature of the metal is raised above the absolute zero, the atoms begin to vibrate about their equilibrium positions in a random way because of the increase in their thermal energy. From these thermal vibrations, the electrons initially in the outer orbit or the highest energy level (valence electrons), will obtain additional energy, enough to be freed from the valence bonds and to move to neighbouring regions in the crystal. These free electrons interact with the vibrating atoms and move randomly as a result of this interaction.

The problem of determining the electronic energies is a statistical one because of the large number of collisions. This means that average values will be taken for the energy transferred after collision, the free path travelled by the electron between two collisions and the free time in which this free distance has been travelled. The energy distribution among electrons has been studied by Fermi[12] and Dirac[10] and the function which satisfies the physical requirements at a given temperature is known as the Fermi–Dirac distribution. This function permits the determination of the number of electrons per unit volume which have an energy $\mathrm{d}\mathscr{E}$ at a temperature T; it states:

$$\mathrm{d}n = f(\mathscr{E})\,\mathrm{d}\mathscr{E} = \frac{4\pi(2M)^{3/2}}{h^2}\,\frac{\mathscr{E}^{1/2}\,\mathrm{d}\mathscr{E}}{1+\exp\left[(\mathscr{E}-\mathscr{E}_\mathrm{F})/kT\right]} \tag{9.1}$$

where $\mathscr{E}_\mathrm{F}$ is the Fermi level of energy, h is Planck's constant and M is the mass of the electron. The total number of electrons per unit volume is:

$$n = \int_0^\infty f(\mathscr{E})\,\mathrm{d}\mathscr{E} \tag{9.2}$$

Equation (9.1) is represented in Figure 9.1 at absolute zero temperature and at a higher temperature T_1. From Figure 9.1 one can see that no electrons can exist above the Fermi level at the absolute zero. Thus absolute zero is the lowest energy state possible for the electrons, and the Fermi energy represents the maximum energy that an electron can have at the absolute zero temperature. It can be seen from equation (9.1) that at the absolute zero temperature one has:

$$\begin{aligned} \mathrm{d}n &= \frac{4\pi(2M)^{3/2}}{h^2}\,\mathscr{E}^{1/2}\,\mathrm{d}\mathscr{E} \quad &&\text{for} \quad \mathscr{E} < \mathscr{E}_\mathrm{F} \\ \mathrm{d}n &= 0 &&\text{for} \quad \mathscr{E} > \mathscr{E}_\mathrm{F} \end{aligned} \tag{9.3}$$

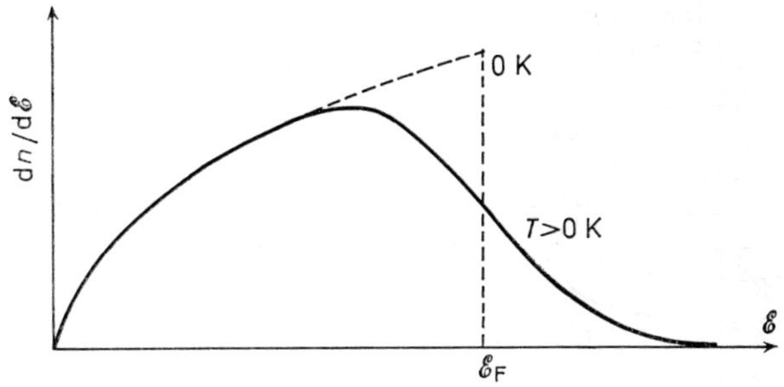

Fig. 9.1. The Fermi–Dirac distribution of electrons at absolute zero temperature and at T > 0 K

Using equation (9.2) one obtains:

$$n = \frac{8\pi}{3} \frac{(2M)^{3/2}}{h^2} \mathscr{E}_F^{3/2} \tag{9.4}$$

or:

$$\mathscr{E}_F = \frac{3}{8M} \frac{h^{4/3}}{(9\pi)^{2/3}} n^{2/3} \tag{9.5}$$

where n is the density of the valence electrons in the metal. It is a function of the type of the metal, and consequently $\mathscr{E}_F$ is also a function of the metal type. At the absolute zero, all the levels of energy up to the Fermi level are filled and all levels above it are empty.

Therefore, an electron free to move in the metal lattice has a certain kinetic energy. To find the value that this energy should reach for the electron to be possibly emitted from the metallic surface, the amount of work for the emission to occur should be calculated.

The surface effects appear as if a barrier were erected at the metal surface over which the electrons must be forced. Inside the metal, the free electrons move in a practically equipotential region. This balance is broken at the surface because of the absence of adjacent atoms. The energy necessary for the electron to cross the surface barrier is called the work function $\mathscr{E}_w$. It is equal to the difference between the internal potential energy $\mathscr{E}_{in}$ and the Fermi energy $\mathscr{E}_F$ still inherent to the electron even at the absolute zero. The surface of a metal presents a roughness at least of the order of the atomic diameter. Consequently $\mathscr{E}_w$ is a function of the metal type and of the conditions of the surface.

When an electron departs from the surface of the metal, the metallic surface will be left with a positive charge and an electric field will be created between the surface and the electron. Since the metal surface is equipotential, it will behave as if it is at mid-distance between two opposite and equal charges as shown in Figure 9.2. An image force will then act on the electron. This force is given by Coulomb's law applied to an electron of charge e at a distance x from the metal surface and its image inside the metal; thus:

$$F(x) = \frac{-e^2}{16\pi\varepsilon_0 x^2}$$

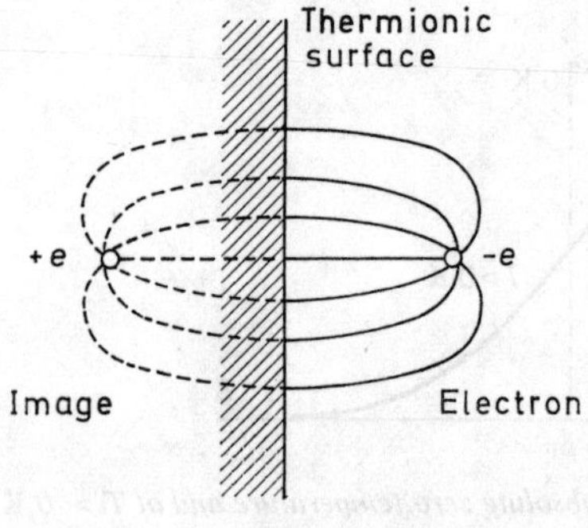

Fig. 9.2. The electric image principle

and the potential of the electron is for $x > r_D$:

$$V(x) = -\int_{r_D}^{x} \frac{F(x)}{e}\,dx$$

Coulomb's law is valid only for distances larger than the Debye radius r_D. For distances smaller than r_D, a constant force is usually postulated[39]; thus for $x < r_D$:

$$V_1(x) = \frac{ex}{16\pi\varepsilon_0 r_D^2} \tag{9.6}$$

and for $x > r_D$:

$$V_2(x) = \frac{-e}{16\pi\varepsilon_0}\left(\frac{1}{x}-\frac{1}{r_D}\right) \tag{9.7}$$

The total work function is:

$$\mathscr{E}_w = eV_1(r_D) + eV_2(\infty)$$

or:

$$\mathscr{E}_w = \frac{e^2}{8\pi\varepsilon_0 r_D} \tag{9.8}$$

Consider two metal surfaces with Fermi levels $\mathscr{E}_{FC}$ and $\mathscr{E}_{FA}$ and work functions $\mathscr{E}_C$ and $\mathscr{E}_A$. At equilibrium, there will be no net current flow from one surface to the other, hence no difference between Fermi levels, and a potential difference eV_{AC}, called the contact potential, appears between the two surfaces; it is:

$$eV_{AC} = \mathscr{E}_C - \mathscr{E}_A \tag{9.9}$$

The current density dJ_s due to electrons having velocities between $v_\perp$ and $v_\perp + dv_\perp$ perpendicular to the surface of a metal is $ev_\perp\, dn_\perp$, where J_s is the saturation current density. $dn_\perp$ is given by equation (9.2) after replacing $f(\mathscr{E})$ by its value in equation (9.1) and using the transformation $4\pi v^2\, dv = dv_\perp\, dv_y\, dv_z$. Integrating from $-\infty$ to $+\infty$ along v_y and v_z in the hodographic space, one obtains:

$$dn_\perp = \frac{4\pi M^2}{h^3} kT \exp(\mathscr{E}_F/kT) \exp(-Mv_\perp^2/2kT)\, dv_\perp \tag{9.10}$$

It should be noted here that 1 has been neglected in the denominator of equation (9.1), which is an acceptable simplification in most thermionic devices. The saturation current density is obtained by integrating equation (9.10) from $v_{\perp} = (2\mathcal{E}/M)^{1/2}$ to infinity; thus:

$$J_s = (4\pi Me/h^3)(kT)^2 \exp(-\mathcal{E}/kT) \tag{9.11}$$

This leads to:

$$J_s = \mathcal{A}T^2 \exp(-\mathcal{B}/T) \tag{9.12}$$

known as the Richardson–Dushman equation, where $\mathcal{B} = 11\,600\mathcal{E}$ K is a constant depending on the type of the material, and $\mathcal{A} = 1{\cdot}20\times10^4$ A/m² K². $\mathcal{A}$ is actually smaller than this value due to reflection effects at the surface of the metal. Richardson plots represent $\ln(J_s/T^2)$ versus $1/T$. As can be seen from equation (9.12), these plots are linear and given by:

$$\ln(J_s/T^2) = 13{\cdot}9 - 1{\cdot}16\times10^4(V/T) \tag{9.13}$$

Experimentally, however, it is found that the work potential V is a linear function of temperature such that $V = V_0 + \mathcal{C}T$ where $\mathcal{C}$ is a constant depending on the type of emitter, V is called the effective work potential, and V_0 is the Richardson work potential.

9.3 VACUUM TUBES[2]

The simplest vacuum tube using thermionic emission is the *diode* which is made up of two elements: a plate and a cathode, the space between the two elements being a vacuum. The cathode acts as a source of electrons and the heat necessary for thermionic emission is generally provided by a filament behind the cathode. The filament is attached to an independent heater circuit. The flow of electrons from cathode to plate is controlled by an externally applied electric field.

When electrons are emitted from the cathode, they leave positive charges behind on the thermionic surface. This surface will tend to attract back the emitted electrons which will tend to remain in the neighbourhood of the cathode, forming a negative space charge. The high electronic density near the cathode creates an electric field which will cause the electrons to be trapped back by the metal surface. An equilibrium is reached when the number of electrons emitted by thermionic emission equals the number of the electrons re-entering the surface.

When a positive voltage is externally applied, the induced electric field will tend to accelerate the electrons towards the anode thus creating a current in the external circuit. This is called forward biasing. With an applied negative voltage, that is reverse biasing, the electrons will be repelled toward the cathode, and practically no current will flow in the external circuit.

The number of emitted electrons will increase with increasing temperature until saturation is reached at a certain temperature T_1, determined by the external voltage. After T_1, the increase of temperature will practically have no effect on the number of emitted electrons. From 0 to T_1, the plate current is then governed by the temperature and is said to be *emission-limited.* It is governed by equation (9.12). After T_1 the dominant effect affecting the plate current becomes the space charge, and the current is said to be *space-charge limited.*

To calculate the voltage–current relationship in the space-charge-limited region let x be the distance from an electron to the emitter. Let d be the inter-electrode distance, and n be the electron density between the electrodes. Since the positive charge carriers are absent, there is a net space-charge density $-n(x)e$ using the one-dimensional form of Poisson's law, one obtains:

$$dE/dx = -n(x)e/\varepsilon_0 \tag{9.14}$$

The current density is:

$$J = -n(x)\,ev(x) \tag{9.15}$$

The velocity $v(x)$ is the result of an acceleration of the carriers by the applied electric field. Its value is:

$$v(x) = [2eV(x)/M]^{1/2} \tag{9.16}$$

where $V(x)$ is the potential difference between the point of abscissa x and the cathode where the initial velocity of the electrons has been neglected. Eliminating $n(x)$ from equations (9.14), (9.15) and (9.16), one obtains:

$$dE/dx = (J/\varepsilon_0)[2eV(x)/M]^{-1/2}$$

This equation can also be written, after multiplying by dV, as:

$$-E\,dE = (J/\varepsilon_0)(M/2e)^{1/2}V^{-1/2}\,dV \tag{9.17}$$

after integrating from 0 to x, one obtains:

$$E^2 = (4J/\varepsilon_0)(M/2e)^{1/2}V^{1/2}$$

or, since $E = -dV/dx$,

$$dV/dx = 2(J/\varepsilon_0)^{1/2}(MV/2e)^{1/4} \tag{9.18}$$

After integrating from 0 to d one obtains for the current density:

$$J = (4\varepsilon_0/9)(2e/M)^{1/2}d^{-2}V^{3/2} \tag{9.19}$$

where V is here the voltage at the distance d from the cathode (plate voltage). More generally equation (9.19) can be written, as:

$$J = \mathcal{K}V^{3/2}$$

which is known as Child's equation, and $\mathcal{K}$, which is called the constant of the tube, is a function of its geometry. The voltage–current characteristic of a vacuum diode is shown in Figure 9.3 for several values of the temperature. Vacuum tubes are normally operated at constant emission temperature in the space-charge-limited region.

Note that very high negatively applied voltages may also lead to an appreciable conduction. This is called 'flashover', and can destroy the tube.

Controlling the flow of electrons by changing the voltage of the main circuit is not very practical. It would be much more interesting if control were achieved by altering the field in the interelectrode space. This would require much less power, for the same amount of flow to be controlled. Indeed, the interelectrode field can be altered by the addition of a control electrode, called a 'grid', thus transforming the diode into a *triode*[5]. The emitted electrons are then accelerated by the anode-to-cathode V_c voltage but their flow is altered by the grid-to-

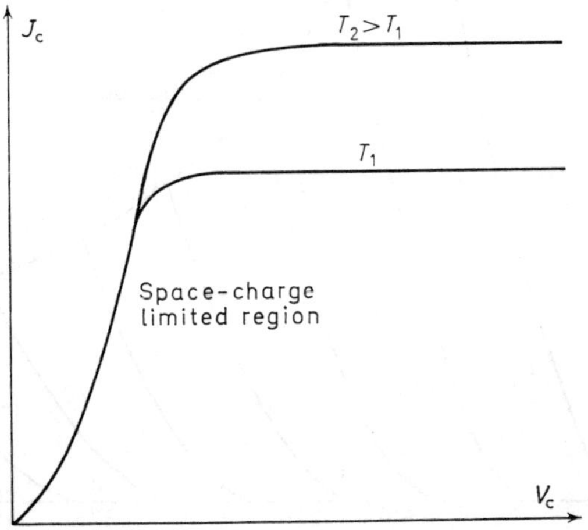

Fig. 9.3. Diode voltage–current characteristics for two different temperatures

cathode potential V_g. For a negative V_g the electrons are decelerated. The current–voltage characteristic of the triode in the space-charge-limited region is given by an equation similar to Child's equation; it is:

$$J = \mathcal{K}'(V_c + m_A V_g)^n \tag{9.20}$$

where $\mathcal{K}'$ is the triode constant, m_A is the amplification factor of the tube, and n is a number not necessarily equal to 1·5 (value corresponding to the theory sketched above). For typical triodes, m_A ranges from 10 to 50 and this shows the effectiveness of the grid control. The resulting fact that in practice V_g is extremely small compared to V_c, shows that the controlling power is often more than a thousand times smaller than the power to be controlled.

The family of curves representing equation (9.20) is shown in Figure 9.4. For a very large negative grid voltage, most of the electrons are returned to the cathode causing the plate current to be very small. The value of the grid voltage under which there is no plate current for a given plate voltage is called the cut-off voltage. For a positive grid, electrons are attracted by the grid itself, increasing markedly the grid current, and the grid acts roughly like a second plate in a diode.

Note that in most applications the operation of the triode is limited by the maximum plate dissipation $(I_c V_c)_{max}$ which is represented in Figure 9.4 by a hyperbola. Under no conditions should operation be allowed above this hyperbola.

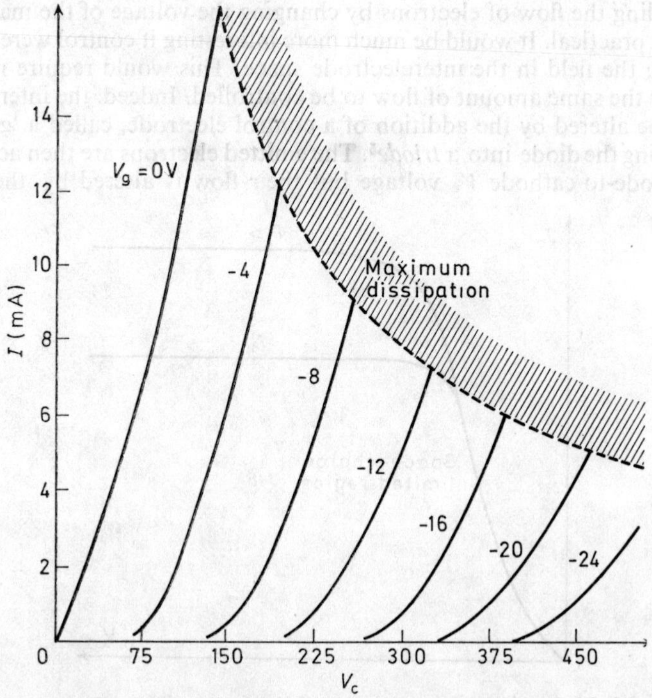

Fig. 9.4. Plate characteristics of a triode (type 6J5) for different values of grid voltage, $V_t = 6{\cdot}3$V

In many applications it is interesting to keep the plate current of the tube independent of the plate voltage. This is achieved by adding to the triode in the space between the grid and the plate a new electrode (also a grid) called a *screen*. The screen is biased in such a way as to keep the screen-to-cathode voltage positive. Because of this positive potential the screen collects some of the flowing electrons. The cathode current will thus be the sum of the screen and plate currents.

The flowing electrons, when they impinge on the surface of the plate, might dislodge new electrons, thus causing a secondary emission. This emission causes a decrease in the plate current. To eliminate the effect of electrons produced by secondary emission, an extra electrode (still a grid) called the *suppressor* is introduced between the screen and the plate. The suppressor is at a negative potential with respect to the anode; it provides an additional acceleration to the electrons coming from the cathode, and repels back to the plate the electrons emitted by secondary emission. The current that might be drawn by the suppressor is sent back externally to the cathode. Such a tube, having five electrodes is called a *pentode*.

Typical voltage–current characteristics of a pentode are shown in Figure 9.5. These characteristics depend on the physical and geometrical properties of a given pentode. They are characterised by a constant plate current for a given grid voltage. Like triodes, pentodes are also limited by the maximum plate dissipation shown by a hyperbola on the figure.

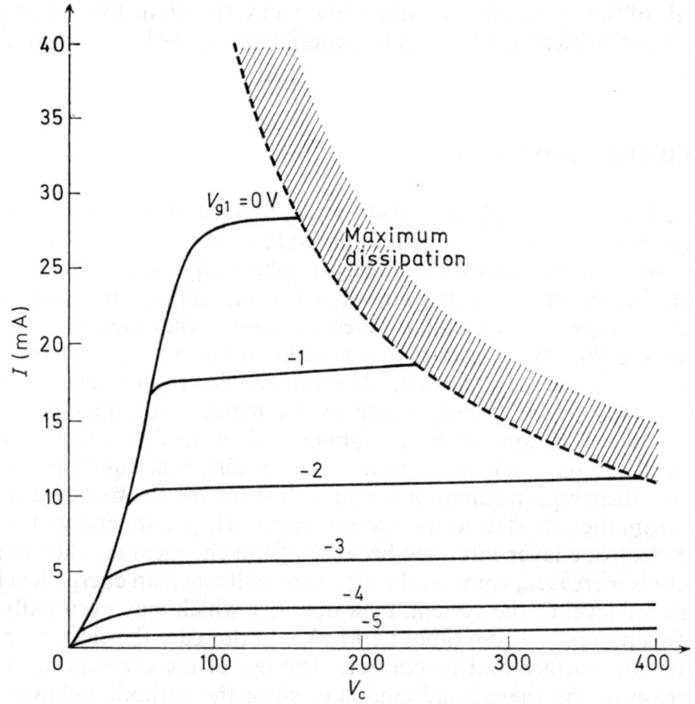

Fig. 9.5. Plate characteristics of a pentode (type 6CB6) for different values of V_{g1} voltage, $V_{g2} = 150$ V

Triodes and pentodes are fundamental electronic devices used (with their semiconductor counterparts) in most areas of computation, communication and control. Diodes are mostly used in rectification of a.c. to d.c. current in electronic applications. Triodes are used in amplifiers for current, voltage, or power, and in oscillators. One should also note that the characteristics of the pentode are very similar to those of the transistor, thus explaining the fact that in most applications pentodes are being replaced by transistors.

The material of the cathode should be a good thermionic emitter, and have both an acceptable lifespan and a good chemical stability. It should have large thermal and electrical conductivities and a low thermal emission. Some good thermionic emitters are tungsten (W), molybdenum (Mo), tantalum (Ta), thorium (Th) and calcium (Ca). Their work functions are 4·54 eV for W, 4·15 eV for Mo, 4·15 eV for Ta, 3·35 eV for Th and 2·24 eV for Ca. In most of the vacuum tubes, especially the larger ones, the cathode is directly heated by a current flowing through the thermionic emitting wires. In this case those wires should be coated by some metallic oxides to avoid evaporation at the working temperatures. The oxide is often a barium oxide with small amounts of strontium oxide to increase the lifetime of the coating. Indirectly-heated cathodes using a heating filament inside the cylindrical cathode are used mostly in smaller tubes. The cathode material is often made of nickel or tungsten.

Materials of the anode should have a low work function, low thermal emissivity and low electrical resistivity. The material having the lowest work function is caesium.

9.4 GAS-FILLED TUBES[23]

When liquid caesium is placed inside an evacuated diode, some atoms are evaporated thus establishing a caesium pressure inside the diode. Each caesium atom has one valence electron and has a spherically symmetrical electronic charge. This symmetry is destroyed when two caesium atoms come close to each other thus creating electric dipoles. Assuming that electrodes are good conductors and that the system is in thermal equilibrium, the caesium dipoles induce 'images' inside the electrodes. The valence electron of the caesium atom on the electrode surface is then shared by the metal inside the electrode. The same reasoning can be applied to caesium-coated electrodes. The caesium layer and the electrode material, say tungsten, are in chemical equilibrium. A new Fermi–Dirac distribution function is established for the electrons moving back and forth from the tungsten to the caesium layer. The Fermi energy is constant across the electrode-layer interface because of the chemical equilibrium. If the temperature is increased, some of the electrons will reach an energy level higher than the Fermi level of the system, by a quantity which may eventually exceed the work function of caesium (about 1·81 eV). In this case the electron becomes free to cross the surface and be emitted. The use of the caesium layer results in an increase of the thermionic emission, since the cathode behaves as if it were made of caesium instead of tungsten (work function about 4·52 eV).

When the temperature is increased again the kinetic energy of the layer of caesium atoms is increased enough to enable the caesium atoms to try to leave the cathode surface. The valence electrons, however, find themselves pulled between two forces, the force of attraction of the caesium atoms and that of the tungsten surface. The electron has to reach the ionisation energy level (3·87 eV) of the caesium atom in order to be freed from the atom and it should reach the work function level of the tungsten in order to leave the tungsten surface. Thus it often occurs that a positive caesium ion is emitted instead of the electron, resulting in what is called a *contact ionisation.*

At equilibrium a plasma is formed in the interelectrode space of the diode by three types of particles: caesium neutral atoms, caesium ions formed by contact ionisation and electrons emitted thermionically. Excess ions or electrons are attracted by the cathode surface leading to a complete neutralisation of the plasma, thus greatly reducing the space-charge limitation described for vacuum diodes.

Since space charge effects are now eliminated, all the electrons emitted from the cathode will reach the anode as soon as the plate voltage is equal to, or less than, the contact potential given by equation (9.9). The diode acts as a constant-current device with a current I_s given by equation (9.12). For an output voltage larger than the contact potential V_{AC}, the variation of the current will be exponential and such that:

$$J = J_s \exp\left[-e(|V| - V_{AC})/kT\right] \quad \text{for} \quad |V| > V_{AC} \tag{9.21}$$

This variation is shown in Figure 9.6. The plasma diode can, in this range, be successfully used as a variable resistance.

If a grid is introduced between the cathode and the anode in the gas-filled diode some control of the anode current would be possible. Such a tube is called a *thyratron*. If, in the absence of plasma, the grid is brought to a negative potential capable of nullifying the effect of the anode potential at the cathode, then no electrons will reach the ionisation potential. When the grid is made less

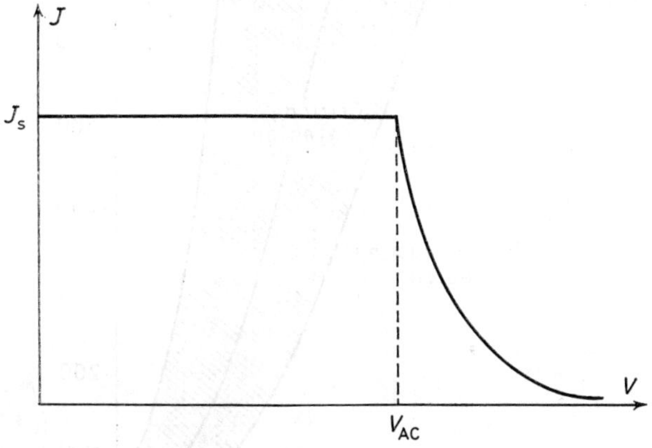

Fig. 9.6. Performance characteristic of a thermionic converter, space-charge neglected

negative, some electrons emitted from the cathode will be accelerated enough to reach an ionising velocity. The phenomenon results in an avalanche until an arc discharge is formed between the cathode and the anode. For a given anode potential, a specific grid voltage is critical in starting the arc. The characteristic of these points at a given temperature is called the critical grid characteristic (Figure 9.7). Once the arc starts, the negative grid attracts the positive ions which form a Debye sheath completely insulating the grid. Changing the grid voltage will then only change the thickness of the sheath and the latter will remain positively charged. Thus after the start of the arc the grid has no more control on the plate current and it cannot stop the arc, which will be stopped only by removing the plate potential.

The current in the tube can be readily calculated if we assume that the gas pressure is raised to a point at which the carrier drift velocity becomes governed by a mobility phenomenon. In this case the velocity $v(x)$ is $\ell E(x)$. Introducing this in equations (9.14) and (9.15), one obtains after eliminating $n(x)$:

$$E\,\mathrm{d}E/\mathrm{d}x = -\frac{J}{\varepsilon_0}\frac{1}{\ell}$$

and after two integrations, and considering the conditions at the limits, one obtains:

$$J = \frac{9\ell\varepsilon_0}{8d^3}V^2 \tag{9.22}$$

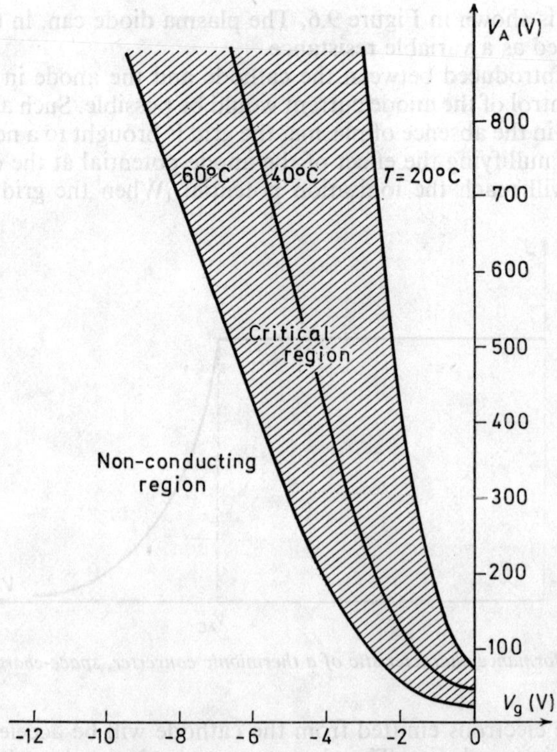

Fig. 9.7. Critical grid characteristic of a thyratron (Type 5559/FG-57)

This result assumes that the mobility coefficient b is independent of the electric field. This is usually a false assumption except for very weak fields. Von Engel and Steenbeck[44] used for moderately slow electrons a $1/\sqrt{E}$ mobility variation and obtained:

$$J = (16\varepsilon_0/9)\,(5/3d^5)^{1/2} V^{3/2} b_1 \tag{9.23}$$

where b_1 is the mobility at unit electric field.

The advent of the thyratron tubes made possible a high efficiency control of d.c. power. The development of ignitrons using a mercury-pool cathode increased the use of electronics in the high power field. Here, one should note that although the mercury arc rectifier might be considered as a device of similar type to the thyratron, the cathode-spot formation and emission is generally not considered of the thermionic type. Langmuir had suggested that it might be a type of high-field emission. This assumption, however, produces its own difficulties.

To reduce the current flowing in the control-grid circuit before the generation of the arc, a shield grid is often introduced around the control grid of a thyratron. The shield grid is biased negatively or positively and is completely isolated from the control grid. For a negative shield, the tube will have a negative grid characteristic, and it will have a positive characteristic for a positive shield. This method improves the reliability of firing of the thyratron for short input pulses. The tube is then often called a *gas tetrode*.

9.5 THERMIONIC CONVERTERS[22]

A thermionic converter converts heat into electricity by using the effect of thermionic emission. No thermionic converter can operate properly without some reduction of the space charge effects. To achieve this condition two possibilities are offered: space charge effects are reduced either by making the distance very small between anode and cathode (in a vacuum diode), or by neutralising the negative charges by recombination with positive charges (in gas-filled diodes).

High vacuum thermionic converters[45]

The use of closely spaced electrodes is a very practical method of reducing space charge effects. The distance does not usually exceed 10 μm. Equation (9.12) is based on a Fermi–Dirac distribution of the electronic velocities in the metal. If now the electrons are assumed to be emitted from the cathode with a Maxwellian velocity distribution f_e, and accelerated after emission toward the anode by an applied electric field, one can follow Langmuir's[25] mathematical analysis. Using a one-dimensional approach of Boltzmann's equation, and neglecting collisions, one might write:

$$v(\partial f_e/\partial x) + \frac{e}{M}(\mathrm{d}V/\mathrm{d}x)(\partial f_e/\partial v) = 0$$

where the variables x and v are separable. The general solution is of the form:

$$f_e(x, v) = \sum_{k=0}^{\infty} \mathcal{A}_k \exp\left\{\mathcal{B}_k\left[\frac{eV(x)}{M} - \frac{1}{2}v^2\right]\right\} \tag{9.24}$$

where $\mathcal{A}_k$ and $\mathcal{B}_k$ are constants determined by the boundary conditions. Langmuir assumed, however, a truncated Maxwellian distribution for the velocity of the electrons in the emitter, since the net current is often larger than the saturation current. He found,

$$J_s = -en_m(2kT_c/\pi M)^{1/2} \exp(-eV_m/kT_c) \tag{9.25}$$

$$J = -en_m(2kT_c/\pi M)^{1/2} \tag{9.26}$$

where n_m is the particle density at the maximum potential barrier at a distance x_m from the cathode. To calculate the electron density of the interelectrode space as a function of the interelectrode distance, it is necessary to solve the integral

$$n(x) = \int_{-v_1}^{\infty} f_e(x, v)\,\mathrm{d}v$$

19

where $f_e(x, v)$ is given by equation (9.24), and v_1 is the velocity necessary for an electron to cross a potential barrier $[V(x)-V_m]$. The result is:

$$n(x)/n_m = \exp(\psi_0)\,[1 \pm \mathrm{erf}(\sqrt{\psi_0})] \qquad (9.27)$$

where:

$$\psi_0 = \frac{Mv_1^2}{2kT_c}$$

Using Poisson's law (equation 1.1), one obtains after integration:

$$\xi_0 = \int_0^{\psi_0} \frac{\mathrm{d}x}{[\exp(x) - 1 \pm \mathrm{erf}(\sqrt{x}) \pm 2\sqrt{(x/\pi)}]^{1/2}} \qquad (9.28)$$

where:

$$\xi_0 = (2\pi Me^2/k^3\varepsilon_0)\frac{|J|^{1/2}}{T^{3/4}}(x - x_m)$$

x in equation (9.28) is an arbitrary variable. Equation (9.23) is plotted in Figure 9.8. J can readily be calculated from the $\psi_0(\xi_0)$ characteristic as a function of the interelectrode distance and the voltage for given values of J_s, T_c and V_{AC}.

Vacuum converters always operate in the space-charge-limited mode and the output current is always smaller than the saturation current, often smaller than 10 per cent of its value in most experiments. Due to the high temperature difference between cathode and anode, close-spacing is a technical difficulty in thermionic converters. Such converters are interesting only for small output power devices not exceeding 10 W.

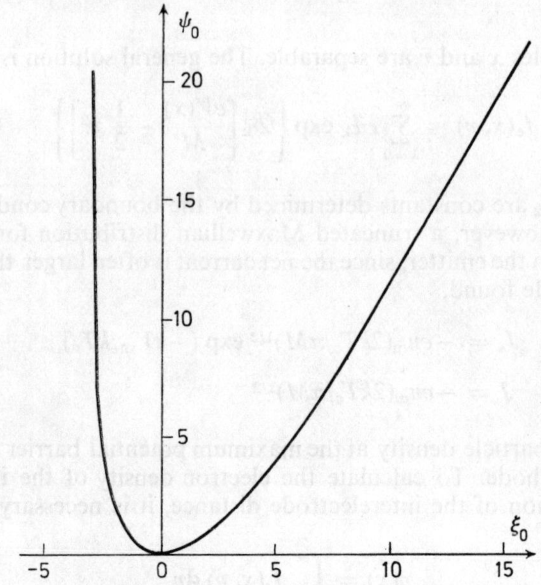

Fig. 9.8. Effect of the distance between the electrodes on the space charge. Collisions neglected

Gas-filled thermionic converters[18]

The difficulties of close-spacing led searchers to finding other ways of reducing space-charge effects in thermionic converters. One of such ways is the neutralisation of the space charge electrons by positive ions in the interelectrode space. Such positive ions can be generated by both surface and volume ionisation as was explained in Section 9.4. The condition for the gas inside the thermionic converter is to have an ionisation potential smaller than the work function of the emitter. For this reason caesium is most often used since it has the lowest

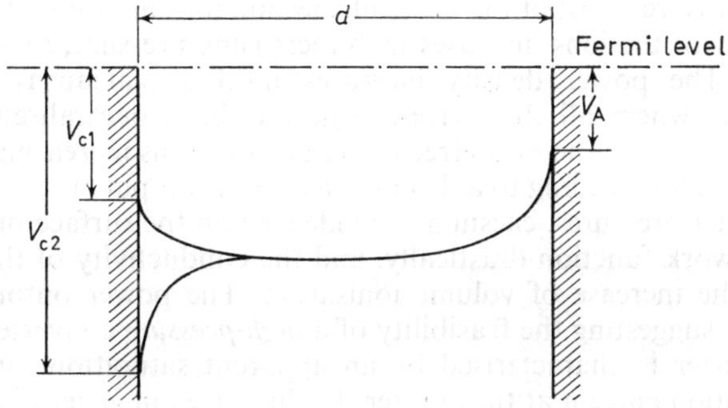

Fig. 9.9. Potential distribution in a low pressure thermionic converter for two values of n_e/n_i

ionisation potential and work function, vapourising it as it condenses on the surface of the anode thus lowering its apparent work function.

Taylor and Langmuir[42] have shown experimentally that the caesium vapour pressure p_{cs} in torr is related to the caesium pool temperature T_{cs} by:

$$p_{cs} = 121{\cdot}2\, T_{cs}^{-1{\cdot}35} \exp(-1755/T_{cs}) \qquad (9.29)$$

Depending on the caesium vapour pressure, one can distinguish three types of converter: (1) low pressure thermionic converters; (2) high pressure thermionic converters; and (3) arc-mode thermionic converters.

In the *low-pressure* converter[40], the mean free path is much larger than the spacing between the electrodes, for caesium pressures of about 10^{-4} mm of mercury. The performance characteristic of the low pressure thermionic converter is identical to that of the gas-filled diode (Figure 9.6). The distribution of the potential in the interelectrode space can be calculated by considering an isothermal metallic enclosure containing neutral caesium atoms, caesium ions, and electrons. The enclosure is assumed to be made-up of two parallel plates of infinite extent and separated by a distance *d*. The particle velocity distribution is Maxwellian in equilibrium. The equations of conservation of momentum are given by Boltzmann's transfer equations for neutral atoms, positive ions, and electrons. These equations are of the form:

$$\frac{kT}{e}\,\frac{\mathrm{d}n_{i,e,n}}{\mathrm{d}x} \pm n_{e,i}\,\frac{\mathrm{d}V}{\mathrm{d}x} = 0 \qquad (9.30)$$

where V is zero for the neutral atoms. This problem has been solved by Auer[3], who obtained $V(x, n_e/n_i)$ which is shown qualitatively in Figure 9.9. By a similar method the effect of the ions on the electric current density can be found[3] without difficulty; its value is:

$$J = J_{ecs} \exp(eV_m/kT_c) + J_{ics} \exp(-eV_A/kT_c) \quad \text{for} \quad V_A > 0 \quad (9.31)$$

where V_A is the anode potential, and V_m is the optimum potential in the interelectrode space; J_{ecs} and J_{ics} are the saturation current densities at the emitter, neglecting the saturation currents at the anode.

The power output of thermionic converters is a function of the caesium pressure. At low pressure, most of the incident caesium ions are ionised at the emitter. The density of the ions increases with increasing pressure, thus decreasing space-charge. The power density increases until a plateau is reached at $p_{cs} \cong 10^{-5}$ torr, where all the space-charge has been neutralised. With the increase of the pressure, the mean free path increases thus increasing the probability of collision and leading to a drop in the power output at $p_{cs} \cong 10^{-1}$ torr. But at still higher pressures, caesium is condensed on the surface of the anode, decreasing its work function drastically, and the conductivity of the plasma is enhanced by the increase of volume ionisation. The power output increases in consequence suggesting the feasibility of a *high-pressure* converter. The high pressure converter is characterised by an apparent saturation much smaller than the saturation current at the emitter. In this case ion generation is mainly due to the resonance ionisation at the cathode, which has a temperature in the 1500–1800°C range.

Higher efficiencies can be obtained in the arc-mode of operation at lower cathode temperatures. The most important mechanism of ion generation is then impact ionisation. The arc converter is thus an externally-heated hot-cathode arc discharge. The cathode becomes space-charge limited, and the performance characteristic rises to much higher current densities. As compared to the ideal space-charge-neutralised converter (Figure 9.6), there is a shift in the output voltage due to the loss of power necessary for the ionisation process. The converter arc-discharge is composed of a Langmuir double-sheath at the cathode and an anode sheath, the rest of the discharge being made of a glow plasma characterised by a space-constant potential and a high electronic temperature.

Other alkali metals than caesium have been considered for gas neutralisation. But, because of the corrosive nature of these metal vapours, noble gases such as argon, xenon, and helium have been suggested. The prospect of those gases does not seem to be encouraging since they have higher ionisation potentials than caesium.

Other space-charge neutralisation methods

Space-charge can be neutralised by the gating action of positive ion sheaths which surround auxiliary grids introduced between the cathode and the anode, as proposed by Johnson and Webster[20]. This method has been improved by Gabor[15], in the device of which the interelectrode space is separated into two regions by the presence of a mesh at zero voltage. The region between the mesh and the auxiliary anode contains a luminous ionising discharge; that in the cells formed by collectors in the space between the mesh and the cathode contains a dark conducting plasma. The electrons flow from the auxiliary anode to the cathode.

Jablonski *et al.*[19] proposed that space-charge neutralisation could be achieved by the use of fission fragments in an argon plasma. Because of the high pressure necessary, it does not seem that this method is feasible.

Another method of space-charge neutralisation is by the use of combined electric and magnetic fields. These so-called magnetic triodes will be studied in the next section.

Losses in a thermionic converter[26]

The detailed energy balance in the plasma diode is approximately represented by Figure 9.10. The thermionic converter operates along the normal lines of a thermal engine, with a hot source and a cold source. The means by which some heat is transferred from the hot source to the cold source, without any interference with the removal of electrons from the Fermi level of the cathode up, can be listed as follows: (1) radiation loss from the cathode; (2) heat convection

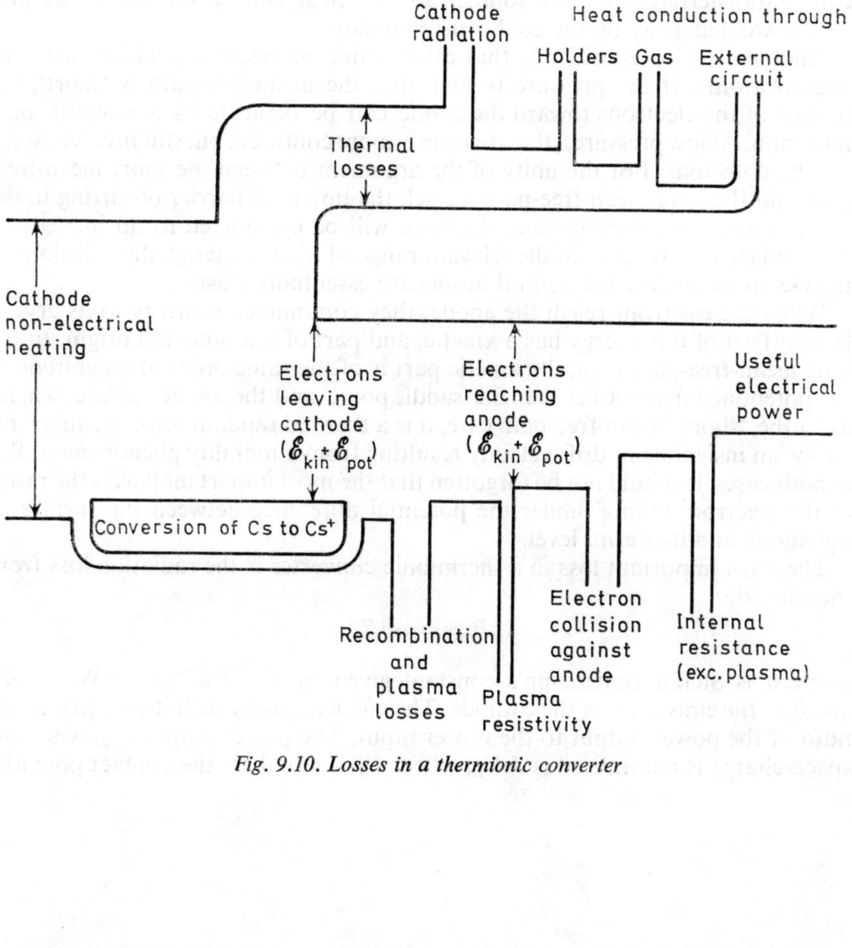

Fig. 9.10. Losses in a thermionic converter

through the neutral gas; (3) heat conduction through the electric circuit; and (4) heat conduction through the insulating holders.

On the average, each time a positive ion is created from cathode bombardment by a neutral atom, some heat is dissipated into the cathode crystal-lattice as a result of the electron transfer from the neutral atom onto the lattice, as specified before. All the heat communicated to the cathode, less the four losses enumerated above appear as potential plus kinetic energy of the electrons, the reference level for the former being the Fermi level inside the cathode. However, only part of this energy ultimately appears as electrical energy in the load.

First, some of the electrons do not reach the anode; they are carried away laterally by ambipolar diffusion or some equivalent mechanism and recombine with positive ions on some surface, where the corresponding energy appears as heat. Since the anode is, under normal conditions, less negative than the bulk of the plasma, only a very small part of this recombination occurs at its surface, owing to the 'saddle point' voltage distribution. Also, if the pressure is sufficiently high,some recombination can occur in the gas where the exact picture would be a balance between carrier generation and carrier annihilation according to Saha's formula, or some analogous picture in a state of non-thermal equilibrium. In both cases, the recombination energy appears as heat or as radiation ultimately converted into heat. Apart from a very small fraction which can be transferred to the hot source, all this heat can be considered as ultimately carried away by the cold source coolant.

Another loss to consider is that attributable to electron collision with the neutral atoms. If the pressure is such that the mean-free-path is 'short', the motion of the electrons toward the anode can be depicted as a mobility phenomenon. At low pressures, the situation is more confused, but still involves some loss by collisions. For the unity of the argument between the short mean-free-path and the long mean-free-path model, the potential barrier occurring in the latter, owing to complete neutralisation, will be considered as an increase in V_{AC} and treated as such. In the relevant range of kinetic energy, the collisions of the electrons against the neutral atoms are essentially elastic.

When the electrons reach the anode, they communicate energy to its crystal lattice. Part of this energy has a kinetic, and part of it a potential origin. In the long mean-free-path case, the kinetic part is of the same order of magnitude as the potential difference between the 'saddle point' and the anode surface, whereas in the 'short' mean-free-path case, it is a thermal random velocity, augmented by an insignificant drift velocity resulting from a mobility phenomenon. But in both cases, it should not be forgotten that the most important part is the result of the electron 'falling' under the potential difference between the surface of the anode and its Fermi level.

The most important loss in a thermionic converter is the radiation loss from the cathode:

$$P_{\text{rad}} = a\theta_c T_c^4$$

where a is Stefan–Boltzmann's constant given by $a = 5{\cdot}67\times10^{-8}$ W/m² K⁴, and θ_c is the emissivity of the cathode. The efficiency is by definition equal to the ratio of the power output to the power input. The power output P_{out} when the space-charge is neutralised is $J_s V_L$, where V_L is equal to the contact potential

V_{AC} minus all the voltage drops in the wires, the ion generation process, and the plasma. The power input is equal, if all the losses are neglected except the loss by radiation, to:

$$P_{in} = J_s V_c + a\theta_c T_c^4$$

Thus the efficiency of the thermionic converter becomes:

$$\eta = \frac{1-(V_A/V_c)}{1+(a\theta_c T_c^4/J_s V_c)} \tag{9.32}$$

This equation is represented in Figure 9.11.

Although the practical values of efficiency for experimental devices are rarely higher than 10 per cent, values as high as 40 per cent can be hopefully considered in the future. A typical high vacuum thermionic cell would produce a power

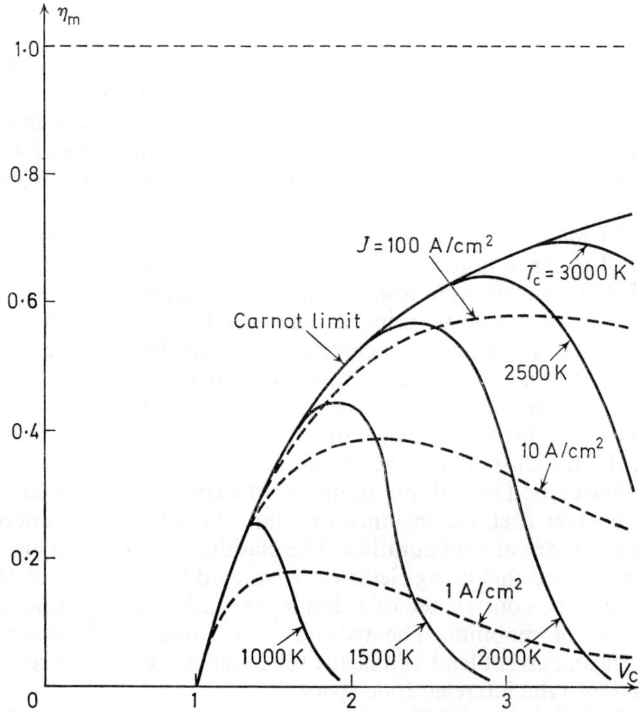

Fig. 9.11. Maximum efficiency as a function of the cathode work function[7]. *The cathode temperature and saturated current densities are taken as parameters,* $\theta_e = 0{\cdot}35$, $V_A = 1$ V

density of 0·4 W/cm² at an output voltage of 0·5 V and a cathode temperature of 1500K. Low-pressure gaseous converters yield an output of 10 W/cm² when the cathode temperature is higher than 2000K.

Thermionic converter systems

Thermionic converters share many of the advantages of other direct energy converters. They are free from moving solid parts and are hence quiet and long lived. This makes thermionic conversion advantageous for space applications and for use in remote, unattended generators. Their high theoretical efficiencies make them useful in conventional power plants mostly as toppers because of their high operating temperatures.

Their high power densities and high powers per unit weight together with the absence of moving parts and their structural rigidity make thermionic conversion ideal for *space*[41] applications. There are three major types of thermionic generators for use in the space field. These are: (1) nuclear-fuelled converters; (2) solar-powered converters; and (3) flame-heated[27] converters. Thermionic converters may be used to convert the heat from high temperature flame into electricity; this has direct applications in missiles where the exhaust flame itself could be used to generate electricity during the flight time. This electricity could be used to run auxiliary equipment operative only during powered flight intervals, or it might be used to charge batteries for use later on in a space mission. One such generator was built[7] and tested to produce power from the heat energy of a rocket exhaust gas. It was of the caesium vapour type with a concentric arrangement of the emitter and the collector with a total emitter area of 150 cm^2 and an emitter temperature of 2250K. This generator achieved 160 W/kg on short term runs.

Design of a *solar*[28] thermionic power system working in the 1–10 kW range and suitable for space applications has been considered[32]. Both vacuum and caesium-filled converters were tested. The most efficient converter was a cylindrical vapour-filled device working at an output voltage of 28 V. The solar energy was used through solar concentrators and the heat was rejected only by radiation. An efficiency of 15 per cent was obtained.

The most promising use of a *nuclear*[6] thermionic converter for space is as an *in-pile* device. The three main parts of the converter are then the reactor, the shield, and the radiator. Each one of the reactor segments is subdivided into unit thermionic cells. The cells are mounted in a tray and are cooled by a liquid metal. The nuclear fuel, for instance uranium dioxide, is introduced into the centre of the cylindrical configuration. The cladding serves as an electron emitter, the encapsulated fuel being electrically isolated by an insulator. The collector is a trilayer tube constructed of a thin layer of aluminium oxide sandwiched between layers of niobium. The trilayer tube simultaneously serves as the collector surface, current lead and heat-exchanger surface. The caesium vapour is introduced into the interelectrode space.

Terrestrial applications of thermionic power generation can be divided into three major classes: (1) use in remote unattended generators[43]; (2) use in motor vehicles; and (3) use in large power plant systems. Thermionic conversion lends itself readily to use in remote areas because it has no moving parts and therefore promises a high reliability. Solar systems, however, require storage capabilities and this is a handicap. Nuclear-fuelled thermionic generators do not have the storage problem and therefore may find a use where relatively large power outputs are needed in remote installations.

One line of development of caesium diodes in motor vehicles aims at fossil-fuelled converters. Such systems require a vacuum-proof envelope capable of being cycled reasonably fast from room temperature to about 2000K. The driving properties of such thermionic cars would be ideal and the ratio of power to weight good, but it would require a leak-proof vacuum envelope capable of operating approximately 2000 h through many cycles. In view of the many problems arising, it does not seem that such a car will be produced for some time yet.

Large fossil-fuelled thermionic plants can be disregarded as impractical. This is because the natural unit of thermionic converters is about a few hundred watts, and even if this can be increased to a few kilowatts it is still not feasible, since present power plants operate in the 500 MW range. Thus about 100 000 thermionic units would have to be used in one single plant. Also commercial fossil-fuelled power plants have efficiencies approaching 40–45 per cent and the slight increase that a thermionic topping would provide might not allay the high capital cost of installation of those units. This method might, however, prove worthwhile on nuclear fuelled power plants, where the overall efficiency is now around only 30 per cent. Assuming an efficiency of 15 per cent in a topping, a thermionic generator would enable one to boost the overall efficiency of a nuclear plus thermionic steam plant above 40 per cent. This increase would presumably justify a fairly high capital cost for the additional equipment involved in the topping.

9.6 MAGNETIC TRIODES[38]

In large thermionic generators, high currents induce magnetic fields, the effect of which is to reduce drastically the efficiency of conversion. The prevention of such induced magnetic fields is possible in small generators. For a high power generator the problem is different. Indeed, since the thermionic converter is a low voltage device, a high-power generator would be by necessity a high-current one. The large current induces a magnetic field in a direction transverse to it. This field deflects some of the electrons back to the emitter.

We have seen that the electrons have a Fermi–Dirac distribution inside the metal. If the surface of the metal is in the x-y plane, the current density $\mathrm{d}J(v_x, v_y, v_z)$ of electrons with initial velocity components between v_x and $v_x+\mathrm{d}v_x$, v_y and $v_y+\mathrm{d}v_y$, v_z and $v_z+\mathrm{d}v_z$, is[29]:

$$\mathrm{d}J = (2eM^3/h^3) \exp\{[-eV_c - \tfrac{1}{2}M(v_x^2+v_y^2+v_z^2)]/kT_c\}\, v_z \,\mathrm{d}v_x \,\mathrm{d}v_y \,\mathrm{d}v_z \quad (9.33)$$

Since the saturation current density is given by equation (9.11), equation (9.33) can be written:

$$\mathrm{d}J = (2J_s/\pi) \exp\{-[(v_x/v_0)^2+(v_y/v_0)^2+(v_z/v_0)^2]\}\,(v_z/v_0)\,\mathrm{d}(v_x/v_0)\,\mathrm{d}(v_y/v_0)\,\mathrm{d}(v_z/v_0) \quad (9.34)$$

The current density at the collector can be calculated by integrating equation (9.24) between the appropriate velocity limits. These limits are given by considering the effect of the magnetic field. Assuming that the caesium vapour in the thermionic converter is of a density low enough to neglect collisions, the velocity of electrons will be:

$$dv/dt = -\frac{e}{M}[(v\times B)+E] \qquad (9.35)$$

Assuming also that the plasma is neutral, one can write for the electric field:

$$E = -(V_{AC}-V_0)/d$$

where d is the distance between the electrodes, and V_0 is the voltage drop in the external circuit. The trajectory of the electron can be calculated; its projection

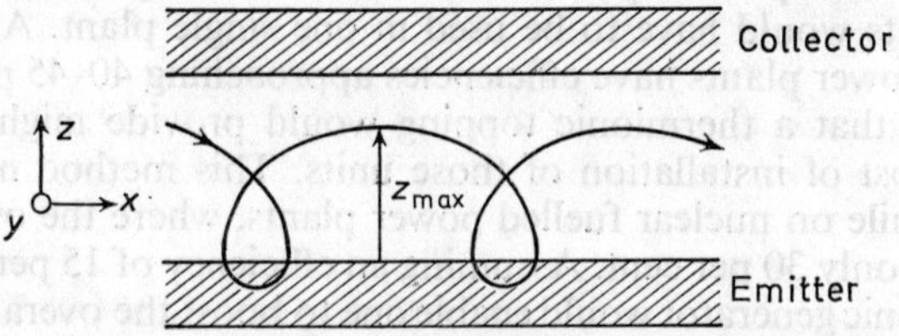

Fig. 9.12. The trajectory of an electron projected in a plane perpendicular to the electrodes under the effect of the induced magnetic field. The trajectory is a trochoid

in the x-y plane is a trochoid as shown in Figure 9.12. The maximum value z_{max} of the electron in the z-direction is:

$$z_{max} = \frac{1}{\omega_c}\left\{\left[\left(v_x-\frac{E}{B}\right)^2+v_z^2\right]^{1/2}+\left(v_x-\frac{E}{B}\right)\right\} \qquad (9.36)$$

The condition for the electon to reach the collector is that $z_{max} = d$, which can occur even for $v_z = 0$, that is for:

$$v \geqslant \frac{1}{2}\omega_c d+\frac{E}{B}$$

In general, one can write:

$$\left(v_x-\frac{E}{B}\right)^2+v_z^2 = \left(\omega_c d - v_x + \frac{E}{B}\right)^2$$

and consequently equation (9.34) must be integrated in two parts between the following limits:

$$-\infty < v_x < \frac{1}{2}\omega_c d+\frac{E}{B}$$

$$-\infty < v_y < \infty$$

$$\omega_c d\left\{1-\left[\left(v_x-\frac{E}{B}\right)\Big/\left(\frac{1}{2}\omega_c d\right)\right]\right\}^{1/2} < v_z < \infty$$

and

$$\frac{1}{2}\omega_c d + \frac{E}{B} < v_x < \infty$$

$$-\infty < v_y < \infty$$

$$0 < v_z < \infty$$

After integration, Schock[38] obtained for the current at the collector:

$$J = J_s \frac{1}{2}\,\mathrm{erfc}\left(\mathcal{B}_1 - \frac{\mathcal{B}_2}{4\mathcal{B}_1}\right) + J_s \frac{1}{2}\,\mathrm{erfc}\left(\mathcal{B}_1 + \frac{\mathcal{B}_2}{4\mathcal{B}_1}\right)\exp(\mathcal{B}_2) \qquad (9.37)$$

with:

$$\mathcal{B}_1 = edB/\sqrt{(8MkT)}$$

and

$$\mathcal{B}_2 = e\Delta V/kT$$

This equation is represented in Figure 9.13 and shows that for a magnetic field different from zero the current density reaching the cathode is smaller than the saturation current density.

To avert this loss of efficiency due to the induced magnetic field, Schock[38] proposes one of two solutions; either choosing special electrode configurations

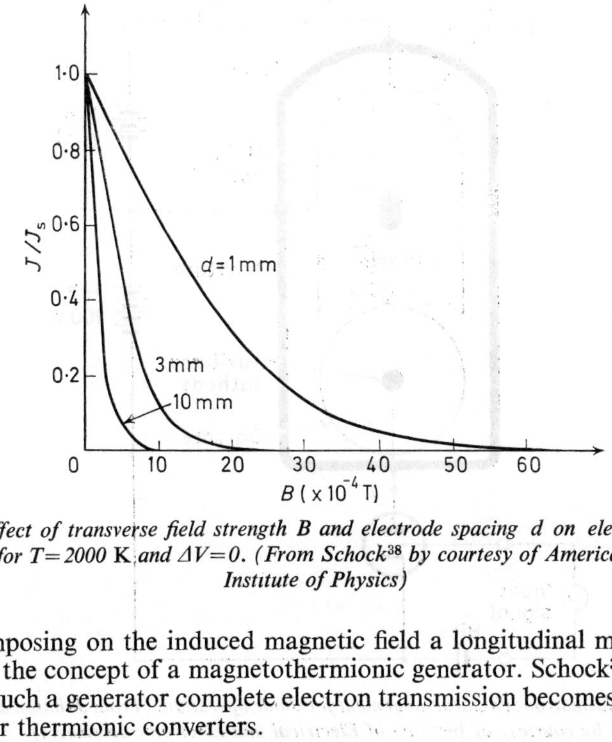

Fig. 9.13. Effect of transverse field strength B and electrode spacing d on electron emission for T=2000 K and $\Delta V=0$. (From Schock[38] by courtesy of American Institute of Physics)

or superimposing on the induced magnetic field a longitudinal magnetic field leading to the concept of a magnetothermionic generator. Schock[38] has shown that with such a generator complete electron transmission becomes possible for high-power thermionic converters.

9.7 THE PLASMATRON

The principles of the plasmatron were outlined by Johnson and Webster[20]. We have seen that the vacuum diode was often used in its space-charge-limited region. It is thus a high impedance device that can control continuously its own current. The gas-filled tubes, such as the thyratron, are low impedance devices in which the grid can only control the starting moment and not the overall current. The idea is to obtain a device that has both the advantages of the vacuum diode and those of the thyratron. Such a device would be able to continuously control large currents at low voltage. The plasmatron has these properties.

The *plasmatron* would be a gas-filled tube in which the plasma is generated by an auxiliary discharge. It is therefore basically different from the thyratron in which the plasma is generated by the discharge current itself. There are two ways of controlling the anode current in the plasmatron, either by changing the intensity of the auxiliary discharge or by the use of a gating grid. The first method is based on the fact that by changing the intensity of the auxiliary discharge, the plasma density changes, thus changing its conductivity and consequently the current density flowing through it. In the second method the positively charged grid acts in a way similar to that of the thyratron. The fact that the plasma is generated independently from the current through the tube

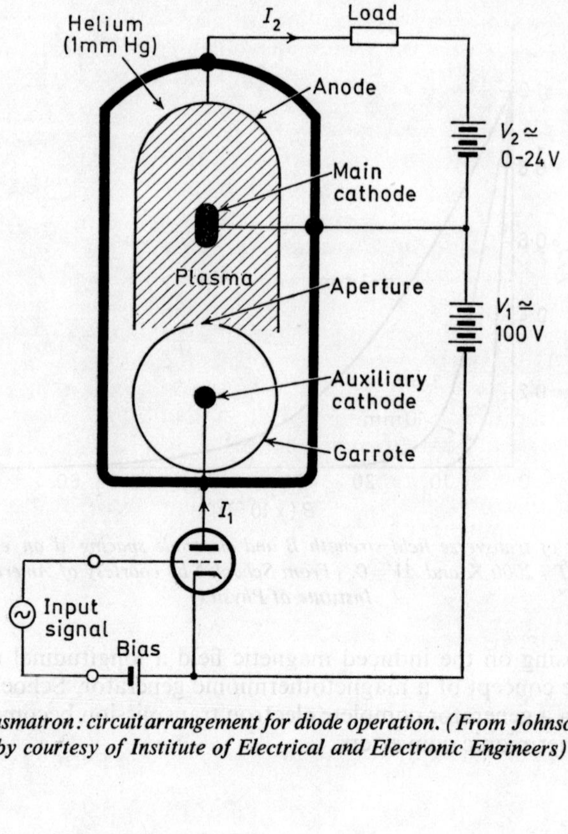

Fig. 9.14. The plasmatron: circuit arrangement for diode operation. (From Johnson and Webster[20] by courtesy of Institute of Electrical and Electronic Engineers)

allows complete control of this current as in a vacuum tube with the extra characteristic of a low impedance operation. Indeed, in a thyratron the grid has two functions; the first one is to ionise the gas and the second one is to establish an electric field that would accelerate the electrons. In the plasmatron these two functions are separated. The first one is provided by the auxiliary circuit and the second by the grid itself.

Figure 9.14 shows an experimental plasmatron as reported by Johnson and Webster[20]. This plasmatron consists of a U-shaped anode, two cathodes, one of which is the auxiliary electrode, and a 'garrote', which is a cylinder having a narrow slit. The circuit consists of a main loop and an auxiliary loop, and the gas used was helium at 1 torr pressure.

The voltage between the main and the auxiliary cathodes is made sufficiently high (about 100 V) to allow the discharge to occur in the auxiliary loop, between the anode and the main cathode. Due to the application of this discharge, large currents will flow if the voltage between these two electrodes is kept small enough to avoid a discharge in the main loop. The function of the garrote is to help increase the ionisation efficiency in the tube.

The current in the auxiliary loop can be varied at will by varying the potential of the grid of a modulator tube. The characteristics of such a plasmatron were reported by Johnson and Webster[20]. For a given auxiliary discharge current, the anode current flows at about -1 V of anode voltage and increases steeply until a saturation value is reached for $V_a \cong 3$ V. This saturation current depends on the value of the auxiliary discharge current. The saturation remains the same when the voltage increases until the ionisation potential of helium is reached (about 25 V). In this case, the auxiliary circuit loses control of the anode current, the latter rising rapidly to higher values.

The current gain is about 40 dB for an anode potential of about 6 V. This gain remains constant when the frequency increases until a value of about 10 kHz is reached. After this value, the gain begins to fall off markedly. This is due to an increase in the time it takes the surplus ionisation to diffuse away from the walls (where the surface recombination takes place).

A plasmatron of the triode type has also been studied yielding results similar to the preceding ones. The frequency response is, however, greatly improved since the modulation of the grid affects only the neighbouring plasma regions, whereas in the diode type the modulation affects the entire plasma. It was found that for an auxiliary current of 5 mA and an anode potential of 6 V, the trans-conductance of the triode was constant at the value 100 dB up to 100 kHz. Then it decreases in an oscillatory manner reaching a new equilibrium at 80 dB for frequencies around 10 MHz.

The use of the plasmatron may be of interest in many applications such as the control of motors and the transformerless drive of loudspeakers. It is, thus, useful for all applications in which a current control is needed at low impedance.

9.8 THE Q-MACHINE

The Q-machine operates on the principle of thermionic ionisation of caesium on the surface of tungsten. Figure 9.15 represents a Q-machine in which a hot tungsten electrode is placed at each extremity of a cylindrical vacuum tube. An

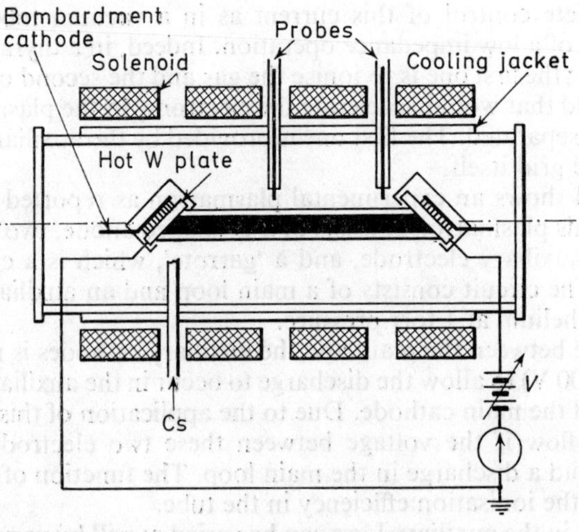

Fig. 9.15. Schematic Q-machine. A hot tungsten electrode is placed at each extremity of the vacuum tube. Thermionic ionisation of caesium atoms occurs at the surface of the electrodes. (From Rynn[36] by courtesy of Gordon and Breach)

atomic beam of caesium is then introduced into the vacuum. When the beam strikes one of the tungsten plates, most of the neutral caesium atoms lose their valence electron and are thus ionised. If the tungsten plate is hot enough (about 2400K) electrons will be emitted from its surface by thermionic emission. Thus a plasma flow is generated, which is made up of the caesium ions and the electrons. This flow moves towards the other end of the tube.

To confine the caesium plasma, a strong axial magnetic field is applied. The tungsten electrode at the other end of the tube is also inclined by a 45° angle. When caesium ions strike this electrode they are re-emitted and the same can be said about electrons if the rate of absorption is equal to the rate of thermionic emission. The result is a neutral plasma confined in the middle of the tube.

Other arrangements can be made for the Q-machine. One such arrangement is by using a single-ended tube. In this case one of the plates is kept cold allowing the electrons to be absorbed. This will cause a continuous flow of neutral plasma along the magnetic lines of force.

Experiments on Q-machines have been performed notably by Rynn[36]. He used a cooling jacket around the tube to permit the removal of any neutral caesium atoms that may remain in the plasma by wall condensation. If the background pressure of the caesium is to be kept at about 10^{-8} torr the wall temperature should be kept at about -20°C. By this method Rynn obtained fully ionised (up to 99 per cent) plasmas with densities as high as 5×10^{12} cm^{-3}. The strength of the confining magnetic field was made to vary from 0·03 to about 0·9 T.

The radius of the tube is chosen by considering many factors. One of them is to obtain as many ion gyration radii as possible. For instance, since caesium has an atomic weight of 133 its gyroradius would be about 1 mm at 1 T. If the radius of the vessel is 1·4 cm, 14 gyrations would be allowed. The length of the tube is determined by the value of the mean free path, the strength of the magnetic field and the plasma density.

It is very important to keep the radius of the incoming atomic flux constant in a Q-machine. Adequate collimators should be used. If no electric field is applied and no current is flowing, the plasma formed by the Q-machine will have a constant density in the axial and azimuthal directions under certain assumptions. The radial variation is:

$$n = K' \exp(-r/2r_0)/r^{1/4} \tag{9.38}$$

where K' is a constant, and r_0 is the radius of the vessel.

The voltage–current characteristic of a Q-machine has been studied by Rynn, whose results have been reported in Figure 9.16. As the voltage is increased, the current increases until it reaches a maximum and then decreases to a constant value. As shown in Figure 9.16, an instability is triggered just after the maximum

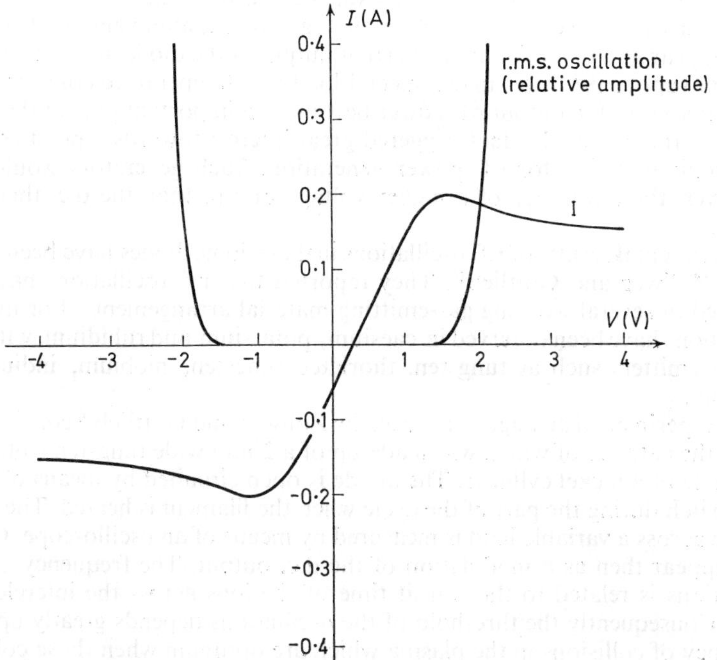

Fig. 9.16. Total current and r.m.s. noise amplitude versus voltage in a Q-machine. (From Rynn[36] by courtesy of Gordon and Breach)

current. Later studies have shown that similar instabilities in the Q-machine may result from several origins: universal collisionless instabilities[24], resistive drift modes[35], collisionless and resistive quasimodes[9]. Experiments on these instabilities have been performed by Chen[8].

It should be noted that combinations other than caesium and tungsten may be used in Q-machines. These are potassium with tungsten, caesium with tantalum, and sodium with rhenium, among others. Chen[8] used potassium with tungsten in his instability experiments.

It should also be noted that it is not always necessary to use atomic beams for injection of the working gas. This gas can be introduced by any method; by heating the walls of the tube, enough neutral atoms will be supplied to the electrodes under the effect of the vapour pressure.

The Q-machine is very important in the production of highly ionised quiescent plasmas. It is thus related very closely to fusion experiments. In this respect the problems of confinement and stability are similar to those encountered in thermonuclear fusion devices.

9.9 GENERATION OF RADIO-FREQUENCY ENERGY[4]

Oscillations in the radio-frequency range (from 100 kHz to 2 MHz) have been observed by many workers[16, 21, 33] in alkali-vapour filled thermionic tubes. These oscillations have been often seen in the case of caesium-vapour filled tubes. They appear as a reduction in the current output of the diode in a given range of caesium vapour pressures under special load and temperature conditions.

In some cases the output r.f. power becomes an important part of the output power of the diode; this fact triggered great interest towards a possible use of thermionic emission for r.f. power generation. Such generators would have, moreover, the advantage of a higher voltage output than the d.c. thermionic diodes.

Experimental results of r.f. oscillations in thermionic diodes have been reported by Zollweg and Gottlieb[46]. They reported that r.f. oscillations have been observed in several working gas-emitting material arrangements. For instance, oscillations have been observed in caesium, potassium and rubidium, with thermionic emitters such as tungsten, thoriated tungsten, niobium, indium and tantalum.

The experimental arrangements used by Zollweg and Gottlieb[46] consisted of a diode, the cathode of which was made up of a 2 mm wide tungsten ribbon and the anode of a nickel cylinder. The anode is open circuited by means of a mercury switch during the part of the cycle when the filament is heated. The output voltage across a variable load is measured by means of an oscilloscope. Oscillations appear then as a modulation of the d.c. output. The frequency of these oscillations is related to the transit time of the ions across the interelectrode space. Consequently the threshold of the oscillations depends greatly upon the frequency of collisions in the plasma which are optimum when those collisions can be neglected. The oscillation threshold depends also upon the level of electron emission from the thermionic emitter. It occurs only if this level is sufficiently high. In the experiments reported, by using a 4 mm interelectrode space the frequency of oscillation was equal to about 200 kHz and increased to about ten times this value by increasing the positive potential of the anode. Those oscillations were practically independent of the temperature of the emitter. These observations have been made at low caesium vapour pressures when collisions are negligible.

At higher pressures, the oscillations vanished but reappeared again at about 0·5 torr, around 1200K. In this range the frequency of oscillation was strongly dependent upon the temperature of the emitter and increased with increasing temperature. The highest a.c. output was obtained for Cs–W systems leading under some load conditions to an almost complete modulation of the d.c. output.

The frequency of oscillation was found to increase linearly with the interelectrode distance of the diode and was found independent of the emitter material. Rocard and Paxton[34] have found that for a given voltage, pressure, and gas density the frequency of oscillation varies as the inverse of the square root of the atomic mass of the working gas in the same manner as ionic oscillations.

Theoretically, the frequency of oscillation can be found by considering an electron-rich plasma region occupying part of the interelectrode space. In this region, caesium ions will oscillate. Assuming that the electron density has a Boltzmann distribution, one can use equation (4.16) for electrons. The ionic density was considered to oscillate around equilibrium with a periodic displacement $\tilde{x}$; thus:

$$n_i = n_{i0}\left(1-\frac{\mathrm{d}\tilde{x}}{\mathrm{d}x}\right)$$

Using Poisson's law and the equation of motion for ions, one obtains:

$$\frac{\mathrm{d}^2}{\mathrm{d}x^2}\left(\frac{\mathrm{d}^2\tilde{x}}{\mathrm{d}t^2}+\frac{e^2 n_{i0}}{\varepsilon_0 M_i}\tilde{x}\right)-\frac{e^2 n_{e0}}{\varepsilon_0 kT}\frac{\mathrm{d}^2\tilde{x}}{\mathrm{d}t^2}=0 \tag{9.39}$$

the dispersion relation of which is:

$$\omega^2 = \omega_i^2/[1+(\lambda/r_\mathrm{D})^2] \tag{9.40}$$

where ω_i is the ion plasma frequency defined in Section 1.2, and r_D is the Debye screening length for electrons given by equation (1.4). The relation between the period of oscillation τ and the distance d between the electrodes is:

$$\tau = \frac{2\pi}{\omega_i}[1+(4d/r_\mathrm{D})^2]^{1/2} \tag{9.41}$$

The importance of the r.f. energy generation cannot be overemphasised. It can be of interest in many applications in communications. Also this discovery makes the realisation of a high voltage a.c. thermionic conversion feasible. Another less important application is the use of r.f. oscillation as a diagnostic method in thermionic devices.

9.10 THERMIONIC AMPLIFIERS[1]

The principle of the thermionic amplifier is based on the electron-wave tube invented by Haeff[17]. This device is based on a new method of generation and amplification of microwave energy, in which amplification occurs as a result of space-charge interaction of electron beams of different velocities. A similar device has been developed independently by Pierce and Hebenstreit[31].

In conventional high-frequency tubes there is an upper limit to the frequency that can be obtained. Indeed at higher frequencies, the dimensions of the resonant circuit become so small that their construction becomes physically impossible. New methods of microwave power generation and amplification were thus sought. One of the alleys of research in this field led to the discovery and development of masers and lasers. The method proposed by Haeff is based upon the theory of beam–plasma interaction.

Haeff's device is shown in Figure 9.17. An electron beam is generated thermionically by two different cathodes placed at one end of a cylindrical tube of metallic material. The emitted electrons are then focused and accelerated towards

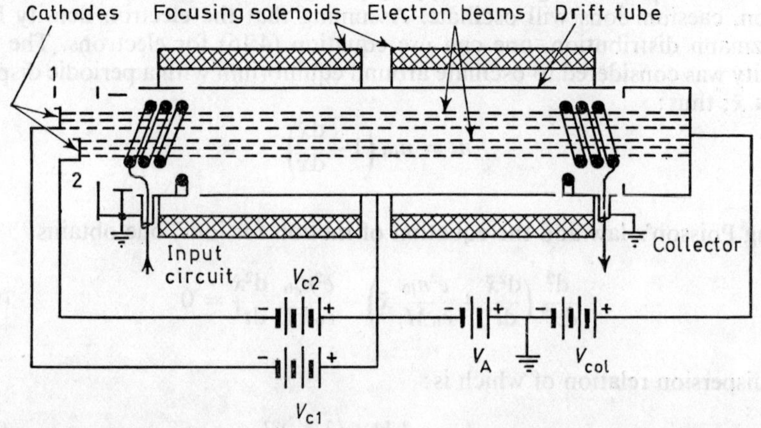

Fig. 9.17. The electron-wave tube. An electron beam is generated thermionically by two different cathodes placed at one end of a metallic tube. The beam is focused by magnetic fields all along the tube. (From Haeff[17] by courtesy of Institute of Electrical and Electronic Engineers)

the centre of the tube and are collected at the other end of the tube by a collector. All along the tube the beam is focused by a magnetic field, the lines of force of which are parallel to the axis of the tube. The input signal is applied at the cathodic end of the tube by a transmission line thus causing a disturbance in the electron beam. Following the theory of the beam–plasma interaction, this disturbance will be exponentially amplified as it moves towards the other end of the vessel. The amplified signal can then be picked up by another transmission line just before the collector end.

To obtain the optimum gain, the potential between the two emitting cathodes had to be kept at such a value as to keep the inhomogeneity factor m_i equal to $\sqrt{3}/2$. The inhomogeneity factor is defined by:

$$m_i = \tfrac{1}{2}(\Delta v/v)(\omega/\omega_{p1}) \tag{9.42}$$

where Δv is the difference in velocity between the two electron beams, v is the average velocity of electrons in the beams, ω is the frequency of oscillation and ω_{p1} is the plasma frequency of one of the beams. Haeff reported the maximum gain in dB of the electron tube as:

$$\eta^* = 1330dJ^{1/2}V^{-3/4} \tag{9.43}$$

where d is the distance between the output and the input of the signal, V is the amplitude of the disturbance and J is the current density in the tube. Gains as high as 70 dB have been obtained by Haeff.

REFERENCES AND BIBLIOGRAPHY

1. ALLEN, M. A. and KINO, G. S., *Beam Plasma Amplifiers*, Microwave Lab. Tech. Rept. No. 833, Stanford University, Stanford, Calif., July (1961).
2. ANGELLO, J. P., 'A. C. Thermionic Converter Operation', *18th Power Sources Conf.*, PSC Publication Committee, Red Bank, N.J., May (1964).
3. AUER, P. L., 'Potential Distribution in a Low-pressure Thermionic Converter', *J. appl. Phys.*, **31**, No. 12, 2096 (1960).
4. BLOSS, W. and SAGGAN, B., 'R. F. Oscillations in Noble Gas Plasma Diodes with Extraneous Ion Supply', *Advd Energy Conversion*, **2**, 639 (1962).
5. BRONNER, G., MURRAY, J. G. and SORRENTINO, J. P., *Evaluation of Arc Damage to Power Triode Grid Wires*, Princeton University, Plasma Lab., Princeton, N.J., MATT-383, Sept. (1966).
6. BUSSE, C. A., CARON, R. and CAPPELLETTI, C. M., 'Operating Experience with an Experimental Nuclear Heated Thermionic Converter', *Advd Energy Conversion*, **4**, No. 11, 121 (1964).
7. CAYLESS, M. A., 'Thermionic Generation of Electricity', *Br. J. appl. Phys.*, **12**, September (1961).
8. CHEN, F. F., 'Oscillations and Diffusion in a Shear Stabilized Q-Machine', *Inter. Conf. Phys. Quiescent Plasmas*, Frascati, Italy, Jan. (1967).
9. COPPI, B. *et al.*, 'Theory of Collisional Drift Modes', *Inter. Conf. Phys. Quiescent Plasmas*, Frascati, Italy, Jan. (1967).
10. DIRAC, P. A., 'Theory of Quantum Mechanics', *Proc. R. Soc.*, **112**, 661 (1926).
11. EDISON, T. A., 'A Phenomenon of the Edison Lamp', *Engineering*, 553, Dec. 12 (1884).
12. FERMI, E., 'Wave Mechanics of Collision', *Z. Phys.*, **40**, No. 5, 399 (1926).
13. FLEMING, J. A., 'Rectification of Electric Oscillations by Means of a Vacuum Valve', *Proc. R. Soc.*, **74**, 476 (1905).
14. FOUAD, A. A. and WALSH, E. M., 'A Cyclic Analysis of the Gabor-type Auxiliary Discharge Thermionic Converter', *Advd Energy Conversion*, **15**, No. 5, 71 (1965).
15. GABOR, D. A., 'New Thermionic Generator', *Nature*, **189**, No. 4768, 868 (1961).
16. GARVIN, H. L., TEUTSCH, W. B. and PIDD, R. W., 'Generation of Alternating Current in the Cesium Cell', *J. appl. Phys.*, **31**, 1508 (1960).
17. HAEFF, A. V., 'The Electron Wave Tube. A Novel Method of Generation and Amplification of Microwave Energy', *Proc. Inst. Radio Engrs*, **37**, 4 (1949).
18. HENQVIST, K. G., 'Analysis of the Arc Mode Operation of the Cesium Vapor Thermionic Energy Converter', *Proc. IEEE*, 748, May (1963).
19. JABLONSKI, F. E., LEFFERT, C. B., SILVER, R., HILL, R. F. and LOUGHBRIDGE, D. H., 'Space Charge Neutralization by Fission Fragments in the Direct Conversion Plasma Diode', *J. appl. Phys.*, **30**, 2017 (1959).
20. JOHNSON, E. O. and WEBSTER, W. M., 'The Plasmatron, a Continuously Controllable Gas Discharge Development Tube', *Proc. Inst. Radio Engrs*, **40**, No. 6, 645 (1952).
21. JOHNSON, F. M., 'Relaxation Oscillations in a Cesium Plasma', *MIT Phys. Electronics Conf.*, Cambridge, Mass., Rept. No. 20, March (1960).
22. KETTANI, M. A., *Direct Energy Conversion*, Chapter 6, Addison-Wesley, Reading, Mass. (1970).
23. KOHL, W. H., *Materials and Techniques for Electron Tubes*, Reinhold, New York (1960).
24. KRALL, N. A. and ROSENBLUTH, M. N., 'Low Frequency Stability of Nonuniform Plasmas' *Physics Fluids*, **6**, 254 (1963).

25. LANGMUIR, I., 'The Effect of Space Charge and Initial Velocities on the Potential Distribution and Thermionic Current Between Parallel Plane Electrodes', *Phys. Rev.*, **21**, 426 (1923).
26. LEWIS, H. W. and REITZ, J. R., 'Efficiency of the Plasma Thermocouple', *J. appl. Phys.*, **31**, No. 4, 723 (1960).
27. MARTINI, W. R., *Flame Heated Thermionic Converter Research*, Atomics International Canoga Park, Calif., AI-8681, Jan. (1964).
28. MENETREY, W. R. and SMITH, A., 'Solar Energy Thermionic Electrical Power Supplies', *J. Spacecraft Rockets*, **1**, 659 (1964).
29. MILLMAN, J. and SEELY, S., *Electronics*, McGraw-Hill, New York (1941).
30. NOTTINGHAM, W. B., 'Thermionic Emission', *Handbuch der Physik*, Vol. 21, Springer-Verlag, Berlin (1956).
31. PIERCE, J. R. and HEBENSTREIT, W. B., 'A New Type of High Frequency Amplifier', *Bell Syst. Tech. J.*, **28**, 33 (1949).
32. REDDEN, E. F. and WALLIS, A. E., *Advanced Solar Thermionic Power Systems*, Rept. No. ASD-TDR-62-877, Dec. (1962)
33. RICHARDS, H. K., 'R. F. Oscillations and Power Output in Alkaline Thermionic Converters', *DOD Symp. on Thermionic Power Conversion*, Colorado Springs, Col., May (1963).
34. ROCARD, J. M. and PAXTON, G. W., 'Relaxation Oscillations in a Plasma Diode', *J. appl. Phys.*, **32**, 1171, June (1961).
35. RUTHERFORD, P. and FRIEMAN, E. A., 'Two Particle Correlation Function for an Unstable Plasma', *Physics Fluids*, **6**, No. 8, Aug. (1963).
36. RYNN, N., 'Cesium Plasma Experiments', *3rd Symp. on Engineering Aspects of MHD*, Rochester, N.Y., March (1962).
37. SCHLICHTER, W., *Die spontane Elektronemission von glühender Metalle und das Glühelektrische Element*, Doctorate Dissertation, Göttingen University (1915).
38. SCHOCK, A., 'Effect of Magnetic Fields on Thermionic Power Generators', *J. appl. Phys.*, **31**, No. 11, 1978, Nov. (1960).
39. SCHOTTKY, W., ROTHE, H. and SIMON, H., 'Physik der Glühelektroden', *Handbuch der Experimentalphysik*', Vol. 13, No. 2, Akademische Verlagsgesellschaft, Leipzig (1928)
40. STEELE, H., 'Energy Converters Using Low Pressure Cesium', *Direct Conversion of Heat to Electricity* (edited by J. H. Welsh and J. Kaye), Wiley, New York (1960).
41. SMITH, A. H., *Thermionics: Auxiliary Power Application for Space*, ASME Paper 63-MD-54, May (1963).
42. TAYLOR, J. B. and LANGMUIR, I., 'Vapor Pressure of Cesium by the Positive Ion Method', *Phys. Rev.*, **51**, 753 (1937).
43. VAN HOOMISSEN, J. E. and CASE, J. M., 'Application of Nuclear Thermionics to Undersea Technology', *ASME Undersea Technol. Conf.*, New London, Conn., May 5 (1965).
44. VON ENGEL, A. and STEENBECK, M., *Elektrische Gasentladungen*, Edwards Brothers, Ann Arbor, Mich. (1944).
45. WEBSTER, H. F., 'Calculation of the Efficiency of a High-Vacuum Thermionic Energy Converter', *J. appl. Phys.*, **30**, 419, April (1959).
46. ZOLLWEG, R. J. and GOTTLIEB, M., 'R.F. Oscillations in Thermionic Diodes', *Proc. IEEE*, **51**, No. 5, 754, May (1963).

10 Masers and lasers

10.1 INTRODUCTION

The development of masers and later of lasers was a result of the long struggle toward the generation of waves at extremely high frequencies. Until 1938 the only available type of amplifier was based on the triode and its extensions, and by its inherent physical characteristics, it was greatly limited in frequency. The development of the resonant cavity led to the first microwave amplifier and oscillator (the klystron), and in 1946 the advent of the travelling-wave tube led to the handling of still higher frequencies. In these devices, electromagnetic waves are made to interact with an electron stream. When the electrons are retarded by the electromagnetic field of the waves, part of their kinetic energy is converted into electromagnetic energy, thus amplifying it. However big the advances made by klystron and travelling-wave tubes in the frequency range, they were still not capable of operating in the range of much shorter wavelengths like the visible or infra-red range.

Theoretically, electromagnetic waves—including visible—can interact not only with electrons but with any elementary particle by exchanging energy with it. This energy might be the kinetic energy of the particle, or even its own internal energy. In this latter case the particle does not have to be charged and might be a simple molecule. The existence of the so-called stimulated emission of radiation was first postulated by Einstein[12] on grounds of simple thermodynamic principles. The *m*icrowave *a*mplification by *s*timulated *e*mission of *r*adiation (MASER) was first proposed in 1953 by Weber[58] at the University of Maryland, Townes and Schawlow[54] at Columbia University, and Basov and Prokhorov in the Soviet Union. The microwave molecular amplifier based on this principle was built by the Columbia group[19] using a transition frequency from ammonia. Solid state masers were proposed by Bloembergen[3] using paramagnetic impurity ions having at least three energy levels*.

The new principle of amplifying electromagnetic energy by using the energy stored in molecules was a revolution by itself in the world of science and tech-

* Dilute impurities in the solid state behave essentially like gaseous atoms (with different parameters).

nology. It opened entirely new horizons in communications and energy concentration. During the last years of the 1950s the operating frequency of the masers was pushed a huge step upward to the optical region, leading to the *optical maser* or what is now more commonly known as *laser*, this name standing for *l*ight *a*mplification by *s*timulated *e*mission of *r*adiation[45]. Maiman[36] reported the successful construction of the first laser in 1960, and since then work has proceeded at a quick pace, both theoretically and experimentally, to perfect the new device.

10.2 ELECTROMAGNETIC WAVE GENERATORS[23]

To understand better masers and lasers, a brief review of microwave theory seems helpful. The 'microwave' region of the electromagnetic spectrum extends over a frequency range between 10^9 and 10^{12} Hz corresponding to wavelengths in the range between a few tens of centimetres to a few tenths of millimetre, i.e. of the same order of magnitude as the circuit components of electronic tube amplifiers. The 'optical' region (from 10^{13} to 10^7 Hz) is intermediate between two important domains: an upper domain dominated by quantum effects, and a lower domain which is the concern of conventional communication sciences.

The simplest wave amplifier is the triode. It is made of three elements, enclosed in an evacuated vessel, a cathode acting as the source of charge carriers, a grid acting as a controlling agent, and an anode acting as a collector. Under the effect of the electric field, the electrons flow from the cathode to the anode. A small voltage applied to the grid may greatly perturb their electron flow. It increases or decreases with the grid voltage. However, since the resulting variations in the anode current are much larger than the corresponding variations in the grid voltage, the amplification of waves has, thus, been obtained. When the anode circuit feeds an external load, the device is said to work as an *amplifier*. When the output is fed back (in part) to the input it is said to be an *oscillator*.

It is possible to carry more information on the wave at higher frequencies, hence their interest. However, the triode could not be made to operate at frequencies in the microwave region, because its electrons are too slow. It works correctly only in the radio-wave region where electromagnetic energy is transported by conventional two-conductor transmission lines. At shorter wavelengths ($\lambda < 10$ cm) the losses in such transmission lines become very large, making it more convenient to use hollow uniform conducting tubes or *waveguides*.

Maxwell's equations yield a detailed and rigorous description of the electromagnetic field arising from arbitrary sources of current in the presence of material bodies. If either $\boldsymbol{E}$ or $\boldsymbol{H}$, is eliminated from Maxwell's equations, a three-dimensional wave equation is obtained. For instance, if $\boldsymbol{H}$ is eliminated,

$$\nabla^2 \boldsymbol{E} - \varepsilon_0 \mu_0 \frac{\partial^2 \boldsymbol{E}}{\partial t^2} = 0. \qquad (10.1)$$

in vacuum, and since $\varepsilon_0\mu_0c^2 = 1$ where c is the velocity of light, equation (10.1) becomes:

$$\nabla^2 \boldsymbol{E} - \frac{1}{c^2}\frac{\partial^2 \boldsymbol{E}}{\partial t^2} = 0 \tag{10.2}$$

If $\boldsymbol{E}$ is eliminated one finds similarly:

$$\nabla^2 \boldsymbol{H} - \frac{1}{c^2}\frac{\partial^2 \boldsymbol{H}}{\partial t^2} = 0 \tag{10.3}$$

It can be easily seen that the general solution of equations (10.2) and (10.3) is of the form $f(x-ct)$. This function represents a disturbance that propagates in the x-direction with a speed c. For sinusoidal time-varying fields equation (10.2) becomes:

$$\nabla^2 \boldsymbol{E} - k^2 \boldsymbol{E} = 0 \tag{10.4}$$

where k is the constant of propagation of the wave, equal to $k = (\omega/c)$. It is related to the wavelength λ by the relation $\lambda = 2\pi/k = c/f$, where f is the frequency. It is always possible to separate solutions of Maxwell's equations in a source-free region into three basic types:

(1) Transverse electromagnetic waves (TEM waves) in which both the $\boldsymbol{E}$-field and the $\boldsymbol{H}$-field vectors lie in a plane perpendicular to the direction of propagation of the wave.
(2) Transverse electric waves (TE waves) where only the $\boldsymbol{H}$-field has a component in the direction of propagation.
(3) Transverse magnetic waves (TM waves) where only the $\boldsymbol{E}$-field has a component in the direction of propagation.

The two basic types of waves in a waveguide are the TE waves and the TM waves; TEM waves cannot propagate if the guide is in one piece*. It turns out that a given mode can propagate only when its frequency is higher than a certain cut-off frequency. It has then a phase velocity v_p and a guide wavelength λ_g greater than the corresponding quantities in free space.

At microwave frequencies the electromagnetic resonant cavity replaces the lumped LC resonant circuit as resonator. Theoretically, any metallic enclosure becomes an electromagnetic resonator when properly excited. Practical resonant cavities are rectangular or cylindrical. Oscillations can be excited from a coaxial line by a small probe or a loop antenna. It can be sustained within the enclosure by a small expenditure of power needed to make up for the loss to the cavity walls. The field solutions in a cavity are readily constructed from the corresponding solutions for the waveguide modes by considering that the cavity is part of a waveguide.

Two important electromagnetic wave generators in the microwave range use electromagnetic resonant cavities in their circuits: the positive-grid oscilla-

* If the guide is not in one piece, a TEM wave is possible, but it is trivial and corresponds to the conventional two-wire line.

tor, and the klystron. The positive-grid oscillator shown in Figure 10.1 consists of a triode tube with the grid at a positive d.c. potential with respect to the cathode and the plate. The tuned circuit is a resonant cavity. An electron space-charge cloud oscillates about the grid plane. This oscillating space-charge induces an alternating current in the load circuit. The voltage drop across the load impedance, in turn, produces an a.c. field in the interelectrode space of

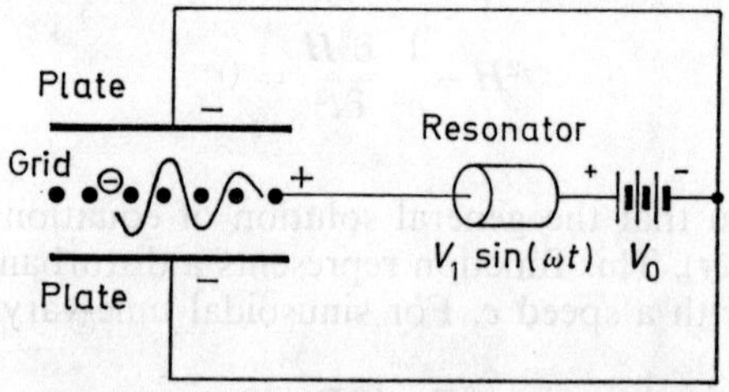

Fig. 10.1. Positive grid oscillator. The tuned circuit is the cylindrical resonator. Both electrodes of the tube are at a negative potential, whereas the grid is at a positive potential

the tube. Most of the electrons are retarded by this alternating field and thus transfer part of their energy to the resonant cavity during each cycle of oscillation.

Klystron oscillators are of two types: double-resonator klystrons, and reflex klystrons. A double-resonator klystron is shown in Figure 10.2. It consists of a cathode, two control grids, two metallic resonators, a 'buncher' near the cathode, a catcher near the anode, and a collecting anode. The electrical connections consist of small coupling loops inside the resonators. The overall assembly is then placed in vacuum. Electrons produced by the cathode are accelerated by a potential V_0. They reach the buncher-grid region with a high initial velocity v_0. The buncher resonator, excited at its resonant frequency by the incoming electromagnetic wave to be amplified, modulates the electrons in velocity. As electrons cross the buncher field, they are either accelerated or

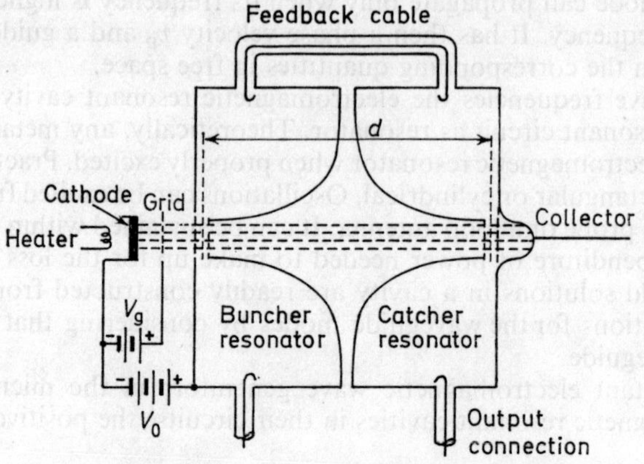

Fig. 10.2. The klystron. Electrons accelerated by V_0 enter the buncher region and are bunched on their way to the catcher region

decelerated depending on the time of their arrival. The accelerated electrons then overtake the decelerated ones in the region between buncher and catcher grids. Thus, electrons are *bunched* as they move toward the catcher, and energy is transferred from the electrons to the catcher field.

The electron transit-time relationships in the klystron can be obtained easily. Indeed, for initial conditions, the kinetic energy of the electrons equals the potential energy of the accelerating field; thus:

$$v_0 = (2eV_0/M)^{1/2} \tag{10.5}$$

The buncher grids are at a potential V_0 and a potential $[V_0+V_1 \sin(\omega t)]$ with respect to the cathode. Consequently, the velocity v_1 of an electron passing through the buncher at time t, will be:

$$v_1 = \left\{\frac{2eV_0}{M}\left[1+\frac{V_1}{V_0}\sin(\omega t_1)\right]\right\}^{1/2} \tag{10.6}$$

after equating the kinetic energy of the electrons to the potential energy of the field. Taking into account equation (10.5), equation (10.6) becomes:

$$v_1/v_0 = \left[1+\frac{V_1}{V_0}\sin(\omega t_1)\right]^{1/2} \tag{10.7}$$

The time τ required for the electron to travel the distance between the buncher and the catcher is:

$$\tau = \frac{d}{v_0}\left[1+\frac{V_1}{V_0}\sin(\omega t_1)\right]^{-1/2} \tag{10.8}$$

The electrons reach the catcher grids at time $t_2 = t_1+\tau$, and for small values of ωt_1 equation (10.8) yields:

$$\omega t_2 = \omega t_1+\alpha\left[1-\frac{1}{2}\frac{V_1}{V_0}\sin(\omega t_1)\right] \tag{10.9}$$

where $\alpha = (\omega d/v_0)$ is the 'transit angle'. Assuming that the buncher is in phase with the catcher and that the catcher voltage is $V_2 \sin(\omega t)$, the energy yielded by the electron to the catcher field will be:

$$\mathcal{E} = -eV_2 \sin(\omega t_2) \tag{10.10}$$

Taking into account equation (10.10), the average energy transferred will be:

$$\mathcal{E}_{av} = -eV_2 J_1(m_b) \sin\alpha \tag{10.11}$$

where $m_b = (\alpha V_1/2V_0)$ is a bunching parameter and $J_1(m_b)$ is the Bessel function of the first kind and the first order.

In the two tubes seen above the electron beam interacts with the modulating field over a short interval in space. In the *travelling-wave* tube the electron beam interacts with the microwave field along most of its trajectory. A helix-type travelling wave tube is shown in Figure 10.3. In this device, the electron beam

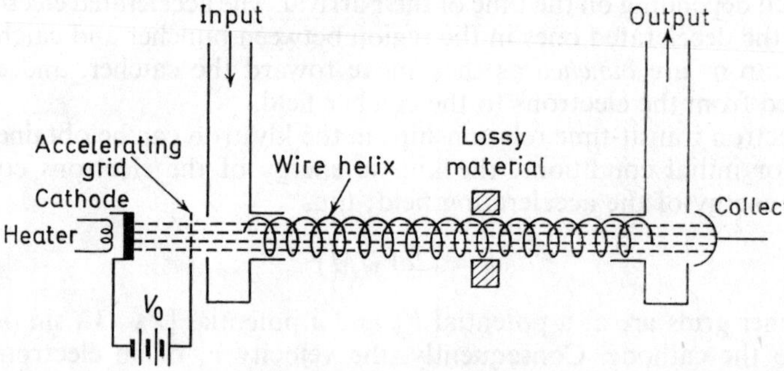

Fig. 10.3. The helix type travelling-wave tube. The wire helix is subjected to the electromagnetic field of the input wave. It bunches the electron beam when it travels through it. The lossy material suppresses the reflected waves

moves in a slow-waveguiding structure and interacts with the electromagnetic field of the wave. As the beam travels, it is bunched into electron clouds and part of its kinetic energy is transferred to the field. The travelling wave is thus amplified in the tube. If part of the output is fed back to the input the travelling-wave tube may work as an oscillator as well.

10.3 EMISSION OF RADIATION

Radiation is characterised by its wavelength λ or its frequency ν. The first treatment of stimulated emission is based on thermodynamic concepts and was given by Einstein. He considered a certain number of molecules in thermal equilibrium in a cavity at an absolute temperature T. Because of the thermal equilibrium, the radiative transitions between the energy levels of the molecules must be of the black-body type. In this case, the energy density is a function of the absolute temperature of the system only. The distribution of this energy over the frequency range is given further by Planck's law:

$$\mathcal{E}_\nu \, \mathrm{d}\nu = \frac{8\pi h\nu^3}{c^3} \frac{\mathrm{d}\nu}{\exp(h\nu/kT) - 1} \tag{10.12}$$

where $\mathcal{E}_\nu$ is the energy density of radiation per unit frequency range near the frequency ν, h is Planck's constant, k is Boltzmann's constant, and c is the velocity of light in vacuum. $\mathcal{E}_\nu$ is related to the total energy density by:

$$\mathcal{E} = \int_0^\infty \mathcal{E}_\nu \, \mathrm{d}\nu$$

Since the molecules are in equilibrium with the radiation in the cavity, Einstein assumed that both the radiation absorbed and emitted by these molecules in their transitions between energy states obey Planck's law. Consider transitions between an upper state m having an energy $\mathcal{E}_\mathrm{m}$ and a lower state n having an energy $\mathcal{E}_\mathrm{n}$. The total probability for the transition (m → n) should equal the total probability for the transitions (n $\rightarrow$ m). That is, the probability of emission of the relevant radiation should, at equilibrium, be equal to the probability of absorption of the same radiation. This radiation consists of quanta:

$$h\nu_{mn} = \mathcal{E}_m - \mathcal{E}_n \tag{10.13}$$

The total probabilities of transition are:

$$\begin{aligned} p_{bm} &= \mathcal{C} \exp(-\mathcal{E}_m/kT) \\ p_{bn} &= \mathcal{C} \exp(-\mathcal{E}_n/kT) \end{aligned} \tag{10.14}$$

where $\mathcal{C}$ is a function of the temperature. Considering first the downward transition (m $\rightarrow$ n), its probability of occurrence in the interval of time dt is $\mathrm{d}p_{bmn} = \mathcal{A}_{mn}\,\mathrm{d}t$, where $\mathcal{A}_{mn}$ is the coefficient of spontaneous emission, similar to a radioactive decay constant. Considering, next, molecules in the lower state in the presence of a radiation energy density $\mathcal{E}_\nu$ per unit frequency, Einstein assumed that the probabilities of transition depend only on $\mathcal{E}_\nu$ in either direction; thus:

$$\begin{aligned} \mathrm{d}p_{bmn} &= \mathcal{E}_\nu \mathcal{B}_{mn}\,\mathrm{d}t \\ \mathrm{d}p_{bnm} &= \mathcal{E}_\nu \mathcal{B}_{nm}\,\mathrm{d}t \end{aligned} \tag{10.15}$$

where $\mathcal{B}_{mn}$ and $\mathcal{B}_{nm}$ are the coefficients of induced emission and of absorption respectively. Because of the thermal equilibrium, the probability of upward transitions should equal the probability of downward transitions; thus from equations (10.14) and (10.15):

$$\mathcal{E}_\nu \mathcal{B}_{nm} \exp(-\mathcal{E}_n/kT) = (\mathcal{E}_\nu \mathcal{B}_{mn} + \mathcal{A}_{mn}) \exp(-\mathcal{E}_m/kT) \tag{10.16}$$

As the temperature increases toward infinity the exponential terms of equation (10.16) tend toward each other and the spontaneous emission coefficient becomes negligible compared to the term $\mathcal{E}_\nu \mathcal{B}_{mn}$. Consequently, at the limit, one must have:

$$\mathcal{B}_{nm} = \mathcal{B}_{mn} \tag{10.17}$$

Using this in equation (10.16) one finds readily:

$$\mathcal{E}_\nu = \frac{\mathcal{A}_{mn}/\mathcal{B}_{mn}}{\exp[(\mathcal{E}_m - \mathcal{E}_n)/kT] - 1}$$

Comparison with equation (10.12) shows that:

$$\mathcal{A}_{mn}/\mathcal{B}_{mn} = 8\pi h\nu^3/c^3 \tag{10.18}$$

and also that $(\mathcal{E}_m - \mathcal{E}_n) = h\nu_{mn}$ [equation (10.14)], known as the Bohr frequency condition. Equation (10.18) shows that the coefficient of spontaneous emission $\mathcal{A}_{mn}$ is proportional to the coefficient of induced emission $\mathcal{B}_{mn}$, with the constant of proportionality equal to the product of the number of radiation modes per unit volume per unit frequency ($8\pi\nu^2/c^3$), and the quantum energy $h\nu$. $\mathcal{B}_{mn}$ is equal to $\mathcal{B}_{nm}$ (the absorption coefficient which is readily measurable). $\mathcal{A}_{mn}$ is obtained through equation (10.18) from $\mathcal{B}_{mn}$ for a given frequency ν.

Equation (10.12) can also be expressed in terms of the wavelength λ, hence:

$$P_\lambda \,d\lambda = 2\pi hc^2 \frac{\lambda^{-5}\,d\lambda}{\exp(hc/\lambda kT)-1} \tag{10.19}$$

where P_λ is expressed in watts per unit volume. Integration of equation (10.19) over all the wavelengths results in the well-known Stefan–Boltzmann law $P = aT^4$, where $a = 5{\cdot}679\times10^{-12}$ W/cm² K⁴.

Differentiation of the Planck equation shows that for a given temperature, there is a wavelength λ_m corresponding to a maximum emitted radiation[11], and one finds:

$$\lambda_m T = 2891 \ \mu\text{m K} \tag{10.20}$$

Equation (10.20) is known as the Wien displacement law. The peak energy $\mathscr{E}_m$ is:

$$\mathscr{E}_m(T) = 1{\cdot}290\times10^{-15}\,T^5 \ \text{W/cm}^2\,\mu\text{m}$$

The energy distribution of black bodies at different temperatures is shown in Figure 10.4, as deduced from equation (10.19).

Planck's equation is approximated for small values of λT by neglecting the unity in the denominator (Wien's approximation). For large values of λT, the exponential can be considered as infinitesimal as compared to unity (Rayleigh–

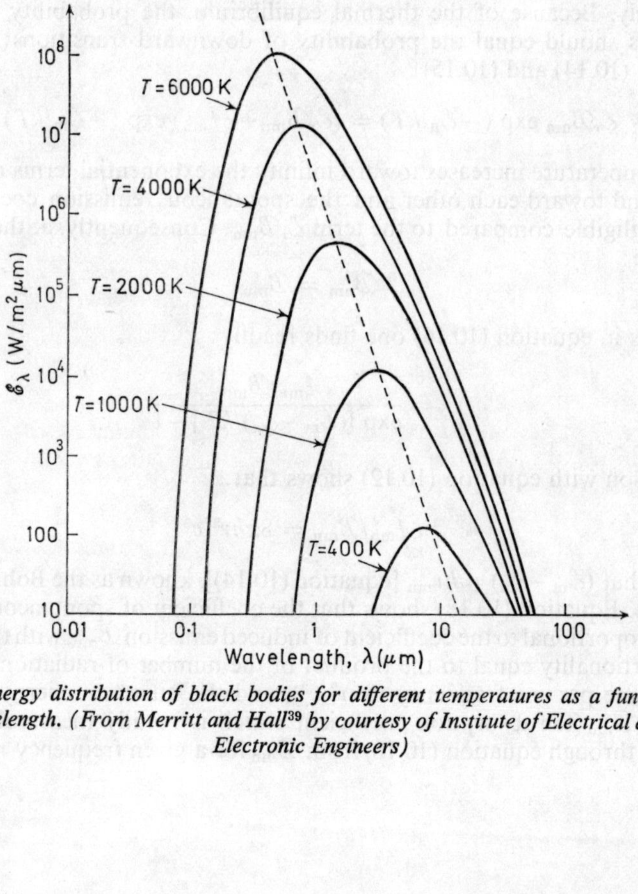

Fig. 10.4. Energy distribution of black bodies for different temperatures as a function of the wavelength. (From Merritt and Hall[39] by courtesy of Institute of Electrical and Electronic Engineers)

Jeans approximation). The black-body radiation is illustrated in Figure 10.5 where the energy has been normalised. From this, it can be noted that one fourth of the total radiated power is always on the short wave side of the maximum in the radiation curve.

The limitations of conventional light sources are numerous: powerful monochromatic sources cannot be obtained since the radiated energy is distributed over a broad spectral region. The collimation of the radiated energy is inversely proportional to its intensity, and no intense collimated beam can be obtained.

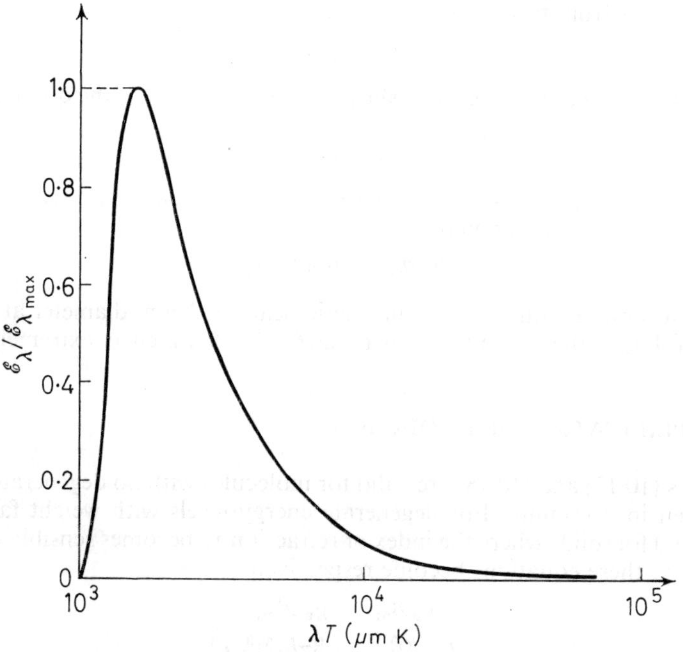

Fig. 10.5. Normalised black body radiation curve. (From Merritt and Hall[39] *by courtesy of Institute of Electrical and Electronic Engineers)*

The radiation emitted by gaseous sources consists of spectral lines determined by their frequencies, their intensities and their widths. At low pressure, the intensity is low and the lines are sharp. As the pressure increases, the brightness of the lines increases, and they become broader and broader until eventually they overlap and lose their discrete character.

However, since light is an electromagnetic radiation, it is also characterised by polarisation and phase. When the light produced by a point source is split into two or more beams, each beam supposedly travelling toward the same destination through a different path (interference), the amplitudes of the beams behave as phasors (they add as vectors having directions depending on the paths travelled). Light radiated from different points of one single source is said to be *incoherent in phase*. If light is made to emanate from one single point source through two different paths, no interference will occur if the difference in paths exceeds a certain critical value. This is because of the fact that the phase of the radiation is lost during the time interval corresponding to the difference in paths. This light is said to be *incoherent in time*. The phase of the light produced by a conventional source undergoes fluctuations proportional in intensity to $\Delta\nu$ of the spectrum. The coherence time is the defined as $\Delta t_c = 1/\Delta\nu$. The light becomes monochromatic and thus coherent only for time intervals much shorter than Δt_c.

A coherent beam of limited intensity can be obtained from a non-coherent source. Let Φ_s be the flux density of the source, and S its circular surface such that $S = \pi r_0^2$, r_0 being its radius. According to Lambert's law, the flux density at a distance r from the radiating surface is:

$$\Phi_p = \Phi_s(r_0/r)^2 \tag{10.21}$$

The Van Cittert–Zernike theorem[5] shows that the radius r_b of the quasi-coherent beam is:

$$r_b = 0{\cdot}08\lambda/\alpha \tag{10.22}$$

where α is the angular radius such that $\alpha = (r_0/r)$. Introducing equation (10.22) into equation (10.21), one finds:

$$(\Phi_p/\Phi_s) = (0{\cdot}08\lambda/r_b)^2$$

In order to obtain a quasi-monochromatic beam of 1 mm diameter at a wavelength of 1 μm, the flux ratio will be $2{\cdot}56\times10^{-8}$, which is extremely small.

10.4 AMPLIFICATION OF RADIATION

Equations (10.17) and (10.18) are valid for molecules with no degenerate energy levels even in a vacuum. For degenerate energy levels with weight factors g_n and g_m, and for solids where the index of refraction ν_R becomes sensibly different from unity, these equations become respectively[31]:

$$\begin{aligned} g_n\mathcal{B}_{nm} &= g_m\mathcal{B}_{mn} \\ \mathcal{A}_{nm}/\mathcal{B}_{nm} &= 8\pi h\nu^3\nu_R^3/c^3 \end{aligned} \tag{10.23}$$

The energy $\mathcal{E}_{abs}$ absorbed by N_n molecules in their transition from the ground state n to state m is:

$$\mathcal{E}_{abs} = N_n h\nu_{mn}\, dp_{bnm} \tag{10.24}$$

and taking into account equations (10.15), equation (10.24) becomes:

$$\mathcal{E}_{abs} = N_m\mathcal{B}_{nm}\mathcal{E}_\nu h\nu_{mn}\, dt \tag{10.25}$$

The energy $\mathcal{E}_{em}$ emitted by N_m molecules in their downward transition from m to n is equal to the sum of the spontaneous $\mathcal{E}_{sp}$ and the stimulated $\mathcal{E}_{ind}$ energies; this total is:

$$\mathcal{E}_{em} = N_m\mathcal{A}_{mn}h\nu_{mn}\, dt + N_m\mathcal{B}_{mn}\mathcal{E}_\nu h\nu_{mn}\, dt \tag{10.26}$$

This relation shows that the spontaneous emitted energy is independent of the incident energy $\mathcal{E}_\nu$ and that the induced emitted energy is coherent with it in phase.

For amplification, the condition $\mathcal{E}_{ind} > \mathcal{E}_{abs}$ should be obtained, and since $\mathcal{E}_{sp}$ is independent of the energy $\mathcal{E}_\nu$ to be amplified, it acts merely as a source of noise in masers, and precautions should be taken to keep it as low as possible. Taking into account equation (10.17), the condition of amplification becomes:

$$N_m > N_n \qquad (10.27)$$

for non-degenerate states, and:

$$N_m/g_m > N_n/g_n \qquad (10.28)$$

when degenerate states exist.

The condition $N_m > N_n$ means that an excess of molecules should somehow be kept in the upper energy state m during the time dt. However, in a thermodynamic equilibrium, the relationship between the populations of the energy levels m and n is given by Boltzmann's law, yielding in this case:

$$N_m/g_m = (N_n/g_n) \exp [-(\mathcal{E}_n - \mathcal{E}_m)/kT] \qquad (10.29)$$

Comparison of equations (10.29) and (10.28) shows that for the amplification to occur, the system should definitely not be in thermodynamic equilibrium, but should contain a *population inversion* corresponding to a state of so-called *negative temperature* (if we are to make Boltzmann's statistics correct). Taking account of equation (10.27) and neglecting the spontaneous emission, the instantaneous power emitted by the molecules becomes then:

$$\mathrm{d}\mathcal{E}/\mathrm{d}t = (N_m - N_n)\mathcal{B}_{mn}\,\mathcal{E}_\nu\, h\nu_{mn} \qquad (10.30)$$

This equation assumes that the spectral lines have an infinite sharpness. Actually, this is not the case, for several reasons, and the frequency response curve always has some broadening. Thus, the right-hand side of equation (10.30) should be multiplied by a factor m_s which is a function of ν and ν_{mn} and which takes into consideration the broadening mechanism.

Experimentally, at a distance x from the surface of an absorbing medium, incident light of constant intensity Φ_0 is absorbed according to the exponential law:

$$\Phi_\nu = \Phi_0 \exp(-k_\nu\, x)$$

where k_ν is the absorption 'constant' which is actually a function of the light frequency ν. k_ν can be obtained experimentally; it is displayed in Figure 10.6 as a function of ν. This function is characterised by its peak value k_m and the width $\Delta\nu$ of the absorption line at $(k_m/2)$.

One of the reasons for the broadening of the spectral lines is related directly to Heisenberg's Uncertainty Principle, which is stated in Section 2.4. The average time spent by a molecule in a certain state is inversely proportional to the spontaneous emission coefficient $\mathcal{A}_{mn}$; hence since the linewidth is:

$$\Delta\nu = \Delta\mathcal{E}/h$$

one finds from the Uncertainty Principle that $\Delta\nu = \mathcal{A}_{mn}/2\pi$. Using the perturbation theory, the function m_s due to this principle is found[20] to be:

$$m_{s1} = \frac{\mathcal{A}_{mn}}{2\pi^2}\,\frac{1}{(\nu-\nu_{mn})^2+(\mathcal{A}_{mn}/2\pi)^2} \qquad (10.31)$$

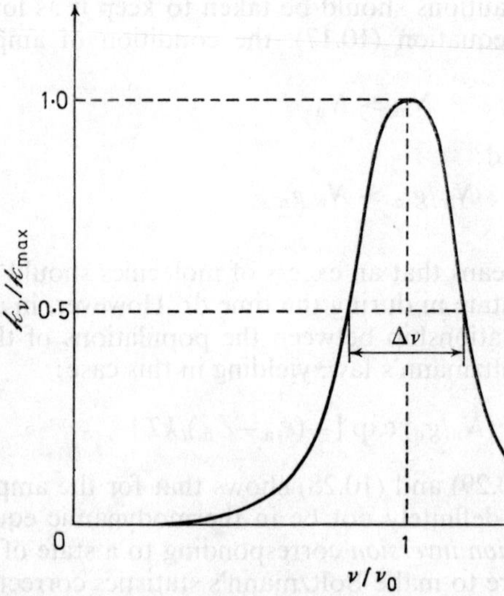

Fig. 10.6. Variation of the absorption coefficient with frequency in an absorption line and definition of the linewidth

Other factors affecting the broadening process are the Doppler effect and collisions. The Doppler effect becomes appreciable whenever the velocity v of the molecules along the axis of observation is not negligible compared to the velocity of light c. Indeed, for $v \ll c$, the observed frequency of radiation is $\nu = \nu_{mn}\,[1 \pm (v/c)]$. The sign of the ratio (v/c) depends on whether the molecule moves toward or away from the observer. The shape factor due to Doppler effect is found to be:

$$m_{s2} = \mathcal{A} \exp\left[-\frac{1}{2}\,\frac{Mv^2}{kT}\,\frac{(\nu-\nu_{mn})^2}{\nu_{mn}^2}\right] \tag{10.32}$$

where $\mathcal{A}$ is a constant. Collisions between molecules and with the walls of the container revert some excited molecules from state m to state n. If ν_c is the collision frequency, the shape factor due to this effect becomes:

$$m_{s2} = \frac{\nu_c}{2\pi^2}\,\frac{1}{(\nu_{mn}-\nu)^2+(\nu_c/2\pi)^2} \tag{10.33}$$

The fact that the system contains a population inversion causes it to 'relax' to its equilibrium position. These relaxation processes are of two types: 'spin–spin' processes and 'spin–lattice' processes. When these processes have the same relaxation time τ, the shape factor is then:

$$m_{s4} = \frac{1}{2\pi^2\tau}\,\frac{1}{(\nu_{mn}-\nu)^2+(2\pi\tau)^{-2}} \tag{10.34}$$

Equation (10.30) is valid only for $\mathcal{E}_\nu$ up to a certain upper limit. After this limit, $d\mathcal{E}/dt$ is no longer linearly proportional to $\mathcal{E}_\nu$. This phenomenon is known as *saturation* and causes a broadening of the spectral lines. Gordy *et al.*[20] reported for the case of molecular collisions along with the saturation effect:

$$m_s = \frac{\nu_c}{2\pi^2} \frac{1}{(\nu_{mn}-\nu)^2+(\nu_c/2\pi)^2+(\mathcal{B}_{mn}\mathcal{E}_\nu/\pi^2)} \tag{10.35}$$

10.5 PARAMAGNETISM AND BLOCH EQUATIONS[21]

Electric polarisation was discussed in Chapter 1 and found to be the result of a negative charge distribution centre being separated from a positive charge distribution centre by a finite distance. The two charge centres constitute an electric dipole and the polarisation vector can be defined as the electric dipole moment vector per unit volume. However, the electron is in motion and moves

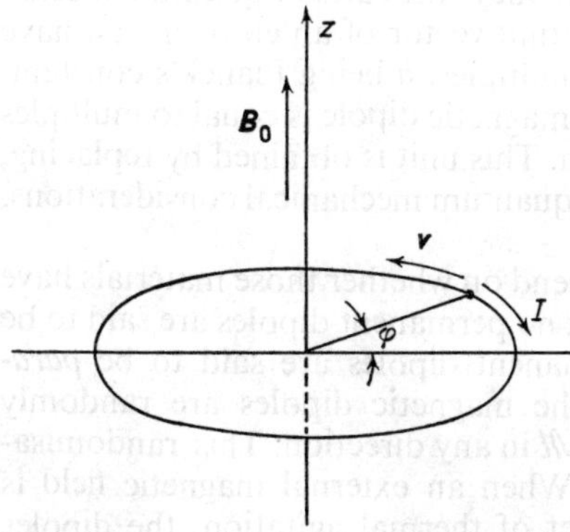

Fig. 10.7. Model of an orbiting electron. The velocity v of the electron is represented by a current I = − ev

around the nucleus of its atom. This electronic motion can be described by defining an angular momentum with each electron.

Consider an electron of rest mass M_0 orbiting around a nucleus at a distance r and with an angular frequency ω (Figure 10.7). The angular momentum of the electron around the axis of rotation z is:

$$\boldsymbol{\mathcal{P}} = \boldsymbol{i}_z M_0 \omega r^2 \tag{10.36}$$

where $\boldsymbol{i}_z$ is the unit vector in the z-direction. The motion of this charge carrier of charge density e gives rise to a current I_ϕ in the ϕ-direction such that:

$$\boldsymbol{I}_\phi = -\boldsymbol{i}_\phi e\omega/2\pi \tag{10.37}$$

where $\boldsymbol{i}_\phi$ is the unit vector in the ϕ-direction. The magnetic moment $\boldsymbol{m}_z$ associated with the electron is then:

$$\boldsymbol{m}_z = -\boldsymbol{i}_z \pi r^2 I_\phi \tag{10.38}$$

Replacing I_ϕ by its value from equation (10.37) and taking into account equation (10.36), one finds for the magnetic moment:

$$\boldsymbol{m}_z = -e\boldsymbol{\mathcal{P}}_z/2M_0 \tag{10.39}$$

Since the angular momentum of an electron is different from zero, its magnetic moment is also different from zero. It can be shown that as far as the magnetic effects are concerned a planar current loop is equivalent to a magnetic dipole. By analogy with the electric dipole seen in Chapter 1, the magnetic dipole consists of two magnetic poles on the axis of rotation of the electrons separated by a distance l, and a pole strength equal to $I\pi r^2/l$ for one pole and to the reverse of this value for the other one. Thus the electron is said to possess a *permanent magnetic dipole*. It is obvious that magnetic moments can also be applied externally by the use of external magnetic fields as is the case in the betatron. This external field induces an orbital motion known as the *Larmor precession* with a Larmor frequency:

$$\omega_L = eB/2M_0 \tag{10.40}$$

where B is the externally applied magnetic field.

The behaviour of each individual electron obeys the rules of quantum mechanics. From these rules, the angular-momentum vector of an electron can have only the values $+h/2\pi$ or $-h/2\pi$ or their multiples, h being Planck's constant. Consequently, the strength of a permanent magnetic dipole is equal to multiples of an atomic unit called the Bohr magneton. This unit is obtained by replacing, in equation (10.38), I_ϕ by its value given by quantum mechanical considerations, $m_B = eh/4\pi M_0 = 0{\cdot}927\times10^{-23}$ A m².

The magnetic properties of materials depend on whether those materials have dermanent dipoles or not. Materials having no permanent dipoles are said to be *diamagnetic*, whereas materials with permanent dipoles are said to be *paramagnetic*. However, in these materials, the magnetic dipoles are randomly oriented and there is no net magnetisation $\mathcal{M}$ in any direction. This randomisation is mainly due to thermal agitation. When an external magnetic field is applied, strong enough to cancel the effect of thermal agitation, the dipoles align themselves in the direction of the field and a net magnetisation is thus obtained. The total magnetic field is:

$$\boldsymbol{B} = \mu_0(H+\mathcal{M}) = \mu_0\mu_r\boldsymbol{H}$$

where μ_r is the relative permeability of the material. One can also write:

$$\mathcal{M} = (\mu_r-1)\boldsymbol{H} = \chi\boldsymbol{H} \tag{10.41}$$

where χ is the magnetic susceptibility of the material. It should be noted that because the external field is in competition with the thermal agitation, the susceptibility will increase with decreasing temperature at a given value of $\boldsymbol{H}$. Also, when spontaneous magnetisation occurs in a material, the material is said to be *ferromagnetic*.

Consider now, an elementary magnetic dipole characterised by a dipole moment $\boldsymbol{m}$ in a constant magnetic field B_0. Under steady state conditions, both the magnetic field and the dipole moment are in the z-direction. Under the effect of en external force the magnetic dipole can be tilted by an angle θ from the z-axis. This introduces a torque, the value of which is:

$$\mathcal{C} = \boldsymbol{m}\times\boldsymbol{B} \tag{10.42}$$

$\mathcal{C}$ is obviously related to the angular momentum $\mathcal{P}$ by $\mathcal{C} = d\mathcal{P}/dt$. The relationship between $\mathcal{P}$ and $\boldsymbol{m}$ is given by equation (10.39), which shows that the dipole moment vector is proportional to the angular momentum $\mathcal{P}$, the constant of proportionality being defined as the gyromagnetic ratio γ_g. Under these circumstances, equation (10.42) becomes:

$$d\boldsymbol{m}/dt = -\gamma_g \boldsymbol{m} \times \boldsymbol{B} \qquad (10.43)$$

If the spectroscopic splitting is taken into account the gyromagnetic ratio takes the value $\gamma_g = em_g/2M_0$, where m_g is called the Landé splitting factor. From the above equations it is seen that $\mathcal{P}$ does not change in magnitude but rotates along the cone of angle θ around the z-axis. It can be easily shown that the angular frequency of precession is equal to the Larmor frequency given by equation (10.40), or if the spectroscopic splitting is considered, $\omega_0 = \gamma_g B_0$.

Consider a volume $\mathcal{V}$ enclosing n dipoles; the resulting magnetisation of these dipoles will be:

$$\mathcal{M} = \boldsymbol{m}/\mathcal{V} \qquad (10.44)$$

Introducing equation (10.44) into equation (10.43), one obtains:

$$d\mathcal{M}/dt = -\gamma_g \mathcal{M} \times \boldsymbol{B} \qquad (10.45)$$

Equation (10.41) gives the relationship between the magnetic field intensity $\boldsymbol{H}$ and the magnetisation $\mathcal{M}$. When $\boldsymbol{H}$ and $\mathcal{M}$ are not in the same direction, their ratio, the magnetic susceptibility, becomes a tensor. The solution of equations (10.45) and (10.41) shows that this tensor is:

$$\chi = \begin{vmatrix} \dfrac{\omega_0 \gamma_g \mu_0 n m_z}{\omega_0^2 - \omega^2} & \dfrac{j\omega \gamma_g \mu_0 n m_z}{\omega_0^2 - \omega^2} \\ \dfrac{-j\omega \gamma_g \mu_0 n m_z}{\omega_0^2 - \omega^2} & \dfrac{\omega_0 \gamma_g \mu_0 n m_z}{\omega_0^2 - \omega^2} \end{vmatrix}$$

The susceptibility tensor is antisymmetrical and has a *resonance*[35, 43] value for $\omega = \omega_0$.

The picture given by equation (10.45) is not, however, realistic, since all the losses were neglected in it. Indeed, the magnetic dipole transfers part of its energy, say, to the host lattice through the phenomena described in Section 10.4. Bloch in a similar study, took account of the spin–spin and the spin–lattice processes*. The resulting Bloch equations are:

$$\left.\begin{aligned} d\mathcal{M}_x/dt &= -\gamma_g(\mathcal{M}\times\boldsymbol{B})_x - (\mathcal{M}_x/\tau_1) \\ d\mathcal{M}_y/dt &= -\gamma_g(\mathcal{M}\times\boldsymbol{B})_y - (\mathcal{M}_y/\tau_1) \\ d\mathcal{M}_z/dt &= -\gamma_g(\mathcal{M}\times\boldsymbol{B})_z - [(\mathcal{M}_z - \mathcal{M}_0)/\tau_2] \end{aligned}\right\} \qquad (10.46)$$

where τ_1 is the spin–spin relaxation time, and τ_2 is the spin–lattice relaxation time. Solutions of equations (10.46) show that the susceptibility has a real component χ_r and an imaginary component χ_i. The real component is the

* Bloch, F., *Phys. Rev.*, **70**, 460 (1946).

result of the reactive power produced by the paramagnet and is called the *dispersive* susceptibility, whereas the imaginary component is due to the average power absorbed by the paramagnet and is called the *absorptive* susceptibility. It is found that when the losses are taken into account, the susceptibility is no longer infinite at resonance and the resonance absorption curve has a finite linewidth resulting in τ_2. Equations (10.46) also show that the average power absorbed by the paramagnet tends toward a saturation value for high magnetic fields resulting in τ_1. A better understanding of τ_1 and τ_2 can only be obtained by a rigorous treatment in terms of quantum mechanics; it is given by Powell and Crasemann[44].

10.6 MASERS[7]

Gaseous Masers

There are two important types of masers: solid-state masers and gaseous masers. Gaseous masers were the first to be developed by the Columbia University group which used the simple ammonia molecule as the working gas[54]. This molecule $^{14}N^{1}H_3$ consists of one nitrogen and three hydrogen atoms arranged

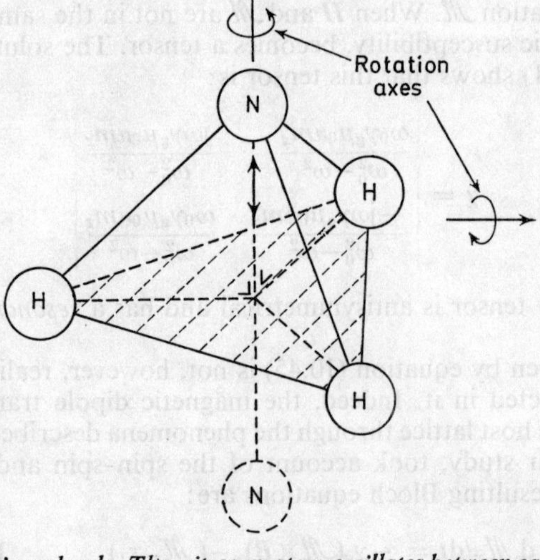

Fig. 10.8. The ammonia molecule. The nitrogen atom oscillates between each side of the hydrogen plane

in a pyramid as shown in Figure 10.8. The three hydrogen atoms may be assumed to form a plane and thus the nitrogen atom will be above or below this plane. The potential energy function and several energy levels of the ammonia molecule are shown in Figure 10.9. as a function of the distance *d* between the hydrogen plane and the nitrogen atom. This potential curve is characterised by one minimum on each side of the hydrogen plane separated by a potential barrier on the surface of the plane. The nitrogen atom vibrates perpendicularly to the hydrogen plane and may cross this plane by a tunnel effect. The probability of finding the nitrogen atom on either side of the plane is the same. For a given energy state, the total Schrödinger wave function should be the linear sum of the wave functions corresponding to it, being on each side. When the spin is taken into account, this leads to a symmetrical and an antisymmetrical solution, with respect to the hydrogen plane. The antisymmetrical solution has the higher energy, and above the potential barrier there will be single levels

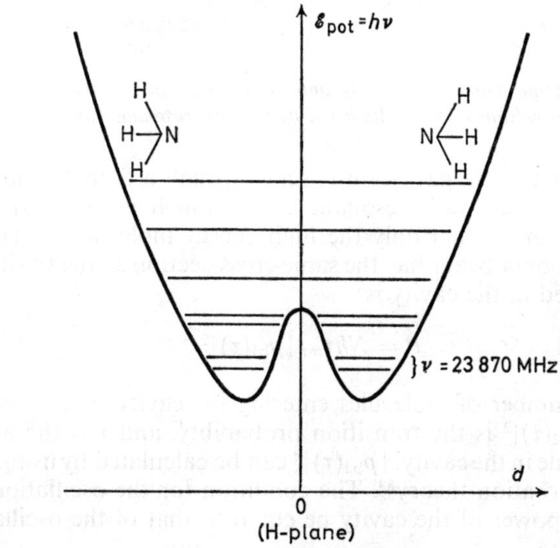

Fig. 10.9. Potential energy as a function of the distance between the H-plane and the nitrogen atom in the ammonia molecule. Some energy levels are shown, with the lowest corresponding to the (3–3) inversion transition

in a single potential well, whereas below the barrier one obtains double levels in a double potential well. An *inversion transition*, that is a transition leading to a population inversion, may occur between two such states. The inversion levels, that is the upper ones, are very fine, since the ammonia molecules rotate around their axis of symmetry and around an axis perpendicular to it. The most probable inversion transition is the one associated with those energy levels in which the two rotational quantum numbers are equal to each other. For microwave technological reasons, the (3 → 3) inversion transition was used in the first experiments. The transition (3 → 3) means that the rotational quantum number was 3 for each axis of rotation. This transition corresponds to a strong spectral line at 23 870 MHz frequency and to a linewidth of about 1 kHz resulting from the natural broadening.

The problem still remains in finding a method to bring about a population inversion. It was known, however, that the reaction of ammonia molecules in uniform electric fields (Stark effect) depends on their energy level. The maser shown in Figure 10.10 takes advantage of this effect to extract the low-energy

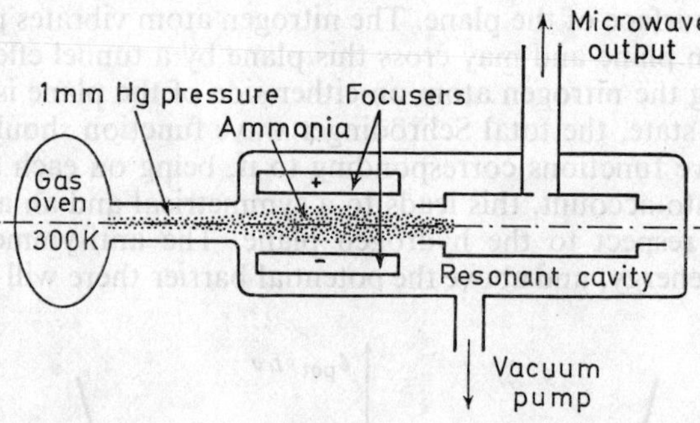

Fig. 10.10. The ammonia beam maser[47]. An ammonia gas beam is sent through a focuser and therefrom into a cylindrical microwave resonant cavity

molecules[47]. An ammonia gas beam is sent through a focuser and from there into a cylindrical microwave resonant cavity which is made to resonate at 23 870 MHz. Assuming that only the high energy molecules enter the cavity, and that the ammonia beam has the same cross-section as the cavity, the average power released in the cavity is:

$$P = Nh\nu_{mn}\,|p_{b1}(\tau)|^2$$

where N is the number of molecules entering the cavity, ν_{mn} is the frequency of resonance, $|p_{b1}(\tau)|^2$ is the transition probability, and τ is the average time spent by a molecule in the cavity. $|p_{b1}(\tau)|^2$ can be calculated by using a quantum mechanical perturbation theory[59]. The condition for the oscillation to start is that the average power of the cavity be equal to that of the oscillation of the gas; this leads to:

$$2\pi\nu_{mn}\mathscr{E}_{st}/Q_L = Nh\nu_{mn}\,|p_{b1}(\tau)|^2$$

where $\mathscr{E}_{st}$ is the energy stored in the cavity and Q_L is the Q-factor of the cavity when it is 'loaded' by the beam. If N_{th} is the number of molecules when the saturation is negligible and N_{sat} their number when the saturation is dominant, it is found that the condition for a sustained oscillation is:

$$\mathscr{E}_{st} = \frac{N_{sat}}{N_{th}}\,hQ_L\,|p_{b1}(\tau)|^2 N_{th}$$

In the first experiments[19] the cylindrical cavity used (made of copper- or silver-plated invar with an unloaded Q of 12 000), was in the TE_{011} mode. The focuser was made to work at 30 kV and its electrodes were filled with liquid nitrogen to help remove the lower state ammonia molecules. The inner faces of the electrodes were hyperboloids, to facilitate the calculation of the non-uniform field. Later, a figure of merit equal to Ql/S was defined, where S is the cross-section of the cavity and l is its length; the TM_{010} mode was used[25]. It was found that the TM_{010} mode required only one third as many molecules as the TE_{011} one to trigger the oscillation. The TM_{011} mode was also investigated and calculations were made on rectangular resonators in the TM_{110} mode[25].

Measurements of the noise temperature factor were made [57] and a large amount of work was done in the field of oscillation stability[41]. The amplification is about 20 dB.

Solid-state masers[46]

The disadvantages of gaseous masers are manifold. The most important ones are the low level of amplification, the existence of a fair amount of noise and worst of all the impossibility of tuning the operating frequency, the latter being determined by the invariable frequency of the ammonia molecule. It should also be noted that the frequency band handled by the gaseous maser is too narrow for many important applications to communications.

The answer to these problems was found in paramagnetic materials. The first solid-state oscillator and amplifier was developed by Scoril *et al.**, using a crystal of hydrated lanthanum ethyl sulphate, with 0·5% of the lanthanum atoms being replaced by triply-ionised gadolinium atoms (Gd^{3+}). This system, of a type described by Bloembergen[3], is a three-energy state one, and coherent stimulated emission from the highest state to the intermediate state can be obtained by saturation of the transition between the highest and the lowest states (Figure 10.11).

The operation of paramagnetic masers is based on the stimulated emission from atoms in a high-energy state. The transition to the lower states will occur

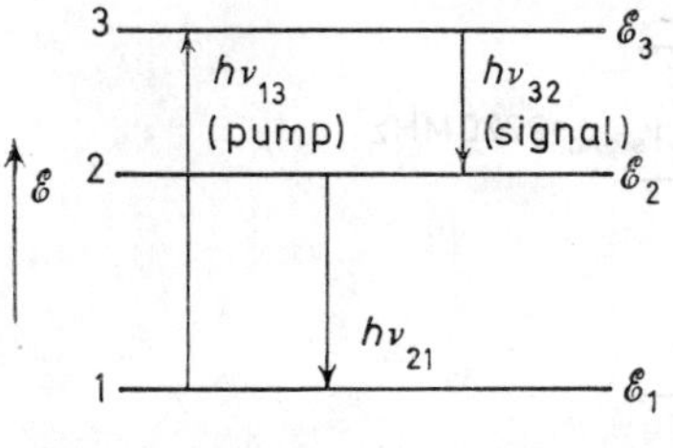

Fig. 10.11. The three-level maser (Blœmbergen type)

as long as there is a net excess of high energy excited atoms in the upper state and will increase with increasing radiation. For continuous operation, a constant surplus of excited atoms should be made available in the upper energy state. This is obtained by the operation called *pumping*, i.e. bringing atoms from the lower to the higher state by supplying a high energy of irradiation of adequate frequency.

Let N be the total change in population from upper level 2 to lower level 1; the spontaneous rate of transition would be, in the absence of external radiation:

$$(\mathrm{d}N/\mathrm{d}t)_{\mathrm{sp1}} = (N_0 - N)/\tau \qquad (10.47)$$

where τ is the relaxation time. In the presence of radiation, a population inversion will occur, as expressed by equation (10.30), leading to a supplement:

$$(\mathrm{d}N/\mathrm{d}t)_{\mathrm{sp2}} = -2\mathscr{B}\mathscr{E}N \qquad (10.48)$$

* Scoril, H., Feher, G., and Seidel, H., *Phys. Rev.*, **105**, 762 (1956).

combining equations (10.47) and (10.48) yields for the total balance:

$$(dN/dt)_{sp} = [(N_0 - N)/\tau] - [2\mathcal{B}\mathcal{E}N]$$

the solution of which is:

$$\frac{N}{N_0} = \frac{1-(N/N_0)\exp[-(\tau^{-1}+2\mathcal{B}\mathcal{E})t]}{1+2\tau\mathcal{B}\mathcal{E}} \tag{10.49}$$

where N_0 is a constant. In a state of equilibrium between pumping and emission, equation (10.49) becomes:

$$N_\infty/N_0 = (1+2\tau\mathcal{B}\mathcal{E})^{-1}$$

Thus if the pumping power is made very large, N_∞ will go to zero. To obtain high values of τ ($\sim$ 10 μs), the paramagnetic material should be cooled to a very low temperature. It would, however, not be too convenient to work between only two levels and both pumping and emitting radiation at the same time. For this reason, multilevel systems (at least three levels) have been chosen.

One of the most successful masers is the ruby (Al_2O_3) maser[36] in which chromium (Cr^{3+}) has been introduced as impurity ions at a 0·1 % concentration. The energy levels of this system are shown in Figure 10.12. Experimentally a

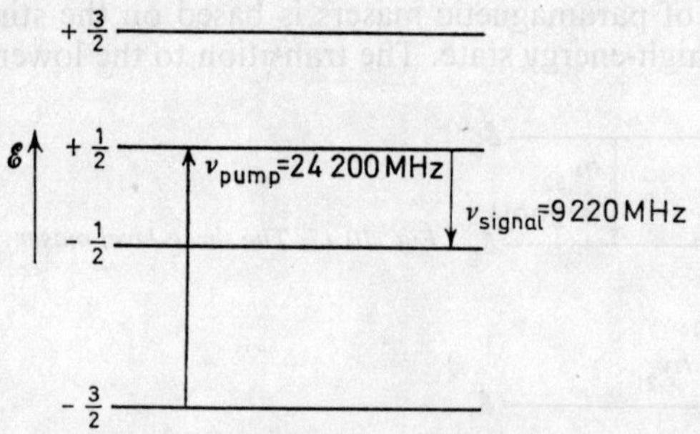

Fig. 10.12. Maser operation in chromium (Cr^{3+}) *doped ruby* (Al_2O_3) *at 0·42* T

ruby crystal is introduced into a cylindrical cavity resonating at both the signal (9220 MHz) and the pumping (24 200 MHz) frequency levels (in the TE_{011} and TE_{114} modes respectively). A klystron supplies the pumping power, whereas the output is amplified by another klystron at the 24 200 MHz frequency. Obviously, the energy of each level is a function of the applied magnetic field. In a typical example, the magnetic field has a 0·42 T strength. Also, in view of the greater value of τ, the crystal must be cooled by immersion in liquid helium.

In the solid-state maser, the transition to be stimulated varies over a certain frequency band. This is due to the fact that the three extra electrons of the chromium atoms are subjected to both the external field and to the influence of the neighbouring atoms, which varies slightly from one to another, and enables the chromium to react to different input frequencies, thus leading to a frequency range of about 25 MHz. Finally, it should be noted that the amplification factor is much higher than in the gaseous maser (about 40 dB).

10.7 LASERS[11, 24, 34]

The advent of the maser was important in its principle, i.e. the achievement of stimulated radiation. The frequencies at which it operated (i.e. microwave), were chosen only because of the technological limitations of the 1950s. There was at that time no acceptable way of manipulating radiation in the submillimetre range. Once the theoretical considerations upon which the maser was based were grasped and advances in handling higher and higher frequencies were achieved, there was no apparent reason for not applying the same principle to amplify waves at higher and higher frequencies. First, the interest was in developing oscillators working in the submillimetre range. But, since the spectra of the materials in this region were barely known, interest grew in the much higher 'optical' frequencies. This led eventually to the development of the laser. Here, three types of lasers can be distinguished: gaseous lasers; solid-state lasers; and liquid lasers.

Gaseous lasers[1, 4]

The most interesting gas laser is the one developed by Javan *et al.*[27]. In this system, two gases, helium and neon, were made to work together, with neon as the amplifying agent and helium the exciting agent. The neon gas was at a partial pressure of 0·1 mm of mercury and the helium gas at 1·0 mm. The overall is in a continuous gas discharge maintained by a source of power at 28 MHz (see

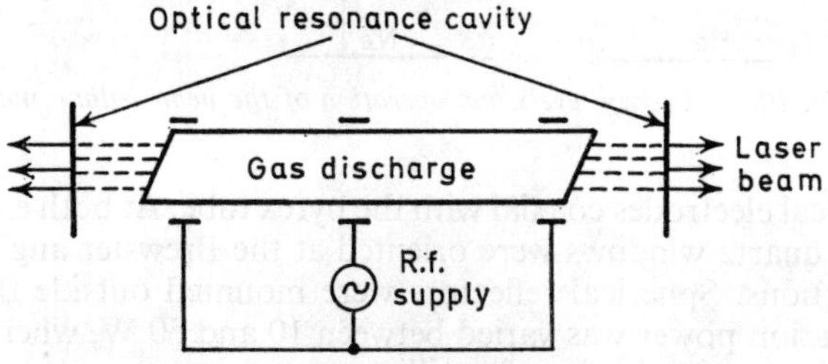

Fig. 10.13. The helium–neon laser. The gas (He–Ne) discharge is inclined at the Brewster angle to avoid reflection

Figure 10.13). The energy levels for the helium–neon maser are shown in Figure 10.14.

Through electron impact in the gas discharge, the helium atoms are raised to the 2^3s state, which is almost at the same energy level as the 2s state of the neon atom. When an excited helium atom collides with a ground state neon atom, most often the helium atom gives up all its energy to the neon atom, thus raising it to the 2s state, the slight difference in energy between the He 2^3s and the Ne 2s states being transformed into kinetic energy of the atoms. Of course, the atoms may also be excited to a lower energy state by electron impact, but since the number of Ne atoms is only a tenth that of the He atoms the excitation of He will be the most probable. Thus a Ne population inversion will build up in the high 2s energy level. The Ne atom then drops to the lower 2p level emitting radiation in the infra-red range. The 2p level contains 10 sublevels, whereas the 2s level contains 4 sublevels. There are altogether 30 possible transitions between the two levels, the strongest being between the $2s_2$ and $2p_4$ sublevels at a 1·153 μm wavelength.

The gas r.f. discharge was contained in a pyrex tube of 1 m length (to obtain an appreciable gain) and 17 mm diameter. The discharge was maintained by

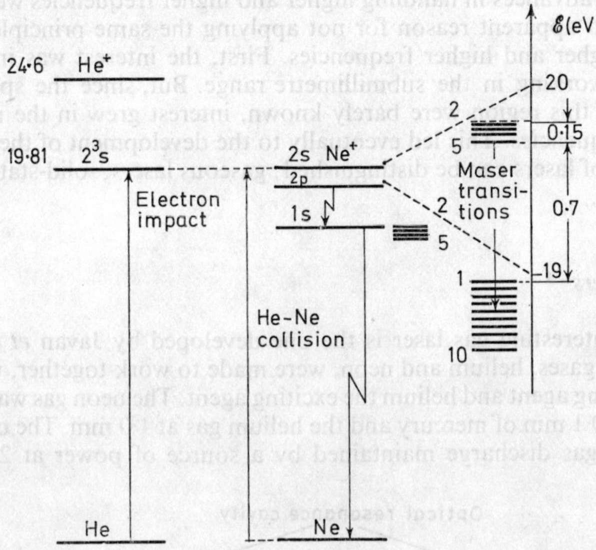

Fig. 10.14. Energy levels and operation of the neon–helium maser

three cylindrical electrodes coaxial with the pyrex tube. At both ends of the tube, optically flat quartz windows were oriented at the Brewster angle to avoid unwanted reflections. Spherical reflectors were mounted outside the pyrex tube. The r.f. excitation power was varied between 10 and 50 W, whereas the output amplifier light was of the order of a milliwatt.

More recently, CO_2 lasers are becoming more and more important in the generation of multi-megawatt pulses[53]. The technique is similar to all other gas lasers: the CO_2 gas is placed in a long cylindrical container inside an optical resonator; it is then excited by an electrical discharge. If Q-switched, this laser can produce high peak power pulses, about 100 ns long. Compared to other lasers, the wavelength of 10·6 μm is very long, resulting in less energetic photons than for the solid-state lasers.

Other gas systems were proposed for laser operation, most of them using an alkali metal vapour[50]. Among these are the potassium vapour and mercury–sodium vapour mixtures, and also the particularly successful helium–caesium mixture[26].

Solid-state lasers

There are two important types of solid-state laser systems: the *ruby laser* and the *semiconductor laser*. These two systems are fundamentally different in the way electrons are used in the amplification of light.

The principle of the ruby laser is similar to that of the ruby maser. In both, impurity chromium ions (Cr^{3+}) are injected into a ruby (Al_2O_3) crystal and photons are injected to pump the chromium atoms to their higher energy levels. The important difference is that for the laser, no magnetic field is used. The energy states of chromium ions in the ruby crystal lattice can be related to those of the free ion by group method theories[22]. When green light (related to the transition from the ground level $^4A_2$ to the upper level $^4F_2$) is applied to a ruby with 0·05% molar Cr_2O_3 at room temperature, the ruby fluoresces by transitions from the sharp levels $^2E(E)$ and $^2E(2A)$ to the ground level. The ^2E state being below the upper $^4F_2$ state, this fluorescence corresponds to the so-called R-lines of the ruby with R_1 (from E) at 6943 Å and with a 4 Å linewidth, and R_2 (from 2A) at 6929 Å and with a 3 Å linewidth. About 70% of the quanta absorbed in the pumping transition ($^4A_2 \rightarrow {}^4F_2$) yield fluorescent quanta, since the non-radiative processes from the upper state to the intermediary state are much more probable than those from the upper state to the ground state. As stimulated emission, the ($^2E \rightarrow {}^4A_2$) transition is predominant and thus only the R_1 line will be observed.

In its transition from the upper level to the intermediary level, the chromium atom gives up its excess energy in the form of heat to the crystal lattice. The lifetime of the atom in the intermediate state (metastable state) is about 0·3 ms. The power absorbed in exciting the atom from the ground level to the upper level can be easily calculated; for the R_1 fluorescent line it is about 1·2 kW/cm³. The minimum power necessary to maintain a population inversion is 15 W/cm³, thus most of the pumping power is spent just to keep the population of the upper state equal to that of the ground state. Because of this, flash tubes were suggested to emit pulsed green light[43]. A cylindrical ruby rod is introduced along the axis of the flash, and is terminated by two mirrors perpendicular to the axis, to reflect the photons. The amplified beam is accepted through one of the mirrors, which is made slightly transparent. Other pumping methods have also been proposed[48].

Lasers using rare-earth elements[40] were considered. These lasers are four-level systems with two working metastable levels. They are more efficient than the three-level ruby lasers. One such device is an yttrium aluminium crystal doped with a neodymium ion. Sunlight has also been successfully used as pumping agent (in a dysprosium-doped crystal with parabolic reflectors).

The *semiconductor laser* is based upon the p–n junction. When this is made in a degenerate semiconductor, a population inversion is established near the junction, provided it is sufficiently biased in the forward direction. The energy-band diagram of a degenerate junction is shown in Figure 10.15 with $\mathscr{E}_{Fn}$ and $\mathscr{E}_{Fp}$ being the quasi-Fermi levels of both regions of the junction. Since at 0 K the conduction-band states are filled up to $\mathscr{E}_{Fn}$ and the valence-band states are empty down to $\mathscr{E}_{Fp}$, a population inversion necessarily exists in the junction area (between planes A and B in Figure 10.15). Radiative recombination will occur when there are enough holes in the valence band and electrons in the conduction band, and will be most intense in a narrow region around the junction itself. In this p–n junction, there are two energy states of interest, an upper one, the conduction band, and a lower one, the valence band. A net radiation emission will occur when there are more transitions from the upper to the lower state than in the opposite direction. If $n_c(\mathcal{E})$ and $n_v(\mathcal{E})$ are the total densities of the conduction and the valence states respectively, and if $n_c^*(\mathcal{E})$ and $n_v^*(\mathcal{E})$ are densities of occupied states having energies between $\mathcal{E}$ and $\mathcal{E}+d\mathcal{E}$, then the condition for radiative emission becomes $(n_c^*/n_c) > (n_v^*/n_v)$. To achieve a population inversion, a non-equilibrium situation should be achieved. To amplify the light output, and thus to sustain the oscillations, a resonant system must be used.

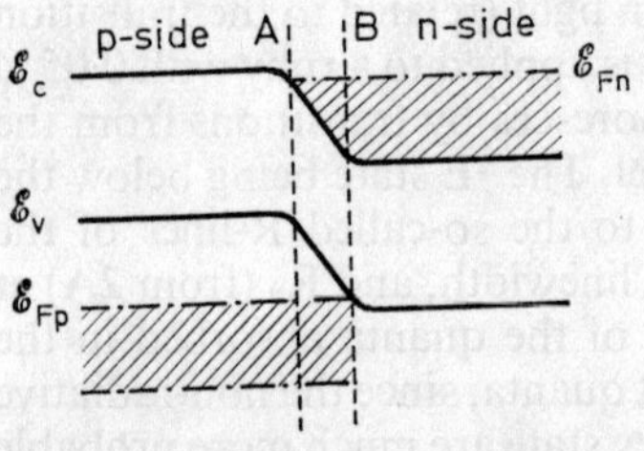

Fig. 10.15. A degenerate p–n junction with a bias in the forward direction

This is most often the Fabry–Perot interferometer[30], which consists of two reflecting surfaces, both parallel to the p–n junction.

Many semiconducting materials have been considered for laser applications, the most frequently used being gallium arsenide[18] (GaAs). The semiconductor laser does not require any separate pumping system, and the junction is excited by a direct current of electrons. The output frequency can be tuned and changed by controlling the temperature and the composition of the junction, whereas the output intensity can be controlled by changing the electron current.

Liquid lasers

In essence, liquid lasers are not different from the two other types of lasers. They are divided into two different types that are basically different in many respects: mineral lasers, and organic lasers[49]. In the first category, various rare-earth ions are incorporated in the liquid. These ions are the active radiators of the laser in the same manner as chromium ions are in an aluminium oxide crystal.

Since 1966, it has been discovered that complex organic dye molecules can be induced to emit laser light. These molecules can be made to work in a regular liquid solvent such as water or alcohol. Nevertheless, one can note that solid plexiglass (polymethyl methacrylate) can have the same function as the liquid solvents with respect to the organic atoms.

The growing interest in organic lasers results from the availability of a large number of organic dyes, each of them emitting at a specific wavelength. The range of these wavelengths is across the entire visible spectrum, in such a way that it would be possible to build an organic-dye laser emitting at any chosen wavelength. Such a laser would also have the advantage of being tunable over a small range of wavelengths. This characteristic is unique to this type of laser.

In the case of the solid-state laser, for instance, the energy of a primary beam has to be converted, in one way or another, into that of a secondary beam, if one needs to shift frequency.

Using the molecules of an organic dye called 7-diethylamino-4 methyl coumarin, Sorokin[49] and his coworkers were able to produce blue laser light. The molecules were dissolved in ethyl alcohol. They also produced an orange laser beam with molecules of another organic dye called rhodamine 6G also dissolved in ethyl alcohol. In both cases the conditions were the same, the first dye solution was simply flushed out from the apparatus, while the second dye solution was pumped in.

10.8 MATERIAL CONSIDERATIONS[52]

Ever since the beginning of this decade, a great effort has been spent in the development of laser materials with the highest possible efficiencies. This led to the investigation of a large variety of gases and crystals (about twenty of each). Some 10 semiconductors were also studied for laser production. For high energy pulsed outputs, neodymium-doped glass has been tested with success. Other materials include liquids and plastics.

The aim of this materials research is to obtain high energy beams at room temperature with the best efficiency possible. Also, it is of great importance to be able to produce amplified light covering the entire optical spectrum. This spectrum presently ranges from ultra-violet to the submillimetre level. Obviously the choice of a certain material after it has been judged satisfactory should depend on the particular application.

The choice of the material depends also on the possible excitation method that can be used with the material. Broadly speaking, the excitation methods can be divided into two categories: continuous and pulsed. In the first case, the molecules in the upper state are segregated from those in the lower state by using the action of the electric field. This type can be called the method of *state segregation* and was used in the early ammonia maser. In the case where the material has more than two energy levels, it becomes possible to excite a higher energy level by pumping the molecules from the ground state at the relevant excitation frequency in order to obtain a population inversion. In this case, the *excitation method* may be either continuous or pulsed. One of these methods is *optical pumping* in which a suitably polarised light irradiates the system of molecules, the wavelength of the light being the same as that of the optical transition of the said molecules from one state to a higher one. This method has been successfully used in several ruby lasers. In the Bloembergen-type of excitation mentioned above, the *Zeeman effect* resulting from the action of a steady magnetic field on a paramagnetic material is used to split the energy levels, of which the material should have three or more. Bloembergen excitation can be either continuous or pulsed. Another excitation method is the *pulse inversion* in which the relaxation time of an energy level in a material should be larger than the exciting pulse duration to allow the population inversion to occur. Finally there is the excitation method of *adiabatic fast passage*, in which the ensemble of molecules is subjected to a strong frequency-varying microwave field having a constant amplitude. The frequency of the field is changed from a very small value, across the frequency of the molecular transition to much higher ones. The passage should be adiabatic but fast compared to the relaxation time of the molecules, and the energy-density of the exciting field must be larger than the radiative energy-density of the molecular ensemble. The functions of excitation and emission can be accomplished by two different materials as in the case of the helium–neon lasers.

Among materials for the gaseous laser, the helium–neon gas mixture seems to be the most frequently used. It produces continuous power between 10 and 100 MW with many lines ranging between 0·59 μm and 132·8 μm in wavelength. When pulsed, this laser produces a 100 W peak power. Below these wavelengths the pulsed nitrogen laser generates ultra-violet light at 0·34 μm whereas the argon ion laser produces a strong emission at 0·488 μm in the vicinity of the maximum of transmission through water. As seen before, still other gas mixtures have been considered, among them mercury–sodium and helium–caesium vapours. Laser operation has been reported for the helium–caesium system at 3·2 μm and 7·18 μm[26].

Among the solid materials the chromium-doped ruby is the most frequently used. It is especially practical for high energy pulsed lasers. It can produce energy in excess of 100 J at room temperature and 250 J in a 3 ms pulse at the temperature of liquid nitrogen. This high pulsed energy output at room temperature is shared by the neodymium-doped glass laser. Neodymium-doped yttrium aluminium garnet pumped optically by a tungsten lamp produces a continuous power output of 250 mW. Among the others are $CaWO_4$ doped with Er^{3+} or Nd^{3+} and CaF_2 doped with U^{3+} or with doubly charged rare earths.

The highest continuous power obtained from a semiconductor laser has been from gallium arsenide doped by diffusion of zinc and tellurium. The power was about 3·1 W at 20 K (with an efficiency of 46 per cent), at 0·84 μm. For ternary semiconducting alloys, the wavelength of the emitted light can be varied over a wide range by changing the percentages of the components. One such material is the gallium–arsenide–phosphide alloy of formula $GaAs_{1-x}P_x$. One should also note that indium antimonide has been considered for laser application at a wavelength of 5·2 μm.

Plastic lasers are also possible. The most interesting one uses a europium-doped chelate material in a substrate of clear plastic. The wavelength is 0·61 μm.

10.9 APPLICATIONS OF THE MASER

The two possible ways of using a maser are as an oscillator and as an amplifier. Since the maser works in the microwave range, it is directly in competition with the conventional oscillators and amplifiers such as klystrons and travelling-wave tubes. However, for frequencies higher than 20 000 MHz, the wavelength becomes too small for those conventional devices to be used. Further, masers have the very desirable characteristic of being low noise devices, this noise being primarily determined by the temperature of the molecules.

As oscillators, masers were used successfully for *frequency standards*, *time measurements*, and for *spectroscopy*. The first application was considered for the ammonia maser because of its extreme phase and frequency stability. Extremely stable beat notes have been obtained[8] by detuning two masers. Because of the small spectral linewidth, masers can be used as standard oscillators, with very small phase modulation, in communication systems. The extremely pure oscillations produced by the maser can be used to measure the absorption and emission spectra of materials at radio and microwave frequencies. Such a spectrometer has been described by Townes[55].

The extremely low noise level of the ruby maser makes it attractive for applications as an amplifier. This characteristic has been used notably in *astronomy* for the detection of extremely weak radio signals such as the so-called 'radio whispers'. Liquid-helium cooled ruby masers have been installed with success in radiotelescopes at several observatories. Radar echoes from the planet Venus were successfully detected by a receiving maser pre-amplifier working at 300 MHz[42] and subsequently at 2388 MHz. In the qrst experiment, the active medium was a triply ionised chromium atom in a $K_2CO(CN)_6$ lattice.

Masers are also considered for *radar* applications. However, the pulse of the transmitter might 'paralyse' the maser and some arrangement should be made to protect it. Because of their low noise and their frequency stability, travelling-wave masers have been successfully used[54] for this application. Here, it should be mentioned that, in the maser radar, low angles with the horizon should be avoided if an optimum performance is desired. Indeed, at such low angles, the atmospheric noise and the thermal noise radiated by the ground become extremely important compared to the main signal. The practical frequency of oscillation should be above 1000 MHz for this application.

Another of the most important applications of masers is that to *satellite communication*. Receivers using travelling-wave masers were used in the project *Echo**. In this project a metallised balloon was placed in orbit around the earth: signals were sent from one side of the U.S. and reflected by the balloon to the other side. A travelling-wave ruby maser was also used in the *Telstar* ground receiver at Andover, Maine. Since the receiving antenna was pointed to the zenith, the noise was kept to a minimum '2·4 K whisper'; the noise of the maser amplifier was not higher than 3·5 K. All noises considered, the total noise of the receiver did not exceed a 32 K level, compared to some 3000 K in the conventional microwave receivers. This helped the Telstar receiver to detect signals of extremely low energy at the radio-frequency. For a bandwidth of 2 kHz corresponding to the range of human voice, incoming signals having only 10^{-18} W could be detected. For smaller bandwidth, still smaller radiated power can be detected.

10.10 APPLICATIONS OF THE LASER[9]

Applications of the laser have invaded almost all the areas of activity of man. This is due to the quality of the beam produced by the laser oscillator. Indeed, the laser beam can be a powerful carrier of energy or of information. Among the many proposed applications, attention will be focused here on four important domains: plasma production, machining and welding, communications, and medical applications.

* *Bell Syst. Tech. J.*, **49**, 975 (1961).

Plasma production[6]

A plasma can be produced by irradiating a small solid target with a high power pulsed laser beam. The development of Q-switched ruby lasers has triggered a great amount of theoretical[13] and experimental[14] work in this field. The Q-switched laser yields Q-spoiled laser pulses having a 4 ns rise time and 8 J energy, thus yielding pulsed power in the gigawatt range. In the experiments at Westinghouse this laser produced a single line at 6943 Å with a 0·06 Å linewidth. These experiments were related to controlled thermonuclear fusion research. The targets used were solid surfaces or thin metallic foils of aluminium or silica fibre. Of importance to fusion is the irradiation of a deuterium gas by the high energy laser beam[15]. By this method, ions were obtained with kinetic energies as high as the keV level. Once the energetic ions are produced, the problem remains of confining them. Indeed for the same reasons discussed in Chapter 7, this plasma cannot be confined by physical walls. The quadrupole field generated by applying a.c. potentials to a cubical array of electrodes has been tried to suspend the ions far from the chamber walls, thus obtaining ionic kinetic energy of about 170 eV. However, the most promising method of confinement remains through the use of magnetic fields. In the experiments reported by Sucov *et al.*[50], plasmas produced by focusing a Q-switched gigawatt laser beam on aluminium spheres, discs and foils were confined in a magnetic mirror. The characteristics of these plasmas were studied in the presence and in the absence of the magnetic field. Experiments using other forms of magnetic confinement such as cusps and stellarators were suggested. With the ever increasing power of the laser beams this field needs much research and is very promising.

Laser beams are also used in the spectroscopic diagnosis[10] of molecules. When a photon hits a molecule, part of its energy may be absorbed by the molecule whereas the remaining part is re-emitted as a photon of lower energy, thus of lower frequency than the incident photon. This is known as *Raman scattering*. Information about the vibrational and rotational states of a molecule can be obtained with great accuracy by comparing the frequency of the emitted photons with the frequency of a strong laser beam. Incidentally, this same Raman effect can be used to produce a beam at a lower frequency than the incident laser beam; the generation of optical harmonics in quartz was successfully demonstrated through this method[16]. Other laser diagnostic techniques have been studied in Section 6.9.

Machining and welding

Lasers can produce extremely highly energetic pulsed beams. For a given energy output, the shorter the pulse the higher is its power. If the pulse is short enough the metal vaporises. By this method and due to the negligibly small scattering in the laser beam, holes of very small diameters can be 'drilled' even in such hard substances as black industrial diamond. This is done by focusing the beam on very small areas. This same vaporisation process can be applied to high precision machining such as that required in horology. Here, however, it should be noted that after a pulse the laser needs some time before it can rebuild its population inversion and then start another pulse. This time is here of the order of the minute.

For too long pulses the bulk of the metal may melt, but this property is used for welding applications, the condition being that enough molten metal should be produced to obtain a strong weld after cooling. This method is extremely helpful and is successful, notably in welding microcircuits of low thermal conductivity.

Communications[38]

Because of the extremely high frequency of the laser beam, a large number of channels of usual bandwidth can be carried simultaneously on the same beam. A laser beam will thus be theoretically able to transmit an enormous amount of information simultaneously. Because of the fact that the beam is highly directional and intense, communication between earth and satellites, and eventually between planets, is possible. On terrestrial applications, laser beams are made especially attractive by the advances in the technology of optical waveguides[29]. Electro-optical modulators and demodulators employing potassium dihydrogen phosphate (KDP) have been developed with success[37]. Photodetectors and photomultipliers using semiconductors have also been considered.

A field of special importance is the use of laser beams in radar (to determine the position and the motion of small distant objects which can be detected by optical rangers using *Q*-switched lasers).

Medical applications

Laser beams can be used in cancer therapy to destroy cancerous cells without damaging the neighbouring ones[17] and have been used successfully in fusing detached retinas in human eyes[28]. The fact that the operation can be done in an extremely short time renders this application especially interesting. There is a problem of safety which should be mentioned here[33]. Intense laser beams are extremely dangerous to different parts of the human body and especially the eyes[51].

One of the most revolutionary applications of laser light is its possible use in cell surgery[2]. It was always possible for a physiologist to identify the function of an organ by removing it from the animal. This is not practical for the cytologist who studies the functions of the organelles of a living cell because of the extremely small dimensions of such organelles, being in the range of the micrometre. This situation is in the process of being changed due to the development of the laser 'scalpel'. The latter consists of a laser whose beam is focused with the help of a microscope and instruments recording the effect of the laser beam on the cell. Thus, the cytologist can destroy chosen organelles to study the effect of their removal on the living cell and consequently to identify the function of the cell. This is helped by the fact that many organelles contain natural chromophores. The latter are molecules that strongly absorb light at a specific wavelength.

Experiments were made on the mitochondria in the myocardial cells of the cardiac muscle of the rat. These myocardial organelles contain chromophores that absorb light at 5200 Å and 5500 Å. An argon laser was used to determine the function of mitochondria in the cardiac cells. Experiments were also performed on the chromosomes of the salamander. Chromosomes are organelles containing the hereditary characteristics of the living cell. This latter experiment would seem to open new horizons into the possibility of cell surgery to determine and possibly cure defects due to abnormal constitution or number of chromosomes.

Other applications

It would be a big task to treat all the domains of application of the laser. Most of them have been listed in *Microwaves* (Vol. 5, No. 10, p. 56, 1966). Among the other ones are the use of lasers for an accurate positioning of machine tools (by interferometric methods). Among the applications to space science, the laser gyro[32] is an important contribution to the advances in aeronautics. This device is not sensitive to sudden accelerations and does not involve prolonged starting times. The ideal gyro must be sensitive to rotations and have an output proportional to increments of angle. The laser gyro combines some properties of general relativity, of optical oscillators, and of lasers to yield a gyroscope with an integrating rate. The laser can also be used as a perfect flash-light for photographing extremely fast moving objects. It can be used to explore the properties of chemical reactions. The laser 'speedometer' using the Doppler shift can measure the velocity of far objects such as artificial satellites. It has been used for a new measurement of the speed of light in vacuum. A laser beam could push a spaceship in a way similar and much more powerful than the so-called 'solar sailing'. The same effect can be used for braking the motion of an artificial satellite. One might think that in the far future power itself could be transported by laser beams. Another example of laser applications in space science is the use of a laser ranging retro-reflector (LRRR) on the Apollo 11 mission that took the first man on the moon in the summer of 1969. The LRRR has been installed on the moon to serve as a reference point on the lunar surface. It was thus hoped to continuously monitor the point to point distances between this lunar reference point and different stations on earth. This was done by the method of short-pulse laser ranging, by measuring the round-trip travel time between the LRRR and, say, the McDonald Observatory. This time has been measured with an experimental error of $\pm 15 \times 10^{-9}$ s or about 2·5 m in the actual one-way distance.

In military applications, a laser proximity fuse is planned to make air-to-air missiles more effective against manoeuvring targets. The optically focused laser beam of the proximity fuse is reflected off the target. It cannot be confused by electronic methods and becomes impossible to detect.

Lasers have entered the business field ever since the second half of the 1960s. As the realm of their industrial applications broadens, large group companies are creating markets for an increasing number of components. New laser types such as the carbon dioxide laser and the YAG laser are being manufactured, bringing the cost of the whole equipment to less than £40. This field is indeed one of the currently fastest growing ones in science and technology.

REFERENCES AND BIBLIOGRAPHY

1. ALLEN, L. and JONES, D. G., *Principles of Gas Lasers*, Butterworths, London (1967).
2. BERNS, M. W. and POUNDS, D. E., 'Cell Surgery by Laser', *Scient. Am.*, **222**, No. 2, p. 98, Feb. (1970.)
3. BLOEMBERGEN, N., *Nonlinear Optics*, Benjamin, New York (1965).
4. BLOOM, A. L., *Gas Lasers*, Interscience, New York (1968).
5. BORN, M. and WOLF, E., *Principles of Optics*, Pergamon Press, New York (1959).
6. BRADLEY, D. J., 'Laser Spectroscopy', *National Atomic and Molecular Physics Congress*, University of Manchester, April (1969).
7. BROTHERTON, M., *Masers and Lasers*, McGraw-Hill, New York (1964).
8. CEDARHOLM, *et al.*, 'New Experimental Test of Special Relativity', *Phys. Rev. Lett.*, **1**, 342 (1958).
9. CHANG, W. S. (Ed.), *Lasers and Applications*, The Ohio State University (1963).
10. DOUGAL, A. A., 'Optical Maser Probing Theory for Magnetoplasma Diagnostics', *4th Symp. on Engng Aspects of MHD*, Berkeley, Calif. (1963).
11. EAGLEFIELD, C. C., *Laser Light*, MacMillan, New York (1967).
12. EINSTEIN, A., *Phys. Z.*, **18**, 121 (1917).
13. ENGELHARDT, A. G., *The Generation of Dense High Temperature Plasma by Means of Coherent Optical Radiation*, Westinghouse Res. Lab., Rept. 63-128-113-R2, Oct. (1963).
14. ENGELHARDT, A. G. *et al.*, *Plasma Production by a High-Power Q-Switched Laser*, Rept. WERL-3472-9, Westinghouse Res. Lab., March (1968).
15. FLOUX, F., 'High Density and High Temperature Laser Produced Plasmas', *Laser Interaction and Related Plasma Phenomena* (edited by H. J. SCHWARZ and H. HORA), Plenum Press, New York (1971).
16. FRANKEN, P. A., HILL, A. E., PETERS, C. W. and WEINREICH, G., 'Generation of Optical Harmonics', *Phys. Rev. Lett.*, **7**, 118 (1961).
17. GOLDMAN, L. *et al.*, 'Laser Radiation of Malignancy in Man', *Cancer*, **18**, 533 (1965).
18. GOOCH, C. H., *Gallium Arsenide Lasers*, Interscience, London (1969).
19. GORDON, J. P., ZEIGER, H. Z. and TOWNES, C. H., 'The Maser. New Type of Amplifier, Frequency Standard and Spectrometer', *Phys. Rev.*, **99**, 1264 (1955).
20. GORDY, W., SMITH, W. V. and TRAMBARULO, R., *Microwave Spectroscopy*, Wiley, New York (1953).
21. GORTER, C. J., *Paramagnetic Relaxation*, Elsevier, Amsterdam, (1947).
22. GRIFFITH, J. S., *The Theory of Transition Metal Ions*, Cambridge University Press (1961).
23. HAMILTON, D. R., KNIPP, J. K. and KUPER, J. B., *Klystrons and Microwave Triodes*, McGraw-Hill, New York (1948).
24. HEAVENS, O. S., *Optical Masers*, Methuen, London (1964).
25. HELMER, J. C., 'Maser Oscillators', *J. appl. Phys.*, **28**, No. 2 (1957).
26. JACOBS, S., GOULD, G. and RABINOWITZ, P., 'Coherent Light Amplification in Optical Pumped Cs Vapor', *Phys. Rev. Lett.*, **7**, 415 (1961).
27. JAVAN, A., BENNETT, W. R. and HERRIOTT, D. R., 'Population Inversion and Continuous Optical Maser Oscillations in a Gas Discharge Containing a He-Ne Mixture', *Phys. Rev. Lett.*, **6**, 106 (1961).
28. KAPANY, N. S., SILBERTRUST, N. and PEPPERS, N. A., 'Laser Retinal Photocoagulator', *Appl. Opt.*, **4**, 517 (1965).
29. KARBOWIAK, A. E., 'A Close Look at Optical Waveguides', *Microwaves*, **5**, No. 7, 37 (1966).
30. KASTLER, A., 'Atomes á l'Interieur d'un Interferomètre Pérot-Fabry', *Appl. Optics*, **1**, 17 (1962).
31. KEMBLE, E. C., *Fundamental Principles of Quantum Mechanics*, McGraw-Hill, New York (1937).
32. KILLPATRICK, J., 'The Laser Gyro', *IEEE Spectrum*, **4**, No. 10, 44 (1967).

33. KOHTIAO, A. *et al.*, 'Hazards and Physiological Effects of Laser Radiation', *N.Y. Acad. Sci.*, **22,** 777 (1965).
34. LENGYEL, B. A., *Lasers*, Wiley, New York (1962).
35. LOW, W., *Paramagnetic Resonance in Solids*, Academic Press, New York (1960).
36. MAIMAN, T. H., 'Stimulated Optical Radiation in Ruby', *Nature*, **187,** 493 (1960).
37. MAKER, P. D., TERHUNE, R. W., NISENOFF, M. and SAVAGE, C. M., 'Effects of Dispersion and Focusing in the Production of Optical Harmonics', *Phys. Rev. Lett.*, **8,** 21 (1962).
38. MARCATILI, E. A. and SCHMEKTER, R. A., 'Hollow Metallic and Dielectric Waveguides for Long Distance Optical Transmission and Lasers', *Bell Syst. tech. J.*, **43,** Part 2, 1783 (1964).
39. MERRITT, T. P. and HALL, F. F. Jr., 'Blackbody Radiation', *Proc. IRE*, **47,** 1435 (1959).
40. METLAY, M., 'How Rare Earth Chelate Lasers Work', *Electronics*, Nov. 15, 67 (1963).
41. MITCHELL, A. M., ROOTS, K. G. and PHILLIPS, G., 'Ammonia Maser Oscillator', *Electron. Technol.*, **37,** 136 (1960).
42. NEWBAUER, J. A., 'Radar Echo From the Planet Venus', *Astronautics*, **4,** 50 (1959).
43. PAKE, G. E., *Paramagnetic Resonance*, Benjamin, New York (1962).
44. POWELL, J. L. and CRASEMANN, B., *Quantum Mechanics*, Addison-Wesley, Reading, Mass. (1961).
45. SCHAWLOW, A. L. and TOWNES, C. H., 'Infrared and Optical Masers', *Phys. Rev.*, **112,** 1940 (1958).
46. SIEGMAN, A. E., *Microwave Solid-State Masers*, McGraw-Hill, New York (1964).
47. SINGER, J. R., *Masers*, Wiley, New York (1959).
48. SINGER, J. R. (Ed.), *Advances in Quantum Electronics*, Columbia University Press, New York (1961).
49. SOROKIN, P. O., 'Organic Lasers', *Scient. Am.*, **220,** No. 2, 30, Feb. (1969).
50. SUCOV, E. W., PACK, J. L., PHELPS, A. V. and ENGELHARDT, A. G., 'Plasma Production by a High-Power Q-Switched Laser', *Physics Fluids*, **10,** No. 9, Sept. (1967).
51. SWOPE, C. H. and KOESTER, C. J., 'Eye Protection Against Lasers', *Appl. Opt.*, **4,** 523 (1965).
52. THORP, J. S., *Masers and Lasers—Physics and Design*, MacMillan, London (1967).
53. TIFFANY, W. B., TANG, R. and FOSTER, J. D., 'Kilowatt CO_2 Gas Transport Laser', *Phys. Rev. Lett.*, **15,** 91 (1969).
54. TOWNES, C. H. and SCHAWLOW, A. L., *Microwave Spectroscopy*, McGraw-Hill, New York (1955).
55. TOWNES, C. H., 'Some Applications of Optical and Infrared Masers', *Advances in Quantum Electronics* (edited by τ. R. SINGER), Columbia University Press, New York (1961).
56. TROUP, G., *Masers and Lasers*, Methuen, London (1963).
57. VUYLSTEKE, A. A., *Elements of Maser Theory*, Van Nostrand, Princeton, N. J. (1960).
58. WEBER, J., 'Application of Microwave Radiation by Substances Now in Thermal Equilibrium', *Trans. IRE, Prof. Group on Electron. Devices*, PGED-3, June (1953).
59. WITTKE, J. P., 'Molecular Amplification and Generation of Microwaves', *Proc. IRE*, **45,** 291 (1957).

11 Solid-state plasmas

11.1 INTRODUCTION[19,23,32]

By definition, a solid-state plasma is a collection of semi-free charge carriers in a crystal susceptible to exhibit at least part of the typical plasma behaviours defined in the first sections of Chapter 1. This field of science was born in the late 1950s, and it has retained the attention of many physicists, both theoreticians and experimenters.

The relevant field of research involves two major areas, with a relatively clear-cut separation. The first area consists of studies based upon the plasma collective behaviour, related to the wave propagation phenomena investigated in Chapter 4. Those are better performed on a solid-state plasma at equilibrium*. In the second area, the solid-state plasma starts with a behaviour related to that of an electrical discharge in a gas, and extends toward the simulation, on an extremely small scale, of 'magnetic bottles' considered for the confinement of the thermonuclear plasma, as discussed in Chapter 7.

In the quasi-totality of gaseous plasma phenomena, it was seen that the motion of particles is described, to an excellent degree of approximation, by the equations of classical mechanics. Few gaseous plasmas are dense enough that degeneracy and non-Maxwellian statistics come into consideration, or hot enough that a relativistic correction has to be taken into account. Particles in gaseous plasmas do not usually find suitable lattices to be diffracted by, and the existence of the de Broglie-associated wavelength may be forgotten, most of the time.

By contradistinction, the very existence of the solid state, let alone its electrical properties, could not be understood *at all* until the wave mechanics of periodic media (crystals) was suitably developed around 1930. However, the theory of solid-state plasmas is not that of wave mechanics. The role of wave mechanics in solid-state media may be easily described by assuming that the semi-free particles are either real particles with modified properties, or are fictitious particles obeying laws quite similar to the previous ones. This short cut is not strictly valid, but no serious discrepancy has been found so far by using such concepts to understand plasma behaviour in the solid state. On the con-

* See the definition below.

trary, it seems that the association of this 'semi-classical' model of the solid state with rigorous plasma equations is likely to explain correctly virtually all the plasma behaviours in the solid state.

Such a formalism, even should it prove ultimately to be insufficient, takes account of a sufficient number of effects with respect to the conventional equations of plasma physics that it deserves to be an object of investigation *per se*. Those effects are mainly related to the three following specific concepts in solid-state physics: (1) clustering of the permitted energy levels in *bands*, with simplified models considering merely a *valence band* and a *conducting band*, separated by a *forbidden band;* (2) effect of anisotropy upon the motion of the real or fictitious semi-free charge carriers; (3) the role played by the degeneracy, notably inasmuch as it justified the consideration of positive fictitious charge carriers or *holes*.

The next section will therefore be a reminder of the said semi-classical model. It will appear merely as a series of rules applying to real or fictitious semi-free particles in solid-state crystals.

11.2 SOLID-STATE PHYSICS[36, 45]

Rule 1

The properties of atoms in the solid state differ significantly from those in the gaseous state, due to the presence in close proximity of other atoms. An interesting model consists of assuming that a crystal is made of a lattice of positive ions, in principle singly ionised ones, sharing in common one 'semi-free' electron per atom; the semi-free electrons behave like a gas; they cannot escape from the lattice as a whole, unless they receive a relatively large amount of energy (a few eV per electron). The positive ions are bound together against their mutual electrostatic repulsion by forces *not* explainable in terms of classical mechanics, which justify their tendency to make regular arrays. The so-called 'exchange forces' are related to the fact that electrons are shared and exchanged among themselves.

Rule 2

Contrary to a possible belief, semi-free carriers are *not* to be considered as scattered by the individual positive ions of the lattice, at least inasmuch as the latter contains no crystallographic defects. Scattering occurs whenever there is a crystallographic defect, an impurity atom, or if the lattice is in thermal motion. The latter, which is treated as a particular transient form of reticular defect, is better described by introducing fictitious scattering particles called 'phonons'. All the scatterers encountered in solid-state physics behave satisfactorily either as hard spheres or as potential (Coulomb) scatterers, carrying the validity of many remarks made in the first two chapters.

Rule 3

The influence of a 'perfect' lattice is better taken into account by assuming a change in the laws of motion between successive scatterings (as described by Rule 2). In the cases of weak degeneracy and anisotropy, the new law of motion, under the influence of *external* electric, magnetic and gravitational fields, differs from the ordinary one by introducing an 'effective mass' differing numerically from the real mass. In anisotropic lattices, the effective mass may vary with the direction of motion and is better described as a tensor. Many cases are encountered where the laws of motion are further complicated by conditions of degeneracy, and require graphical constructions in the configuration 'wave-vector' space (Section 11.3).

Rule 4

Degeneracy is said to occur whenever the number of particles to accommodate comes too close to the number of allowed states, according to the rules of quantum mechanics. Semi-free electron gases in metals are strongly degenerate. Semiconducting materials, depending upon the case, offer a wide range between negligible degeneracy as it is achieved in most of the gaseous plasmas, and strong degeneracy almost as complete as in metals. When, in an energy band of the solid, the number of particles equals the number of available energy levels, there is, strictly speaking no possibility of transport phenomena, and the solid is said to be saturated. Everything happens as if the mobility* were zero. This is among the most important properties of solid-state physics.

Rule 5

A situation involving an almost complete saturation is better described by focusing the attention on the few vacant sites and regarding them as populated by fictitious particles called 'holes'. The holes obey essentially the same laws of motion as the electrons, but their electric charge is regarded as positive. By lifting an electron from an almost saturated energy band to an almost empty one, an electron–hole pair is generated[33, 44]. By contradistinction with electron–positron pairs in a vacuum, the energy required is only of the order of 1 eV. It may come from a photon, from a phonon, or from the kinetic energy of a third charge carrier of any nature. Annihilation of pairs occurs as the reverse process. Apart from their order of magnitude, the holes share most of the properties that positrons could have.

* Sections 1.9 and 2.5.

11.3 MOTION OF CHARGE CARRIERS IN THE ABSENCE OF SCATTERING

(a) Negligible degeneracy

In a non-degenerate solid-state plasma, hosted by a lattice of negligible anisotropy, the only change with respect to the law of motion [equation (2.1)], is the replacement of the mass with its usual value by an 'effective mass' M^*, a function of the lattice parameters. No scattering occurs, as long as the lattice is regarded as perfect. Equation (2.1) may therefore be generalised as:

$$M^* \, \mathrm{d}\boldsymbol{v}/\mathrm{d}t = q(\boldsymbol{E}+\boldsymbol{v}\times\boldsymbol{B}) \tag{11.1}$$

($q = -e$ for an electron and $+e$ for a hole), $\boldsymbol{E}$ and $\boldsymbol{B}$ are 'external' fields disregarding entirely the field generated by the ions of the host-lattice. Gravitational forces are the object of a small debate, largely academic. Is the weight of the particle $M\boldsymbol{g}$ or $M^*\boldsymbol{g}$, where $\boldsymbol{g}$ is the acceleration due to the gravitational field? We think that the weight, as a force, is not influenced by the host lattice and is therefore $M\boldsymbol{g}$ ($-M\boldsymbol{g}$ for a hole). Gravity is unimportant in solid-state physics, but inertial forces may be treated as a pseudo-gravity. They should be treated as the acceleration term in equation (11.1), hence with M^* instead of M, if they represent motions of the plasma with respect to the host-lattice, and similar to the weight (hence with $\pm M$) if they represent motions of the host-lattice with respect to the laboratory. Another important refinement to equation (11.1) is the consideration of anisotropy, which confers a tensor nature to the mass. All the refinements together lead to an amended equation (11.1) of the form:

$$\frac{\mathrm{d}}{\mathrm{d}t}[\boldsymbol{M}^* . (\boldsymbol{v}+\boldsymbol{v}_\mathrm{d})] = q(\boldsymbol{E}+\boldsymbol{v}\times\boldsymbol{B}) \pm M(\boldsymbol{g}+\boldsymbol{g}') \tag{11.2}$$

In this equation, $\boldsymbol{v}_\mathrm{d}$ represents the drift velocity of the plasma portion under consideration with respect to the host-lattice, $\boldsymbol{v}$ is the velocity of the particle in a frame of reference travelling with the said plasma portion, $\boldsymbol{g}$ is the true acceleration of the gravitational field (generally negligible), whereas $\boldsymbol{g}'$ is a pseudo-gravity acceleration generated by the inertia in the motion of the host-lattice with respect to the laboratory.

Most of the difficulties involved in the anisotropy can be eliminated by considering a vector $\boldsymbol{p} = \boldsymbol{M}^* . \boldsymbol{v}$ or a vector $\boldsymbol{k} = \hbar \boldsymbol{p}$, with $\hbar = h/2\pi$, h being Planck's constant. Then, disregarding the inertial and gravitational refinements, the law of motion is formally identical to that of a particle in a vacuum. In particular, in a magnetic field alone, one obtains, after replacement of $\boldsymbol{v}$ by $\mathrm{d}\boldsymbol{r}/\mathrm{d}t$:

$$\mathrm{d}\boldsymbol{p}/\mathrm{d}t = -(q\boldsymbol{B}\times\mathrm{d}\boldsymbol{r}/\mathrm{d}t) \tag{11.3}$$

By a proper choice of the origin (made coincident with the instantaneous centre of gyration) this integrates into:

$$\boldsymbol{p} = q\boldsymbol{B}\times\boldsymbol{r} \tag{11.4}$$

The motion of the vector ***p*** follows that of the vector ***r***, save for a change in scale and a rotation by 90° around the direction of ***B***, and vice versa. The same is true for the vector ***k***. In many cases, motions are best represented in a configuration space, with ***k*** plotted instead of ***r***. In particular, in the relevant case one should plot in the configuration space a surface of constant energy. Any orbit possible, with a given energy in a given magnetic field, is obtained by intersecting the relevant surface of equal energy by a plane perpendicular to the magnetic field [Figure 11.1(a)]. Since a case of negligible degeneracy combined with appreciable anisotropy gives ellipsoidal surfaces of equal energy[45], the intersecting curves are in general ellipses instead of circles. This ellipsoidal

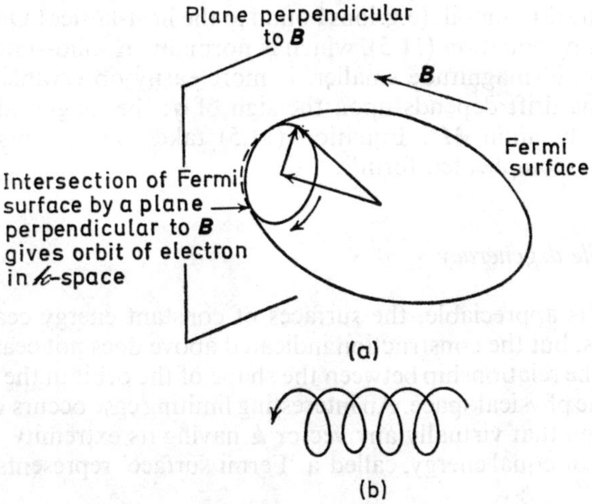

Fig. 11.1. Motion of charge carriers in anisotropic and degenerate solid-state plasmas. In (a), *the motion in the configuration space is obtained by intersecting a surface of constant energy, an ellipsoid if degeneracy is negligible, some more sophisticated surface if it is not, by a plane perpendicular to the magnetic field. In* (b), *the corresponding motion in the physical space, made of a gyrating portion given by the magnetic field and a non-orbiting part given by the non-magnetic forces*

motion should be combined with the usual drift along the magnetic line of force. Here, however, it should be remembered that for the non-orbiting part of the motion [Figure 11.1(b)], ***p*** (or ***k***), *not* ***v***, is parallel to ***B***. In general, the instantaneous gyration centre does not drift parallel to ***B***; it drifts in such a way that the dot product of its velocity with the mass-tensor has the same direction as ***B***.

The problem of cross-drifts in a dominant magnetic field combined with one or several non-magnetic forces is fairly simple in its principle. The cross-drift in an electric field, given by equation (2.8), does not involve the mass and should not be influenced by any variations thereof, provided they are symmetrical around the orbit, which is actually the case with a tensor mass. To obtain the drift induced by any other non-magnetic force, including the force resulting from the orbital magnetic moment placed in a gradient of the magnetic field, replace ***E*** by $\boldsymbol{F}/q$ in equation (2.8), whatever the nature of the force ***F*** is.

An important example is provided by the 'polarisation drift' associated with the inertial force $-(q/\boldsymbol{B})(\partial E_{\perp}/\partial t)$ corresponding to the motion of particles in a constant and uniform magnetic field upon which is superimposed a rapidly variable (e.g. oscillating) electric field. The relevant drift is given by simply inserting this inertial force:

$$v_p = -M^*(\partial E_{\perp}/\partial t)/qB^2 \tag{11.5}$$

In a plasma, it is important to remark that the drift given by equation (2.8) is independent of the nature of the particle; a plasma drifts as a whole, and if positive and negative mobile charges are compensated, there is no resulting current. The current becomes, however, observable if at least part of the particles of one sign are immobile (e.g. ions bound to the host-lattice). On the contrary, the drift given by equation (11.5), which is normally in solid-state plasmas two or three orders of magnitude smaller, is more easily observable as a current. The sign of the drift depends upon the sign of q; the magnitude of the drift depends notably upon M^*. Equation (11.5) takes, in an anisotropic host-lattice, a rather complicated form[45].

(b) Appreciable degeneracy

If degeneracy is appreciable, the surfaces of constant energy cease in general to be ellipsoids, but the construction indicated above does not cease to be valid. Neither does the relationship between the shape of the orbit in the configuration space and in the physical space. An interesting limiting case occurs when degeneracy is so strong that virtually any vector $\boldsymbol{k}$ having its extremity inside a particular surface of equal energy, called a 'Fermi surface' represents an occupied

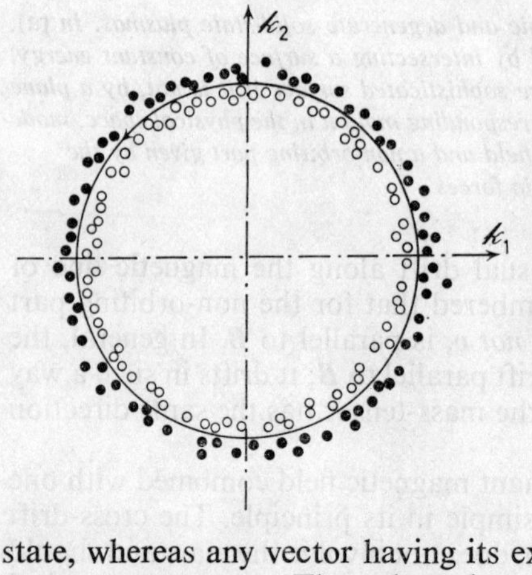

Fig. 11.2. Very strong degeneracy: when degeneracy is almost complete, all the energy states corresponding to points (in the $\boldsymbol{k}$-space) inside a particular surface of constant energy (Fermi surface) tend to be occupied and all the ones outside to be empty. There are just a few holes left inside and a few electrons outside, close to the surface. Any motion significant from a transport point of view is represented by a motion on the Fermi surface, as shown by the arrow

state, whereas any vector having its extremity outside the same surface represents an empty state. Then, the only motions significant from a transport point of view are represented in the configuration space by motions on the Fermi surface (Figure 11.2). In particular, motions in a uniform time-independent magnetic field will be obtained from intersections of the Fermi surface by planes, as above.

For a given material, the Fermi surface is given in shape and size. It may vary slightly with the temperature, as a result of the thermal expansion of the host-lattices. Its shape is quite often complicated and the relevant complication is carried by the gyrating orbits. A case of special interest is that of *connected* Fermi surfaces relevant to adjacent Brillouin zones[10]. Then, whereas many orbits are still topologicaly equivalent to ellipses, planes of accurately defined orientation may intersect the Fermi surface along open curves corresponding to 'open orbits' both in the configuration and in the physical space (Figure 11.3). The advent of open orbits, when the orientation of $\boldsymbol{B}$ is changed, corresponds to the re-establishment, suddenly, of a transverse mobility in a strong magnetic

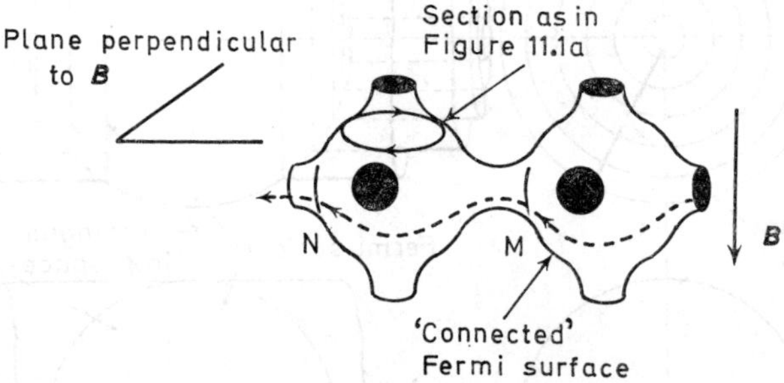

Fig. 11.3. Open orbits: application of the construction of Figure 11.1 to a case of strong degeneracy with 'connected' Fermi surface. While most of the orbits are still closed as in Figure 11.1, certain orientations of the magnetic field permit open orbits to occur. The transposition to the physical space is as in Figure 11.1(b)

field comparable to the longitudinal one. Magnetic breakdown, that is breakdown in a given moderate electric field simply by tilting a magnetic field, may occur.

When the magnetic field is very strong*, a condition is eventually achieved in which the gyrating part of the orbit must be quantised, following the modified Bohr–Sommerfeld formalism, with half integer numbers instead of integers. The surface area S, embraced by the orbit in the physical space, must then satisfy the condition:

$$qBS = (n+\tfrac{1}{2})h \tag{11.6}$$

The surface area S_0 of the orbit in the k-space will also be quantised. Since lengths are in the ratio $qB/2\pi h$ then $S_0/S = q^2B^2/4\pi^2h^2$ and from equation (11.6)

$$S = \left(n+\frac{1}{2}\right)\frac{qB}{4\pi^2 h} \tag{11.7}$$

Thus the gyrating part of the kinetic energy is restricted to quantised values:

$$\mathscr{E}_{kg} = (n+\tfrac{1}{2})h\nu_g \tag{11.8}$$

with $\nu_g = \omega_c/2\pi$ being the gyration frequency.

* In practice, a superconducting magnet is needed.

When the size of the quantised orbits in the k-space becomes of the same order as that of the Fermi surface, even for small ns, the effects of this quantisation become experimentally observable. In Figure 11.4, a simplified version is shown in which the orbits are treated as circles and the Fermi surface as a sphere; actual cases are more complicated, but topologically equivalent. The

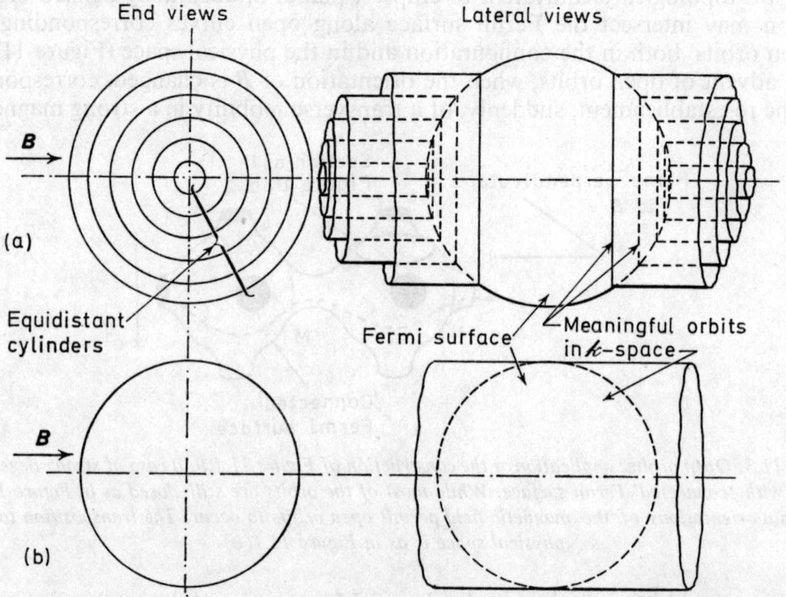

Fig. 11.4. Quantisation of closed orbits leading to 'quantum oscillations'. When the magnetic field is very strong, the area embraced by the orbit is quantised. The only orbits significant from a transport point of view lie at the intersection of the Fermi surface (here represented by a sphere for the sake of simplicity) with a sequence of cylinders. When the magnetic field increases, less and less cylinders intersect, until there is only one left; then, when further expanding, it carries the Fermi surface tangentially with it. The successive expulsions correspond to quasi-cyclic variations in a number of measurable parameters versus B, designated as 'quantum oscillations'

quantised orbits are on the cylinders, but only states represented by $\boldsymbol{k}$-vectors having their extremities inside (or on) the Fermi surface represent occupied energy levels. Even more so, only motions *on* the Fermi surface are significant in transport phenomena, and correspond to experimentally observable facts. Therefore, the only orbits which are both possible and meaningful lie at the intersections of the quantised cylinders with the Fermi surface. As $\boldsymbol{B}$ increases, there are fewer and fewer cylinders small enough to intersect the Fermi surface, and each time the number is reduced by one unit, this is acknowledged by suitable variations in measurable parameters such as the magnetic susceptibility of the sample, or the Landau damping of plasma waves*. Those variations are termed 'quantum oscillations'. Actually, nothing oscillates but the value of a measurable parameter, as the magnetic field increases.

* See Section 11.8.

11.4 MOTION OF CHARGE CARRIERS IN THE PRESENCE OF SCATTERING

On this topic, there is little new to say because solid-state plasmas obey essentially the same laws as gaseous plasmas. In both cases, the basic types of scattering are the same: there are hard-sphere scatterers, and Coulomb scatterers, and the mathematical equations do not differ from those seen in Chapter 1.

At the risk of repetition, *the normal positive ions of a perfect lattice are not scatterers*. Their role is entirely taken into account by the substitution of M^* for M in the equation of motion between the scatterings. Most of the scattering events have actually a co-operative character. This is a situation involving hundreds of displaced atoms on which they act as *one* scattering centre, unless they are excited by the de Broglie-associated wavelength.

In ordinary metals, the most important scatterers are the phonons; they are fictitious particles describing, for that particular purpose and several others, the role of the departure from 'perfect' regularity introduced by the thermal agitation, each phonon describing a 'principal' mode of acoustic vibration involving in fact the totality of the lattice. The resistivity of ordinary metals may be calculated approximately by applying the mobility concept in conditions when the scattering of the electrons results exclusively from phonons in number density proportional to the absolute temperature and behaving like heavy, constant cross-section hard-sphere scatterers. This leads, at room temperature, to collision times τ_c of the order of 10^{-15} s (mobility about 10^{-4} m²/Vs). This time is very short with respect to the period of gyration of the charge carriers in feasible magnetic fields, and with respect to the period of typical waves that may propagate through the solid-state plasma. This time is at most of the same order of magnitude as the period of the plasma oscillation in the solid. Therefore, an important criterion of plasma behaviour is not satisfied. To eliminate this shortcoming, the temperature of the host-lattice must be lowered until the density of phonons is below that of the other scatterers. This means that most of the solid-state plasma experiments are performed at low or very low temperatures: 77K (liquid nitrogen), 4K (liquid helium), or even less. In such conditions, charge carriers are scattered by reticular imperfections, which behave essentially like heavy hard spheres if they imply no change in the number density of lattice ions, but like negative or positive scatterers if they imply vacancies or interstitial radicals, or by impurities. In the degrees of purity required for solid-state plasma experiments, the atoms of impurity are sufficiently remote from each other to behave essentially as gaseous atoms, save for one quantitative difference: in estimating the size of quantised orbits or the magnitude of the Coulomb force (e.g. in the expression for the cross-section), the dielectric constant of the host-lattice should be inserted instead of that of a vacuum. A dielectric constant of about 100, as for bismuth, may introduce a drastic change in the order of magnitude of the final result (four decimal orders in Gvosdover scattering effects). Another significant change may be introduced when M^* is inserted instead of M.

Since these parameters intervene in the calculation of the ionisation potential, some impurities ionise very readily once they are in the crystal (donors and acceptors, with ionisation potentials in the 0·001–0·01 V range). In normal conditions, they should behave as Coulomb scatterers; this statement is, however, less than true, because in strongly degenerate plasmas subjected to strong magnetic fields, fairly fast charge carriers may play a predominant role, especially in the conditions of quantised gyration described at the end of the last section. Fast carriers like this fail to 'feel' appreciably the distant Coulomb field of the ion, further reduced with respect to a vacuum in the ratio of the dielectric constant, and are scattered by the ion itself as if it were a hard sphere.

A typical material for solid-state plasma experiments, indium antimonide, may be purified and cooled down to such a point that the mobility of its carriers increases to about 10^2 m^2/Vs, to compare with 10^{-4} m^2/Vs quoted for metals under many conditions. In applying the equations of mobility, ambipolar diffusion, Gvosdover effect, etc., to solid-state plasmas, one more point should be kept in mind. When the plasma becomes degenerate, the Maxwellian distribution of velocities, which intervenes in the calculation of the statistical coefficients, becomes replaced by the Fermian one. When the degeneracy is nearly complete and the particles may be regarded in the k-space as moving *on* the Fermi surface (Figure 11.2), it is generally assumed that they all have the same speed, called the 'Fermi speed', v_F. Of course, other speeds are still possible, but only particles having the Fermi speed are in a position to contribute to transport phenomena, a reminder of the fact that a fully saturated band does not do so at all. In the following pages, this property, which simplifies significantly many calculations, will be frequently used without comment. It notably gives a standard size to the radius of a cyclotron orbit, $r = Mv_F/qB$.

11.5 GENERATION AND ANNIHILATION OF CHARGE CARRIERS

On the question of the principles, the problems of generation and of annihilation of charge carriers in the solid-state plasma are strongly related to their equivalents in the gaseous plasma. Because of the different orders of magnitude of some parameters, however, some important effects are eventually appreciably different. The concept of 'plasma at equilibrium' is the most important of them.

(a) Plasma at equilibrium

A gaseous plasma cannot exist without a continuous input of energy from some external source. Most of the plasmas of astrophysical or geophysical interest rely upon an external source of energy too, this source being one or several stars. However, in the stars themselves, which are gravitationally-confined quasi-steady state thermonuclear reactors, only an insignificant portion of the total volume is energy producing.

In gaseous plasmas, the concept closest to that of 'plasma at equilibrium' is that of 'plasma in local thermodynamic equilibrium' (L.T.E.), in which the plasma density is a function of the initial chemical composition of the gas and of the thermodynamic parameters (i.e. pressure and temperature) only. In the simplest case of a monatomic gas generating singly ionised ions, the percentage of ionised atoms is given as a function merely of the ionisation potential, the temperature and the pressure, by Saha's equation [equation (2.38)].

In the solid-state plasma, Saha's equation is not generally valid because of the role played by the acceptor and donor impurities and because of degeneracy. The important fact is that the quantity which now plays the role of the ionisation potential is the width of the forbidden gap[36]. Sometimes this role is played by the modified ionisation potential of the impurity, that is the ionisation potential that it has when it is embedded in the host-lattice. Whereas a typical value for the ionisation potential of a gas or a vapour is 10 V, a typical value for the forbidden gap width in a semiconductor is 1 V; as for the ionisation potential of donor or acceptor impurities embedded in the lattice, it can be of the order of 0·01 V or less. Since the relevant figure in Saha's equation or its substitute, whatever it is, has an exponential dependency, this results in the plasma density, at room temperature or below, being of an entirely different order of magnitude. Whereas gases in such conditions are essentially un-ionised, solid-state plasmas in intrinsic or doped semiconductors offer plasma densities typically of the order of 10^{16} cm^{-3} or above, maintained without any energy requirement. Most metals have semi-free electron densities of the order of 10^{23} cm^{-3}. Another interesting case is that of semi-metals like bismuth. They contain electrons and holes in equal number densities, the range of which is between that of metals and of semiconductors.

Metals and semi-metals have in principle carrier or carrier-pair densities fixed by the properties of the host-lattice itself, and largely insensitive to the presence of impurities or a variation in temperature. The character of a semi-metal depends upon the existence of an overlap between valence and conduction bands[11]. If this overlap is small with respect to the equivalent in volts of the temperature, the semi-metal exhibits some of the properties of an intrinsic semi-conductor. This case, of relatively little practical interest, will be disregarded.

An intrinsic semiconductor hosts a plasma made of an equal number of electrons and holes. Designating the number density of electrons by n and the number density of holes by p, one has:

$$np = n_n n_p \exp(-eV_F/kT) \tag{11.9}$$

in which V_F is the width of the forbidden gap, playing here the role of V_i, n_n and n_p are respectively the number of effectively available energy levels for electrons or holes, playing more or less the role of a number density of gaseous neutral atoms available for ionisation. In equation (11.9), set $n = p$, and substitute for n_n the following:

$$n_n = \frac{(M_x^* M_y^* M_z^*)^{1/2} (2\pi)^{1/2} (kT)^{3/2}}{2\pi^2 h^3} \tag{11.10}$$

in which the various symbols have been defined before. Note the intervention of the geometric mean of the effective mass in all three crystallographic directions; n_p is given by a similar expression, the hole effective masses replacing the electron effective masses. The fact that the quantities playing the role of the atom number density vary with temperature is not entirely unknown in gaseous plasmas. This is similar to a situation, achieved in most electrical discharges through metallic vapours, in which the amount of metal in the vaporised state varies with the temperature. Equation (11.10) is valid only if the the value of n_n

thus obtained is small with respect to the number density of states available for the band as a whole; the latter is of the same order as the electron density in true metals.

Doped semiconductors generally have charge carriers of only one sign. Quasi-neutrality is achieved by combining either donors and electrons or acceptors and holes. Let n_d be the number density of donor impurities in the first case, V_d being their ionisation potential in the lattice. Then, the equivalent of Saha's equation is:

$$\frac{2n^2}{(n_d - n)} = n_n \exp(-eV_d/kT) \tag{11.11}$$

Because of the extremely low values of V_d, this reduces to $n = n_d$ (a constant) over most of the range of interest. In the vicinity of the absolute zero, equation (11.11) indicates a departure from this asymptotic value with n going to zero together with T. When the temperature increases, the electron number density given by equation (11.11), reduced in principle to the simplified form $n = n_d$, tends to intercept the curve corresponding to equation (11.9) (with $n = p$) at some critical temperature. In the vicinity of that temperature, a more complicated expression is applicable, but the gap in temperature between the ranges of validity of the two simplified equations is about 20K in most practical cases, and may be disregarded in practice by assuming an angular point at the critical temperature.

In a p-type semiconductor, with an acceptor number density p_a having an ionisation potential V_a in the lattice, an entirely similar equation applies:

$$\frac{2p^2}{(p_a - p)} = n_p \exp(-eV_a/kT) \tag{11.12}$$

The remarks made on equation (11.11) are also applicable to equation (11.12) with n in equation (11.11) and p in equation (11.12).

Some semiconductors may be prepared in extremely pure form, and donor or acceptor impurities introduced at a perfectly controlled level. Then, in the range when $n = n_d$ or $p = p_a$, which occurs in practice at lower temperatures than usual, the plasma density can be as low as 10^{12} cm^{-3}. This is just about as low as one can go with solid-state plasmas. The upper limit is represented by electrons in metals, but here the density is so high that the important criterion of plasma behaviour that the Debye sheath should contain enough particles for it to behave like a continuum, fails to be met. The practical range of number densities for solid-state plasmas in equilibrium with the host lattice and requiring no external source of power, goes therefore from 10^{12} cm^{-3} to slightly below 10^{23} cm^{-3}.

(b) Non-equilibrium generation of charge carriers in the plasma volume

In the so-called non-equilibrium solid-state plasma, a certain plasma density is generated somehow above and beyond that pre-existing at the equilibrium conditions. Then, the plasma has most of the properties of a gaseous plasma, and if only generating processes occurring in the plasma volume are considered, there will be an almost perfect correspondence with processes in gaseous plasmas.

Extra carriers may be generated by raising locally the temperature of the host-lattice above that at equilibrium. This increase can be a result of the heat dissipated by the plasma phenomenon itself, in which case a situation is achieved which parallels exactly that in high-pressure arcs, and that inside stars.

Irradiation is a commonly used means of generating hole–electron pairs. It is in fact considerably easier than in the gaseous state, because, once more, of a forbidden gap of the order of 1 V taking the place of an ionisation potential some ten times greater. Radiations of relatively long wavelengths, up to the infra-red region, may ionise the crystal efficiently.

Electron or hole impact on selected crystallographic defects leads to the generation of electron–hole pairs; in an electric field of sufficient intensity (200 V/cm and more), breakdown by avalanche becomes possible, as in gases. The only major new effect is that positive charge carriers ionise just about as efficiently as negative ones.

Eventually, the role played in electrical discharges in gases by the *metastable atoms* is provided here by energy levels occurring in the forbidden gap. Various forms of step-ionisation can be made possible that way.

(c) Introduction of charge carriers at the plasma boundaries (Figure 11.5)

In a gaseous plasma, it often happens that part of the charge carriers are introduced by one or several electrodes (cathodes or anodes); sometimes 'emitters' giving off electrons and ions at the same time. A similar, but not entirely parallel role, is played here by suitable *contacts*, that is junctions with either the same material differently doped (homo-junctions) or with another material (hetero-junctions).

The theory of junctions is extremely complicated. Here, only a few basic facts will be given. Consider an intrinsic or slightly doped semiconductor. Any welding between it and an *n*-type semiconductor or a metal, effected with care so that the reticular defects and impurities are kept to a minimum, will provide, for a suitable bias, an electron emitter capable of playing essentially the same role as the cathode in a gaseous discharge. The case is better compared to a field-emitting cathode, although considerably lower values of the field will be needed in practice. Of course, holes may be injected with the same ease by means of a properly biased junction to a *p*-type semiconductor [Figure 11.5(c)].

In gaseous discharges, space-charge limitation effects should theoretically arise whenever particles of one sign (e.g. electrons) are injected without compensation. In practice, such a limitation is avoided by the action of an entirely different mechanism: each time an electron–ion pair is generated, the electron tends to move away much more rapidly than the ion. Gaseous discharges have in practice no difficulty in building up a positive space charge in front of a cathode rather than a negative one.

The same favourable circumstance cannot be expected in solid-state plasmas; whenever an electron–hole pair is generated, both components tend to move away at just about the same speed; the continuous injection of charge carriers

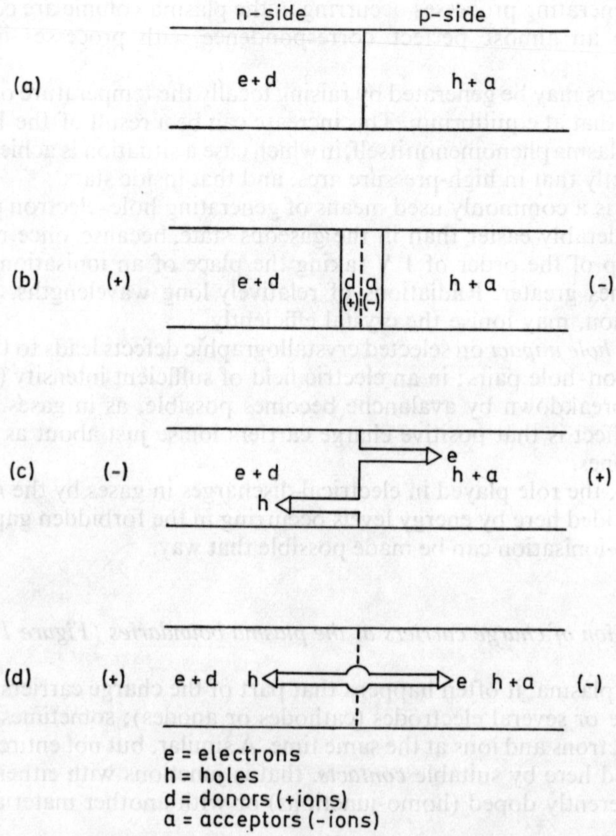

Fig. 11.5. Rectifying and ohmic junctions: under proper bias, junctions can be used to inject particles into solid-state plasmas. In (a), *rectifying junction at rest; sheaths of the order of the Debye screening length are neglected. In* (b), *the same under reversed bias: theoretically, no injection. In* (c), *the same under direct bias: injection of 'minority carriers'. In* (d), *ohmic junction, generating pairs of carriers, and injecting also under reverse bias*

of only one sign tends to build up space charges of that sign, and the latter tend to limit further injection.

This is prevented by making use, in addition to injecting contacts, of ohmic contacts. Here, the rectifying properties that the junction naturally has are wilfully jeopardised by introducing a large density of centres favouring the generation of electron–hole pairs [Figure (11.5d)]. Such junctions inject at will carriers of one or the other sign, and tend to adapt automatically their rate in order to make up for any excess of space charge resulting from the injecting contact(s). Therefore, an injecting contact–ohmic contact pair in the solid state does not parallel exactly a cathode–anode pair in a gaseous discharge, but for reasons of convenience, it is the most practical way of simulating a phenomenon of electrical discharge in a gas. Of course, the injecting contact becomes positive if hole injection is at stake.

(d) Annihilation of charge carriers in the plasma volume

As in gaseous plasmas, carrier pairs may disappear by recombination processes in the plasma volume itself. This is generally catalysed by the intervention of impurities and/or defects in the host-lattice.

In principle, the recombination of a carrier pair, whether spontaneous or catalysed, is a quadratic process. Assuming a hole–electron plasma with $n = p$, one would have, for a free and uncompensated recombination:

$$\mathrm{d}n/\mathrm{d}t = -\varrho_c n^2 \tag{11.13}$$

in which ϱ_c is a recombination coefficient playing an analogous role to that in gaseous plasmas. As there, the solution is of the form:

$$\frac{1}{n} = \frac{1}{n_0} + \varrho_c t \tag{11.14}$$

and is characterised by a linear plot of $(1/n)$ versus t (Section 2.8). But this is less general than in gaseous plasmas with dominant volume recombination, because of the role played either by the plasma density at equilibrium or by the donor or acceptor atoms.

When a hole–electron plasma is exactly in equilibrium, carrier pairs are continuously generated and annihilated, but the generation rate and the annihilation rate must cancel each other. Besides, by virtue of the well-known principle of detailed balancing in statistical physics, the cancellation is not only global, it must also hold good for each generating process which must be balanced by the annihilating process which is its exact counterpart. This principle is not valid in systems which are in a steady state without being really in thermodynamic equilibrium, e.g. in steady-state low-pressure plasmas. Then a generating process may be balanced by an event or a chain of events which differ from the reverse process. If $n_e = p_e$ is the plasma density at equilibrium [equation (11.9)], generations and annihilations must compensate each other, which can be expressed as:

$$\left(\frac{\mathrm{d}n}{\mathrm{d}t}\right)_g = -\left(\frac{\mathrm{d}n}{\mathrm{d}t}\right)_a = \varrho_c n_e^2 \tag{11.15}$$

Paradoxically, this leads to an expression of the rate of generation as a function of the recombination coefficient. Now, if a departure from equilibrium is made, the rate of generation, a function of the lattice and thermodynamic parameters, may be assumed invariant, whereas the rate of recombination is given by equation (11.13). Therefore, the plasma density will obey the following equation:

$$\frac{\mathrm{d}n}{\mathrm{d}t} = \varrho_c(n_e^2 - n^2) \tag{11.16}$$

If n is very large by comparison with n_e, one has essentially the situation described above. Another case of interest is that of a *small departure from equilibrium* in which $(n^2 - n_e^2)$ may be approximated as $2n_e(n - n_e)$ leading to:

$$\mathrm{d}(n - n_e)/\mathrm{d}t = -2\varrho_c n_e(n - n_e) \tag{11.17}$$

which corresponds to an exponential decay with an average lifetime of

$$\tau^* = (2\varrho_c n_e)^{-1} \tag{11.18}$$

This quantity is fairly easy to determine experimentally[10].

In a doped semiconductor, the same type of exponential decay prevails in similar conditions. Consider a p-type semiconductor with p_a acceptors per unit volume. In equilibrium conditions, $p_e = p_a$, $n_e \cong 0$, with, however, equation (11.9) being satisfied by their product. In conditions departing from equilibrium $(n-n_e) \cong n = (p-p_e)$ and:

$$\frac{dn}{dt} = \frac{dp}{dt} = -\varrho_c pn \tag{11.19}$$

For a small departure, $p = p_a$ still holds and:

$$dn/dt = -\varrho_c p_a n \tag{11.20}$$

and τ^* is now equal to $\varrho_c p_a$ ($\varrho_c n_d$ in an n-type semiconductor). The experimental determination of τ^* is once more straightforward.

(e) Annihilation of charge carriers by surface effects

Eventually, the annihilation of carrier pairs by recombination at the walls, which is known to prevail in low-pressure gaseous plasmas, may be simulated in solid-state plasmas if the density of crystallographic defects favouring the recombination in the volume is kept sufficiently low. The boundaries of the crystal are normally the locus of an enhanced density of crystallographic defects and of impurities. In some practical cases, the crystal suffers a preliminary treatment, such as superficial attack, welding, or injection of foreign substances, which accentuates the effect. Then, boundary recombination becomes dominant; the plasma density obeys a diffusion equation and goes to zero at or near the boundaries. In a case of pure surface recombination, the lifetime becomes as in a low pressure gaseous plasma:

$$\tau^* = \Lambda^2/\mathcal{D}_a \tag{11.21}$$

where $\mathcal{D}_a$ is the ambipolar diffusion coefficient, and Λ is a diffusion length which is a function of the size and the shape of the sample.

11.6 COLLECTIVE PHENOMENA

The chief similarity between gaseous and solid-state plasmas is that the collection of charge carriers may behave collectively the same in both cases, if the relevant criteria are met. The criteria are essentially the same, although some precautions are necessary, because either some parameters are widely different, or degeneracy and/or anisotropy must be taken into account.

The plasma frequency is given by equation (1.12). It is essentially a function of the plasma density and may range up to 10^{15} Hz in metals, in which case the energy quantum of plasma oscillation, or plasmon, $\hbar\omega_p$, may play a role. A typical value in metals is as high as 25 eV. In calculating numerical values of the plasma frequency in semiconductors, it should not be forgotten that the dielectric constant is involved, and may introduce a change of one order of magnitude.

The Debye screening length in non-degenerate plasmas is given by equation (1.4). But this equation is less general than the previous one, although the generalisation is easily obtained by stipulating that $r_D \cong v/\omega_p$, where v is the average random speed of the particles. If the Fermi velocity distribution has to be substituted for the Maxwell one, the relevant average velocity should be used. In particular, when the degeneracy is nearly complete, v should simply be replaced by the Fermi speed v_F. The expression 'Thomas–Fermi screening length' is then preferable. In general, the screening length is still shorter in solid-state plasmas than it is in gaseous plasmas. In true metals, it comes below the lattice spacing, which invalidates one of the criteria of plasma behaviour: the Debye (or Thomas–Fermi) sphere, that is, a sphere having the screening length as a radius, contains on the average less than one particle (against 10^3–10^6 in typical gaseous plasmas). Some typical plasma properties, and notably the ability to propagate waves, are nevertheless conserved. Because of the ease with which they are purified, true metals like sodium and potassium are excellent experimental subjects.

As in gaseous plasmas, the key to a rigorous treatment of plasma collective phenomena is Boltzmann's general equation [equation (3.19)]. In general, it is written as such (with M^* simply substituted for M), meaning that anisotropy and degeneracy are neglected. Plain anisotropy is fairly simple to take into account, in principle, at least: the factor $\boldsymbol{F}/\boldsymbol{M}$ representing the acceleration, is then replaced by $\boldsymbol{M}^{*-1}.\boldsymbol{F}$, $\boldsymbol{M}^{*-1}$ being a second order tensor such that $\boldsymbol{M}^* : \boldsymbol{M}^{*-1} = \mathbf{1}$ and whose components in a given frame of reference are calculable from those of M^* by Cramer's rule. The right-hand side of Boltzmann's equation is satisfactorily expressed by the classical Boltzmann collision integral if the scatterers are essentially hard spheres. If Coulomb or both types of scatterers are at stake, both the use of the right-hand side of the Fokker–Planck type or a sum is indicated.

Very few problems have been solved so far with the form taken by Boltzmann's equation when the mass becomes anisotropic. Very little attention has been given also to extensions to degenerate cases, save for almost complete degeneracy. The application to transport phenomena, inasmuch as it differs from the case of gaseous plasmas can be found in the literature[1, 36, 45].

In practice, wave phenomena are investigated chiefly by approximate methods such as writing Maxwell's equations and expressing the current density in the plasma as a result of a drift or a mobility phenomenon. If there is a dominant uniform magnetic field, which is generally the case, the mobility has a tensor character and because of the oscillating character of the electric field in the wave, complex numbers are also used. The accuracy of dispersion relations written that way is in general excellent, at least if the damping is weak. The calculation of the terms representing damping mechanisms may require certain precautions, but the use of Boltzmann's rather than Maxwell's formalism would not necessarily solve the problem.

Because of their extremely high number density, solid-state plasmas generally lend themselves to excellent interpretations in MHD language, such as the concept of frozen magnetic lines of force (see Section 3.9). Here, however, the line of force is frozen in the mobile plasma, *not* in the host-lattice. More generally, the same is true of any application of MHD concepts to solid-state plasmas.

Table 11.1 summarises the various modes of approach of solid-state plasma problems.

Table 11.1 MODES OF APPROACH TO SOLID-STATE PLASMA PROBLEMS

Orbital approach

General equation of motion in an external electromagnetic field:

$$d(\boldsymbol{M}^*\cdot\boldsymbol{v})/dt = q(\boldsymbol{E}+\boldsymbol{v}\times\boldsymbol{B})$$

Adiabatic approximation: gyration plus drifts of the instantaneous centre, notably:

Hall drift: $\boldsymbol{v}_H = (\boldsymbol{E}\times\boldsymbol{B})/B^2$

Polarisation drift: $\boldsymbol{v}_p = -\boldsymbol{M}^*(\partial\boldsymbol{E}/\partial t)\times(\boldsymbol{B}/qB^3)$

Mobility equation $\boldsymbol{v}_d = \mathcal{b}\cdot\boldsymbol{E}$ or in a more general condition $\boldsymbol{v}_d = \mathcal{b}\cdot\boldsymbol{F}_{nm}/q$ where $\mathcal{b}$ is a tensor of components (if the magnetic field is along the z-axis):

$$\left|\begin{array}{ccc} \dfrac{(1/\tau)+j\omega}{[(1/\tau)+j\omega]^2+\omega_c^2} & \dfrac{-\omega_c}{[(1/\tau)+j\omega]^2+\omega_c^2} & 0 \\ \dfrac{\omega_c}{[(1/\tau)+j\omega]^2+\omega_c^2} & \dfrac{(1/\tau)+j\omega}{[(1/\tau)+j\omega]^2+\omega_c^2} & 0 \\ 0 & 0 & \dfrac{1}{(1/\tau)+j\omega} \end{array}\right| : \boldsymbol{M}^{*-1}$$

$\boldsymbol{F}_{nm}$ is any non-magnetic force, including the effect of non-uniformities in the magnetic field. The drifts are used to calculate drift currents, which are, for instance, inserted into Maxwell's equations.

Collective approach

Boltzmann's general equation, for each particle species:

$$(\partial f/\partial t)+(\boldsymbol{v}\,.\,\nabla)f+[(\boldsymbol{M}^{*-1}.\boldsymbol{F})\,.\,\nabla_v]f = (\partial f/\partial t)_{inter}$$

f is a function of seven variables: $x, y, z, v_x, v_y, v_z, t$, from which macroscopic parameters like current densities are deduced by integration over the v's; they are inserted in Maxwell's equations.

MHD approach

The anisotropy of the mass is often neglected, making the equations identical to those in gaseous plasmas. Only the density and motions of the plasma, not the host-lattice, are important.

11.7 SPECIFIC CHARACTERISTICS OF SOLID-STATE PLASMAS

The collections of semi-free carriers present in semiconductors, semi-metals, and, under already specified restrictions, true metals, deserve the name 'plasma' because they are the locus of typical behaviours considered by scientists as typical of the plasma state. Of particular interest is the so-called collective behaviour, including the screening mechanism against external influences, and the possibility of a variety of waves, and related phenomena. Also of interest is the fact that phenomena like electrical discharges in gases can be easily reproduced in samples hosting a solid-state plasma, as soon as the said sample is provided with adequate equipment simulating the electrodes. As an extension thereof, it is possible, by avalanche or otherwise, to generate non-equilibrium plasmas and to confine them by miniature 'magnetic bottles', in order to investigate, for instance, the conditions of confinement of a thermonuclear reactor.

Solid-state plasmas are far from being identical to gaseous plasmas. Some differences lie merely in the fact that the common parameters have an entirely different order of magnitude. But others are more profound and closely related to the basic properties of the solid state, like the band structure, the anisotropy and degeneracy within bands, and the quantisation of some phenomena not normally found in gaseous plasmas.

Table 11.2 gives a comparison between orders of magnitude tipically assumed by some key parameters both in gaseous and in solid-state plasmas. As already shown, the typical solid-state plasma has a higher number density than its gaseous relative: slightly doped semiconductors (one donor or acceptor impurity atom per 10^7 basic atoms) hosts a plasma of some 10^{16} cm^{-3}. This is already equal to the plasma density in the heart of a fairly intense arc. An immediate result is that the plasma frequency in the solid state is typically greater by several orders of magnitude than that in the gaseous state. The usual smallness of the samples authorises the use of extremely high magnetic fields, making the cyclotron frequency typically greater than in gaseous plasmas. This is especially true for the positive particles, since the positive ions are replaced by holes, having the same order of mass as the electrons. The collision frequency is normally high; in fact, the plasma behaviour does usually require that it is lowered by four or five orders of magnitude, in order that the criteria be met.

The temperature is an important factor. Solid-state plasmas are often in thermal equilibrium with their host-lattice, and this means, by virtue of what was said above, at temperatures of 77K, 4K and even less. In degenerate plasmas, however, electrons and holes involved in transport phenomena move approximately at the Fermi speed, and this would be quite a typical random velocity in a gaseous plasma. Drifts, however, are much slower because of lower mobility and diffusion coefficients, even when electric fields or concentration gradients may be several orders of magnitude greater. To conclude, the effect of the generally higher dielectric constant of the host-lattice may not be negligible.

To summarise: solid-state plasmas are usually denser, hosted by very small samples (on the millimetre scale), and live for a much shorter time (down to the nanosecond scale). The existence of the plasma at equilibrium may be regarded as merely due to a difference in the orders of magnitude. The replacement of positive ions by holes may have consequences more far reaching than just the variation in mass.

Table 11.2 PARAMETERS AND THEIR ORDERS OF MAGNITUDE IN GASEOUS AND IN SOLID-STATE PLASMAS

Parameter	*Unit*	*Gaseous plasma*	*Solid-state plasma*
Plasma density	m^{-3}	10^{6}–10^{24}	10^{18}–10^{29}
Mass of negative carrier	Electron mass	1 (except negative ions)	10^{-2}–1
Mass of positive carrier	Electron mass	10^{3}–10^{5}	10^{-2}–1 (except uncompensated plasma)
Plasma temperature	K	10^{2}–10^{9}	0–10^{3} (though degeneracy is possible)
Plasma frequency	Hz	10^{3}–10^{13}	10^{8}–10^{15}
Debye screening length	m	10^{-9}–10^{3}	10^{-5} to below lattice spacing (when it becomes meaningless)
Negative carrier cyclotron frequency (for 'usual' magnetic field)	Hz	0–10^{10}	10^{10}–10^{13}
Positive carrier cyclotron frequency	Hz	0–10^{7}	Same as for negative carrier (except in uncompensated plasma)
Dielectric constant at low frequencies	Dimensionless	Essentially 1	1–10^{2}
Average time interval between collisions	s	10^{-10}–10^{7}	10^{-14}–10^{-10} (plasma phenomena often round about highest value)

To come to more profound differences between solid-state and gaseous plasmas, all the major facts of solid-state physics itself should be considered. The existence of the so-called 'uncompensated plasmas' is one of them; since it has led to the discovery of at least one new type of wave, it is one of the most important features of this chapter. The uncompensated plasma is one in which particles of one sign only are mobile, their space charge being neutralised by particles rigidly bound (apart from thermal agitation) to the host-lattice. Doped semiconductors fall in this category. Either a gas of semi-free electrons has its space charge neutralised by positively ionised donors, or a gas of holes has its space charge neutralised by negatively ionised acceptors. True metals, still with the specified restriction relevant to the Debye sphere, also provide uncompensated plasmas. Semi-metals and intrinsic semiconductors have on the contrary equal number densities of electrons and holes. Partially compensated plasmas are also achievable by setting the temperature close to the transition between intrinsic and impurity conduction, as well as by low level irradiation or weak avalanches, but this has not received much attention.

The anisotropy of the effective mass modifies appreciably the motion of the particles in the solid-state plasma. Extensions of this to the collective level have been incompletely explored. Gaseous plasmas in strong magnetic fields are known to have anisotropic properties with respect to, say, the propagation of waves; here a case of double anisotropy is faced, implying the possibility of fairly complicated combinations by orienting variously the magnetic field with respect to the crystallographic axes. Degeneracy and its consequences, such as open orbits and quantisation of gyration orbits, open the door to hitherto unknown behaviour.

All this implies that solid-state plasmas do not merely duplicate, on a different time and space scale, the behaviour of gaseous plasmas. They have also a few separate properties of their own, sometimes so complicated that an accurate mathematical prediction may be difficult. Of course, conditions may also be selected so that the solid-state plasma simulates more or less accurately its gaseous counterpart and contributes to its knowledge by experiments performed on smaller and cheaper devices. But the majority of the investigators prefer to stress the differences rather than the similarities. This may either provide original information relevant to the host-lattice (e.g. the shape of its Fermi surface) or suggest entirely new types of macroscopically observable phenomena which may have practical applications.

11.8 WAVES IN SOLID-STATE PLASMAS

Wave phenomena in solid-state plasmas differ from those in gaseous plasmas mostly because of the two following features: the time scale is different, and the existence of uncompensated plasmas creates new properties, imperfectly simulated by gaseous plasmas, even those in which the ions are so sluggish that the electrons may be regarded, to all intents and purposes, as the only mobile component. The physical reason lies in the fact that the $\boldsymbol{E}\times\boldsymbol{B}$ drift is independent of the mass of the particles; positive ions in a gaseous plasma, disregarding how heavy, drift in equal fields with exactly the same electron velocity, and a compensated plasma carries no $\boldsymbol{E}\times\boldsymbol{B}$ current. On the contrary, if particles of one sign are bound to a host-lattice, any drift is made impossible for them, and the drift of the particles of the other sign carries an $\boldsymbol{E}\times\boldsymbol{B}$ current.

In solid-state plasmas, the order of magnitude of the characteristic frequencies brings any type of wave with a low-frequency cut-off into an unattractive range of frequencies. Another type, the iono-acoustic wave, is possible in gaseous plasma down to the lowest frequencies, merely because of the large disparity between ionic and electronic masses. The so-called MHD waves are thus left; this is the form taken in plasmas by electromagnetic waves at low frequencies. But the concept of 'low-frequency' in a solid-state plasma has not the usual meaning, and extends well into the GHz range. This is due to the entirely different order of magnitude assumed by the cyclotron frequency of the positive charge carriers (see *Table 11.2*).

In the presence of the restrictions as to the number of wave types, a general theory of waves in solid-state plasmas based, say, upon Boltzmann's general equation, is relatively unrewarding. Here, we will use rather an orbital approach *(Table 11.1)*, with Maxwell's equations, neglecting the displacement current, but including the current(s) generated by the plasma drift(s).

Two drifts are here of relevance, although not at the same time, except in the little investigated partially-compensated plasma. They are the $\boldsymbol{E}\times\boldsymbol{B}$ drift [equation (2.8)], and the polarisation drift [equation (11.5)]. Because of the

MHD character of the wave, it is assumed that there is no space charge and that the fields pertaining to the wave are purely transverse. The magnetic field involved in the said equations is a uniform magnetic field B_0 much greater than the magnetic field pertaining to the wave. Replacing $\partial E_\perp/\partial t$ by $j\omega E$ in equation (11.5) and introducing the cyclotron frequency $\omega_c = qB/M^*$, one sees that the magnitudes of the two drifts are in the ratio (ω/ω_c). This is supposedly a small number, since the very concept of MHD wave requires it to be much smaller than unity. However, in compensated plasmas, the $\boldsymbol{E}\times\boldsymbol{B}$ drift carries no current since it is shared equally by both components. In an uncompensated plasma, both currents exist with the mobile kind of particle, but the polarisation drift is negligible. Therefore, the current density $\boldsymbol{J}$ assumes the following values; for a compensated plasma:

$$\boldsymbol{J}_c = \Sigma \boldsymbol{J}_p = \Sigma(nM^*/B_0^2)(\partial \boldsymbol{E}/\partial t) = (\varrho/B_0^2)(\partial \boldsymbol{E}/\partial t) \tag{11.22}$$

and for an uncompensated plasma:

$$\boldsymbol{J}_u = n_a q \boldsymbol{E}\times\boldsymbol{B}_0/B_0^2 \tag{11.23}$$

This assumes implicitly a wave propagation along the magnetic lines of force. Oblique propagations will be briefly mentioned later. Either value of $\boldsymbol{J}$ should now be inserted into equation (3.4) in which the displacement current is neglected. The other equations of the electromagnetic field theory should be taken into account. In relations between fields and inductions, the anisotropy of the host-lattice is in general not explicitly taken into account. By eliminating all the vectors but one in essentially the same way as in the classical theory of electromagnetic waves one obtains equation (4.41) for compensated plasmas; for uncompensated plasmas, one has:

$$\nabla^2 \boldsymbol{E} - (\mu n_a q/B_0^2)(\partial \boldsymbol{E}/\partial t)\times\boldsymbol{B}_0 = 0 \tag{11.24}$$

In equation (4.41), there has been introduced, on this level of approximation, no character specific to the solid-state plasma. Apart from the order of magnitude of the key parameters, this equation shows no difference from its homologue for gaseous plasmas. It is easily recognised as the equation for the normal Alfvén wave, propagating with a frequency-independent speed given by equation (4.42).

Despite the fact that ϱ is the mass density of the plasma only, the numerical values may be fairly high with respect to gaseous plasmas, leading to extremely slow waves (down to less than 1 m/s) which might lead to interesting applications.

In extremely strong magnetic fields, the quantisation of orbits mentioned at the end of Section 11.3 plays an important role. It introduces slight discrepancies with respect to equation (4.42). These discrepancies have an oscillatory character when B_0 is treated as the independent variable. This is another example of 'quantum oscillation'.

Equation (4.41) may be projected on any axis perpendicular to the direction of propagation without losing its structure; physically, this shows that the wave may be plane-polarised. Because of the presence of the cross-product, this ceases to be true for equation (11.24). The cross-product introduces a screw effect, well depicted by the name 'helicon' given to this type of wave. The projection of the extremity of $\boldsymbol{E}$ onto a plane perpendicular to $\boldsymbol{B}_0$ is seen to describe a circle with a uniform angular velocity equal to ω. In three dimensions, this combines with the wave propagation to give a helix. Only one sign of helicity is possible; it is reversed when B_0 is reversed. It corresponds to the direction of gyration of the particles themselves, although the angular velocity is different. Both cases are illustrated by Figures 11.6(a) and 11.6(b).

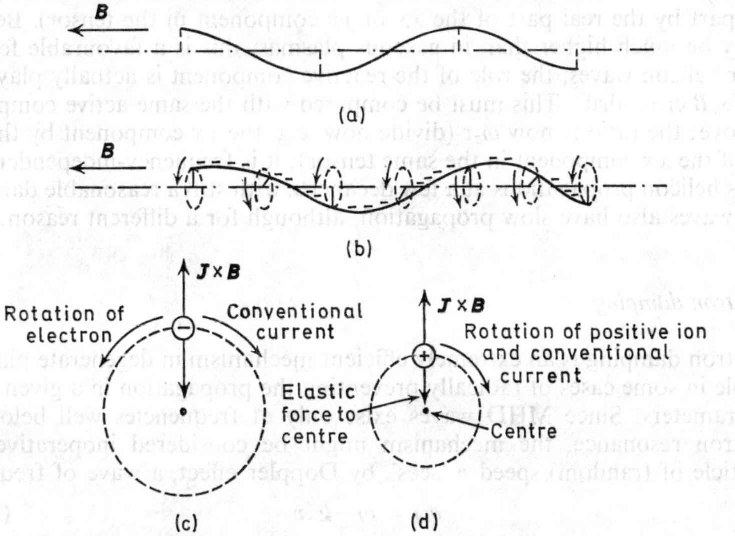

Fig. 11.6. Image representation of Alfvén and helicon waves. In (a), *the Alfvén wave, which may be plane-polarised, can be visualised as the oscillation of a magnetic line of force, as a string subjected to a tension B_0^2/μ and having a mass ϱ per unit length. In* (b), *analogous representation of a helicon. The string is electrically charged and wound as a helix, with only one direction of helicity acceptable, and rotating around the axis. In* (c), *the balance of forces in this rotation: the $\boldsymbol{J}\times\boldsymbol{B}$ force must balance the restoring force resulting from the tension B^2/μ. Because of the lower frequency, the inertia is now negligible. In* (d), *the balance of forces in a circularly-polarised transverse acoustic wave. The two waves have in common an azimuthal current, enabling them to become coupled when at resonance*

The dispersion relation is obtained as usual by setting jω for $\partial/\partial t$ and $-\mathrm{j}\boldsymbol{k}$ for ∇; furthermore, the cross-product is here treated as an ordinary product. The result comes out as:

$$k^2 = \omega\omega_p^2/\omega_c c^2 \qquad (11.25)$$

leading to a strongly frequency-dependent phase-velocity

$$v_p = c(\omega\omega_c)^{1/2}/\omega_p \qquad (11.26)$$

and at equal frequency, one sees that $v_g = 2v_p$.

Damping

The experimental investigation of Alfvén waves in laboratory plasmas is strongly handicapped by the fact that the damping is usually significant. Collision damping is normally predominant; it is easily investigated by replacing the collisionless polarisation drift by mobility drift, by treating the mobility coefficient as a tensor with complex components *(Table 11.1)*. The number of wavelengths leading to a damping by a factor e is proportional to the ratio of the 'reactive' to the 'active' components; this ratio turns out from *Table 11.1* to be equal to $\omega\tau$ where τ is the already defined collision time (divide the imaginary part by the real part of the xx or yy component in the tensor). Because ω may be much higher than in gaseous plasmas, this is a favourable feature.

For helicon waves, the role of the reactive component is actually played by the $\boldsymbol{E}\times\boldsymbol{B}$ cross-drift. This must be compared with the same active component as above; the ratio is now $\omega_c\tau$ (divide now, e.g. the yx component by the real part of the xx component in the same tensor); it is frequency-independent and allows helicon propagations at a few decahertz with still a reasonable damping. Such waves also have slow propagation, although for a different reason.

Cyclotron damping

Cyclotron damping is an extremely efficient mechanism in degenerate plasmas, capable in some cases of radically preventing the propagation in a given range of parameters. Since MHD waves exist only at frequencies well below the cyclotron resonance, the mechanism might be considered inoperative, but a particle of (random) speed $\boldsymbol{v}$ 'sees', by Doppler effect, a wave of frequency

$$\omega_d = \omega - \boldsymbol{k}.\boldsymbol{v} \qquad (11.27)$$

In strongly degenerate plasmas, virtually all the particles which are efficient, from the transport point of view, have essentially the same speed v_F. If $\boldsymbol{k}.\boldsymbol{v}_F = \omega_c$, and since $\omega \ll \omega_c$, then an exceptionally large fraction of these particles will be in a range useful for acceleration by the cyclotron mechanism, and will acquire very large gyration orbits. It is usually an accepted fact that somehow the energy carried by such orbits is dissipated (collisions, radiation, etc.).

When gyration orbits become quantised, as explained at the end of Section 11.3, the kinetic energy of the particles becomes a function of $\boldsymbol{B}_0$, and the latter intervenes, although in a complicated and imperfectly understood manner, in the value of the damping. Landau damping is inexistent when $\boldsymbol{k}$ is exactly aligned with $\boldsymbol{B}_0$, but appears in oblique propagations, in a slightly modified form with respect to that described in Chapter 4.

The influence of *open orbits* in combinations of degenerate plasmas together with connected Fermi surfaces (Section 11.3) may strongly affect the damping. Usually, the mobility coefficient orthogonal to the magnetic line of force is small; an electric field in that direction pushes little or no current, hence there is little or no power involved.

If the particles start gyrating in the magnetic field and follow large orbits, which in practice happens when the direction of the magnetic field with respect to the crystallographic axes falls within certain limits permitting such orbits according to the construction of Section 11.3, then they suddenly acquire extremely large mobility coefficients in the direction of the electric field of the wave. The 'active' component of the current, hence the damping, is increased.

At the same time as the gyration disappears, the $\boldsymbol{E}\times\boldsymbol{B}$ drift disappears too. As a consequence, the propagation of a helicon, which rests upon the existence of an uncompensated cross-drift, becomes impossible. The wave, even in an uncompensated medium, reverts to the properties of an Alfvén wave, but one with extremely strong damping.

Collisions between helicons and ultrasounds

Because of the frequency-dependency of the phase velocity expressed by equation (11.26), helicon waves have a resonance with acoustic waves, at a frequency obtained by equating this velocity to the velocity of sound, v_s. No coupling occurs with the ordinary longitudinal sound waves, but the host-lattice is also capable of propagating circularly polarised acoustic waves. In the case of an uncompensated plasma the host-lattice is electrically charged. It carries a circular current analogous to that of the helicon and provides the coupling element, since in equation (11.24), one must now include the sum of the two currents [see Figure 11.6(c)]. The dispersion relation becomes then:

$$\left(k^2 - \frac{\omega_p^2 \omega}{c^2 \omega_c}\right)(k^2 v_s^2 - \omega^2) = \frac{(n_a^2 e^2 \omega^2)}{k^2 \varepsilon \varrho c^2}. \tag{11.28}$$

The two parentheses on the left-hand side of equation (11.28) represent the separate contributions of the helicon and sound wave dispersion-relations. The right-hand side characterises the intensity of coupling. The effect on the phase velocity is shown in Figure 11.7. This coupling enables us, in theory, to generate helicons from an ultrasonic transducer or ultrasounds from a helicon trans-

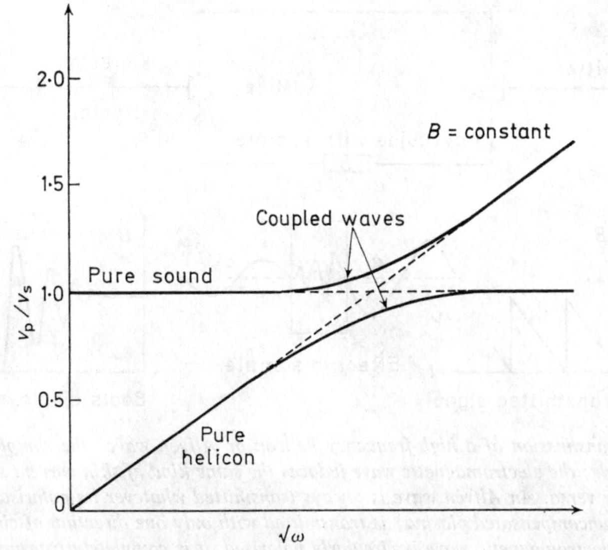

Fig. 11.7. Coupling between acoustic and helicon waves: the phase velocity is plotted versus the square root of frequency. That of an acoustic wave is frequency-independent at v_s; that of the helicon is, after equation (11.26), linear in $\sqrt{\omega}$. The solid line shows the effect of the more complete equation (11.28)

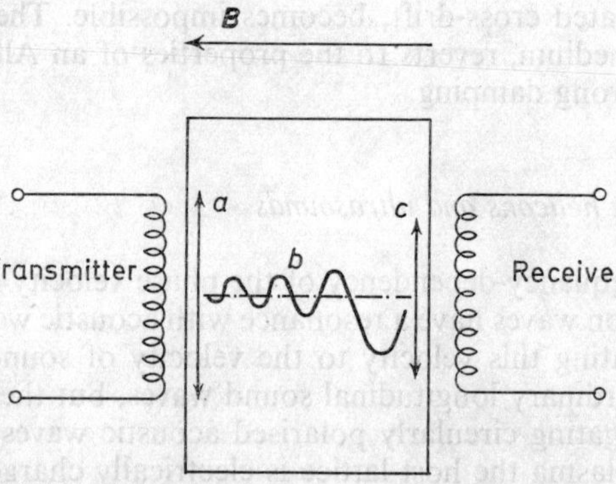

Fig. 11.8. Transmission of a low-frequency helicon: the transmitting coil induces skin-currents, which trigger a helicon propagation. At the receiving end the process is reversed. Even in short samples, the damping is generally significant

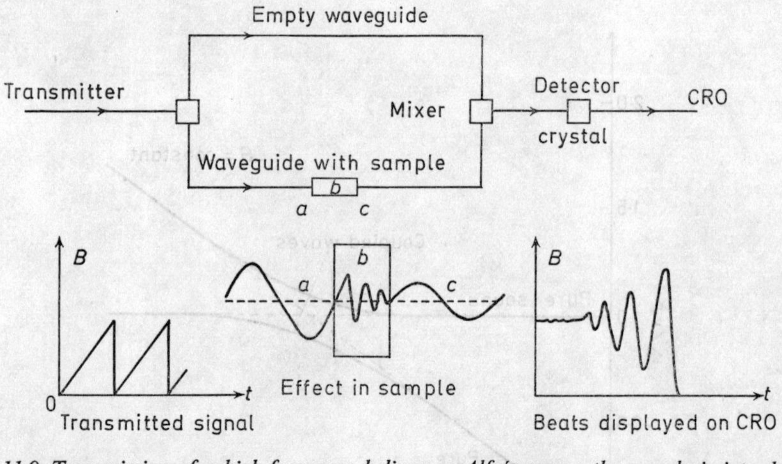

Fig. 11.9. Transmission of a high-frequency helicon or Alfvén wave: the sample is introduced in a waveguide; the electromagnetic wave induces the same kind of skin-current shown in Figure 11.8 and vice versa. An Alfvén wave is always transmitted whatever its polarisation. A helicon wave (in an uncompensated plasma) is transmitted with only one direction of circular polarisation. If the electromagnetic wave is circularly polarised, it is completely transmitted (save for the attenuation in the sample) or completely stopped. This device as well as that of Figure 11.8 may be made part of a Mach–Zehnder interferometer. If the magnetic field varies as a sawtooth, beats of increasing amplitude due to a weakening of the attenuation, may be obtained at the crystal detector and displayed on a cathode-ray oscilloscope

ducer, which may be a simple coil with its axis parallel to the sample surface (Figure 11.8), or a plane surface with a medium in which a circularly-polarised electromagnetic wave exists (Figure 11.9). The coupling has been observed experimentally, but requires a high frequency and a strong magnetic field.

Oblique propagation

The form taken by the preceding waves has also been investigated when $\boldsymbol{k}$ is oblique, and even perpendicular to $\boldsymbol{B}_0$. The theory of oblique Alfvén waves has no new features with respect to gaseous plasmas. When the two vectors are mutually orthogonal, the transition is smoothly effected to the magneto-acoustic wave, also mentioned in Chapter 4, and which has recently been observed also in solid-state plasmas.

The dispersion relation of the undamped helicon wave becomes:

$$k^2 = \frac{\omega_p^2 \omega}{c^2 \omega_c} \cos \varphi \tag{11.29}$$

implying a phase velocity at a given frequency decreasing as $\cos^{1/2} \varphi$ increases and becoming zero in the orthogonal case. Wave-like effects relevant to the periodic character of cyclotron orbits at the speed v_F may still be observable in degenerate plasmas. They look like a degenerate, non-propagating form of helicon.

More important is the fact that a new damping mechanism, proportional to $\sin^2 \varphi$, appears as φ departs from zero; it is the Landau damping, but the axial bunching force is not now electric as in Chapter 4; it is magnetic (mirror effect) and related to the fact that the magnetic field of the wave is not purely transverse any more; the combination of its longitudinal component with $\boldsymbol{B}_0$ gives alternate maxima and minima giving rise to the said mirror effect.

The Landau damping proves to be exceedingly sensitive to the phenomenon of 'quantum oscillation' related to the quantisation of gyration orbits. Instead of a few per cent, the 'oscillations' become 'giant' (50 per cent or more; Figure 11.10).

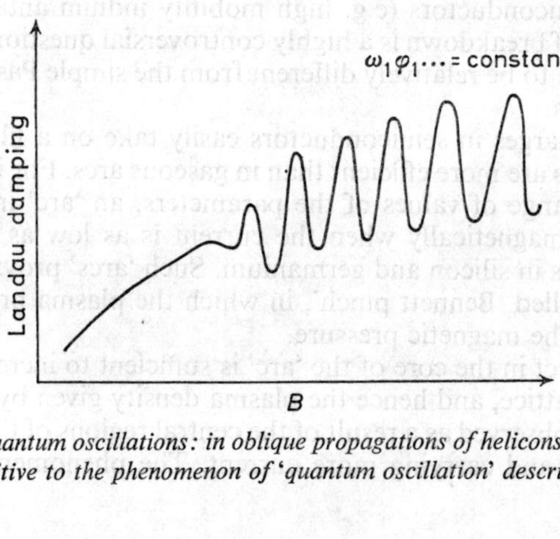

Fig. 11.10. Giant quantum oscillations: in oblique propagations of helicons, the Landau damping is especially sensitive to the phenomenon of 'quantum oscillation' described by Figure 11.4

11.9 NON-EQUILIBRIUM SOLID-STATE PLASMAS

The possibility of simulating, with solid-state plasmas, behaviour such as electrical discharges in gases, including several mechanisms of pinch, as well as confinements by magnetic bottles, fascinates certain investigators in the field[3, 17]. The diagnostics, however, is relatively crude, and consists almost entirely of conduction measurements between pairs of auxiliary contacts (Figure 11.11), giving a qualitative idea of the plasma density in between; the accuracy seems to be much better if the problem is to measure a time-constant of plasma

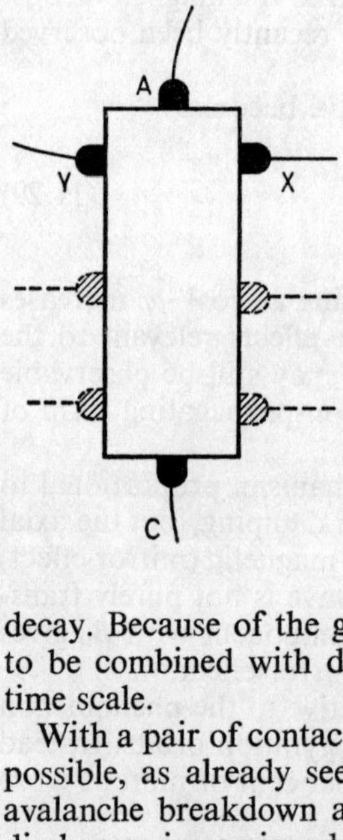

Fig. 11.11. Solid-state sample with equipment simulating the electrodes (A, C) and auxiliary appendages for diagnostics (X, Y, etc.)

decay. Because of the generally fast evolution of solid-state plasmas, this has to be combined with delicate sampling methods operating on a nanosecond time scale.

With a pair of contacts simulating the electrodes of a gaseous discharge, it is possible, as already seen, to reproduce various processes of breakdown, by avalanche breakdown after diffusion of an injected plasma, and steady-state discharges in semiconductors (e.g. high mobility indium antimonide at 77K). The mechanism of breakdown is a highly controversial question; when analysed in detail, it proves to be relatively different from the simple Paschen mechanism occurring in gases.

Electrical discharges in semiconductors easily take on a filamentary shape; pinch mechanisms are more efficient than in gaseous arcs. For instance, because of the different range of values of the parameters, an 'arc' in InSb at 77K is already pinched magnetically when the current is as low as *one ampere*[17, 25], about ten amperes in silicon and germanium. Such 'arcs' provide good illustrations of the so-called 'Bennett pinch', in which the plasma pressure is entirely counteracted by the magnetic pressure.

If the Joule effect in the core of the 'arc' is sufficient to increase the temperature of the host-lattice, and hence the plasma density given by equation (11.9), more pinching is observed as a result of the central regions of the 'arc' becoming more conducting and carrying more current. The phenomenon may become catastrophic and lead to the melting, vaporisation, and even explosion of the host-lattice, if the equilibrium mentioned above requires an excessive temperature.

The application of a longitudinal magnetic field to the discharge enables one to reproduce well-known phenomena of gaseous plasmas, such as the Hoh and Lehnert effect[27], which is the reduction in the axial electric field of an arc as a result of a better lateral confinement away from the boundaries, itself the result of a decreased coefficient of ambipolar diffusion transverse to the magnetic field; also the Kadomtsev instability[4], which is a helical deformation of the discharge axis at a certain critical magnetic field, marking the prelude toward plasma turbulence.

Eventually, such discharges may be used to generate a plasma inside a miniature magnetic confinement system, such as a magnetic mirror machine, with or without multipolar parallel bars (Figure 11.12), achieving a 'minimum-B' configuration; or a stellarator, or forms taken by the latter in order to conform

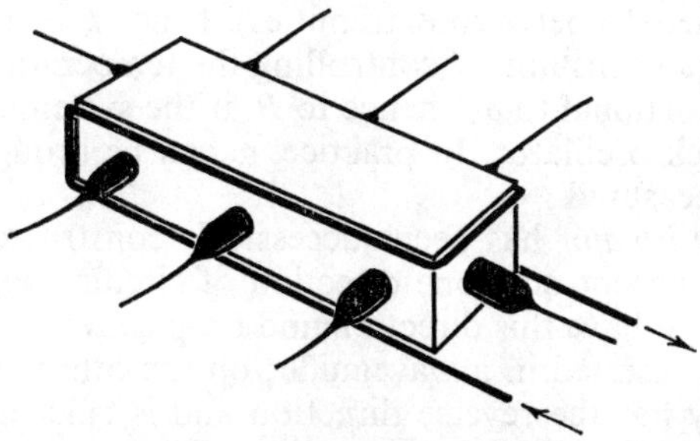

Fig. 11.12. Solid-state sample with a magnetic quadrupole. This confinement may be completed by two coils forming a mirror machine, achieving the simplest 'minimum-B' configuration

to the criteria of 'maximum $\int (dl/B)$' of Furth and Rosenbluth. A typical experiment consists then of comparing time constants for plasma decay with and without such a confinement.

Interesting as they are, such experiments might, however, be of limited scientific value because an electron–hole plasma in a host-lattice is basically an imperfect means of simulating an electron–ion plasma in a vacuum. Whether or not they will actually contribute something significant to the progress of the thermonuclear field is impossible to say at present.

11.10 POSSIBLE ENGINEERING APPLICATIONS

It is impossible to predict at this time what kind of engineering future lies ahead for solid-state plasma physics. Such a future is handicapped by the need for unprecedented purities and crystallographic perfections, low temperatures, and in most of the cases very strong magnetic fields. To find practical applications in which such handicaps are non-existent or tolerable is not easy. Otherwise, there is no reason why solid-state plasma devices should not share the general and well-known advantages of all the solid-state devices, and notably the absence of delicate parts such as filaments, the compactness and lightness, extended life and reliability.

The possibility of propagating extremely slow waves is attractive for *memory banks*, e.g. in computers. Because of the slowness of the wave, a larger number of bits may be stored per unit length than in conventional delay lines. But the attenuation must still be reduced before this can become commercially exploitable.

The coupling between helicons and ultrasounds (Section 11.8), is attractive because of two well-known limitations of quartz transducers: they are limited in the high frequency side well below gigahertz, and they do not offer any interesting possibilities of wide frequency modulation. But the possibility of practical application of the sound coupling to a *high-frequency* or a *frequency-modulated ultrasound generator* is still considered as relatively remote.

A *fluxmeter*, efficient at about 10 T, has been successfully built and used based upon a helicon standing wave. By applying equation (11.25) for the case of a standing wave in a sample of given size, one sees that the wavelength is then given (e.g. $\frac{1}{2}$ wavelength between extremities), hence k is given. If the plasma density is maintained constant by controlling the temperature, ω_p is also given, and ω is then proportional to ω_c, hence to B, if the system is kept resonant by means of a feedback oscillator. In practice, ω can be brought into the audio range and easily measured.

A *high-frequency isolator* has been successfully constructed on the principle that helicon waves accept only one direction of circular polarisation; the forward wave is made to have this direction and propagates easily through a semiconducting sample inserted in a waveguide; on the other hand, the backward wave automatically has the reverse direction and is efficiently blocked by the same semiconducting sample; the conversion of the polarised electromagnetic wave into a helicon and vice versa at the interface between air and a semiconductor has been proved experimentally to be straightforward (see Figure

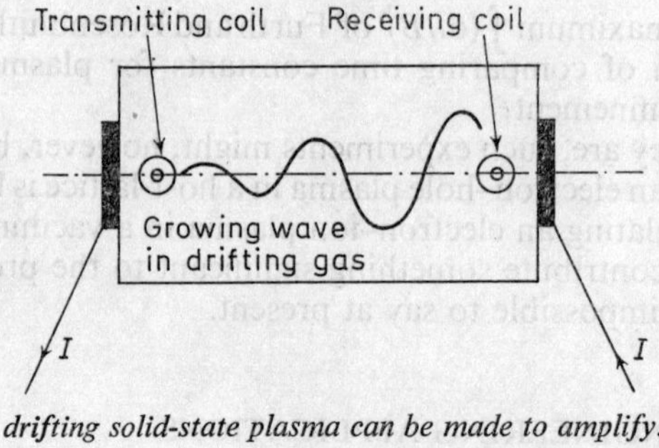

Fig. 11.13. A wave in drifting solid-state plasma can be made to amplify. The drift results from the current injected

11.13). Changing the direction of the magnetic field changes the permitted direction of propagation. This opens new possibilities, notably in the field of two-way radios.

Several models of *amplifiers*, loosely based upon the *travelling wave principle*, have been attempted on the laboratory scale. Actually, the physical principle is closer to that of the plasma–plasma interaction mentioned in Chapter 5.

In a compensated plasma (e.g. in a semi-metal like bismuth), the application of an electric field parallel to the magnetic field guiding an Alfvén wave sets the electrons drifting one way and the holes drifting the other way, thus creating excellent conditions for such an interaction. If it occurs, the amplitude of the Alfvén wave acquires a negative damping coefficient, and hence undergoes amplification. But this is observed experimentally to occur only at prohibitive current densities (in the 10^3 A/cm^2 range) and the Joule effect is then a conspicuous nuisance, especially in view of the low temperature requirement. There is virtually no possibility of building a device on these principles which is technologically acceptable.

In an uncompensated plasma (e.g. suitably doped indium antimonide), there is only one component. Amplification based on the travelling wave principle would, however, be possible if the drift velocity of the electrons or holes could match the phase velocity of the helicon. This is not possible without precautions, since the helicon is supported by the drifting electron or hole gas shift, and changes its phase velocity with the drift of the said gas. A simple law of addition does not hold, because the helicon in the drifting medium has its frequency changed by Doppler effect; this, according to equation (11.26), also changes the phase velocity in a drifting frame of reference.

An ingenious suggestion was made to overcome this difficulty by slicing the semiconducting sample longitudinally and inserting insulating layers in between[40]. The electric field causing the drift is only applied, for example, to every second layer, whereas the wave permeates the whole sample quite easily by tunnel effect. In such conditions, a situation may be attained in which the drifting electrons or holes in the active layers exactly match the phase velocity of the helicon, the latter 'feeling' an average electron or hole gas made of active (drifting) and inactive (non-drifting) layers. Amplification has not yet been experimentally demonstrated, however, as far as we know. If successful, it would have possible applications to millimetre and submillimetre waves (and even down in the infrared) where there is an unsatisfied demand.

Still in the same field, there is a definite possibility of utilising characteristic frequencies of solid-state plasmas, like the cyclotron and the plasma frequency, to operate *masers or lasers*. Such frequencies are fixed in given conditions, but tunable by adjusting the plasma parameters upon which they depend, whereas 'optical' frequencies used so far in masers and lasers are not.

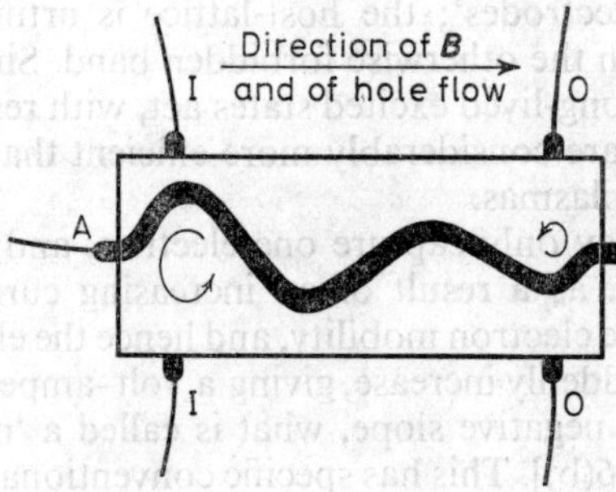

Fig. 11.14. The wave is replaced by a helical instability of a discharge column in the solid-state plasma. A, C, 'electrodes' of the discharge, II input, OO output

Still another type of amplifier utilises the principle of the travelling wave. This amplifier is more successful so far but also less interesting because of its lower frequency in the megacycle range. The wave in question is, however, not a true plasma wave, it is a helical instability of a plasma column generated by striking an arc through a semiconductor. The helical instability is triggered in the vicinity of one 'electrode' and made to propagate with negative damping in the direction of the other one where the amplified signal is picked-up (Figure 11.14). The same principle may be used to build a successful oscillator, called the *oscillistor* (Fig. 11.15). The oscillistor is tunable over a wide range by controlling the

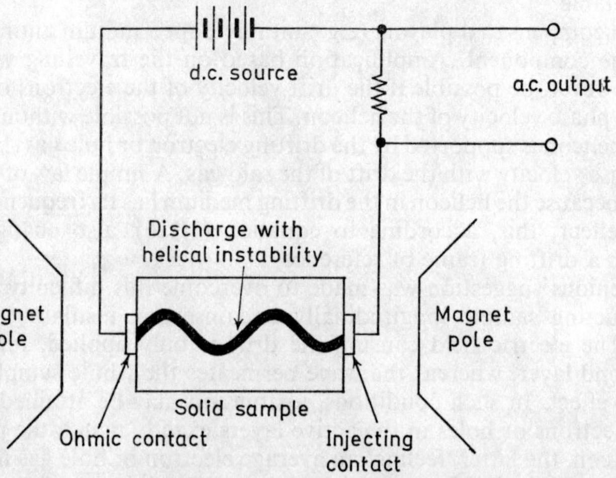

Fig. 11.15. Principle of the oscillistor: the same type of helical instability as in Figure 11.14 is here made to oscillate freely and steadily. An oscillating component appears in the discharge current and is easily picked-up

'arc' parameters, but neither the frequency, nor the power nor the efficiency, are in an especially useful range, and both devices will have to overcome much competition before they have a real technological future.

The same is true of an interesting device, or more exactly, a principle applicable to several devices, called a *madistor* (*ma*gnetic *d*eflection of an *i*njected plasma on *s*aturated *t*raps). Here also, a discharge is struck in a semiconducting medium between two 'electrodes'; the host-lattice is artificially spiked with metastable energy levels in the otherwise forbidden band. Since they are bound to the host-lattice, these long-lived excited states act, with respect to the mobile electrons, as 'traps', and are considerably more efficient than the formation of negative ions in gaseous plasmas.

Each trap, however, may only capture one electron, and when the electron population increases (e.g. as a result of an increasing current) they may all become saturated; then the electron mobility, and hence the electrical conductivity of the sample, may suddenly increase, giving a volt–ampere characteristic in which some parts have a negative slope, what is called a 'negative resistance' [Figures 11.16(a) and 11.16(b)]. This has specific conventional applications.

So far, the 'magnetic deflection' mentioned before plays no role, and the device may and does operate as such without it. It is employed if some regions of the sample, e.g. in the vicinity of the boundaries, have a higher density of traps. Their efficiency may be increased or decreased if a magnetic field perpendicular to the discharge axis is able to cause the electron gas to collect on them [Figures 11.16(c) and 11.16(d)].

Also, imagine a 'cathode' facing at least two identical 'anodes', each of them in front of a region of the sample having the same density of traps, and connected to the same voltage source through identical resistors. All the anodes are therefore on an equal footing, and if the arc is struck on one of them, it will stay there

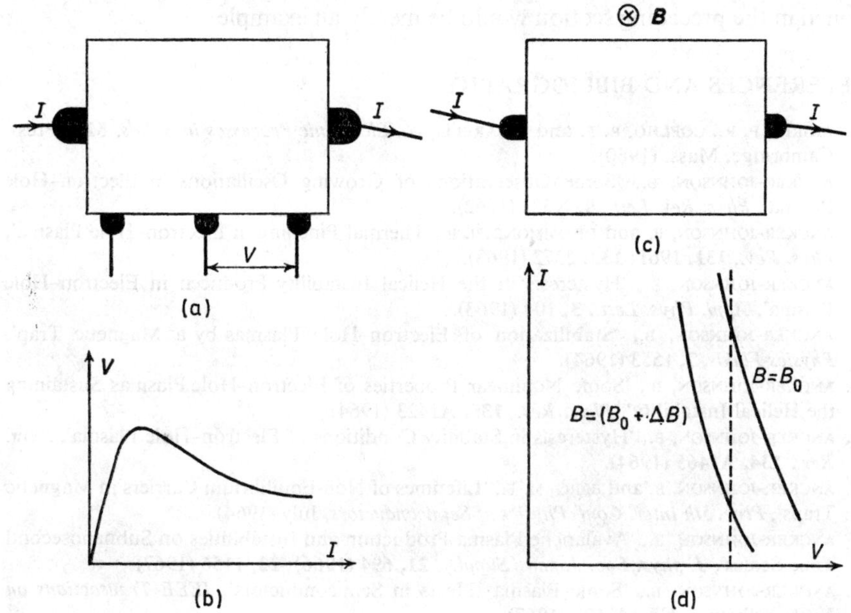

Fig. 11.16. Two forms of the madistor: in (a), *a sample as in Figure 11.11 but spiked with electron traps. In* (b), *the traps are saturated, and the voltage drop V decreases when the current increases. In* (c), *the same sample as in (a) in the presence of a magnetic field instead of a voltage; this magnetic field controls the current very efficiently*

because the corresponding traps are already saturated, making the resistivity locally smaller. This is a kind of stability reminding one of the 'flip-flop' and its equivalents in electronic circuits. Commutation may be induced by several methods, especially by the application of a transverse magnetic field. If the 'arc' is forced into a new location for any time exceeding the lifetime of the traps, it is the traps corresponding to the new, not to the old location, which are saturated and make the resistivity smaller. Decimal counting may be achieved easily with ten anodes around a small slotted disc, the cathode being located in a central hole. In a certain strength magnetic field, perpendicular to the plane of the disc, the commutation can be induced by signals; at higher fields, it becomes spontaneous and the 'discharge' rotates, generating periodic signals.

Magnetic deflection may be applied to transistors, giving them an extra controlling device, making them equivalent to tetrodes. There are also possible applications to switching, thereby eliminating the problem of contact wear. But technological progress is slow.

Solid-state plasmas also have indirect applications; for instance, if a solid-state plasma experiment provides information about a solid-state material (e.g. the accurate shape of its Fermi surface) and this information suggests a technologically valid application which would not have otherwise been considered. Another possibility exists if, for instance, solid-state plasmas can be made to emulate satisfactorily on small scale models, behaviour exhibited only in bulky and costly gaseous plasma experiments, of which magnetic confinement mentioned in the preceding section would be merely an example.

REFERENCES AND BIBLIOGRAPHY

1. AIGRIN, P. R., COELHO, R. P. and ASCARELLI, G., *Electronic Processes in Solids*, MIT Press, Cambridge, Mass. (1960).
2. ANCKER-JOHNSON, B., 'Some Observations of Growing Oscillations in Electron–Hole Plasma', *Phys. Rev. Lett.*, **9**, X523 (1962).
3. ANCKER-JOHNSON, B. and DRUMMOND, J. E., 'Thermal Pinching in Electron–Hole Plasma', *Phys. Rev.*, **131**, 1961; **132**, 2372 (1963).
4. ANCKER-JOHNSON, B., 'Hysteresis in the Helical Instability Produced in Electron–Hole Plasma', *Appl. Phys. Lett.*, **3**, 104 (1963).
5. ANCKER-JOHNSON, B., 'Stabilization of Electron–Hole Plasmas by a Magnetic Trap', *Physics Fluids*, **7**, 1553 (1964).
6. ANCKER-JOHNSON, B., 'Some Nonlinear Properties of Electron–Hole Plasmas Sustaining the Helical Instability', *Phys. Rev.*, **135**, A1423 (1964).
7. ANCKER-JOHNSON, B., 'Hysteresis in Stability Conditions of Electron–Hole Plasma', *Phys. Rev.*, **134**, A1465 (1964).
8. ANCKER-JOHNSON, B. and BERG, M. F., 'Lifetimes of Non-Equilibrium Carriers in Magnetic Traps', *Proc. 5th Inter. Conf. Physics of Semiconductors*, July (1964).
9. ANCKER-JOHNSON, B., 'Avalanche Plasma Production and Instabilities on Subnanosecond Time Scales', *J. phys. Soc. Japan (Suppl.)*, **21**, 694 (1966); **22**, 1156 (1967).
10. ANCKER-JOHNSON, B., 'Some Plasma Effects in Semiconductors', *IEEE Transactions on Nuclear Science*, NS-14, 627 (1967).
11. ANCKER-JOHNSON, B., 'Microwave Emission from Magnetic-Field-Free Electron–Hole Plasmas', *Appl. Phys. Lett.*, **10**, 279 (1967).
12. ANCKER-JOHNSON, B., 'Gigahertz Radiation from Magnetic-Field-Free Electron–Hole Plasma', *Phys. Rev.*, **164**, 1050 (1967).
13. ANCKER-JOHNSON, B., 'Plasma Effects in Semiconductors', *Proc. 9th Intern. Conf. Physics of Semiconductors*, October (1968).
14. ANCKER-JOHNSON, B., 'Magnetoconductance of Non-Equilibrium Plasmas in InSb', *J. Phys. Chem. Solids*, **29**, 1127 (1968).
15. ANCKER-JOHNSON, B., 'Microwave Emission from Non-Equilibrium Plasmas in InSb Subject to Magnetic Fields', *J. appl. Phys.*, **39**, 3365 (1968).
16. ANCKER-JOHNSON, B., 'The Spectrum of Microwave Emission from InSb', *Proc. IEEE*, **56**, 154 (1968).
17. BOK, J. (Ed.), 'Plasma Effects in Solids', *Proc. 7th Inter. Symp. Semiconductor Physics*, Academic Press, New York (1966).
18. BOWERS, R., LEGENDY, C. and ROSE, R., 'Oscillatory Galvanometric Effect in Metallic Sodium', *Phys. Rev. Lett.*, **7**, 339 (1961).
19. BOWERS, R., 'Plasmas in Solids', *Scient. Am.*, **209**, 46 (1963).
20. BOWERS, R. and STEELE, M., 'Plasma Effects in Solids', *Proc. IEEE*, **52**, 1105 (1964).
21. BUCHSBAUM, S. J. and CHYNOWETH, A. G., 'Plasmas in Solids', *Inter. Sci. & Technol.*, **48**, 40 (1965).

22. BUCHSBAUM, S. J. and WOLFF, P. A., 'Effect of Open Orbits on Helicon and Alfvén Wave Propagation in Solid-State Plasmas', *Phys. Rev. Lett.*, **15,** 505(c) (1965).
23. BUCHSBAUM, S. J., 'Plasmas in Solids', *Advances in Plasma Physics* (edited by A. SIMON) Wiley, New York (1968).
24. CHYNOWETH, A. G. and BUCHSBAUM, S. J., 'Solid State Plasmas', *Physics to-day*, **18,** No. 11, 56 (1965).
25. DUMKE, W. P., 'Theory of Avalanche Breakdown in InSb and InAs', *Phys. Rev.*, **167,** 783 (1968).
26. GANTMAKHER, V. F., 'A Method of Measuring the Momentum of Electrons in a Metal', *Soviet Phys. JETP*, **15,** 982 (1962).
27. GLICKSMAN, M., 'Instabilities of a Cylindrical Electron Hole Plasma in a Magnetic Field', *Phys. Rev.*, **124,** 1655 (1961).
28. GRIMES, C. C. and BUCHSBAUM, S. J., 'Interaction Between Helicon Waves and Sound Waves in Potassium', *Phys. Rev. Lett.*, **12,** 356 (1964).
29. GRIMES, C. C., ADAMS, C. and SCHMIDT, P. H., 'Observation of the Effect of Open Orbits on Helicon-Wave Propagation', *Phys. Rev. Lett.*, **15,** 506(c) (1965).
30. GUREVICH, V. L., SKOBOV, V. G. and FIRSOV, Y. A., 'Giant Quantum Oscillation in the Acoustical Absorption by a Metal in a Magnetic Field', *Soviet Phys. JETP*, **13,** 552 (1961).
31. HOYAUX, M. F., 'Plasma Phenomena and the Solid State', *Contemp. Physics*, **9,** No. 2, 165 (1968).
32. HOYAUX, M. F., *Solid State Plasmas*, Pion, London (1970).
33. HUNTER, L. P., *Handbook of Semiconductor Electronics*, McGraw-Hill, New York (1962).
34. KANER, E. A. and SKOBOV, V. G., 'Theory of Resonance Excitation of Weakly Decaying Electromagnetic Waves in Metals', *Soviet Phys. JETP*, **18,** 419, 1964.
35. KIRSCH, J. and MILLER, P. B., 'Doppler-Shifted Cyclotron Resonance and Alfvén Wave Damping in Bismuth', *Phys. Rev. Lett.*, **9,** 421 (1962).
36. KITTEL, C., *Introduction to Solid State Physics*, Wiley, New York (1956).
37. KONSTANTINOV, O. V. and PEREL, V. I., 'The Behavior of Fermian Spin in Elastic Scattering', *Soviet Phys. JETP*, **11,** 117 (1960).
38. LIBCHAKER, A. and VEILEX, R., 'Wave Propagation in a Gyromagnetic Solid Conductor—Helicon Waves', *Phys. Rev.*, **127,** 774 (1962).
39. LONGINI, R. L. and ADLER, R. R., *Introduction to Semiconductor Physics*, Wiley, New York (1962).
40. LUPATKIN, W. L. and NANNEY, C. A., 'Magnetosonic Waves in Bismuth', *Phys. Rev. Lett.*, **20,** 212 (1968).
41. MAXFIELD, B. W., *Helicon Waves in Solids*, Atomic and Solid State Physics Lab. Rept. 14850, Cornell University, Ithaca, N.Y., Sept. (1968).
42. PINES, D. and BOHM, D., 'A Collective Description of Particle Interactions', *Phys. Rev.*, **83,** 221 (1951); **85,** 338 (1952).
43. ROBBINS, W. P., LANTZ, R. M. and ANCKER-JOHNSON, B., 'Nanosecond Rise Time, High Power, Variable Delay Double Pulser with Application to Electron–Hole Plasma', *Rev. Scient. Instrum.*, **39,** 69 (1968).
44. SHOCKLEY, W. W., *Electrons and Holes in Semiconductors*, Van Nostrand, Princeton, N.J. (1950).
45. SINNOTT, M. J., *The Solid State for Engineers*, Wiley, New York (1956).
46. WALSH, W. M., Jr., GRIMES, C. C., ADAMS, G. and RUPP, L. W. Jr., 'Extremal Dimensions of Cyclotron Orbits in Tungsten', *9th Inter. Conf. Low Temperature Physics*, LT9 (Part B), 765, Plenum Press, New York (1965).
47. WILLIAMS, G. A. and SMITH, G. E., 'Alfvén Wave Propagation in Bismuth: Quantum Oscillations of the Fermi Surface', *IBM J. Res. Dev.*, **8,** 276 (1964).
48. WILLIAMS, G. A., 'Alfvén Wave Propagation in Solid State Plasmas. (I) Bismuth', *Phys. Rev.*, **139,** A771 (1965).
49. ZIMAN, J. M., *Principles of the Theory of Solids*, Cambridge University Press (1964).

12 Astrophysics and space sciences

12.1 INTRODUCTION[1]

The applications of plasma physics treated up to this point have been on a laboratory scale or an industrial scale. In this chapter, the realm of much vaster configurations in the plasma state is entered, including the cosmic scale itself.

On the cosmic scale, the plasma state is vastly more prevalent than the three other states of matter: solid, liquid, and gas. The solid state is encountered only in planet nuclei, in cometary heads, meteorites and interstellar dust; the gaseous state is encountered in planetary atmospheres and un-ionised interstellar gases; the liquid state is much more exceptional, since formations like the terrestrial oceans, not only represent a small percentage of the mass of the planet, but also exist only on exceptional planets. In contrast with this, all the remaining matter in the universe, including the stars, part of all the planetary atmospheres, an important fraction of the interplanetary and interstellar gases, and a fraction of the cometary tails, are in the state of partial, or even more frequently, of total ionisation. The fraction of matter in the universe which is ionised by far outweights the total of the three other states of matter.

Except for the electrical conductivity, which in most cases, does not appear as such in astrophysics and space science applications, there would be no major difference between the behaviour of this cosmic ionised matter and that of the ordinary gases, were it not for the existence, in the cosmos, of very significant magnetic fields, their significance being related either to their actual value or to their vast extent. Consequently, the physics of ionised gases in strong magnetic fields has become an increasingly important field of astrophysics.

Most of the field to be reviewed is still in a preliminary state of investigation. In general, no theory is universally accepted, even though some theories have been worked out with remarkable depth.

12.2 THE EARTH'S ATMOSPHERE[20, 46]

The layer of gas that surrounds the earth is defined as the *atmosphere*. This atmosphere is described by a certain number of characteristics, such as the temperature, the density, the chemical composition, and the degree of ionisation.

These properties change with increasing altitude. The study of the properties of the upper atmosphere has developed in a separate science known as *aeronomy*.

The temperature of the particles forming the atmosphere varies with increasing altitude as shown by Figure 12.1. This temperature depends also on the latitude and on the time of year. Up to a distance of about 16 km from the earth's surface, the temperature decreases steadily with a *lapse rate* of about 6·5°C/km down to a minimum value of −60°C. This layer is known as the *troposphere* and the minimum temperature region, the *tropopause*. After the tropopause, the temperature rises continuously up to an altitude of 30 km reaching about −40°C. This region is called the *stratosphere* and the maximum, the *stratopause*. From 30 to 80 km, the temperature increases up to 10°C, and then decreases to an overall minimum of −90°C or less depending on the season

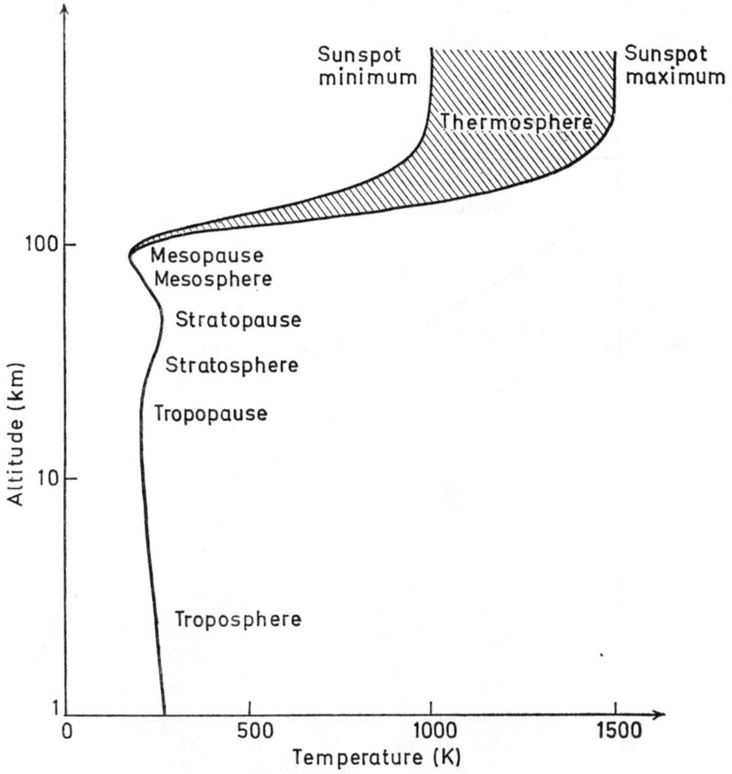

Fig. 12.1. Atmospheric temperature as a function of altitude

and the latitude. This region is known as the *mesosphere* and the overall minimum, the *mesopause*. Above an 80 km altitude, the temperature increases steadily reaching a saturation value of 750°C at the minimum sunspot, 1250°C at the maximum sunspot, and over 1700°C under solar flare conditions, at an altitude varying between 320 and 400 km. This region of the atmosphere is known as the *thermosphere*.

Most of the weather phenomena occur in the troposphere whereas most of the ionisation processes induced by the radiations emanating from the sun occur in the mesosphere. The temperature of the thermosphere depends almost exclusively upon the solar activity.

The density of the atmosphere decreases continuously with increasing altitude. However, this decrease changes continuously with time due to many factors. The most important of these factors are variations in light and corpuscular radiation and changes in the earth's magnetic field, both a result of solar activity. An approximate picture of this atmospheric density variation is shown in Figure 12.2. At very high altitudes, atmospheric densities become extremely small leading to a molecular mean free path of 1·6 km at about 300 km, and much higher values above this altitude.

The molecular density and temperature in the atmosphere are related to each other by the barometric equation. Indeed, if a small atmospheric region, in

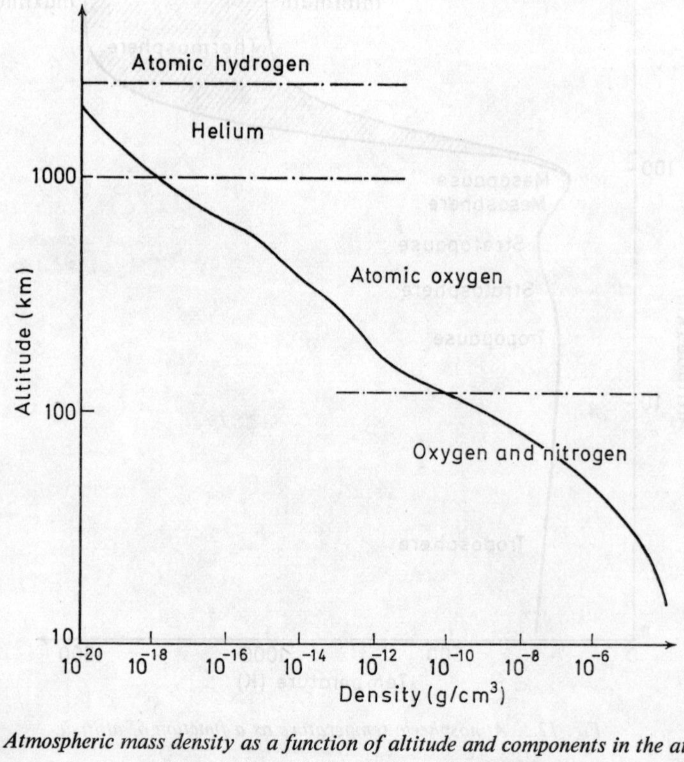

Fig. 12.2. Atmospheric mass density as a function of altitude and components in the atmosphere

which the temperature T is assumed constant, is considered, it can be shown that the relationship between the density n_1 at the altitude z_1 and the density n_2 at the altitude z_2 is:

$$\ln (n_1/n_2) = -(Mg/kT)(z_1-z_2) \tag{12.1}$$

where M is the average mass of the atmospheric molecules, and g is the gravitational acceleration. Since this acceleration varies with altitude, it is convenient to define a geopotential altitude as:

$$z' = r_0 z/(r_0+z) \tag{12.2}$$

where r_0 is the radius of the earth, and z the geometric altitude. If g_0 is g at sea level, equation (12.1) becomes:

$$(n_1/n_2) = \exp(-d/z_m) \tag{12.3}$$

where $d = h_1'-h_2'$, and $z_m = kT/Mg_0$, z_m being defined as the geopotential scale height. Equation (12.3) gives good values for the molecular density up to an altitude of 100 km. Above this altitude, both the molecular weight and temperature change with altitude.

The composition of the atmosphere also changes with altitude as indicated in Figure 12.2. Up to an altitude of about 100 km, the composition and the average molecular weight remain constant, because of the effect of the circulation and mixing of the air. This region is called the *homosphere*. It is constituted, apart from water vapour, of the following elements: molecular nitrogen (N_2), 78·08 volumetric %; molecular oxygen (O_2), 20·9 per cent; argon (Ar), 0·93 per cent and carbon dioxide, 0·033 per cent. The remainder is made up of about 15 minor constituents: noble gases especially helium, or gases produced mostly by biological processes (nitrous oxide, methane), or industrial exhausts (nitric oxide, nitrogen dioxide, carbon monoxide, sulphur dioxide); other constituents of importance are ozone and hydrogen.

Above 100 km there is a region of diffusive equilibrium where the effect of air circulation is negligible. In this region, called *heterosphere*, the molecules are distributed over a range of altitudes, depending on their molecular mass, the heaviest (N_2, O_2) being the lowest because of gravity, and the lightest (O, He, H) are the highest. Also, the effect of solar radiation is important in this region; for instance oxygen atoms are produced by photochemical dissociation of molecular oxygen. Up to about 230 km the atmosphere is then constituted of molecular nitrogen, and molecular and atomic oxygen. From 230 km to 600 km, atomic oxygen is predominant. From 600 to 2500 km, helium predominates, and above 2500 km, atomic hydrogen becomes the most important constituent.

The upper region of the heterosphere is called the *exosphere* and is of special importance for neutral particles. Indeed, some light particles, which are fast enough, can escape altogether from the earth's gravitational field.

The measurement of temperatures, densities, and composition in the atmosphere were first made with instruments carried by balloons, mainly for the troposphere. Later more accuracy could be achieved by rocket-borne instruments and radiosondes. The most accurate methods are those based on indirect measurements, such as measuring the velocity of sound by the rocket grenade. For very high altitudes, above 160 km, good results can be obtained by measuring such quantities as the aerodynamic drag on artificial satellites. Satellite-borne mass spectrometers and other measuring instruments are often used.

Many important effects appear in the atmosphere as the result of the different chemical reactions occurring in it. One such effect, the *greenhouse effect*, is due to the existence of ozone (O_3), carbon dioxide and water vapour in the atmosphere. Indeed, solar radiation penetrating the atmosphere is absorbed by the earth's surface which, as a result of its heating, emits infra-red radiation. The effect of ozone, carbon dioxide, and water vapour is to strongly absorb this radiation and thus help the earth's surface to retain its mild temperatures as in a greenhouse. The proportion of ozone in the atmosphere never exceeds 0·001 volumetric %.

Another phenomenon of interest in the atmosphere is the airglow that seems to exist at all times but most especially on dark nights, and is thus called *night-glow*. It is characterised by a faint luminosity in the sky which becomes rather strong at about 10° above the horizon and is the faintest at the zenith. The spectrum of this light reveals the existence of atomic oxygen, thus placing the origin mostly at an altitude of 100–160 km. Part of the radiation energy of the nightglow is believed to be due to the formation of atomic oxygen by absorption of the solar ultra-violet radiation:

$$O_2 + h\nu \rightarrow O + O$$

The two oxygen atoms could recombine in an excited oxygen molecule:

$$O + O \rightarrow O_2^*$$

This excited molecule is very unstable. It reverts to the ground state, liberating radiation in the green line (5577 Å) and the red lines (6300 Å, 6364 Å, 6391 Å) of its ground state.

12.3 THE IONOSPHERE[38]

From experiments on radio transmission, it has been concluded for a long time that ionised layers, conducting electric currents and reflecting radio waves, should exist in the upper atmosphere. These layers are termed the *ionosphere*. Accurate sounding of these layers is possible by using the radio-echo technique. As a first approximation, both terrestrial magnetic field and the energy loss by collisions can be neglected, and the most important property is then the cut-off at the plasma electron electrostatic resonance frequency. In a collisionless plasma without magnetic field, an electromagnetic wave propagates without attenuation, if its frequency is above this resonance frequency, and it is totally reflected if it is below it.

A concentration of n particles/cm³ each of charge e and mass M, if subjected to an electric field $E = E_0 \sin(\omega t)$, gives rise to a current density J_c such as:

$$J_c = -(ne^2/M\omega)\, E_0 \cos(\omega t) \tag{12.4}$$

The displacement current density is:

$$\partial \boldsymbol{D}/\partial t = \varepsilon_0 \omega E_0 \cos(\omega t) \tag{12.5}$$

The total current density is equal to the conduction current and the displacement current densities; thus, from equations (12.4) and (12.5):

$$\boldsymbol{J} = [\varepsilon_0 - (ne^2/M\omega^2)]\, \mathrm{d}E/\mathrm{d}t \tag{12.6}$$

The effect of the motion of charges is then to reduce the dielectric constant from ε_0 to ε'; the value of ε' can be written as a function of the characteristic plasma frequency ω_p:

$$\varepsilon' = \varepsilon_0[1 - (\omega_p/\omega)^2] \tag{12.7}$$

This will give rise to an increase in the phase velocity of the waves, the value of which is $v_p = 1/\sqrt{(\mu\varepsilon')}$. Therefore, a wave incident at some angle on the surface of an ionised layer will propagate more and more away from its initial direction of propagation. This is the basis of *Larmor's theory* of reflection of radio waves from the ionosphere. If the number density, thus ω_p, decreases with increasing altitude, the direction of propagation of the wave will become more and more horizontal as it proceeds upward, until it is eventually reflected completely as illustrated in Figure 12.3. The normal refractive index $\nu_R = c/v_p = \sqrt{(\varepsilon'/\varepsilon)}$

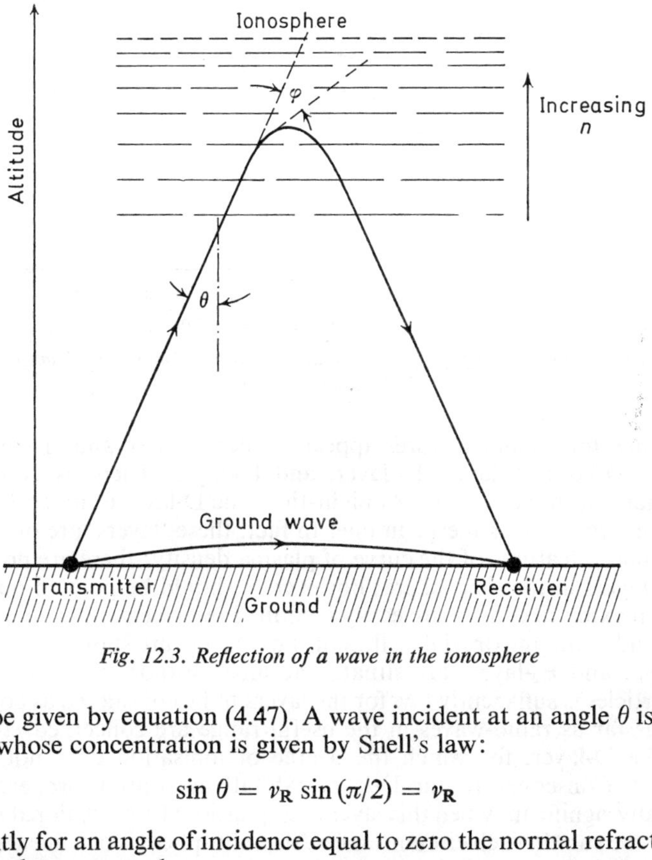

Fig. 12.3. Reflection of a wave in the ionosphere

will then be given by equation (4.47). A wave incident at an angle θ is reflected at a level whose concentration is given by Snell's law:

$$\sin\theta = \nu_R \sin(\pi/2) = \nu_R \tag{12.8}$$

consequently for an angle of incidence equal to zero the normal refractive index will be equal to zero, and $\omega = \omega_p$.

A transmitting plasma has a refractive index different from unity and its 'optical path' can be different from the geometrical one and frequency-dependent. This leads to the conclusion that echo-sounding is not straightforward and needs simplification using a curve of the 'apparent' or 'virtual' height as a function of frequency. The result is a curve of the plasma density as a function of the height such as shown in Figure 12.4. A significant difference exists in this curve between daytime and night-time, which is responsible for the well-known diurnal differences in the range of radio stations. Also, abnormal solar phenomena generate abnormal states in the ionosphere.

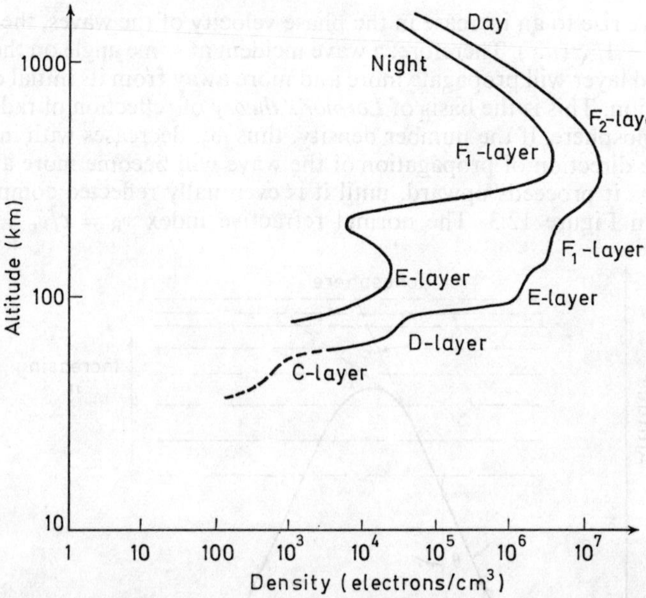

Fig. 12.4. Carrier density as a function of the altitude in the ionosphere (sunspot maximum, 45° latitude, summer)

In daytime, four major features appear in such a curve and are respectively termed the D-layer, E-layer, F_1-layer, and F_2-layer. There is sometimes a C-layer underneath the D-layer. At night-time, the D-layer is much less signifi-ant and the two F-layers merge in one. In fact, these 'layers' are only more or less pronounced features of the curve of plasma density; there are no distinctly separated layers in the ionosphere; however, it should be pointed out that the regions of maximum density are easily identified by the echo-sounding method.

The altitudes and maximal densities of those layers are shown in Figure 12.4. The F-, F_1-, and F_2-layers are situated at such altitudes that the density of neutral particles is sufficiently low for the layers to be considered as collisionless plasmas, as far as radio-waves in the useful range are concerned. This is not true for the D-layer, for which the degree of ionisation does not normally exceed 10^{-9}. Consequently, the D-layer exhibits absorption properties which are especially significant when this layer is emphasised by exceptional radiations emitted by the sun. In such circumstances, the ionosphere is detrimental to radio wave propagation.

Factors affecting the ionosphere

Radiation from the sun has a major influence on the density and structure of the ionosphere. Not only are these significantly different at night-time from what they are in daytime, but they also depend on the latitude, season, and even more so on typical solar phenomena such as sunspots and flares. During total solar eclipses the carrier density falls markedly in all the layers. However, the theory of the ionosphere is still far from being complete; other ionising agents than those emitted by the sun play a significant role.

One way to study the ionosphere is to consider the different ionising agents coming from outside the earth and to determine to what depth they can penetrate the atmosphere; in general, an ionising agent under such conditions becomes most effective close to the end of its path. In this respect, far ultra-violet rays should be responsible for the F_2-layer, and medium ultra-violet for F_1. Soft X-rays have a path terminating close to the E-layer, whereas quite a number of ionising agents contribute to the D-layer, including ultra-violet and moderately fast protons and electrons. In addition, it is suspected that an important part of the normal D-layer is due to small meteorites, which ionise the air through the shock-wave that they develop in crossing it. The influence of cosmic rays is the most pronounced in the tenuous C-layer. The ionosphere cannot be related to the radioactivity of the ground or of the air; both are too weak and too short-ranged to account for the ionospheric phenomenon. However, an important contribution to the ionisation of the upper atmosphere can be traced to the radiation leaking downward from the upper earth radiation belts.

Sounding the ionosphere

Ground-based echo-sounding is practical only as a means of detecting deviations in the rising part of the curve of the carrier density versus height. Electric probe measurements and other techniques carried out by ionospheric rockets and artificial satellites are used to determine the decreasing part of the plasma density above the layers. The kind of probe used is not different from the Langmuir probe, but it has some of the features of a double probe, the second member of the pair being the vehicle body itself. This kind of probe measurement is not easy to perform. Not only does the vehicle perturb the gaseous medium around it, but also it creates its own ionising track through its shock wave. Moreover, the vehicle velocity, the effect of which is to superpose a drift in relative motion on the supposedly Maxwellian distribution of velocities, is not negligible with respect to the particle velocity, especially in the case of ions. The probe shape and probe location must be such as to minimise such effects. More convenient, but less direct, is the method of echo-sounding as applied, not to the layers, but to the vehicle. The method consists essentially in comparing, at the same time, two 'optical paths' between the same space vehicle and the same tracking station, at two different frequencies, taking advantage of the property that the refractive index is a function of carrier density and of frequency. Thanks to such methods, it is now known that the F-layer extends very far, up to 1000 km under normal conditions, with exceptional bulges, especially in the area above the magnetic equator, of some 5000 km in height.

Dynamo effects in the ionosphere[22]

The atmosphere, like the oceans, is subjected to tidal effects caused by lunar and solar gravitational attraction. However, the solar gravitational field is more effective than the lunar one, owing to some resonance effect. Also, the influence of thermal convection is much more pronounced. Owing to the combination of these two effects, atmospheric tides are basically a 24 h period phenomenon rather than a 12·5 h period one as for the oceans.

The ionosphere can be depicted as a structure of ionised layers essentially carried up and down by the surrounding non-ionised gas. In this motion, the conducting layers move with respect to the lines of force of the terrestrial magnetic field, hence the ionosphere is a gigantic MHD generator under short circuit. The terrestrial magnetic field is less than 10^{-4} T in strength and the ionosphere is not a very good conductor (40 mho/m), but owing to the overall dimensions, currents of the order of 100 000 A circulate even when the sun is perfectly quiet.

The solar component of the ionospheric tide can be depicted as comprising four rings, circulating respectively around the points having their local time equal to approximately 11 and 18 h, and their latitude equal to $\pm 40°$. If the MHD effect were of pure gravitational origin, the period would be 12 h instead of 24 h; but the effect of heating during the day profoundly modifies the picture. Moreover, the ionosphere is less conducting by night than by day.

The lunar component of the ionospheric tides is about a quarter as strong. Since it is essentially the result of gravitational attraction, the period is identical to that of sea tides. However, the lunar component itself has a 24 h modulation, owing to the variation in ionospheric conductivity between day and night. Neglecting this modulation, the resulting pattern is made up of eight loops encircling the points $\pm 40°$ of latitude, for which the lunar local time is about 3, 9, 15 and 21 h. Both components show additional irregularities owing to the different locations of the magnetic poles and the geographic poles, and because of the obliquity of the terrestrial axis of rotation (seasonal effect).

Currents of such an order of magnitude generate non-negligible magnetic fields; thus the reaction of these currents on the terrestrial magnetic field is significant. However, it is much less significant than spurious effects, connected with abnormal solar activity through hydrodynamic shock waves and particle streams, the results of which are designated as 'magnetic storms'. They are in general accompanied by partial or total fading of long distance broadcast communications.

Artificial creation of ionospheric layers

With the advent of the fission bomb, the creation of artificial ionospheric layers is definitely not beyond man's grasp. Such creation could be undertaken either to help in understanding the way the natural layers are established or as a substitute for the natural 'radio-mirrors' when the latter are fading. However, such experiments should be conducted with extreme care. Until more is known about the natural ionospheric process, there is always some probability that a badly designed or badly conducted fission explosion might permanently damage the ionosphere, or perturb it in a detrimental way for a time comparable to the human lifetime.

Aurorae[13]

Apart from the nightglow and the normal luminosity in daylight (blue sky), which is the result of particle scattering, some strongly luminous phenomena appear sometimes in the upper atmosphere, especially in areas separated about 25° in latitude from both magnetic poles, where exceptionally luminous phenomena, designated as *aurorae*, appear. These are due to swarms of charged particles, in the form of plasmas rather than unipolar clouds, striking the upper atmosphere. The swarms are not directly responsible for the luminous phenomena, the latter are the result of secondary radiations.

There are two major types of aurorae, the *normal aurora* and the '*sunlit*' *aurora*. The first one appears as an independent phenomenon, resulting only from the interaction of a particle swarm with the upper atmosphere, and can exist even in complete darkness. The second one necessitates the intervention of a third feature, illumination by the sun. In this case, it can be assumed that some kind of ionic and molecular compound is generated in the auroral phenomenon, which is excited and rendered luminous only under solar radiation.

The normal aurora appears generally at an altitude of 100–120 km; the sunlit aurora is, on the average, two to three times higher, and much less concentrated at the one altitude. Aurorae and magnetic storms are connected through a common origin, abnormal solar phenomena. Also, an aurora in the northern hemisphere is always accompanied by a symmetrical one in the southern hemisphere just about at both ends of the same lines of force of the terrestrial magnetic field.

The spectrum of the aurorae comprises mostly forbidden lines and bands of atomic oxygen, atomic and molecular nitrogen, and atomic excited hydrogen (Balmer lines), the latter being widely displaced by Doppler effects. This is an indication that protons play a major role in the initial triggering, and are major constituents in the plasma swarms sent by the non-quiescent sun. The peculiarities of the sunlit aurora have been attributed to the molecular nitrogen.

Whistlers[43]

When a radio receiver is connected to an antenna, a whistle is sometimes heard at very low frequencies. This whistle starts at a high pitch and then falls in about one second. It may be followed by fainter whistles of longer duration. When those whistles have rising tones they are called *nose whistlers*, with the lower frequency being called the *nose frequency*.

These whistlers are the result of lightning occurring at very high altitudes (about 30 000 km), the discharge of the lightning being accompanied by electromagnetic radiation over a wide range of frequencies including those corresponding to the audible sound waves.

12.4 THE EARTH'S RADIATION BELTS

Since the 1940s, it was suggested that particles might be trapped in the terrestrial magnetic field, like satellites, in equilibrium between the gravitational pull, the centrifugal and the magnetic forces. However, it was only after the launching

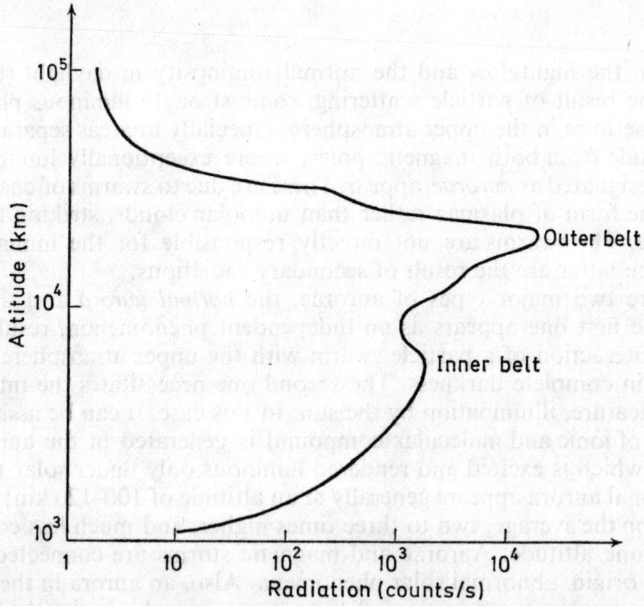

Fig. 12.5. Response of radiation counter in the Van Allen belts on the spacecraft Pioneer III (Dec. 6, 1958)

of the first artificial satellites in 1958 that the true belts of electrically charged particles were discovered around the earth, and named the *Van Allen belts*, after their discoverer. These belts are of approximately toroidal form and can be divided into two important belts as shown by Figure 12.5. In the inner belt, about 3000 km above ground, the most significant component consists of fast protons; in the outer belt, about 15 000 km above ground, fast electrons are the most conspicuous. However, the method of detection used in artificial satellites (Explorers, Pioneers, Sputniks, Lunas) are essentially based upon Geiger–Müller tubes or photoscintillators, and emphasise the role played by fast particles whereas particles essentially at rest remain unnoticed. It is suspected that the density of the latter is rather high and that the radiation belts are essentially plasmas, in which exceptional classes of particles move at much higher speeds than the average. The flux of fast particles is four or five orders of magnitude above the normal intensity of cosmic rays in the interplanetary range. Shielding against such radiation appeared as a difficult problem for manned spaceships.

The simplified model of the Van Allen belts as plasma rings more or less similar to the rings of Saturn, except that the magnetic force intervenes significantly in the equilibrium between the gravitational pull and the centrifugal force, is far from being appropriate for a quantitative description. A model involving a trapping between two magnetic mirrors represented by the two magnetic poles is much more satisfactory.

The terrestrial magnetic field, outside the inner metallic core, can be depicted, to a rather high accuracy, as a dipole field. It is well known that the field of

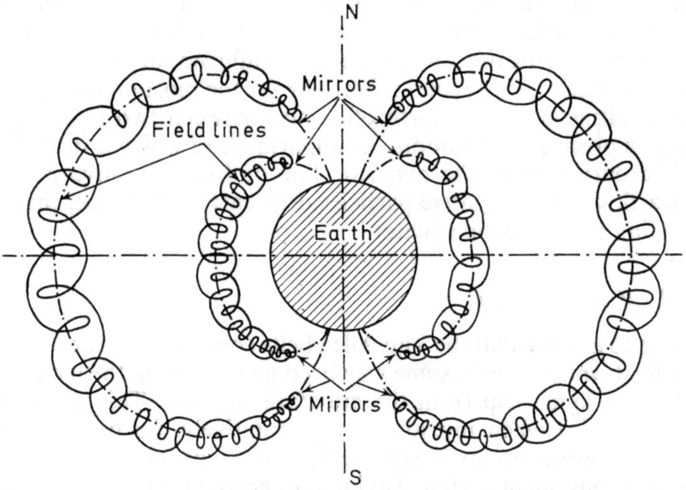

Fig. 12.6. Motion of charge-carriers in the Van Allen belts. They are trapped between two magnetic mirrors

a uniformly magnetised sphere outside its boundary is identical to that of an ideal dipole in its centre, and the model is probably good enough if restricted to the inner core of the earth.

Figure 12.6 shows that both magnetic poles act as mirrors. As a first approximation, the particle orbits can be viewed as helices of variable diameter and curved along the magnetic lines of force, with magnetic reflection at both ends. It is thus possible to confine particles in a manner equivalent to that of the mirror machine, with the only difference that the confined zone will be shaped in a kind of torus. A cross-section of this torus is shown by the idealised proton flux contours of Figure 12.7.

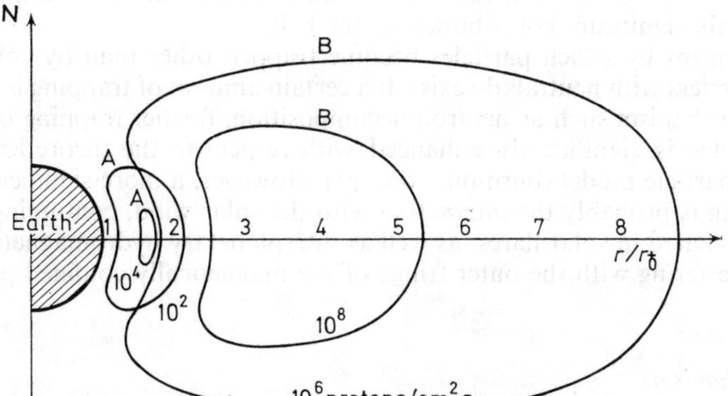

Fig. 12.7. Idealised proton flux countours in the Van Allen belts. Curves A: 'energetic' protons ($\mathcal{E} > 30$ MeV). Curves B: 'non-energetic' protons ($0{\cdot}1 < \mathcal{E} < 5$ MeV)

To refine the model, several effects need to be taken into consideration. They are: the gravitational pull of the earth, the centrifugal force caused by the curvature of the magnetic lines of force, and the influence of the plasma pressure. For energetic particles, the second effect is by far the most predominant.

Particles orbiting back and forth off centre between two magnetic mirrors also undergo a precessional motion, the importance of which is directly proportional to the curvature of the magnetic line of force. Positive particles will drift westward, whereas negative particles drift eastward; hence, there is a net current circling the magnetic equator.

Origin of the Van Allen belts

It is tempting to assume that the Van Allen belts comprise particles from outside the earth's field which have in some way become trapped in the terrestrial magnetic field, with some help from the gravitational field. This model becomes much less attractive when investigated quantitatively. Such trapping is only possible if some event occurs which modifies the magnetic moment while the particle is in a convenient region. Taking into account the density of the interplanetary gas, the probability that such an event is simply a collision between a charge carrier and a neutral atom is several orders of magnitude too small.

Some of the particles in the belts are injected as neutrals from below. Cosmic rays interfering with the terrestrial atmosphere can generate neutrons. Part of those neutrons are sent upward (albedo neutrons), and since the neutron is an unstable particle, they decompose spontaneously into proton–electron pairs. A major argument in favour of this theory is that the particle velocity distribution within the belts is highly inexplicable, unless the particles have been released from the decomposition of neutrals with a given energy emission. The fast proton component in the inner belt is mostly due to such an origin. It is less certain, however, that the fast electrons in the outer belt have the same origin. Whether or not the slow components arise from the deceleration of the fast components without jeopardising the trapping conditions is not yet clear. However, the majority of investigators believe that neutron decomposition is not the only significant contribution to the belts.

Mechanisms by which particles become trapped other than by collision of charge carriers with neutrals do exist. If a certain amount of trapping is originated in a mechanism such as neutron decomposition, further trapping of incoming particles is significantly enhanced with respect to the theoretical results of a one-particle model (burn-out concept). However, a more significant cause of trapping is probably the interaction with the solar wind, continuing plasma blobs originated in solar flares, as well as interplanetary hydromagnetic shockwaves interfering with the outer fringe of the magnetically confined plasma of the belts.

Loss mechanisms

At the low particle densities present, recombination is a highly improbable event and can be neglected. The particles can be lost downward, by interference with the terrestrial atmosphere, and upward by a defect of the magnetic confinement.

In a one-particle model of the magnetic mirrors and in a vacuum, the turning point should always be at the same height above the northern or southern magnetic pole. However, if a plasma of non-negligible density is trapped in between, the probability that, in a given lapse of time, the magnetic moment changes by some kind of interaction is not negligible. Moreover, since the density of the terrestrial atmosphere decreases regularly with increasing height, the curved part of the path, where the magnetic lines of force plunge toward the magnetic poles, occurs in a denser gas. On the average, the particle loses more momentum than it gains and its turning point comes closer and closer to a magnetic pole, until it reaches a region where the atmosphere has enough density to annihilate it in an auroral-like phenomenon.

Similarly, in a strictly collisionless plasma, there would be no mobility phenomenon in the direction perpendicular to the magnetic lines of force. But if interactions do occur, there will be a leakage; for fast particles, the centrifugal force caused by the magnetic line of force curvature is distinctly predominant with respect to the gravitational pull and the pressure gradient so that the dominant leakage is upward.

It is suspected that the leakage is enhanced during magnetic storms, when the magnetic lines of force of the earth are more or less 'shaken' by the intrusion of plasma blobs from the sun. During such periods, not only might the rate of leakage be significantly changed, but also the rate of capture.

Artificial belts

It has been demonstrated experimentally that Man could create artificial belts of density not negligible with respect to that of the natural ones. Once more, the source of particles is a fission explosion occurring at a sufficient height. In the so-called 'Starfish' blast of July 9th 1962, about 10^{25} fast electrons, originating from beta activity of fission fragments, were semi-permanently injected between the magnetic mirrors of the terrestrial field. The corresponding belt differs from the Van Allen belts in altitude, composition and velocity, but the overlapping is considerable in all three parameters. Moreover, the decay time-constant proves to be much longer than expected, so it is considered that, unless some unexpected phenomenon intervenes, the perturbation of the radiation belts will persist for as long as 20 years. Under such conditions, it can be questioned whether the benefit from such experiments exceeds the perturbation created in the normal study of the natural belts. In any case, much care is necessary if future tests of that kind are to be conducted.

12.5 THE MAGNETOSPHERE AND THE SOLAR WIND[6, 37]

The solar corona is in a state of dynamic, rather than static equilibrium. At a temperature of one million degrees, an important fraction of the particles is above the velocity of escape and the observational evidence suggests that an interplanetary plasma is continually streaming radially away from the sun. The solar corona is replenished from underneath. The total amount of matter involved is tremendous compared to human standards, but negligible with respect to the mass and probable lifespan of the sun.

At the distance of the earth's orbit this '*solar wind*' has a density of 5–10 carrier pairs/cm^3 (mostly ionised hydrogen) and the velocity is estimated between 300 and 600 km/s. During the solar flares, the density increases by one order of magnitude or more, and the velocity also increases significantly. It appears that such increases are in the form of plasma blobs with trapped magnetic lines of force. Although the dense part of the Van Allen belts is reasonably symmetrical about the terrestrial magnetic equator, the remote areas, at some 10–20 earth radii, show definite asymmetry because of the pressure exerted by the solar wind.

The solar wind cannot mix with the earth's magnetic field because of the frozen flux concept. A cavity in the interplanetary plasma therefore exists around the earth, called the *magnetosphere*. At the surface of this cavity, the pressure of the solar wind is in equilibrium with the magnetic pressure of the terrestrial field. Reciprocally the magnetic lines of force are pushed back towards the earth on the sunlit side, whereas there is a kind of trail on the dark side. There is normally no interpenetration between the solar wind and the Van Allen belts, and the former is thus kept away from the earth. Probing satellites have observed the existence of the boundary of the magnetosphere, the results being shown in Figure 12.8.

Whenever a solar flare occurs, plasma blobs in the shape of magnetic bottles are sent away as local reinforcements of the plasma wind. When they strike

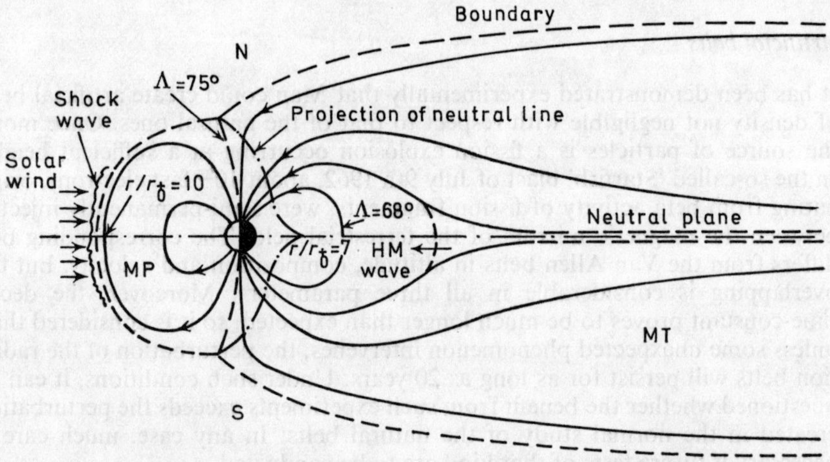

Fig. 12.8. The geomagnetic field lines of force in the magnetosphere. The dashed line is the boundary of the magnetosphere. (From Piddington[37] by courtesy of Pergamon Press)

the surface of separation of the Van Allen belts, the equilibrium between the pressure of the solar wind and that of the terrestrial magnetic field is locally broken. The magnetic lines of force are shaken (magnetic storm) and particles may become trapped. Magnetic storms are generally preceded by smaller disturbances; the latter have been attributed to hydromagnetic shock waves travelling through the interplanetary plasma and interacting with the magnetosphere surface.

Magnetosphere theories[8,18]

Chapman and Ferraro[16,17] were the first to develop a theory of the magnetosphere from data received through magnetic storms and aurorae. This theory was based on the assumption that the solar wind exerts a normal pressure on the geomagnetic field which is confined to a cavity. In spite of the fact that a ring current[42] was introduced into the model, the latter remained unsatisfactory in explaining most of the components of the geomagnetic storms as well as the aurorae. Piddington[36] suggested that a tangential solar wind pressure also exists leading to the concept of a frictional interaction resulting from the instability of two interpenetrating ion streams. One of these streams is made of fast electrons (solar wind), the other of slow protons (Van Allen belts). Both streams flow on the magnetosphere surface causing sudden changes in the magnetic field. The frictional interaction results also in stretching the magnetic lines of force from the day side of the magnetosphere into the night side, thus forming a magnetospheric tail as shown in Figure 12.8. Some of those lines comprise an open tail, that is, they are frozen away far from the orbit of the earth.

The above model overlooks many properties of the magnetosphere. Mead and Beard[34] suggested the assumption of specular reflection. The latter, however, does not explain the existence of a magnetospheric tail. An image dipole model has been also suggested[33] by using the Biot–Savart law, and integrating over the boundary of the magnetosphere. This model is satisfactory for fields near the boundary. The ground-level field, however, will depend upon the current induced by the solar wind, the latter flowing in a sheet around the polar cap from dawn to dusk.

Description of the magnetosphere

Depending upon the activity of the sun, the boundary of the magnetosphere in the day region varies in distance from 8 to 10 earth radii. In the night region, there exists an open magnetic tail. In this tail the field lines above 75° on the noon meridian and above 68° on the midnight meridian are frozen away from the earth's orbit. The part of the magnetosphere in the day region is called magnetosphere proper (MP), whereas in the night region it is called magnetospheric tail[7] (MT). Between the two a neutral line of zero magnetic field exists. At this line there is an abrupt change in the direction of the magnetic field. This line constitutes the weakest region of the magnetosphere as far as its shielding capability from an external plasma is concerned. The MP has a pair of 'wings' stretching away near the dusk and dawn meridians. These wings stretch in the MT, joining in a second neutral line. They create a motion of the magnetic lines of force transferring a large amount of magnetic energy. They then contract back into hydromagnetic waves, along the night auroral zone and then to the MP surface. An intermittent shock wave occurs on the surface of the MP heating the electrons to energies higher than 40 keV.

It should be noted that the rotation of the earth around its axis does not lead to any twisting of the magnetosphere. Indeed, the latter is insulated from the ionosphere into which the field lines are not frozen.

The region between the intermittent shock wave and the boundary of the MP (the magnetopause) is the seat of a strong magnetic turbulence. In this region, 50 keV electrons are trapped. One should note, finally, that there is a marked asymmetry about the midnight–noon meridian. Indeed, the intermittent shock wave is evident only in the dawn hemisphere. It degenerates into a strong hydromagnetic wave toward the dusk hemisphere. In the latter there seems to be no evidence of a shock or even of trapped energetic electrons.

12.6 SOLAR PHYSICS

Until the early 1940s, solar physics was mainly considered as the extrapolation of conventional physics and conventional spectroscopy. The concept that the solar behaviour is essentially magnetohydrodynamic was first introduced by Alfvén[1], who pointed out that magnetohydrodynamic waves were likely to play a major role in the sun as well as in all the stars. Progress in the quantitative field has, however, been slow. Theories based on analogies with phenomena at the laboratory scale or at the scale of the terrestrial atmosphere fail to explain the observational facts by several orders of magnitude, when numerical values are introduced. Two reasons can be invoked to explain such discrepancies. Either our knowledge of the fundamental laws at the solar scale is too crude, or essential MHD phenomena are still unknown. If a fundamental phenomenon like the terrestrial magnetic field could be explained without necessitating any correction to conventional equations, by working out a successful model, it could be considered as an indication that no correction is needed in solar physics either. At first sight, the enormous dimensions of the sun, and its high temperature, should introduce simplifications. For instance, most of the solar matter can be considered as ideally conductive so that the magnetic lines of force are frozen. Also, the compensation between positive and negative space charges should be perfect.

Description of the sun

The different parts of the sun and the solar atmosphere are shown in Figure 12.9. The visible part of the sun is called the *photosphere*. Its external shape is remarkably spherical due to the gravitational field. The photosphere looks darker and cooler close to the edge, because oblique rays reach opacity at a smaller depth, hence in cooler layers. Ultra-high speed photography shows that the brightness of the solar surface is remarkably heterogeneous and suggests an intense state of turbulence. Exceedingly dark spots, the 'sunspots', appear on the photosphere, associated with exceedingly bright areas, the 'facules'. From the spot motion, it has been inferred that the sun rotates on its axis in about 25·5 days at the equator. However, this rotation is not the same at all latitudes, Doppler measurements in the polar areas indicate a period of rotation of some 38 days.

Above the photosphere is a tenuous atmosphere called the '*chromosphere*', which is visible only during total solar eclipses, but which can be rendered visible at all times using special spectrographic techniques. The chromosphere is 3000–5000 km thick and has exceptional bulges and arches termed '*prominences*' extending up to heights of the same order as the solar radius. Whereas the photosphere radiates like a black body, the chromosphere radiates like a rarefied gas.

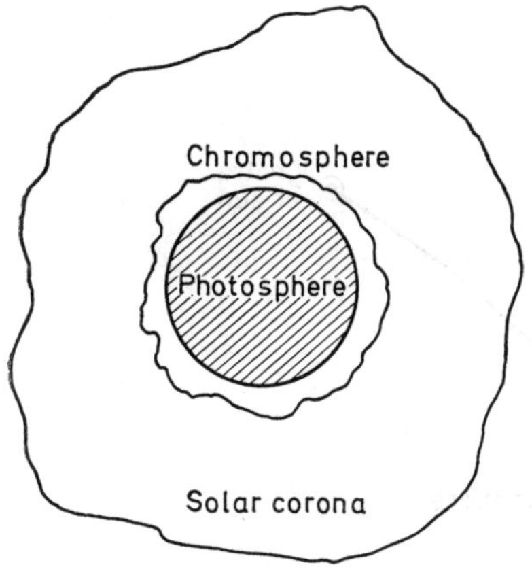

Fig. 12.9. The sun and the solar atmosphere

Above the chromosphere is an extended atmosphere designated as the 'corona'. The emission of spectral lines corresponding to ions which have lost more than 10 electrons from the solar corona shows that the temperature of the corona is about 100 eV, and is therefore much higher than that on the solar surface.

To determine the number density n and the temperature T in the corona, the properties of electromagnetic waves propagating through it have been used. Consider a star emitting radio-waves; when these waves cross the corona, they

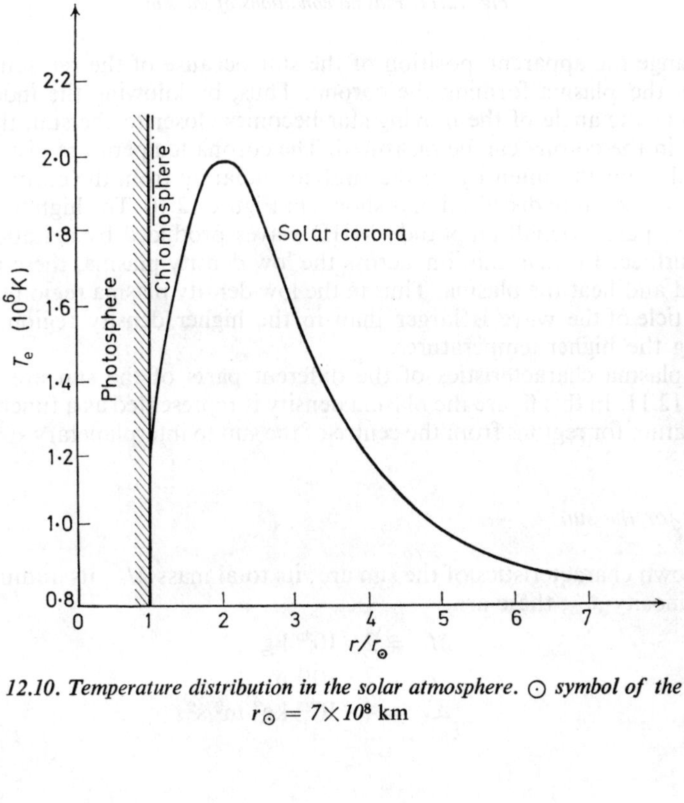

Fig. 12.10. Temperature distribution in the solar atmosphere. ⊙ symbol of the sun; $r_\odot = 7\times10^8$ km

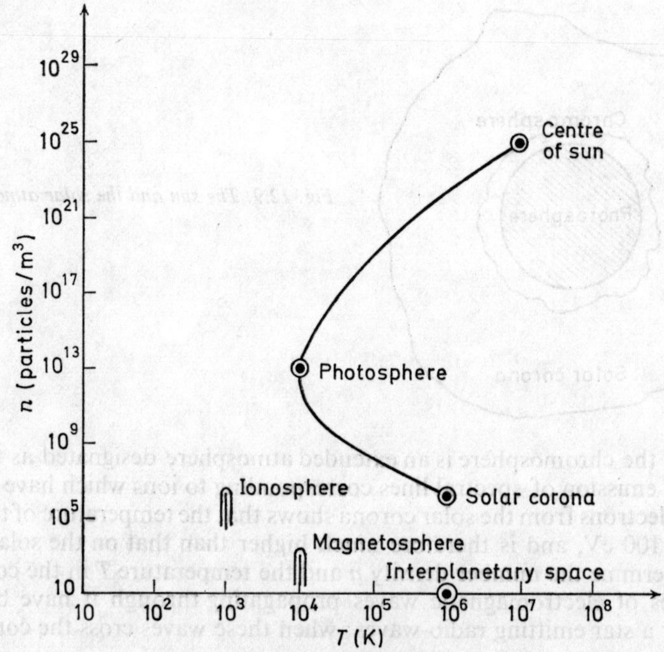

Fig. 12.11. Plasma conditions of the sun

will change the apparent position of the star because of the refraction of the wave in the plasma forming the corona. Thus, by knowing the index of refraction as the angle of the moving star becomes closer to the sun, the plasma density in the corona can be measured. The corona temperature distribution is deduced from the intensity of the emitted radiation from the corona. The resulting temperature distribution is shown in Figure 12.10. The high temperature is due in part to small amplitude MHD waves produced by agitation on the solar surface. In their motion across the low density plasma, these waves are damped and heat the plasma. Thus in the low density plasma region the energy per particle of the wave is larger than in the higher density region, thus explaining the higher temperatures.

The plasma characteristics of the different parts of the sun are shown in Figure 12.11. In this figure the plasma density is represented as a function of the temperature for regions from the centre of the sun to interplanetary space.

Models for the sun

The known characteristics of the sun are: its total mass $M_\odot$, its radius $r_\odot$, and its luminosity $\mathcal{L}_\odot$; these are:

$$M_\odot = 2\times 10^{30} \text{ kg}$$
$$r_\odot = 7\times 10^{11} \text{ m}$$
$$\mathcal{L}_\odot = 4\times 10^{25} \text{ kg}^2 \text{ m}^2/\text{s}^2$$

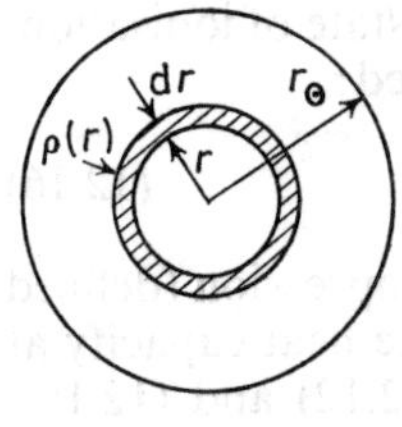

Fig. 12.12. A model for the sun

It is proposed to calculate the density distribution $\varrho(r)$, the pressure, the temperature, and the chemical composition. Assume first that the sun is a symmetrical sphere in hydrostatic equilibrium (Figure 12.12). At a point r inside the sun, the spherical symmetry assumption yields:

$$dM(r) = 4\pi r^2 \varrho(r)\, dr \tag{12.9}$$

Because of the hydrostatic pressure p_r, one has:

$$dp_r = -g\varrho(r)\, dr \tag{12.10}$$

where g is the gravitational acceleration. One has also:

$$g = GM(r)/r^2 \tag{12.11}$$

where G is the universal constant of gravity; thus:

$$dp_r = -GM(r)\,\varrho(r)\, dr/r^2 \tag{12.12}$$

Eliminating dr between equations (12.9) and (12.12), yields:

$$dp_r = -GM(r)\, dM/4\pi r^4 \tag{12.13}$$

and the pressure at the centre of the sun will be:

$$p_{rc} = -\frac{G}{4\pi}\int_{r_\odot}^{0} \frac{M(r)\, dM}{r^4} \tag{12.14}$$

The variation of the mass as a function of the radius is unknown. However, one can use a limit method to solve equation (12.14), by noting that $r \leqslant r_\odot$; one can thus write:

$$p_{rc} \geqslant \frac{G}{4\pi}\int_{0}^{r_\odot} \frac{M\, dM}{r_\odot^4} = \frac{GM_\odot^2}{8\pi r_\odot^4} \tag{12.15}$$

Replacing G, $M_\odot$, and $r_\odot$ by their values, leads to:

$$p_{rc} \geqslant 4{\cdot}4\times10^8 \text{ atm } (4{\cdot}4\times10^{13} \text{ N/m}^2)$$

Following the same method, the temperature at the centre of the sun is found to be:

$$T_c \geqslant 2\times10^6 \text{ K}$$

Now assume that the interior of the sun is a gas in a complete state of ionisation. In this case the equation of state of adiabatic gases can be used:

$$p_r = K\varrho^\gamma \tag{12.16}$$

where K is a constant, and γ is the coefficient of adiabatic compression (defined as the ratio of the heat capacity at constant pressure over the heat capacity at constant volume). The combination of equations (12.9), (12.12) and (12.16), yields:

$$\frac{K\gamma}{4\pi G}\frac{1}{\varrho(r)}\frac{\mathrm{d}}{\mathrm{d}r}\left[r^2\varrho^{\gamma-2}\frac{\mathrm{d}\varrho(r)}{\mathrm{d}r}\right]+1 = 0 \tag{12.17}$$

The solution of this equation gives the density distribution inside the sun.

Actually at a temperature of 2×10^6K, there are two important reactions inside the sun:

$$^6\mathrm{Li}+{}^1\mathrm{H} \rightarrow {}^3\mathrm{He}+{}^4\mathrm{He}$$
$$^7\mathrm{Li}+{}^1\mathrm{H} \rightarrow 2\,{}^4\mathrm{He}$$

The assumption of adiabatic conditions is no longer valid and equation (12.16) does not hold. The flow of energy from inside the sun is by *radiation* only. Thus assuming that the sun radiates as a black body one can write for the pressure:

$$\mathrm{d}p_r(r) = (a/c)\,d(T^4) \tag{12.18}$$

where a is the radiation constant and c is the velocity of light in vacuum. From the luminosity of the sun, one can also write:

$$\mathrm{d}p_r(r) = \frac{\mathcal{L}(r)}{4\pi r^2}\frac{m_A(r)}{c}\varrho(r)\,\mathrm{d}r \tag{12.19}$$

where $m_A(r)$ is the mass absorption coefficient. Equating equation (12.18) to equation (12.19) yields:

$$\frac{16\pi a}{m_A(r)}\frac{r^2T^3}{\varrho(r)\mathcal{L}(r)}\frac{\mathrm{d}T}{\mathrm{d}r}+1 = 0 \tag{12.20}$$

which is the equation of radiative equilibrium. In this equation there are three unknowns: $T(r)$, $\varrho(r)$, and $\mathcal{L}(r)$, therefore two other equations are needed to solve the system. The luminosity is given as function of the energy production $\mathcal{E}(r)$ by:

$$\mathrm{d}\mathcal{L} = 4\pi r^2\varrho(r)\,\mathcal{E}(r)\,\mathrm{d}r \tag{12.21}$$

$m_A(r)$ and $\mathcal{E}(r)$ are functions of the density ϱ, the temperature T, and X, Y, Z the fractional composition per weight for hydrogen, helium and heavy elements respectively. The second equation is obtained by making the total pressure equal to the sum of the gas pressure and the radiation pressure:

$$p_r(r) = (K'/\mu_{e1})\varrho T+(a/c)T^4 \tag{12.22}$$

K' being a constant and

$$\mu_{e1} = (2X+\tfrac{3}{4}Y+\tfrac{1}{2}Z)^{-1} \tag{12.23}$$

After solving these equations one finds at the centre of the sun:

$$p_{rc} = 8 \times 10^{10} \text{ atm } (8 \times 10^{15} \text{ N/m}^2)$$
$$T_c = 15{\cdot}3 \times 10^6 \text{ K}$$
$$\varrho_c = 125 \text{ g/cm}^3$$

As for the composition of the sun, it was found at the centre:

$$X_c = 0{\cdot}49$$
$$Y_c = 0{\cdot}49$$
$$Z_c = 0{\cdot}02$$

and at the outer surface:

$$X_o = 0{\cdot}74$$
$$Y_o = 0{\cdot}24$$
$$Z_o = 0{\cdot}02$$

The different types of energy transport in the interior of the sun are summarised in Figure 12.13. It is convective at the centre of the sun and at up to 9 per cent of its radius, then it becomes radiative up to 71 per cent of its radius, and then convective again.

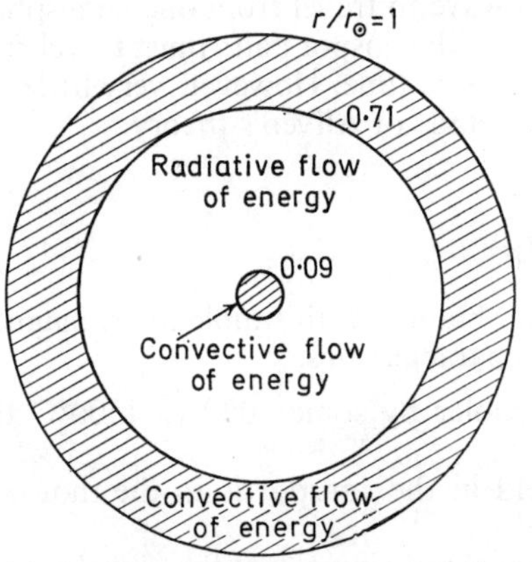

Fig. 12.13. Outward flow of energy from the centre of the sun

The differential rotation of the sun

The differential rotation problem has received no satisfactory solution. The theoretical aspects are strongly dependent upon whether or not there is a general, dipole-like, magnetic field inside the sun. If there is, since the magnetic lines of force are essentially frozen, the internal distribution of angular velocity will be closely related to the pattern of magnetic lines of force.

In any case, viscosity would be amply sufficient to eliminate any kind of differential rotation pattern if there was no powerful motor to maintain it. The only possible motor is the transmission of heat from the inner thermonuclear core toward the external regions. A possible model is then to consider such a transmission in a sun rotating as a rigid body, and to show that this is an unstable configuration. So far, the results from this approach have been unconvincing. It is unlikely that substantial progress can be made before the problem of the general magnetic field is solved.

The solar cycle

The number of sunspots on the solar surface varies significantly over an 11-year cycle. Moreover, the average latitude at which the sunspots appear follows a regular pattern. Luminous phenomena in the chromosphere follow the same pattern. The visual aspect of the corona also is dependent upon the phase in the sunspot cycle. The total emission of heat as well as that of visible and near-visible light is almost constant, having a variation between 1 and 2 per cent, but short waves (UV and X-rays) as well as long waves (radio waves) are strongly dependent upon the phase. If magnetic polarities are taken into account, the actual cycle turns out to be a 22-year one.

It seems beyond doubt that the sun is some kind of a gigantic hydromagnetic resonant cavity. Alfvén[1] has shown that, if reasonable assumptions are made about the permanent magnetic field of the sun, 11 years is just about the time acquired for an Alfvén wave to travel from one hemisphere to the other. Some kind of rings associated with sunspot pairs must travel from one hemisphere to the other in exactly the same time. However, very little confidence is generally credited to the further details of Alfvén's theory.

The theory of sunspots

Some success has been obtained with simple hydromagnetic models explaining the following sunspot characteristics.

(1) The sunspot is cooler by some 1000 to 1500°C than the neighbouring areas;
(2) the magnetic field in the sunspot is in the thousands of gauss (10^{-4} T) range;
(3) sunspots appear generally in pairs at the same latitude; they have opposite magnetic polarities; the leading and trailing poles with respect to the sun's rotation are related to the 22-year cycle of the sunspots;
(4) the time constants, lasting several hours, are rather short, taking into account the dimensions of the sunspots, which are several terrestrial radii;
(5) the sunspots are definitely not vortices; the motion of matter, if any, is rather radial than azimuthal;
(6) the sunspots look slightly depressed;
(7) the sunspots, in spite of being cooler, are actually centres of activity, in relation for instance with chromospheric phenomena.

The fundamental phenomenon is the existence of the magnetic field. The time constants observed in sunspots are six or seven orders of magnitude below those required to establish such magnetic fields. Therefore, the field must be assumed to be permanent underneath the solar surface (for a period at least equal to the human lifespan). Also, the birth of a sunspot should be assumed to be an interference of the pre-existing magnetic field with the surface of the photosphere. Because of the solenoidal character of the magnetic induction, the magnetic field pattern is made up of a number of rings or topologically equivalent structures. The exact shape and dimensions of the rings is still a matter of pure speculation.

Whenever such a ring interferes with the photosphere surface, the presence of a vertical magnetic field locally reduces the convection of heat, hence the sunspot is cooler. Moreover, the magnetic pressure is of the same order of magnitude as the gas pressure on the photospheric level, hence the spot may be in approximate equilibrium with a smaller gas pressure. Consequently the density at the spot location is not necessarily greater than in the surroundings. This explains why the matter does not tend to sink, which would normally be the case for cooler matter in the absence of magnetic fields.

The plasma near the solar surface is turbulent and energy is transported outwardly through turbulent convection. The magnetic field increases the velocity of the plasma flow, inhibiting the convection and decreasing the temperature, thus creating a sunspot. The expansion and contraction of a sunspot can be explained by the snow-plough model (Chapter 7). Indeed, the sunspot has the major features of a theta pinch, but in a reversed configuration similar to that of the hard-core pinch.

The theory of solar flares

A flare is an explosion on the solar surface resulting in a sudden increase of intensity of the emitted light, especially above the solar limb in the Hα region. Plasma jets are accelerated up to velocities of about 500 km/s. The velocity of these jets increases reaching a maximum in a few minutes and then decreases until it is reversed bringing the jet back to the sun in about half an hour.

Flares occur near sunspots and their frequency is closely related to the frequency of sunspots, reaching a maximum of several occurrences per day. An increase in intensity of the cosmic rays is observed to accompany these flares, as well as other emissions, notably visible light and ionised particles.

First, it was believed that a flare occurrence could be explained in terms of a strong electrical discharge guided by a longitudinal magnetic field. However, this simple model with particles travelling along the magnetic lines of force under the influence of a longitudinal electric field fails to conform with observations by a considerable number of orders of magnitude in the time constants. The fact that the flare jets are observed to follow rather complicated paths suggests that the ejected plasma contains frozen magnetic lines of force. When a flare of moderate intensity travels through the corona, it emits radio waves by exciting the plasma oscillations in the corona. In this excitation, non-linear effects are far from being negligible; the second and third harmonics of the plasma frequency have been observed. Large flares are followed by disturbances in the earth's magnetic field after one or two days delay. This results from a shock wave travelling through the interplanetary space and compressing the earth's magnetosphere and, consequently, its magnetic field.

12.7 THE SOLAR SYSTEM[2, 12]

The solar system is characterised by the overwhelming influence of the sun itself. Indeed, the volume of the sun is 743 times larger than the volume of its planets taken together, and the farthest planet Pluto is at a distance no larger than 4000 diameters of the sun.

All the planets orbit around the sun in a counter-clockwise direction when observed from the north of the ecliptic, and the inclination of their orbits from the ecliptic is no larger than 17° (for Pluto). The characteristics of the planets are shown in *Table 12.1* as compared to earth. It is not obvious that Pluto is the farthest planet and it is possible that other planets may exist, but they are undetectable with the present technological capabilities. Apart from the planets, other solid bodies roam in all directions in the solar system and interact in different ways with the interplanetary plasma. They vary in diameter from a few kilometres to a few microns and are called meteors and comets. The total mass of meteors does not exceed that of earth, and as for comets they are mostly confined to the farthest regions of the solar system.

Plasma phenomena similar to those on the earth have been detected or are suspected on other planets. Several planets emit radio waves, the most conspicuous sources being Venus, Mars[39] and Jupiter. Nothing of special interest has been discovered in the case of Venus and Mars, where the black-body radiation, in spite of being not exclusive, is very intense. The atmospheric component does not suggest a structure essentially different from that of the terrestrial atmosphere. The ionised layers of Jupiter, by contradistinction, seem to be of special interest. It has been known that the interaction of the jovian atmosphere with the solar ultra-violet radiation is especially conspicuous, and that the coloured belts around Jupiter are due to frozen free radicals generated photonically. Consequently, it is likely that the ionising events are also particularly numerous. Indeed, the jovian ionosphere seems to be at least one order of magnitude denser than that of the earth. Moreover this ionosphere not only reflects radio waves but it is also a powerful emitter. The origin of the emission does not seem to be related to lightning. Other mechanisms involving plasma electrostatic oscillations triggered from the lower layers by shock waves have been tentatively suggested.

The moon is devoid of any atmosphere, magnetic field or radiation belt. Only the four major planets, Jupiter, Saturn, Uranus and Neptune have a significant probability of having Van Allen-like radiation belts, provided the mechanism of belt creation is still operative at such distances from the sun. Evidence has been obtained that part of Jupiter's radio noise originates not in the planet itself, but in extended radiation belts trapped in a magnetic field about one order of magnitude above that of the earth. As for Venus, Mars and Mercury, they do not have any inherent magnetic field and any radiation belts.

Cometary tails[3] have been long attributed to the effect of radiation pressure on small solid particles. However, this mechanism fails to explain several exceptionally long and straight tails, designated as class I tails. Biermann *et al.*[9] has shown that the aspect of such tails points towards an explanation by the solar wind. More specifically, he assumes that the tail is a plasma, which is accelerated owing to its interference with the solar wind. A major argument in

Table 12.1 THE SOLAR SYSTEM

Characteristics	*Sun*	*Mercury*	*Venus*	*Earth*	*Mars*	*Asteroids*	*Jupiter*	*Saturn*	*Uranus*	*Neptune*	*Pluto*
Diameter	109	0·39	0·97	1	0·53	0·0001–0·07	11·2	9·2	3·925	4·225	1 ?
Mass	333×10^3	0·06	0·82	1	0·11	0·001	318	95	15	17	0·1–0·8 ?
Orbit diameter		0·387	0·723	1	1·524	2·8	5·2	9·54	19·2	30	39·5
Revolution (years)		0·241	0·615	1	1·88	4·7	11·9	29·5	84	165	248
Axial rotation (days)	25–27	59	13	1	1·03	0·2–0·5	0·41	0·43	0·45	0·53	6·39
Number of satellites		0	0	1	2		12	10	5	2	0
Average temperature in °C	5500	188	66	15	−39	−101	−147	−180	−207	−220	−227

favour of this theory is that such tails exhibit features which can be related, in terms of solar wind velocities, to known solar flares and terrestrial magnetic storms. Using a hypersonic collision free flow model, Finson and Probstein[23] developed a theory of the dust comets. In this model, dust particles are released from the nucleus of the comet due to solar heating. Drag forces resulting from the expansion of the comet head gas accelerate those dust particles outwardly. After leaving the head region, they are subjected to significant gravitational and radiation pressures from the sun. The effect of the solar wind is to cause a Lorentz force to act on those particles. The gravitational force is:

$$F_g = GM_{\odot}\varrho_d \pi d^3/6r^2 \quad (12.24)$$

where G is the universal gravitational constant, ϱ_d is the density of the dust particles and d is their diameter. The radiation force is:

$$F_r = (\eta_p/c)(\Phi_s/4\pi r^2)(\pi d^2/4) \quad (12.25)$$

where η_p is called the scattering efficiency and is close to unity, and Φ_s is the mean total solar radiation ($3{\cdot}93 \times 10^{20}$ MW). The Lorentz force is:

$$F_L = 2\pi\varepsilon_0 v_s BVd \quad (12.26)$$

where V is the potential on the charge, v_s is the solar wind velocity, and B is the magnetic field.

The distribution of meteorites has been studied by many workers. The model taken comprises a gravitational centre acting on an infinite parallel stream[41]. Some of these studies were applied with relative success to the meteorite distribution near the earth.

12.8 THE MILKY WAY[14, 15]

Examination of the universe through large telescopes shows that it is made up of units called galaxies. A galaxy is an ensemble containing billions of stars more or less similar to our sun, the overall having a definite shape and being flooded in a plasma and in dust. The galaxy in which we are located is called the *Milky Way*.

The Milky Way has a disc-like form with a nucleus, to which spiral arms are attached as shown in Figure 12.14, its total mass being equal to about 10^{11} solar masses. Around the Milky Way are a multitude of smaller groups of stars called *globular clusters.* These clusters behave as tiny galaxies containing from thousands to millions of stars. They are spherical in shape and vary in diameter from 70 to 600 light-years*. They form a halo around the galaxy. Other clusters of stars exist inside the galaxy and are called *galactic clusters.* The radius of the Milky Way as a whole is about 47 000 light-years, with the sun situated in one of the galactic spiral arms at a distance of 33 000 light-years from its centre. The nucleus of the Milky Way has a radius of about 16 300 light-years and a thickness of 10 000 light-years. This thickness decreases to about 4000 light-years at

* The light-year is the distance travelled by light in one year.

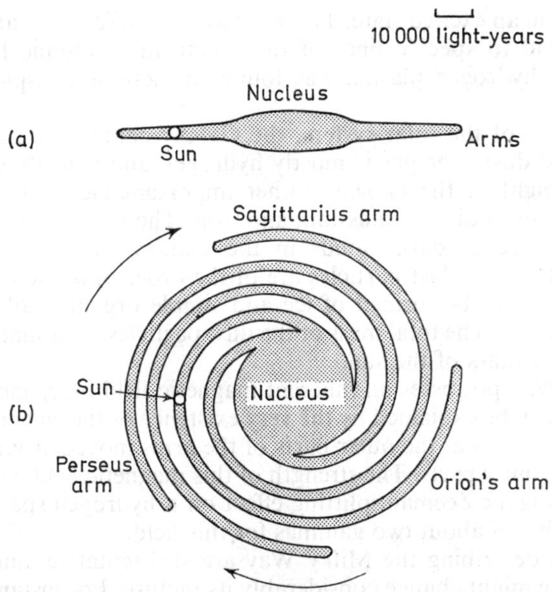

Fig. 12.14. Schematic views of the Milky Way: (a) *side view;* (b) *view from the north galactic pole*

the location of the solar system. The exact number of spiral arms attached to the nucleus is not known, three are so far identified; these are the Orion arm (containing the sun), the Sagittarius arm and the Perseus arm. Other arms have been tentatively identified by radio-astronomy. It is suggested that the width of these arms is about 2500 light-years and that they are seprated from each other by about 6500 light-years.

The Milky Way is also characterised by a rotation around its centre. This rotation is determined by applying Kepler's third law to the relative rotational velocities of stars at different distances from the centre of the Galaxy. Kepler's third law states:

$$M = r_a^3/\tau_p^2 \tag{12.27}$$

where M is the mass of the centre of the Galaxy, r_a is the orbital radius of a star in astronomical units and τ_p is the period in years. The frequency of rotation of the Milky Way at the sun level is about 7×10^{-16} rad/s corresponding to about 220 km/s. However, the stars in the Galaxy do not orbit with the same frequency around its centre. Thus, the idea of a peculiar velocity, which is the velocity of a star relative to its neighbours, is introduced. The peculiar velocity of the sun is about 20 km/s.

Distances between stars in the Milky Way are of the order of several light-years. The nearest star to the sun, Alpha Centauri, is at about 4·3 light-years.

One important characteristic of the Milky Way is the existence of a hydrogen plasma in the interstellar space, especially at the arms of the Galaxy. This gas has been detected by radio-astronomical methods. When a proton recombines with an electron, part of the energy liberated is retained by the resulting neutral which remains in an excited state. This excited neutral radiates, and these radiations correspond to specific lines of the spectrum of atomic hydrogen. The density of this hydrogen plasma was found to increase conspicuously in the spiral arms.

As in the case of the solar system, the Galaxy contains a great amount of neutral gas and dust. The gas is mostly hydrogen and constitutes about 2 per cent of the weight of the Galaxy. Other important gases in the interstellar space are calcium, sodium, potassium and iron. The presence of dust is shown by the appearance of dark clouds in the Galaxy called *dark nebulae*[5, 10]. The dimensions of the dust particles are of the order of a micron, on average. They are believed to be formed of ice and simple organic molecular crystals (NH_3, CH_4, etc...). The total mass of the dust particles is estimated to be about 2×10^7 times the mass of the sun.

The Milky Way possesses an inherent magnetic field independent of that of the stars. This can be explained by the very existence of the galactic spiral arms. Also, it was noticed that the outer parts of the arms moved around the Galaxy slower than the inner parts. The strength of this magnetic field has been measured by observing the Zeeman splitting effect on a hydrogen spectral line. This leads to the value of about two gammas for this field.

The theories describing the Milky Way are still tentative, and new experimental evidence might change considerably its picture. For instance, a balloon-borne instrument package containing a telescope and an infra-red detector was sent on July 1968 to an altitude of about 36 km. Infra-red radiation equalling the energy of 7×10^8 suns was detected coming from a region at about 1500 light-years from the sun. It was proposed that either the Galaxy has many more stars than ever predicted or that somehow its centre pumps a huge amount of energy towards itself. Or maybe, an interaction between the cool stars and the interstellar dust causes this region to radiate strongly[30].

12.9 THE WORLD OF GALAXIES[11]

The known universe is made up of a collection of galaxies between which there is presumably an almost complete vacuum*. To obtain an idea of the particle densities in the different parts of the universe, see *Table 12.2*.

However, galaxies themselves are grouped into ensembles, or clusters (possibly some kind of supergalaxy). The cluster in which the Milky Way is situated is called the *Local Group*. About 19 galaxies are known in this group. Its diameter is believed to be about $3{\cdot}2\times10^6$ light-years. In this group individual galaxies are several hundred thousand light-years apart from each other. The galaxy closest to the Milky Way, the Large Magellanic Cloud, is at a distance of 160 000 light-years from it. The Local Group, like all the other clusters of galaxies, is a bound physical system that moves as one unit in the universe.

Hubble[29] was first to classify galaxies in accordance with their shapes. Two main classes can be distinguished, the elliptical galaxies and the spiral ones. The latter are divided into two super-classes, the normal spirals (such as the Milky Way) and the barred spirals (in which the nucleus is much smaller).

* Neutral hydrogen might possibly exist and escape detection.

Table 12.2 PARTICLE DENSITIES IN THE UNIVERSE

Region	*Density* (kg/m³)
Air at sea level	$8{\cdot}4\times10^{-2}$
Inner ionosphere	10^{-6}
Solar wind	10^{-14}
Interstellar space	10^{-22}
Intergalactic space	$<10^{-35}$

Beyond the Local Group are recognised other clusters of galaxies such as the Group M81 at a distance of about 8×10^6 light-years. This group contains more than a thousand known galaxies. The difference in shape of the galaxies can be interpreted in terms of collisions having occurred between them. Galaxies in a cluster, orbit about their centres and may collide among themselves. These collisions have been observed through the relevant intense radio emission. Collisions of this kind are essentially a plasma physics phenomenon, since the stars have much too small cross-sections for collisions among themselves to be likely. The so-called collision is actually an interpenetration, having a duration of some 10^6 years during which the stars are left essentially unaltered, whereas the interstellar gases strongly interfere. Their kinetic energy is converted into heat and temperatures in the million degree range can be locally attained. Intense radio emission, essentially of bremsstrahlung origin, results. Elliptical galaxies are produced from spiral ones when they lose their arms after some collisions.

At extremely large distances (billions of light-years), it becomes extremely difficult to observe the frontiers (for man's knowledge) of the universe. Some very remote objects are still recognised as radio sources due to their extremely great intensity. They were discovered in 1966 and named *quasars*. Quasars supposedly have ceased to exist by the time their light reaches us and they are known to have intensities some 100 times those of the Milky Way.

12.10 EVOLUTION OF THE UNIVERSE[21, 24, 25]

The sun, like all the other stars, has a specific lifespan. The age of a star can be estimated from two parameters: (1) its brightness, measured as the amount of visible light emitted per unit time, and (2) its surface temperature, observed from the wavelength distribution of the light emitted. A diagram, along the axes of which are plotted these two parameters is known as a Hertzsprung–Russel (H–R) diagram. The position of any star in the H–R diagram indicates also its mass and its composition. Stars born by the contraction of the interstellar plasma at the same time as the sun will have the same initial composition as the sun but might have different masses. These stars are observed to lie on a *main sequence* in the H–R diagram, as shown in Figure 12.15. After they are born, stars produce energy following two different cycles; either through

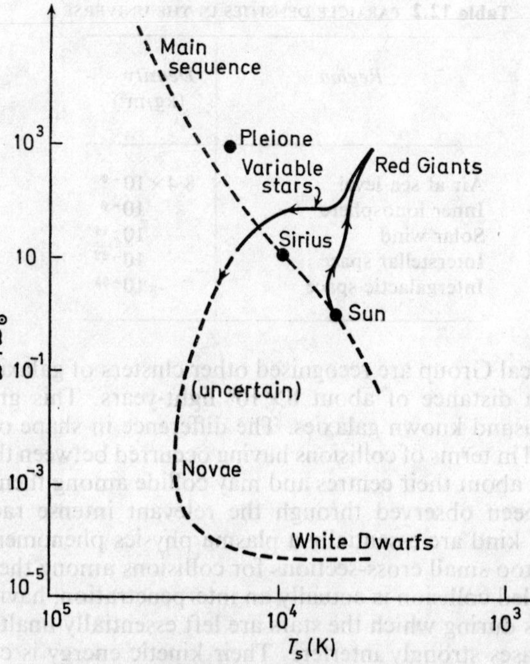

Fig. 12.15. Evolution of the stars. The main sequence is for the mixed stars, whereas the other curve is for the unmixed stars

a proton-chain or through a carbon–nitrogen cycle. In both cases hydrogen is transformed into helium. If the helium spreads throughout the star, it is likely that it will depart only slightly from the main sequence. But if it stays where it is produced, a new sequence is followed as shown by Figure 12.15. The sun, about 5×10^9 years old, is relatively young and is at the beginning of the main sequence. It has burned only about 5% of its hydrogen. A departure from the main sequence occurs when the star has used about 10% of its hydrogen, thus having its active core mostly transformed into helium. As it grows older, the gravitational force shrinks the inner helium core, and the star as a whole expands and becomes a Red Giant. Under the effect of this compression, all the nuclei in the core become stripped of their electrons and the gas is said to become *degenerate**. Eventually, the star will become an almost totally degenerate, dead, and highly condensed cold star called a White Dwarf. In some cases there is a final explosion (during about one year or two) called a supernova, the result of which is a superdense neutron star called a *pulsar*.

The different theories of evolution of the entire universe agree on one point: at the present time the universe is expanding. This expansion is observed by examination of the spectral lines emitted from very distant galaxies as well as the quasars. By measuring the red shift of the quasar 3C 273, for instance, it was found that it is moving away with a speed equal to 15% of the speed of light.

* As in solid-state plasmas. See Chapter 11.

This observed expansion led to two different theories, the *steady-state theory* and the *big-bang theory*. The first, supported by Hoyle[28], assumes that the universe has always been expanding. Furthermore, it has looked the same at any point in time, so as galaxies move away, newer galaxies are created out of the intergalactic hydrogen, and the density of the universe remains constant. Hoyle's universe is therefore infinite in space and in time. The big-bang theory advanced by Gamow[24], assumes that the universe was created about 10 million years ago in a huge explosion of densely packed matter. The fragments of that explosion (galaxies) continue to expand away from each other. In this theory, the universe tends toward infinity in space. Others believe, however, that gravitational attraction will eventually overcome this expansion and will shrink the universe back to its original state of densely packed nuclear matter[26]. From this to the *oscillating theory* of Sandage[35] there is only one step: assume that the phenomenon will repeat itself with a period of some 80 million years. In this latter theory the universe, is, at any time, a finite entity.

REFERENCES AND BIBLIOGRAPHY

1. ALFVÉN, H., *Cosmical Electrodynamics*, Clarendon, Oxford (1950).
2. ALFVÉN, H., *On the Origin of the Solar System*, Clarendon, Oxford (1954).
3. ALFVÉN, H., 'On the Theory of Comet Tails', *Tellus*, **9,** 92 (1957).
4. ALFVÉN, H. and FÄLTHAMMER, C. G., *Cosmical Electrodynamics*, Oxford University Press (1963).
5. ALLER, L. H., 'Chemical Compositions of Selected Planetary Nebulae', *Astrophys. J.*, **125,** No. 1, 84 (1957).
6. AXFORD, W. I., 'The Interaction Between the Solar Wind and the Earth's Magnetosphere', *J. geophys. Res.*, **67,** 3791 (1962).
7. AXFORD, W. I., PETSCHEK, H. E. and SISDOE, G. L., 'Tail of the Magnetosphere', *J. geophys. Res.*, **70,** 1231 (1965).
8. BARTHEL, J. R. and SOWLE, D. H., 'A Mechanism of Injection of Solar Plasma into the Magnetosphere', *Planet. Space Sci.*, **12,** 209 (1964).
9. BIERMANN, L., BROSOWSKI, B. and SCHMIDT, H. V., 'The Interaction of the Solar Wind with a Comet', *Solar Phys.*, **1,** 254 (1967).
10. BOHM, D. and ALLER, L. H., 'The Electron Velocity Distribution in Gaseous Nebulae and Stellar Envelopes', *Astrophys. J.*, **105,** No. 1 (1947).
11. BONDI, H., *The Universe at Large*, Doubleday, New York (1960).
12. BRIDGE, H. S., DILWORTH, C., LAZARUS, A. J., LYONS, E. F., ROSSI, B. and SCHERB, F., 'Direct Observation of the Interplanetary Plasma', *Kyoto Conf. on Cosmic Rays and the Earth Storm*, Japan (1961).
13. CHAMBERLAIN, J. W., *Physics of the Aurora and Airglow*, Academic Press, New York (1961).
14. CHANDRASEKHAR, S., *An Introduction to the Study of Stellar Structure*, University of Chicago Press (1939).
15. CHANDRASEKHAR, S., *Principles of Stellar Dynamics*, University of Chicago Press (1942).
16. CHAPMAN, S. and FERRARO, V. C., 'A New Theory of Magnetic Storms. Part I, the Initial Phase', *Terr. Magn. atmos. Elect.*, **36,** 77, 171 (1931).
17. CHAPMAN, S. and FERRARO, V. C., 'A Theory of Magnetic Storms. Part II, the Main Phase', *Terr. Magn. atmos. Elect.*, **38,** 79 (1933).
18. DESSLER, A. J. and WALTERS, G. K., 'Hydromagnetic Coupling Between Solar Wind and Magnetosphere', *Planet. Space Sci.*, **12,** 227 (1964).
19. DUNGEY, J. N., *Cosmic Electrodynamics*, Cambridge University Press (1958).

20. ECKART, C., *Hydrodynamics of Oceans and Atmospheres*, Pergamon Press, New York (1960).
21. EDDINGTON, A., *The Expanding Universe*, MacMillan, New York (1933).
22. ELSASSER, W. M., 'Hydromagnetic Dynamo Theory', *Rev. mod. Phys.*, **28**, 135 (1956).
23. FINSON, M. L. and PROBSTEIN, R. F., *A Theory of Dust Comets*, Fluid Mechanics Lab. Publ. No. 67-4, MIT, Cambridge, Mass., Aug. (1967).
24. GAMOW, G., *The Creation of the Universe*, Mentor, New York (1952).
25. GAMOW, G., 'The Evolutionary Universe', *Scient. Am.*, Sept. (1956).
26. HARRISON, B. K., THORNE, K. S., WAKANO, M. and WHEELER, J. A., *Gravitation Theory and Gravitational Collapse*, University of Chicago Press (1965).
27. HOYLE, F., *Frontiers of Astronomy*, Mentor, New York (1955).
28. HOYLE, F., 'The Steady State Universe', *Scient. Am.*, Sept. (1956).
29. HUBBLE, E., *The Realm of the Nebulae*, Dover, New York (1958).
30. JOHNSON, H. L., 'Infrared Stars', *Science*, **157**, No. 3789, 635, Aug. 11 (1967).
31. KUIPER, G. P. (Ed.), *The Solar System*, University of Chicago Press (1961).
32. MACKIN, R. J. and NEUGEBAUER, M. (Eds.), *The Solar Winds*, Pergamon Press, New York (1966).
33. MEAD, J. D., 'Deformation of Geomagnetic Field by Solar Wind', *J. geophys. Res.*, **69**, 1181 (1964).
34. MEAD, J. D. and BEARD, D. B., 'Shape of the Geomagnetic Field Solar Wind Boundary', *J. geophys. Res.*, **69**, 1169 (1964).
35. ÖPIK, E. J., *The Oscillating Universe*, Mentor, New York (1960).
36. PIDDINGTON, J. H., 'Geomagnetic Storm Theory', *J. geophys. Res.*, **65**, 93 (1960).
37. PIDDINGTON, J. H., 'The Magnetosphere and its Environs', *J. Space Sci.*, **13**, 363 (1965).
38. RATCLIFFE, J. A., *The Magneto-Ionic Theory and Its Applications to the Ionosphere*, Cambridge University Press (1959).
39. SAGAN, C. and VEVERKA, J., 'Martian Ionosphere: A Component Due to Solar Protons', *Science*, **158**, No. 3797, 110, Oct. (1967).
40. SCARF, F. L., 'Plasma in the Magnetosphere', *Advances in Plasma Physics* (edited by A. SIMON), Interscience, New York (1968).
41. SHELTON, R. D. *et al.*, 'Some Aspects of the Distribution of Meteoric Flux About an Attractive Center', *Space Res. in U.K. Univ.*, **4**, 875 (1964).
42. SINGER, S. F., 'Particles in the Magnetosphere', *J. phys. Soc. Japan*, **17**, Suppl. A2, 609 (1962).
43. STOREY, L. R., 'An Investigation of Whistling Atmospherics', *Phil. Trans. R. Soc., London*, **246A,** 113 (1953).
44. UREY, H. C., *The Planets, their Origin and Development*, Yale University Press (1952).
45. WHEELER, J. A., 'Our Universe, the Known and the Unknown', *Am. Scient.*, **56,** No. 1, 1 (1968).
46. WILKES, M. V., 'Oscillations of the Earth's Atmosphere', Cambridge University Press (1949).

Glossary of symbols used

a	acceleration (m/s^2)
a_r	unit vector in r-direction
a	Stefan–Boltzmann constant $= 5{\cdot}679 \times 10^{-8}$ W/m^2 K^4
A	mass number of an atom
$\boldsymbol{A}$	magnetic vector potential (Wb/m)
$\mathcal{A}$	unitless function
$\boldsymbol{\mathcal{A}}$	tensor
$\mathcal{A}''$	constant
$\mathcal{A}_{mn}$	coefficient of spontaneous emission
$\mathcal{b}$	mobility (m^2/V s)
$\boldsymbol{B}$	magnetic field (Wb/m^2)
$\mathcal{B}$	constant
$\boldsymbol{\mathcal{B}}$	tensor
$\mathcal{B}_{k\omega}$	matrix element
$\mathcal{B}_{mn}$	coefficient of induced emission
$\mathcal{B}_{nm}$	coefficient of induced absorption
c	velocity of light in vacuum $= 2{\cdot}997\,93 \times 10^8$ m/s
C	root mean square velocity (m/s)
C_f	capacitance (F)
C_v	specific heat at constant volume (J/K)
C_p	specific heat at constant pressure (J/K)
$\mathcal{C}$	constant
$\mathcal{C}_1$	constant (C m/F)
$\mathcal{C}_2$	constant (C m/F)
$\mathcal{C}_{k\omega}$	instability function
$\mathcal{C}_{\varphi}$	constant
$\mathcal{C}_{\mu}$	probability that a single collision is an annihilation process
d	distance (m)
$\mathcal{D}$	diffusion coefficient (m^2/s)
$\mathcal{D}_a$	ambipolar diffusion coefficient (m^2/s)
$\mathcal{D}_v$	diffusion with respect to velocity (m^2/s)
e	modulus of electronic charge $= 1{\cdot}602 \times 10^{-19}$ C
e_s	average energy per unit mass of particles (J/kg)
$\boldsymbol{E}$	electric field (V/m)

$\mathcal{E}$	energy (J)
$\mathcal{E}_F$	Fermi level of energy (J)
$\mathcal{E}_m$	peak energy (J)
$\mathcal{E}_\nu$	energy density per unit frequency range near the frequency ν (J s)
$f(\boldsymbol{v})$	distribution function
f_p	plasma frequency (1/s)
$\boldsymbol{F}$	force (N)
$\dot{\boldsymbol{F}}$	surface force (N/m²)
$\mathcal{F}$	Helmholtz free energy (J)
g	weight factor
g_i	degeneracy of the *i*th level
$\boldsymbol{g}$	acceleration due to gravity (m/s²)
G	universal constant of gravity $= 0{\cdot}667\times10^{-10}$ m³/s² kg
h	Planck's constant $= 6{\cdot}625\times10^{-34}$ J s
$\hbar$	$h/2\pi$
$\boldsymbol{H}$	magnetic field vector (A/m)
$\mathcal{H}$	enthalpy (J)
$\boldsymbol{i}$	unit vector in the *x*-direction
I	electric current (A)
I_p	probe current (A)
I_r	radiation intensity (W/m²)
j	$\sqrt{-1}$
$\boldsymbol{j}$	unit vector in the y-direction
$\boldsymbol{J}$	current density (A/m²)
J_H	Hall current density (A/m²)
J_p	probe current density (A/m²)
J_r	emission coefficient (W/m³)
$\mathcal{J}$	action (m²/s)
k	Boltzmann's constant $= 1{\cdot}38\times10^{-23}$ J/K
$\boldsymbol{k}$	unit vector in the *z*-direction
$\boldsymbol{k}$	propagation constant (1/m)
k_ν	absorption constant (1/m)
K	constant
K'	constant
K_1	constant
K_c	constant (C²m/F)
$\mathcal{K}$	electric field ratio
l	length (m)
L	inductance (H)
$\mathcal{L}$	luminosity (kg²m²/s²)
m_A	mass absorption coefficient; amplification factor
m_B	magnetic field ratio
m_F	ratio of the magnetic body force to the fluid inertial force
m_g	Landé splitting factor
m_i	inhomogeneity factor
m_k	pinch ratio
m_p	unitless ratio
m_s	saturation factor
m_α	ratio

$\boldsymbol{m}$	magnetic moment (A m^2)
M	mass (kg)
M^*	effective mass (kg)
M^-	electronic mass = $9{\cdot}108\times10^{31}$ kg
$\mathcal{M}$	Mach number
$\boldsymbol{\mathcal{M}}$	magnetisation (A/m)
n	charged particle density (m^{-3})
n^+	positive charge density (m^{-3})
n^-	negative charge density (m^{-3})
n_e	plasma density at equilibrium (m^{-3})
n_{sc}	density of scatterers (m^{-3})
$\mathscr{n}$	number
$\mathscr{n}_f$	number of degrees of freedom in a gas
$\mathscr{n}_\theta$	integral value
N	total number of particles; particle current density (1/m s)
N_n	number of molecules initially in state n
N_m	number of molecules initially in state m
$\mathcal{N}$	Avogadro's number = $60{\cdot}2\times10^{25}$ particles/mol
p	density of holes (m^{-3})
p_a	acceptor number density (m^{-3})
p_b	probability
p_{cs}	caesium vapour pressure (kg/m s^2)
p_r	pressure (kg/m s^2)
$\boldsymbol{p}$	momentum vector (m kg/s)
$\mathscr{p}$	dipole moment (m C)
P	power (W)
P_λ	power density at wavelength λ (W/m^3)
$\mathcal{P}$	polarisation moment (C/m); angular momentum (kg m^2/s)
q	electric charge (C)
$\mathscr{q}$	Hamiltonian
Q	scattering cross-section (m^2)
Q_L	Q-factor of a loaded cavity
Q_t	total scattering cross-section (m^2)
$\mathcal{Q}$	heat flux (J)
r	distance, radius (m)
r_a	radius of the atom (m)
r_c	curvature radius of magnetic field lines (m)
r_D	Debye screening length (m)
r_h	Larmor radius (m)
R	resistance (Ω)
$\mathcal{R}$	universal gas constant (8·314 J/mol K)
$\mathcal{R}_n$	Reynolds number
S	surface area (m^2)
$\dot{S}$	angular momentum (m^2/s)
$\mathcal{S}$	entropy (J)
t	time (s)
T	temperature (K)
$\mathcal{T}$	torque (kg m^2/s^2)
U	internal energy (J)

$\boldsymbol{v}$	velocity (m/s)
$\bar{v}$	average speed (m/s)
v_A	Alfvén speed (m/s)
$\boldsymbol{v}_d$	drift velocity (m/s)
v_F	Fermi speed (m/s)
$\boldsymbol{v}_g$	group velocity (m/s)
v_i	speed of incident particle (m/s)
v_0	most probable speed (m/s)
$\boldsymbol{v}_p$	phase velocity (m/s)
$\boldsymbol{v}_r$	radial velocity (m/s)
v_s	speed of sound in a medium (m/s)
$\boldsymbol{v}_t$	tangential velocity (m/s)
$\boldsymbol{v}_\phi$	shock velocity (m/s)
V	voltage (V)
V_i	ionisation potential (V)
V_p	probe voltage (V)
$\mathcal{V}$	volume (m^3)
x	distance (m)
x_b	impact parameter (m)
X	shear stress (N/m^2)
$\mathcal{Y}$	unspecified variable function
Z	atomic number; partition function
Z_p	figure of merit of propulsor
α	degree of ionisation
α_e	ratio of excited molecules
α_L	ratio
α_r	absorption coefficient
α_s	departure coefficient
β	expansivity
γ	ratio of specific heats
γ_g	gyromagnetic ratio
γ_t	second Townsend coefficient
Γ_r	random flux ($1/m^2$ s)
δ	radius ratio small compared to unity
δ_B	ratio of magnetic fields small compared to unity
δ_p	pressure ratio small compared to unity
δ_v	velocity ratio small compared to unity
ε	permittivity of a medium (F/m)
ε_0	permittivity of free space = $(1/36\pi)\times10^{-9}$ F/m
ε_g	relative permittivity of glass
ε_r	relative permittivity
ξ_0	thermionic function
η	efficiency
η^*	gain
η_t	first Townsend coefficient per volt (1/V)
θ	angle (degrees)
Θ	emissivity
$\varkappa$	thermal conductivity (J/K m s)
λ	mean free path (m); wavelength (m)

λ_g guide wavelength (m)
λ_m wavelength corresponding to a maximum emitted radiation (m)
λ_t first Townsend coefficient per metre (1/m)
Λ diffusion length (m)
μ permeability of a medium (H/m)
μ_0 permeability of free space $= 4\pi\times10^{-7}$ H/m
μ_r relative permeability
ν radiation frequency (1/s)
ν_c frequency of collision (1/s)
ν_i frequency of ionisation (1/s)
ν_k kinematic viscosity (m^2/s^3)
ν_R refractive index
Π pressure gradient (N/m^3)
ϱ mass density (kg/m^3)
ϱ_c recombination coefficient
ϱ_e space charge density (C/m^3)
σ electric conductivity (mho/m)
τ mean collision time (s)
τ_B Bohm diffusion time (s)
τ_c average time flight (s)
$\boldsymbol{v}$ viscosity coefficient (kg/m s)
φ angle (degrees)
Φ magnetic flux ($m^{-2}\,s^{-1}$)
Φ_s mean total solar radiation $= 3{\cdot}93\times10^{20}$ MW
χ magnetic susceptibility
χ_a ratio
χ_m ratio of masses
ψ Schrödinger function; electric susceptibility
ψ_0 thermionic function
Ψ pressure tensor (kg/m s^2)
ω radial frequency (rad/s)
ω_c gyrofrequency of a particle (rad/s)
ω_L Larmor frequency (rad/s)
ω_p characteristic plasma frequency (rad/s)
ω_{pi} ion plasma frequency (rad/s)
Ω solid angle (steradians)
Ω_p overall plasma frequency (rad/s)

Problems

CHAPTER 1

1. Consider an infinitely long cylindrical region of radius a in which electrons and ions are uniformly distributed, the overall region being neutral. Now the electrons are assumed to be displaced through a small distance x perpendicular to the axis of the cylinder, leaving the ions fixed.

   (a) Derive the equation of oscillation under such a condition.
   (b) Do a similar calculation for a spherical region.
   (c) Prove that in both cases the oscillation frequency is generally different from ω_p.

2. Prove equation (1.14).

3. Solve equation (1.9) for the potential V, assuming the following boundary conditions: the potential goes to zero for r going to infinity and it is reduced to a Coulomb potential $V_C = e/(4\pi\varepsilon_0 r)$ for small values of r.
   (a) Deduce the value of the Debye radius r_D and plot $V(r)$.
   (b) The sphere of radius r_D is known as the Debye Sphere. Find the total number of electrons in such a sphere.

4. Find the potential along the axis of a uniformly charged disc of radius a.

5. In a gas, an atom could be considered as consisting of a positive point charge (nucleus) surrounded by a spherically symmetric cloud of electrons. If an external electric field is applied, then the nucleus will be displaced slightly resulting in an 'electronic' polarisation. Calculate the electronic polarisability and the dielectric susceptibility in such a case.
   Hint: equate the force exerted by the external field to the Coulomb restoring force.
   Result: $\psi = 4\pi n_a r_a^3$ (r_a = radius of the atom, n_a = number of atoms per cubic metre)

6. Use equation (1.28) to find the scattering cross-section of the two hard spheres shown in Figure 1.8.

7. Show that in general

$$Q_m/Q_t = \langle 1-\cos\theta \rangle$$

where the $\langle\,\rangle$ symbol stands for average.

8. $Q(\theta)$ is defined in the centre-of-mass system of co-ordinates. If one defines $Q'(\alpha)$ as the differential scattering cross-section in the laboratory frame of co-ordinates, then $dQ'_t = 2\pi Q'(\alpha)\sin\alpha\, d\alpha$, where α is the deflection angle in the laboratory system. Find the ratio $Q'(\alpha)/Q(\theta)$ for two hard-spheres having the same mass.
Hint: show first that $\tan\alpha = \sin\theta/(1+\cos\theta)$.

9. Show that for a central force having a potential proportional to r^{-n}, the differential cross-section is indeed proportional to $\mathcal{E}^{-2/n}$

10. Verify the values of $\bar{v}$, v_0 and C given by equations (1.49).

11. Give a physical meaning to the difference in the value of the diffusion coefficient between equation (1.56) and equation (1.61).

12. Show that for a Coulomb interaction the electronic conductivity of a plasma is:

$$\sigma = 3\times 10^7\,(kT_e)^{3/2}$$

13. At what temperature does the conductivity of the plasma of Problem 12 equal that of silver?

14. According to thermodynamics the pressure $p_r(T, \mathcal{V})$ of a gas satisfies the relation:

$$\frac{\partial}{\partial T}\left(\frac{P_T}{T}\right) = \frac{1}{T^2}\frac{\partial U}{\partial \mathcal{V}}$$

Show that equations (1.77) and (1.80) do not contradict this relation.

15. Prove equation (1.88).

CHAPTER 2

1. Find the path of an electron which, initially at rest, is subjected to the constant fields $\boldsymbol{E}$ in the x-direction and $\boldsymbol{B}$ in the y-direction. What is the name of this path? What is the wavelength of the motion?

2. *(a)* Describe the motion of a particle in the field of a straight current-carrying wire in the presence of a uniform electric field,
   (b) What happens to the motion when the current in the wire is switched off suddenly?
   (c) What happens when the electric field is switched off suddenly?
   (d) What happens when both the current and the electric field are switched off suddenly?

3. Calculate the average velocity of an electron in a sinusoidally-varying electric field when:
   (a) The field is varying in space but constant in time;
   (b) the field is constant in space but varying in time.

4. Consider a current I flowing in a loop enclosing an area S, show that (IS/c) is an adiabatic invariant.

5. Using Faraday's law:

$$\frac{\mathrm{d}\Phi}{\mathrm{d}t} = \oint \boldsymbol{E}.\mathrm{d}\boldsymbol{r}$$

show that the magnetic flux through any closed contour moving with a perfectly conducting fluid is a constant.

6. The initial velocities $\boldsymbol{v}_1$ and $\boldsymbol{v}_2$ and final velocities $\boldsymbol{v}_1'$ and $\boldsymbol{v}_2'$ of colliding particles of masses M_1 and M_2 may be written as:

$$\boldsymbol{v}_1 = \boldsymbol{v}_{\mathrm{gn}} + \frac{M_2}{M_1+M_2}\boldsymbol{v}_{\mathrm{gnr}} \qquad \boldsymbol{v}_2 = \boldsymbol{v}_{\mathrm{gn}} - \frac{1}{M_1+M_2}\boldsymbol{v}_{\mathrm{gnr}}$$

$$\boldsymbol{v}_1' = \boldsymbol{v}_{\mathrm{m}}' + \frac{M_2}{M_1+M_2}\boldsymbol{v}_{\mathrm{mr}}' \qquad \boldsymbol{v}_2' = \boldsymbol{v}_{\mathrm{m}}' - \frac{M_1}{M_1+M_2}\boldsymbol{v}_{\mathrm{mr}}'$$

where $\boldsymbol{v}_{\mathrm{m}}$, $\boldsymbol{v}_{\mathrm{m}}'$ are the initial and final centre-of-mass velocities, and $\boldsymbol{v}_{\mathrm{mr}}$, $\boldsymbol{v}'_{\mathrm{mr}}$ are the initial and final centre-of-mass relative velocities. Using the laws of conservation of energy and momentum show that $\boldsymbol{v}_{\mathrm{m}} = \boldsymbol{v}_{\mathrm{m}}'$ and $|\boldsymbol{v}_{\mathrm{mr}}| = |\boldsymbol{v}_{\mathrm{mr}}'|$.

7. Prove equation (2.25).

8. The constant in equation (2.38) is given by $\mathcal{A} = 3{\cdot}16\times10^{-7}$ when the temperature is expressed in Kelvin and the pressure in atmospheres. Plot the degree of ionisation for argon as a function of temperature at one atmosphere pressure. Here, the ionisation potential is $V_{\mathrm{i}} = 15{\cdot}755$ V.

9. Find the mobility in the case where the equation of motion is given by:

$$M\frac{\mathrm{d}\boldsymbol{v}_{\mathrm{d}}}{\mathrm{d}t} = q(\boldsymbol{E}+\boldsymbol{v}_{\mathrm{d}}\times\boldsymbol{B}) - M\boldsymbol{v}_{\mathrm{d}}\nu_{\mathrm{c}}$$

and where $\boldsymbol{B}$ is a uniform magnetic field in the z-direction, $\boldsymbol{E}$ being uniform and in the x-direction.

10. Show that if $\Gamma^- = \Gamma^+$ then a constant magnetic field would have no effect upon diffusion.

11. In the presence of a uniform magnetic field $\boldsymbol{k}B_0$ and when no electric field is present, one can write for a very slow diffusion:

$$\boldsymbol{\Gamma} = -\mathcal{D}^-\nabla n + \frac{\omega_{\mathrm{c}}}{\nu_{\mathrm{e}}}\boldsymbol{\Gamma}\times\boldsymbol{k}$$

Using equation (1.54), find the value of the diffusion constant $\mathcal{D}$ in this case.

12. Integrate equation (2.61) to find the number of photons having an energy larger than eV_{i}.

13. Discuss the possibility of deionisation by diffusion. Investigate the problem by using a simple one-dimensional model.

14. The presence of an electric field changes the effective potential V_i of a cathode to a new value V_i' in equation (2.70). Show that if $E_i(x) = e/4x^2$, then this new function will be:

$$V_i' = V_i - \sqrt{(eE)}$$

15. Prove equation (2.77).

CHAPTER 3

1. From Maxwell's equations establish Poynting's theorem:

$$-\iiint_{\mathcal{V}} (\boldsymbol{E}.\boldsymbol{J})\,\mathrm{d}\mathcal{V} = \iiint_{\mathcal{V}} \left(\boldsymbol{E}.\frac{\partial \boldsymbol{D}}{\partial t} + \boldsymbol{H}.\frac{\partial \boldsymbol{B}}{\partial T}\right)\mathrm{d}\mathcal{V} + \iint_{S} (\boldsymbol{E}\times\boldsymbol{H}).\mathrm{d}\boldsymbol{S}$$

Give to this theorem a physical meaning.

2. Knowing that an electric current is defined as the rate of change of electric charges, prove equation (3.10).

3. Find the Hamiltonian and the general form of the distribution function $f(x, v_x, v_y, v_z)$ of non-interacting particles moving in a uniform magnetic field $\boldsymbol{B} = \boldsymbol{k}B_0$. The distribution function is assumed stationary and independent of y and z.

4. Write Boltzmann's equation in cylindrical co-ordinates and show that $f = f\left(\frac{1}{2}Mv^2, Mr^2\frac{\partial \varphi}{\partial t}, qrA_\varphi\right)$ is a stationary solution for an azimuthally symmetric magnetic field. The vector potential $\boldsymbol{A}$ is given by $\boldsymbol{B} = \nabla\times\boldsymbol{A}$.

5. Prove that the transformation from the $(\boldsymbol{r}, \boldsymbol{v})$ space to the phase space is volume preserving.
Hint: show that the Jacobian of this transformation is a constant.

6. Prove equation (3.10) by using equations (3.30a) and (3.30b).

7. Show that in the absence of collisions Boltzmann's equation can be written in the form of a conservation equation.

8. Show that equation (3.41) can be written in the form of a conservation equation in which the derivative with respect to time of the total momentum density plus the gradient of the total momentum flux equals zero.

9. Show that the sum of the derivative of the total energy density and the gradient of the total energy flux is equal to zero.

10. Find the adiabatic equation of state of a collisionless gas where the motion of particles in the z-direction is independent of the motion in the xy plane.

11. If $Y(v)$ is any scalar property of the particles, then the flow of Y is:

$$\boldsymbol{F}(\boldsymbol{r}) = \iiint_{\text{all } \boldsymbol{v}} f(\boldsymbol{v})\,Y(\boldsymbol{v})\boldsymbol{v}\,\mathrm{d}\boldsymbol{v}$$

Knowing this, calculate the heat flow $\boldsymbol{Q}$ from Boltzmann's equation.

12. Show that if τ_m and n are constants and f_0 is a modified Maxwell–Boltzmann distribution function such as

$$f_0(\boldsymbol{r}, \boldsymbol{v}) = n\left[\frac{M}{2\pi kT(\boldsymbol{r})}\right]^{3/2} \exp\left[\frac{-Mv^2}{2kT(\boldsymbol{r})}\right]$$

then the heat flow can be written as

$$\boldsymbol{Q} = \varkappa(n, T)\,\nabla T(\boldsymbol{r})$$

where $\varkappa$ is the thermal conductivity, and assuming that no external fields are present. Find the value of $\varkappa(n, T)$.

13. Prove equations (3.54) and (3.55).

14. Prove equation (3.58).

15. In the case of an electric field varying sinusoidally with time with a frequency ω:

(a) show that neglecting the displacement current with respect to the conduction current consists in having

$$(\varepsilon_0\omega/\sigma) \ll 1$$

(b) show that neglecting the flow current due to the transport of charges with respect to the conduction current consists in having

$$(\varepsilon_0/\tau\sigma) \ll 1$$

where τ is some characteristic time;

(c) show that the MHD approximations are valid in a plasma where $\sigma = 100$ mho/m and $\tau = 10^{-6}$ s, even at microwave frequencies.

CHAPTER 4

1. A non-relativistic point-charge moves in the direction of an exponentially decreasing electric field, such that:

$$\boldsymbol{E} = \boldsymbol{i}F_0 \exp(-kx)$$

Calculate the total energy radiated by this charge when it starts moving from infinity at time $t = -\infty$ toward the origin with a velocity $\boldsymbol{v}$.

Answer: $$\mathcal{E}_t = \frac{e^2kv^3}{9\pi\varepsilon_0 c^3}$$

2. Prove equation (4.14).

3. Solve equation (4.21). Under which conditions would this equation have exponential solutions only? Deduce the characteristic rising lengths in such a case and compare with the Debye radius.

4. Find an equation similar to equation (4.27) when the oscillation of positive ions is taken into account.

5. A plasma stream moves with a velocity v inside a homogeneous and isotropic medium having a relative dielectric constant $\varepsilon_r > 1$.
   (a) Find the dispersion relation of a plane wave propagating transversally to the stream.
   (b) Calculate both the phase and the group velocities of such a wave.
   Answer: (a) $c^2k^2+\omega_p^2-\varepsilon_r\omega^2 = 0$
6. Show that Alfvén waves are circularly polarised if they propagate in the direction of the magnetic field.
7. Prove equation (4.44). Deduce from it the dispersion relation of the wave and calculate its phase and group velocities.
8. Consider a hydromagnetic plane wave travelling along the magnetic field in an adiabatic fluid. Find the dispersion relation in such a case. Comment on this result physically.
9. Prove that there is equipartition of magnetic and kinetic energy in an Alfvén wave, for small amplitudes.
10. Calculate the damping attenuation length of a hydromagnetic wave propagating in a viscous fluid having a kinematic viscosity ν_k.
11. Find a relationship between the damping factor and the power given to the plasma by a wave.
12. Solve equation (4.67), by setting

$$v_{01} = v+\delta$$

$$v_{02} = v-\delta$$

$$k = \frac{\omega}{v}+\alpha$$

where v is the average velocity of electrons and 2δ is the difference in velocity of the two streams. Explain your results physically.
Hint: Compare the factors $(\delta\omega/v\omega_1)$ and $(\gamma v/\omega_1)$.

13. Solve equation (4.67) for the case when $\omega_1^2 = \omega_2^2 = \omega_p^2$. Discuss your results.
14. Prove equation (4.73).
15. Show that if the entropy increases across a shock then the pressure and density increase as well.

CHAPTER 5

1. Prove equation (5.10).
2. Consider a system having n_f degrees of freedom. How many types of equilibria does this system have? How many of them are stable?
3. Prove that when the magnetic field containing a plasma is concave towards the plasma, then any disturbance in this field will be amplified.

4. Prove equation (5.21).
5. Study the stability of the flow between parallel walls in the presence of a uniform magnetic field parallel to the walls. The velocity of the fluid is taken to be that of a laminar flow.
6. A longitudinal wave whose field is

$$E_x(x, t) = E_0 \sin (k_x x - \omega_r t)$$

is established in a plasma.
   (a) Show that charge carriers moving with a velocity approximately equal to the phase velocity of the wave will oscillate with an angular frequency:

$$\omega_0 = \left(\frac{qE_0 k_x}{M}\right)^{1/2}$$

for small displacements.
   (b) Under which conditions would these charge carriers be trapped by the wave?
7. Consider two intersecting beams of electrons. Assume that each beam is uniform, has the same number density n_0, and travels with the same velocity v, but in opposite directions on the x-axis. Assume also that the distribution function of each beam is given by:

$$f_0 = \tfrac{1}{2} n_0 \delta(v_x)\, \delta(v_z)\, [\delta(v_x - v) + \delta(v_x + v)]$$

   (a) Show that the dispersion equation for small-amplitude longitudinal waves is given by:

$$1 = \frac{\omega_p^2}{\frac{1}{2} n_0} \iiint \frac{f_0 \, dv_x \, dv_y \, dv_z}{(kv_x - \omega)^2}$$

where ω is a complex quantity.
   (b) Find the dispersion relation of this two-stream problem.
   (c) Under which conditions would the wave be growing? This is called a two-stream instability.
8. Prove that the growth rate of the fastest two-stream instability is $\omega_p/2$, and that the maximum in the growth rate occurs at

$$\omega_r = (\tfrac{3}{8})^{1/2} \omega_p$$

9. Find the same results as in Problem 7 *(b)*, from the macroscopic cold plasma equations for two beams of electrons having densities:

$$n_{1,2} = \tfrac{1}{2} n_0 + \tilde{n}_{1,2} \exp [j(kx - \omega_r t)]$$

velocities:

$$v_{1,2} = \mp v + \tilde{v}_{1,2} \exp [j(kx - \omega_r t)]$$

and an electric field:

$$E_x = \tilde{E} \exp [j(kx - \omega_r t)]$$

ω_r being the real part of ω, its imaginary part being neglected in this case.

10. Prove equation (5.39).
11. The magnetic field in a toroid is given by:

$$\boldsymbol{B} \sim \frac{\boldsymbol{a}_\theta}{r}$$

where $\boldsymbol{a}_\theta$ is the unit vector in the θ-direction. Show that it is not possible to contain a plasma in such a magnetic field. Is this magnetic configuration unstable in the outer or in the inner limits of the toroid? Prove your statement.
12. Prove equation (5.42).
13. What would be the effect on equation (5.46) of the injection of a weakly modulated beam into the plasma?
14. Study the convection of a weakly ionised plasma in an inhomogeneous toroidal magnetic field. Show that at a certain critical value of the magnetic field, the plasma becomes convectively unstable. Assume that the plasma is located between two ideally conducting cylinders of radii r and $r+\delta$ (with $\delta \ll r$), and that the density in the plasma is n at the inner cylinder and $n+\Delta n$ at the outer one (with $\Delta n \ll n$).
15. Discuss the stability of an inhomogeneous plasma in a magnetic field. Show that if this plasma is always unstable, then this instability must lead to a turbulent motion of the plasma with a velocity of the same order of magnitude as the drift velocity[8].

CHAPTER 6

1. One method of measuring the ratio (e/M) of an electron is by studying its motion in a magnetic field. Consider a cathode-ray tube placed in a longitudinal magnetic field $\boldsymbol{B}$. Electrons are ejected in this tube with velocities $\boldsymbol{v}_0$ as a result of an accelerating potential V_a. When the beam is defocused, electrons will strike the screen, at a distance l from the anode, at different locations, forming a smudge. As $\boldsymbol{B}$ is increased, the smudge will eventually become a tiny dot on the screen. The beam has then been focused.
   (a) Show that the field B_c at which the beam is focused occurs when the pitch of the helical path equals the anode-screen distance.
   (b) Show that under these conditions

$$\frac{e}{M} = \frac{8\pi^2 V_a n^2}{l^2 B_c^2}$$

   where n is an integer.
   (c) Calculate (e/M) for $V_a = 400$ V, $l = 20$ cm, and $B_{c\,min} = 21$ Wb/cm^2.
2. Find a method of measuring the velocity of light in vacuum from the observation of Jupiter's satellites.

3. Consider a simple pendulum and assume that the kinetic energy of the bob of the pendulum at the mid-point of its swing is equal to its potential energy at the end of the swing. Assume also that the velocity of the bob at mid-point is equal to the velocity of a point moving uniformly on a circle whose radius is equal to the amplitude of vibration of the pendulum and with an equal period. If n_s is the number of swings of the pendulum per unit time show that the acceleration due to gravity is:

$$g = \frac{4\pi^2 n_s^2}{l}$$

where l is the length of the pendulum.

4. Generally, the path of a particle in an electric field perpendicular to a magnetic field is a cycloid.
   (*a*) Show that if $\boldsymbol{v} = i v_0$, $\boldsymbol{B} = kB$, and $\boldsymbol{E} = jE$, then for $v_0 = E/B$ the motion is a straight line.
   (*b*) How can this situation be exploited to design a mass spectrometer?
5. Show that if the plasma density is small enough, the current density $I_p(V_p)$ at the probe is:
   (*a*) constant for the ideal plane probe;
   (*b*) a parabola for the cylindrical probe;
   (*c*) a linear function for the spherical probe.
6. Prove equations (6.3) and (6.4).
7. Consider two large copper plane parallel electrodes P_1 and P_2 with 1 cm × 1 cm measuring sections A_1 and A_2, and two meters M_1 and M_2 which are ideal current-measuring oscilloscopes with zero impedance (see Figure 6.17). A sudden pulse of perfectly parallel gamma radiation produces an instantaneous ionisation of 10^{12} pairs/cm³ in a plane region 1 cm thick at the centre of the space between the plates. The gas is N_2 at atmospheric pressure. Attachment of electrons to molecules can be neglected. The mobility of the positive ions is 2 cm²/s V and the mobility of electrons is 10^3 cm²/s V. Plot current versus time as read on meters M_1 and M_2.

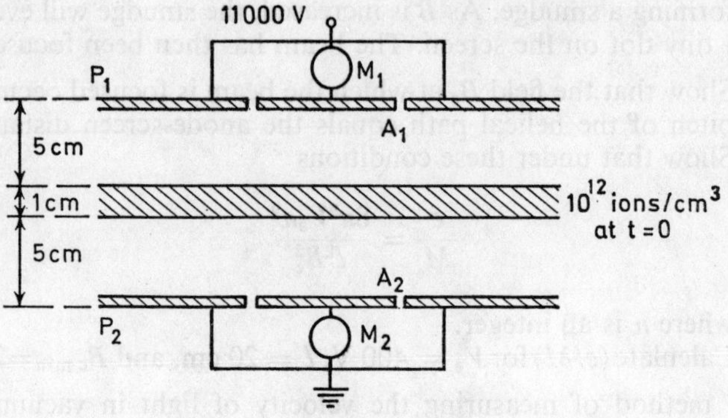

Fig. 6.17

8. Prove equation (6.7).

9. Consider a waveguide operating in the TE_{21} mode.
   (a) Find the cut-off frequency of the waveguide if it is filled with air and has a radius equal to 1 cm.
   (b) Calculate to which value this cut-off frequency will change if the waveguide is filled with a plasma having $\sigma=0{\cdot}1$ mho, $\varepsilon_r=2$, and $\mu_r=1$?

10. Prove equation (6.25).

11. A Mach–Zehnder interferometer is a device in which a beam of light is split into two different paths and recombined again to form a fringe system. Find a way of measuring the electron densities of a plasma by using such an interferometer and a pressure-driven shock tube.

12. Prove equation (6.30).

13. Prove equation (6.33).

14. Consider a hydromagnetic capacitor made up of an annular disc of fully ionised plasma, with a potential V applied between its inner and outer radii. A uniform magnetic field $\boldsymbol{B}$ is then applied along the axis of the disc.
   (a) Calculate the total energy stored in the plasma.
   (b) Calculate the effective dielectric permittivity of the plasma.
   (c) How can this capacitor be used as a diagnostic device? To measure what?

15. A fully ionised plasma is placed in uniform fields such that $\boldsymbol{B} = \boldsymbol{j}B$ and $\boldsymbol{E} = \boldsymbol{i}E$.
   (a) Show that the resulting motion of the particles would lead to an average charge separation and an electric dipole polarisation given by:

$$\mathcal{P} = \frac{n(M^- + M^+)}{B^2} \boldsymbol{E}$$

   where $n^- = n^+ = n =$ number density of particles, M^- is the electronic mass, and M^+ is the ionic mass.
   (b) Is there any way of using this property for the measurement of the number density of a plasma?

CHAPTER 7

1. What is the magnetic field strength necessary to confine a plasma having a density of 8×10^{15} particles/cm³ at temperatures of 1 keV, 10 keV and 100 keV respectively?

2. Calculate the diffusion time of a thermonuclear plasma having an average energy equal to 20 eV and a density of 10^{15} particles/cm³. The overall length of the plasma is 8 cm and the magnetic field strength is 120 Wb/cm².

3. Calculate the temperature at which the rate of thermonuclear energy release for Reaction (3) equals that for Reactions (1) and (2) *(Table 7.1)*.

4. As the current increases in a linear pinch, the temperature increases, leading to a loss of energy by radiation. Therefore, one should expect a limit to the heating of the discharge and to the current. If the maximum value of this current is 10^6 A, find the maximum temperature possible for densities of 10^{20} particles/cm³, the cross-section of the pinch being 10 cm².

5. Prove that the pressure-balance for the pinch configuration is:

$$\frac{d}{dr}\left[(n^+ + n^-)kT + \frac{B_\theta^2}{2\mu_0}\right] + \frac{B_\theta^2}{\mu_0 r} = 0$$

6. Consider a particle confined along the axis of a mirror machine between the turning points $\pm z_0$. Assume that the magnetic field along the axis of the machine is given by $B_z = \Phi_0 z^2 + B_0$, with $\Phi_0(t)$ and $B_0(t)$ slowly varying with time.
   (a) What is the variation of $v_\perp$ and $v_\parallel$ at $z = 0$?
   (b) What is the motion of the turning points?

7. The picket-fence geometry is made up of an 'infinite' number of straight wires. Currents varying periodically with space are then passed through these wires. Describe the motion of charged particles injected against the fence (see Figure 7.21).

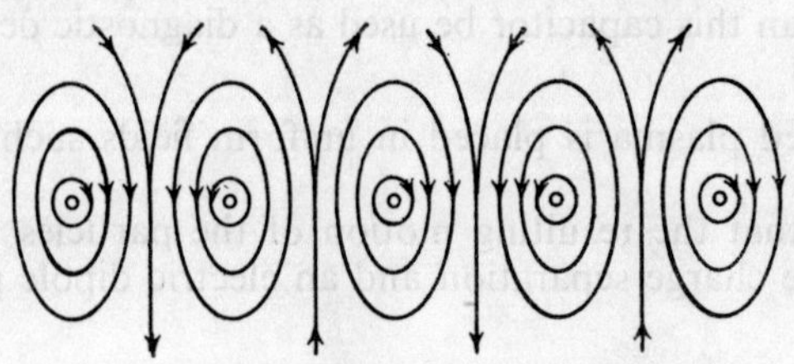

Fig. 7.21

8. Determine the configuration of the magnetic field in the cusp geometry shown in Figure 7.14. Assume that $\boldsymbol{n}.\boldsymbol{B} = 0$, where $\boldsymbol{n}$ is a unit vector normal to the surface, that $B = B_0$ near both exterior conductors and that there is a constant magnetic pressure. (Characteristic numerical values for the cusp: radius $= r_0 =$ 10 cm; electron temperature $= T =$ 10 eV; magnetic field $=$ $B =$ 1 Wb/m², confinement time $=$ 10 ms.)

9. Find the solution for the vector potential $\boldsymbol{A}$ in the case of a cylindrical astron device, using Maxwell's equations.

10. Calculate the equilibrium pinch radius of a zeta-like machine. Assume that there is no toroidal curvature and that the current flows in a thin layer at the pinch surface. Numeral example: $r_0 = 0{\cdot}5$ m, $B = 0{\cdot}02$ Wb/m², $I = 10^5$ A, $p_{\text{gas}} = 10^{-4}$ mm D_2.

11. (a) What are the advantages of using an octahedron geometry to confine a plasma?
    (b) Show that a regular octahedron can be wound along all its edges with a simple wire, making the curvature of the magnetic lines of force inside the geometry favourable for stability.

12. Study the motion of charged particles inside a 'baseball'-machine.
13. Find the mean free path for capture of thermal neutrons in ^{6}Li and in natural Li; what is the effect of high neutron energy on the mean free path?
14. Determine the electronic temperature and density of a 500 MW thermonuclear fusion reactor, having a 1 per cent efficiency. Use the most efficient reactions.
15. A deuterium plasma radiates as much cyclotron energy as bremsstrahlung energy. Find:
    (a) the average energy of the electrons;
    (b) the radiated energy
    for $n_D = 10^{-3}$ particles/m^3 and $\omega_c = 10^{22}$ Hz.

CHAPTER 8

1. Solve the equations of continuity, motion, and conservation of energy for a unidimensional, compressible, but non-viscous flow.
2. Give a complete analysis of a linear MHD generator operating at a constant Mach number.
3. Give a complete analysis of a linear MHD generator in the case of a constant velocity operation.
4. Design a linear constant Mach number MHD generator producing 300 MW power and using 6 Wb/m^2 field strength. The conditions of the plasma at the inlet are:

$$v_0 = 700 \text{ m/s}$$
$$T_0 = 2600 \text{ K}$$
$$p_0 = 4{\cdot}5 \text{ atm}$$

   Choose the type of segmentation and make reasonable assumptions whenever necessary.
5. In the a.c. MHD generator of Figure 8.6, assume that the walls are rigid boundaries and calculate the perturbed values of the Hall and Faraday currents.
6. Consider a constant area propulsor in which the working fluid is a perfect gas. Using a one-dimensional approach, calculate the ratio E/B for maximum acceleration of the plasma.
7. In the source-flow model of Figure 8.8, a radial flow is established if one assumes continuous electrodes.
    (a) Write the equations of motion, conservation of mass, and conservation of energy, for such a model.
    (b) Deduce the value of the radial current flow in terms of the radial velocity. Is this value different from the one that would be obtained if one sets $\nabla . \boldsymbol{J} = 0$? Why?
    (c) Find the value of the pressure gradient due to the radial current.

8. Consider a constant area MHD linear accelerator having the following characteristics:

Magnetic flux density	1 Wb/m²
Applied voltage	2000 V
Electrical conductivity	100 mho/m
Average velocity of the flow	2000 m/s
Interelectrode distance	0·3 m
Area of each electrode	0·4 m²

(*a*) Calculate the rate at which electrical energy is delivered to the moving fluid and the amount of mechanical energy obtained.

(*b*) Calculate the efficiency of such an accelerator.

9. In the accelerator described in Problem 8, calculate the field intensity at which a maximum amount of power is converted into mechanical power, and find the value of this maximum power.

10. In a plasma rocket one would have two types of accelerations, one due to ions and one due to electrons, such as:

$$a^+ = -\frac{kT^+}{BM^+}\frac{\partial B}{\partial z}$$

$$a^- = -\frac{kT^-}{BM^-}\frac{\partial B}{\partial z}$$

Show that when the same acceleration is imparted to electrons and ions in the presence of an electric field, then the overall acceleration becomes:

$$a = -\frac{k}{B}\frac{\partial B}{\partial z}\left(\frac{T^- + T^+}{M^- + M^+}\right)$$

Assume that $\partial B/\partial z$ is constant in the moving frame of reference.

11. Deduce the results of Figure 8.16.

12. In Section 8.7 an MHD magnetic amplifier has been described. Could you find a way of devising an MHD amplifier amplifying the electric field only?

13. Prove equation (8.43).

14. In ordinary gas dynamics, Mach one occurs at the geometric throat of a diverging nozzle. Is this still true in an MHD device?

15. Write down the basic equations necessary for the solution of the MHD stirring problem. Choose a realistic model.

CHAPTER 9

1. The saturation current from a tungsten filament at 1900 K is 140 μA. Find the current emitted by a thoriated-tungsten filament operating at the same temperature, and having the same area. The constants are:

for tungsten

$$\mathcal{A} = 60{\cdot}2\times10^4 \text{ A/m}^2\text{ K}^2$$

$$\mathcal{B} = 5{\cdot}243\times10^4 \text{ K}$$

for thoriated-tungsten

$$\mathcal{A} = 3{\cdot}0\times10^4 \text{ A/m}^2\text{ K}^2$$

$$\mathcal{B} = 3{\cdot}051\times10^4 \text{ K}$$

2. Plot the current density versus temperature for tungsten, thoriated tungsten, and an oxide-coated cathode whose constants are

$$\mathcal{A} = 0{\cdot}01\times10^4 \text{ A/m}^2\text{ K}^2$$

$$\mathcal{B} = 1{\cdot}16\times10^4 \text{K}$$

3. For the plane-diode treated in Section 9.3, the perveance of the tube is given by:

$$\mathcal{K} = \frac{4}{9}\,\varepsilon_0\left(\frac{2e}{M}\right)^{1/2}\frac{1}{d^2}$$

Find the perveance of a diode having a cylindrical symmetry, the anode being a cylinder and coaxial with a cylindrical cathode.

4. Show that in the case of a bipolar discharge in a vacuum, the overall current density is given by:

$$J = \frac{9}{4}\,\varepsilon_0\left[2e\left(\frac{1}{M^-}+\frac{1}{M^+}\right)\right]^{1/2}\frac{V^{3/2}}{d^2}$$

where M^- is the mass of the electrons and M^+ is the mass of the ions.

5. Calculate the current density of a bipolar discharge in a gas with carrier generation in the gas. Assume that the electric field between anode and cathode is weak enough for volume recombination to occur.

6. Consider a thermionic close-spaced high-vacuum diode in which the cathode is made of barium-impregnated tungsten ($V_c = 1{\cdot}7$ V) and the anode of Cs/WO ($V_a = 0{\cdot}71$ V). Calculate:

   (a) the saturation currents of the anode and the cathode;
   (b) the power output per unit area.

$$T_a = 600°\text{C}, \quad T_c = 1200°\text{C}, \quad d = 14\ \mu\text{m}$$

7. Plot equation (9.29). Find the caesium pool temperature for a pressure equal to 10^{-4} torr.

8. Design a flame-heated thermionic converter receiving 3 kW of heat energy and yielding an optimum output voltage of 1 V. Choose the type of converter, the materials necessary for the electrodes, the interelectrode distance, and the operating temperature.

9. Design a nuclear-power thermionic system with the cylindrical nuclear source having the following characteristics:

$$\text{Heat produced} = 0{\cdot}3 \text{ MW}$$
$$\text{Surface area} = 0{\cdot}1 \text{ m}^2$$
$$\text{Surface temperature} = 1200°\text{C}$$

Choose the type of converter and type of materials necessary for the electrodes. Calculate the output voltage, current and efficiency, as well as the number of unit cells, their mode of staging, and their dimensions.

10. Calculate the minimum cathode temperature for complete ionisation in the caesium of a gas-filled thermionic converter*.
11. Prove equation (9.36).
12. Calculate the current gain as a function of frequency in a plasmatron.
13. Plot the radial variation of the plasma density in a Q-machine. Comment on your result.
14. Prove equations (9.40) and (9.41).
15. Analyse Haeff's device and prove equation (9.43).

CHAPTER 10

1. Solve Equation (10.4) for the TE modes in a circular waveguide.
   (a) Find the value of the cut-off frequency below which no wave can propagate. Calculate the value of this frequency when the wave guide is air-filled, has a 1 cm radius, and operates in the TE_{21} mode.
   (b) Find the radius of a circular waveguide filled with a dielectric medium having $\sigma = 0$, $\mu_r = 1$, and $\varepsilon_r = 3$, operating in the same mode and having the same cut-off frequency.

2. The Q-factor of a resonant system is a parameter which measures the sharpness of resonance. If the resonant frequency is ω_0, and $\Delta\omega$ is the range of frequency in which the energy of oscillation is higher than 50 per cent of its resonant value, then $Q_L = \omega_0/\Delta\omega$.
   (a) Show that the Q-factor of a cubical resonator working in the

   TE_{101} is $$Q_L = \frac{1}{3}\sqrt{\left(\frac{\pi\sigma}{2\varepsilon f}\right)}$$

   (b) Calculate this value for a silver plated cavity working at a frequency equal to 10 000 MHz.

3. Prove the Stefan–Boltzmann law from equation (10.12).

* See Nottingham, W. B. 'Emitter Materials for High Temperature Energy Conversion', *Inter. Symp. High Temp. Technology*, Butterworths, London, 369 (1963).

4. Show that for a given temperature the wavelength corresponding to a maximum emitted radiation is given by $\lambda_m T = 2891\ \mu$K.

5. Ruby crystals present two metastable levels, both of them are two-fold degenerate, the $\bar{E}$ level, 14 418 cm^{-1} above ground, and the $2\bar{A}$ level, 29 cm^{-1} above the $\bar{E}$ level. Find the ratio N_E/N_{2A} of their populations at room temperature.

6. A magnetisable sphere is placed in a uniform external magnetic field. Show that inside the sphere the magnetisation is given by:

$$\mathcal{M} = 3\frac{\mu_r - 1}{\mu_r + 2}\boldsymbol{H}_0$$

where $\boldsymbol{B}_0 = \mu_0 \boldsymbol{H}_0$ is the external magnetic field and $\boldsymbol{B} = \mu_0\mu_r\boldsymbol{H}$ is the total magnetic field.

7. Solve equations (10.41) and (10.45) to obtain the magnetic susceptibility tensor.

8. Show that the TM_{010} mode requires only one third as many molecules as the TE_{011} one to trigger oscillation in a gaseous maser.

9. For a small average time spent by the molecule in the cavity of a maser and a small r.f. field strength, the transition probability is shown to be

$$|P_{b1}(\tau)|^2 = (\pi \wp E_1 \tau/h)^2$$

where $\wp$ is the electric dipole moment and E_1 is the amplitude of the electric field in the cavity. Show that under the circumstances, the minimum number of molecules per second required to maintain oscillation becomes:

$$N = \frac{\varepsilon_0 h v^2 S}{\pi \wp^2 l Q_L}$$

where v is the average molecular velocity.

10. *(a)* Calculate the minimum number of molecules per second required to maintain oscillation in a maser in the following case.

$$v = 10^5 \text{ cm/s}$$
$$l = 10 \text{ cm}$$
$$S = 2{\cdot}5 \text{ cm}^2$$
$$Q_L = 12\,000$$
$$\wp = 6 \times 10^{-30} \text{ C m}$$

(b) If the molecule considered is ammonia, find the oven temperature.

11. Show that the resonance condition in a cylindrical laser rod is given by:

$$\left(\frac{J_{nm}}{r}\right)^2 + \left(\frac{\pi n}{l}\right)^2 = \left(\frac{2\pi v_R}{\lambda}\right)^2$$

where l is the length of the laser, r is its radius, J_{nm} is the mth root of the Bessel function of order n, n is an integer, ν_R is the refractive index of the laser material, and λ is the wavelength.

12. Consider the simpler case where the radiation in the laser rod is a uniform plane wave passing axially between the reflecting ends. Calculate the separation in wave numbers between consecutive axial mode resonances in the case of a ruby laser at 0·69 μm wavelength.

$$\nu_R = 1{\cdot}76; \quad l = 10 \text{ cm}$$

Answer: 0·028 cm^{-1}

13. Plot equations (10.31)–(10.35) qualitatively as a function of ν and determine the maximum value of m in each case.

14. Deduce the population inversion per unit volume Δn in a laser from equation (10.28).

15. Calculate the number of ions that need to be brought to the metastable state to achieve the threshold condition, in terms of the total number of active ions in the crystal:

    (a) in a three-level laser; *(b)* in a four-level laser.

    Assume that the final and initial states of the laser transition have equal degeneracy.

CHAPTER 11

1. Consider the Coulomb nature of the potential energy of an electron inside a metal as a function of its distance from a nucleus.
   (a) Draw the potential energy curve resulting from two nuclei.
   (b) Draw the potential energy distribution inside the metal and at its surface.

2. A 400 V potential is applied between the faces of a silicon slab in the direction where $M^* = 0{\cdot}19\ M_0$ (M_0 = mass of an electron).
   (a) What is the time of flight of an electron in the slab along a distance of 3 cm? The initial velocity of the electron is assumed to be negligible.
   (b) Solve the same problem in free space, and compare the two flight times.

3. From equations (11.6) and (11.7) find a relationship between the surface area embraced by the instant orbit of an electron in physical space and that in the k-space. Comment on your result.

4. The mobility of holes in germanium is 1900 cm^2/V s. In silicon it is 3800 cm^2/V s. Find the diffusion constant of holes in both materials at 300 K.

5. In an intrinsic semiconductor the Fermi level is half-way between the conduction band and the valence band. From this statement deduce expressions for n and p separately, similar to equation (11.9).

6. Radiation $h\nu$ is emitted when an electron transfers from a stationary state having an energy level $\mathcal{E}_2$ to another stationary state having an energy level $\mathcal{E}_1$. Show that if the ionisation potential V_i is given in volts, the wavelength λ in angströms of the emitted radiation would be:

$$\lambda = \frac{12\,400}{V_i}.$$

7. If one assumes a self-decaying plasma with no recombination within its volume, then

$$\frac{dn}{dt} = -\mathcal{D}_a \nabla^2 n$$

where $\mathcal{D}_a$ is the ambipolar diffusion constant. Solve the above equation for a cylindrical rod and show that the value of Λ given in equation (11.21) is $(r_0/2{\cdot}405)$ where r_0 is the radius of the rod.

8. Calculate the plasma frequency and the Debye radius of a solid-state plasma having a density of 10^{20} electrons/m³ at 'room temperature'.

9. Find the average random speed of the electrons in the plasma of Problem 8. Comment.

10. Extrinsic semiconductors are 'doped' by impurities having 5 or 3 valence electrons whereas the host lattice of the intrinsic semiconductor has only 4 valence electrons. Comment on the possible use of extrinsic semiconductors in solid-state plasma applications.

11. Compute the Alfvén wave speed in a solid-state plasma having a density of 10^{20} particles/cm³ in a 0·1 Wb/m² magnetic field. Charge carriers are holes with a mass equal to $0{\cdot}04\,M_0$, M_0 being the mass of the electron.

12. Under the conditions of maximum coupling between helicons and ultrasounds, the two waves become indistinguishable. Show that under these conditions the ratio of the magnetic field over the drift velocity of the positive ions in the sound wave is $c(\varepsilon\varrho)^{1/2}$.

13. Rewrite equation (11.28) such that use could be made of the variables V_p and $(\omega\omega_c)^{1/2}$. Plot V_p as a function of $(\omega\omega_c)^{1/2}$ and comment.

14. Prove equation (11.29).

15. Calculate the temperature in a solid-state plasma pinch having the following characteristics:

$$\left(\frac{r}{I}\right) = 6\ \mu\text{m/A}$$

$$n = 10^{23}\ \text{particles/m}^3$$

What would be the temperature for $(r/I) = 4\ \mu$m/A?

CHAPTER 12

1. Consider a simplified model for the atmosphere with an exponentially decreasing density as one goes up. Ionising particles (e.g. electrons) come vertically from above and are all assumed to terminate their paths at the same altitude. Above this altitude, each electron creates a number of charge carrier pairs per unit length proportional to the local density of the air. Carrier pairs are recombined in the gas and the recombination coefficient is proportional to the local air density. Diffusion from regions of higher concentration to regions of lower concentration should be taken into account. Calculate the plasma density as a function of altitude, including that part of the atmosphere where the ionising particles do not penetrate.
2. Calculate the speed of an Alfvén wave in the ionosphere for $n = 10^5$ cm^{-3} and $B = 10^{-4}$ Wb/m^2, the positive charge carriers being nitrogen ions.
3. Using Figures 12.1 and 12.4 sketch the variation with height of the Debye radius and of the plasma frequency in the ionosphere.
4. Assume that the terrestrial magnetic field is identical to that of a dipole having a moment of $8{\cdot}1\times10^{22}$ A m^2, calculate the drift velocity of 10^5 eV electrons having a pitch angle of 60° at the magnetic equator, at a distance of 4 earth radii from the earth's centre.
5. Using Figure 12.4, sketch the variation with height of the Alfvén speed in the ionosphere. Assume that the average ionic molecular weight is approximately equal to 16.
6. *(a)* Calculate the mean radius of gyration of an electron in the terrestrial magnetic field at a distance of 4 earth radii at the equator. Assume the energy of the electrons to be constant and equal to 0·009 eV.

   (b) Plot this radius as a function of altitude between 2 and 9 earth radii at the equator.
7. Describe the motion of charged particles in the Van Allen belts. Show that the surface swept by the guiding centre of a charged particle is closed. Assume an azimuthal symmetry for the magnetic field.
8. Equating the solar wind pressure and the geomagnetic field pressure, show that the radius of the magnetosphere proper is given by:

$$r_{\mathrm{Mp}} = \left(\frac{2B_0^2}{\mu_0 \varrho_s v_s^2}\right)^{1/6} r_0$$

   where r_0 is the earth radius, ϱ_s is the mass density of the solar wind, v_s is its undisturbed speed, and B_0 is the surface value of the undisturbed magnetic field of earth.
9. Compute the Alfvén wave speed in the solar corona, assuming $n = 10^{12}$ m^{-3} and $B = 10^{-3}$ Wb/m^2, the positive charge carriers being protons.

10. Considering the effect of a vertical magnetic field on the equilibrium of a conducting medium, study the equilibrium of sunspots. Use a cylindrical configuration, and assume the field is constant along a magnetic line of force but varies from one line to another.

11. Explain the differential rotation of the sun in terms of magnetic rigidity. Show that the magnetic lines of force should be frozen in the solar material and that the magnetic field should be symmetrical about the sun's axis of rotation.

12. Explain the occurrence of solar prominences by assuming the prominence to be an arch along a magnetic line of force. Criticise this model.

13. Compute the Alfvén wave speed in the interstellar space, assuming $n = 10^7$ m^{-3} and $B = 10^{-7}$ Wb/m^2, the positive charge carriers being protons.

14. Some theories suggest that the occurrence of spiral arms in the Galaxy is due to the presence of magnetic fields whose lines of force lie along the spiral arms. Compare these theories to the case of Problem 12.

15. Criticise the different theories of evolution of the universe. Could this evolution be explained in terms of an MHD model?

Index